NONLINEAR CLIMATE DYNAMICS

This book introduces stochastic dynamical systems theory to synthesise our current knowledge of climate variability. Nonlinear processes, such as advection, radiation and turbulent mixing, play a central role in climate variability. These processes can give rise to transition phenomena, associated with tipping or bifurcation points, once external conditions are changed. The theory of dynamical systems provides a systematic way to study these transition phenomena. Its stochastic extension also forms the basis of modern (nonlinear) data analysis techniques, predictability studies and data assimilation methods. Early chapters present the stochastic dynamical systems framework and a hierarchy of climate models to study climate variability. Later chapters analyse phenomena such as the North Atlantic Oscillation, the El Niño/Southern Oscillation, Atlantic Multidecadal Variability, Dansgaard-Oeschger Events, Pleistocene Ice Ages and climate predictability. This book will prove invaluable for graduate students and researchers in climate dynamics, physical oceanography, meteorology and paleoclimatology.

HENK A. DIJKSTRA is Professor of dynamical oceanography at the Institute for Marine and Atmospheric Research, Department of Physics and Astronomy, Utrecht University, The Netherlands. His main research interests are in the application of dynamical systems methods to problems in climate variability. He is the author of *Nonlinear Physical Oceanography, Second Edition* (2005) and *Dynamical Oceanography* (2008). He is a member of the Dutch Royal Academy of Arts and Sciences. In 2005, he received the Lewis Fry Richardson medal from the European Geosciences Union and in 2009 he was elected a Fellow of the Society for Industrial and Applied Mathematics.

NONLINEAR CLIMATE DYNAMICS

HENK A. DIJKSTRA
Utrecht University

CAMBRIDGE
UNIVERSITY PRESS

University Printing House, Cambridge CB2 8BS, United Kingdom

One Liberty Plaza, 20th Floor, New York, NY 10006, USA

477 Williamstown Road, Port Melbourne, VIC 3207, Australia

314-321, 3rd Floor, Plot 3, Splendor Forum, Jasola District Centre, New Delhi - 110025, India

79 Anson Road, #06-04/06, Singapore 079906

Cambridge University Press is part of the University of Cambridge.

It furthers the University's mission by disseminating knowledge in the pursuit of education, learning and research at the highest international levels of excellence.

www.cambridge.org
Information on this title: www.cambridge.org/9780521879170

First published 2013

A catalogue record for this publication is available from the British Library

Library of Congress Cataloging in Publication data
Dijkstra, Henk A.
Nonlinear climate dynamics / Henk A. Dijkstra, Utrecht University.
pages cm
Includes bibliographical references and index.
ISBN 978-0-521-87917-0 (hardback)
1. Climatology – Statistical methods. 2. Dynamic climatology. I. Title.
QC874.5.D55 2013
551.601′175–dc23 2012044078

ISBN 978-0-521-87917-0 Hardback

Contents

Preface

Dynamical systems theory is an extremely powerful framework for understanding the behavior of complex systems. Its concepts apply to many scientific fields, and hence its language provides a multidisciplinary and unifying communication tool. The theory provides a systematic approach for assessing the sensitivity of a mathematical model of a particular phenomenon to changes in parameters and initial conditions. As such, it finds application in stability problems, transition behavior and predictability studies. In addition, techniques and concepts from dynamical systems theory have led to the development of a diverse set of nonlinear methods of time series analysis.

For many phenomena, existing models cannot resolve all relevant spatial and temporal scales, and hence small-scale features are often represented as 'noise'. As a result of the increase in computational power, solutions of the resulting stochastic partial differential equations are now within reach. Although stochastic dynamical systems are difficult to deal with, in recent years, the theory of stochastic dynamical systems has matured and is ready to be applied to many scientific areas.

This book developed from a course on climate dynamics that I taught at Colorado State University in 2005 and a course on stochastic climate models that I taught at Utrecht University in 2008. My main motivation in writing this book was to provide both an introduction into stochastic dynamical systems theory and to show the application of these methods to problems in climate dynamics.

The book is, therefore, logically divided into two parts. In the first part, Chapters 2 through 5, introductions to dynamical systems (Chapter 2), stochastic calculus (Chapter 3) and random dynamical systems (Chapter 4) are given. With the level of mathematical detail provided, this material should be accessible to graduate students in climate physics. Chapters 7 through 11 provide a description of how stochastic dynamical systems theory has been used to understand particular phenomena in climate physics. Chapter 6 illustrates the hierarchy of climate models used in the description of climate variability and climate change. Chapter 12 shows the application of dynamical systems theory to predictability studies. The choice of all the material is quite a personal one and I apologise for certainly having forgotten to cite and discuss results of additional very relevant publications.

In the present scientific world in which everybody can be a 'Google' specialist on any topic in fifteen minutes, and where the focus appears to be more and more on concepts that are appealing rather than precise, it is good to know that there are frameworks available that can be used to fit pieces of a complex puzzle. The material in this book forms the basic material for such a framework related to problems in climate variability. I hope it finds its way to the younger generation of climate scientists.

Acknowledgements

The writing of this book has been a pleasure as a result of the interaction with many colleagues in the fields of climate dynamics and dynamical systems.

First, I thank Will de Ruijter (IMAU, Utrecht) for his unfaltering support of my research activities. He and the rest of the faculty, staff, postdocs and students at IMAU have been responsible for creating the environment in which this book could be written. In addition, I thank my colleagues at the Royal Dutch Meteorological Institute, Geert Jan van Oldenborgh and Sybren Drijfhout, for the nice discussions over the years. Special thanks go to Fred Wubs (RUG) for his collaboration in many joint projects and his contributions to the underlying numerical methodology to handle high-dimensional dynamical systems.

Joint work with Michael Ghil (LMD, Paris, France) and Eric Simonnet (INLN, Nice, France) has been very important for the development of the material in this book. Chapter 2 is based on a joint paper that we wrote for the *Handbook of Numerical Analysis* (Vol XIV). Chapter 4 is based on Eric's notes on random dynamical systems, which we discussed extensively during my many enjoyable visits to INLN (for which I thank the CNRS for support). I thank Eric Deleersnijder (Louvain-la-Neuve, Belgium) for inviting me to give several lectures on stochastic dynamical systems, which eventually led to Chapter 3. Most of the book was written during a four-month sabbatical at the Australian National University in 2011. I thank Andy Hogg and Ross Griffiths for a quiet and great time in Canberra and the ARC Centre of Excellence for Climate System Science for the generous support.

I thank members of my current group at IMAU (Elodie Burrillon, Andrea Cimatoribus, Leela Frankcombe, Lisa Hahn-Woernle, Anna von der Heydt, Dewi Le Bars, Maria Rugenstein, Matthijs den Toom, and Jan Viebahn) for providing feedback on early versions of this book. I want to mention in particular Matthijs den Toom and Jan Viebahn for essential contributions and Anna von der Heydt for creating several of the figures used in Chapter 11. Daan Crommelin is thanked for providing very useful comments on a first version of Chapter 7. I thank many colleagues for providing me with original files for many of the figures in this book (see permissions).

The generous support from the Netherlands Organization for Scientific Research (NWO) over the years is much appreciated. This support has been essential in carrying out the research on the applications of dynamical systems theory in physical oceanography and climate dynamics. The support in supercomputing time on the machines at the Academic Computing Center in Amsterdam (SARA) through the projects from the National Computing Facilities Foundation (NCF) is also much acknowledged.

Matt Lloyd at Cambridge University Press has been very supportive and patient in waiting for the final version of the manuscript. I also thank Amanda O'Connor for going through the text and for providing useful comments on layout and style. Peggy Rote (Aptara, Inc.) did a fantastic job in the copyediting of the book; all remaining errors are my sole responsibility.

Finally, I thank Julia for always being around.

1

Climate Variability

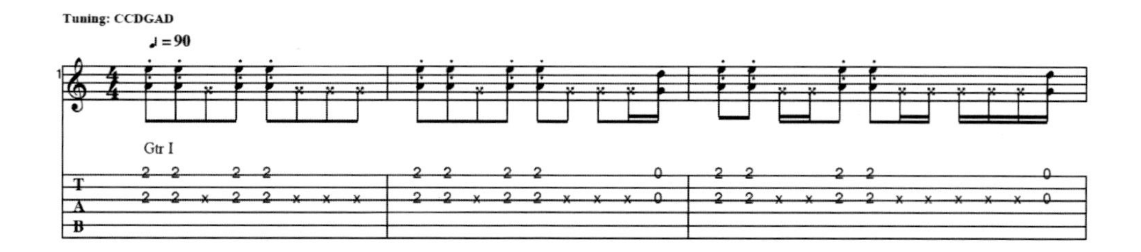

Complex motions on the sphere.
CCDGAD, Aerial Boundaries, Michael Hedges

Human life is possible because of the specific conditions of the fluid envelopes surrounding the Earth. These fluid envelopes and the processes affecting their behavior are usually grouped into one system: the climate system. Quantities in the climate system, such as temperature and precipitation, vary on many time scales, and these variations are highly relevant for many aspects of human life, such as food production and safety.

There are many very good textbooks containing a description of the components of the climate system (Peixoto and Oort, 1992; Ruddiman, 2001), the relevant processes (Hartmann, 1994) and the modelling of the development of this system (McGuffie and Henderson-Sellers, 2006; Neelin, 2011). Many of these books first introduce the radiation balance with all the physical, chemical and biological processes affecting it. Next, the large-scale atmospheric circulation and ocean circulation are considered, followed by the smaller-scale processes in these components of the climate system. Finally, the role of the biosphere and cryosphere is discussed.

This is a book in which variability in the climate system is viewed from a stochastic dynamical systems framework. After an introduction into the observational database in Section 1.1, typical phenomena of climate variability are presented in Section 1.2. In Section 1.3, we focus on the dynamical organization of the climate system, which is followed by an introduction into the stochastic dynamical systems framework in Section 1.4. An overview of the contents of the book is given in Section 1.5, together with specific reading paths.

1.1 The observational database

Climate research is a data-poor science considering the questions it attempts to answer (Wunsch, 2010). There are two sources of data: the instrumental record and the palaeorecord. The instrumental record teaches us about the transient behavior of the climate system under a changing forcing (solar, greenhouse gases) since the early 1860s. It consists of the following different databases:

- Local routine measurements at certain stations. Examples are radiosonde observations, tide gauge data and data from ocean weather ships. The length of the time series varies enormously from location to location, and the spatial coverage is usually poor (e.g., only a few locations in the Southern Ocean region).
- Databases from different international monitoring programs. In several of these programs, such as the World Ocean Circulation Experiment (WOCE), a global (but coarse) coverage was achieved.
- Remotely sensed data. Satellite data have been collected since the early 1980s, providing near global coverage of the Earth.
- Data from drifting platforms. Since the early 2000s, floats have been released in the oceans, which monitor quantities such as temperature and salinity as they move with the currents.

Much of our knowledge on climate variability is based on observations of which data are stored in several databases around the world. These databases are continuously updated with new observational records from in situ (atmospheric, oceanic) measurements and with satellite data. The observational data sets are composed of time series of particular so-called climate indices and of three-dimensional (longitude, latitude, time) or four-dimensional (longitude, latitude, height/depth, time) fields of many quantities, such as temperature and precipitation. Most of the instrumental climate data are available through the Royal Netherlands Meteorological Institute's Climate Explorer (http://climexp.knmi.nl), where they can also be easily visualised and analysed. An important example is the time series of the sea surface temperature anomaly in the eastern Pacific, the so-called NINO3 time series (Fig. 1.1a).

In addition to the instrumental record, there is also the palaeorecord, containing (mainly) data that are not direct measurements of climate variables but of a quantity

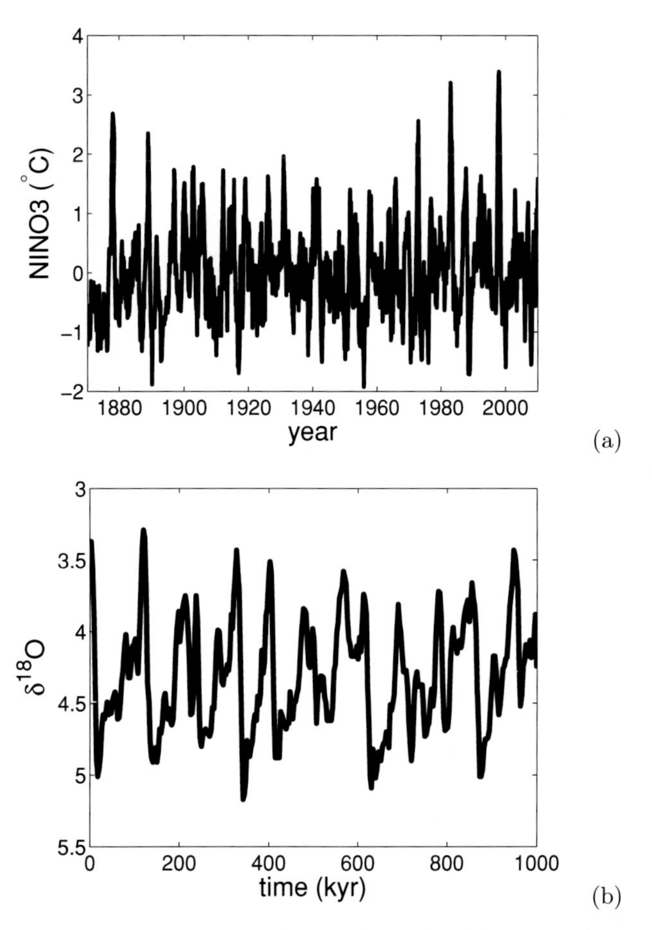

Figure 1.1 (a) Time series over the period 1870–2010 of the sea surface temperature anomalies (with respect to the mean seasonal cycle) in the Eastern equatorial Pacific averaged of the domain [150°W-90°W] × [5°S-5°N], the so-called NINO3 box, as provided in the HadISST1 data set (Rayner et al., 2003). The time series is low-pass filtered with a cut-off value of 24 months. (b) Time series of δ^{18}O at Ocean Drilling Program Site 677 (1°N, 83°W) at a depth of about 3 km over the last one million years providing an indirect measure of temperature fluctuations in the deep ocean (a high [low] value of δ^{18}O indicates a low [high] temperature).

(a proxy) that can be related to a climate variable. An example is the oxygen isotope ratio δ^{18}O determined from water molecules in ice cores, which can be related to the temperature at the time of deposition of the snow. Also δ^{18}O derived from carbonate containing marine organisms provides information on the changes in the deep ocean temperature (Fig. 1.1b). Most of the palaeodata are available through the palaeoclimate Web site at the National Oceanic and Atmospheric Administration (NOAA) http://www.ncdc.noaa.gov/paleo/paleo.html.

1.2 Phenomena: temporal and spatial scales

An artist's view of climate variability on 'all' time scales is provided in Fig. 1.2. The first version of this figure was produced by Mitchell (1976), and many versions thereof have circulated since. The figure is meant to summarize our knowledge of the spectral power $S = S(\omega)$, that is, the amount of variability in a given frequency band, between ω and $\omega + \Delta\omega$; here the frequency ω is the inverse of the period of oscillation, and $\Delta\omega$ indicates a small increment. This power spectrum is not computed directly by spectral analysis from a time series of a given climatic quantity, such as (local or global) temperature. There is no single time series that is 10^7 years long and has a sampling interval of hours, as the figure suggests. Instead, Fig. 1.2 includes information obtained by analysing the spectral content of many different time series, for example, those in Fig. 1.1.

Between the two sharp lines at one day and one year lies the synoptic variability of mid-latitude weather systems, concentrated at 3–7 days, as well as intraseasonal variability, that is, variability that occurs on the time scale of 1–3 months. The latter is also called low-frequency atmospheric variability, a name that refers to the fact that this variability has lower frequency, or longer periods, than the life cycle of weather systems. Intraseasonal variability comprises phenomena such as the Madden–Julian oscillation of winds and cloudiness in the tropics or the alternation between episodes of zonal and blocked flow in mid-latitudes.

Immediately to the left of the seasonal cycle in Fig. 1.2 lies interannual (i.e., year-to-year) variability. An important component of this variability is the El Niño phenomenon in the Tropical Pacific: once about every four years, the sea-surface temperatures (SSTs) in the Eastern Tropical Pacific increase by a few degrees over a period of about one year. This SST variation is associated with changes in the trade winds over the tropical Pacific and in sea-level pressures (Philander, 1990); an east–west seesaw in the latter is called the Southern Oscillation. The combined El Niño/Southern Oscillation (ENSO) phenomenon arises through large-scale interactions between the equatorial Pacific and the atmosphere above. Equatorial wave dynamics in the ocean play a key role in setting ENSO's time scale. The time series of the NINO3 index as shown in Fig. 1.1a is a measure for the variability in the SST in the eastern Pacific.

On slightly larger time scales, decadal-to-multidecadal variability appears, which is particularly prominent in the North Atlantic because of a phenomenon called the Atlantic Multidecadal Oscillation (AMO). The AMO is associated with basin-wide changes in sea surface temperatures on a 20- to 70-year time scale, with most of the action occurring in the northern North Atlantic. Multidecadal variability was also found in the 335-year-long record of the Central England temperature time series (Ghil and Vautard, 1991).

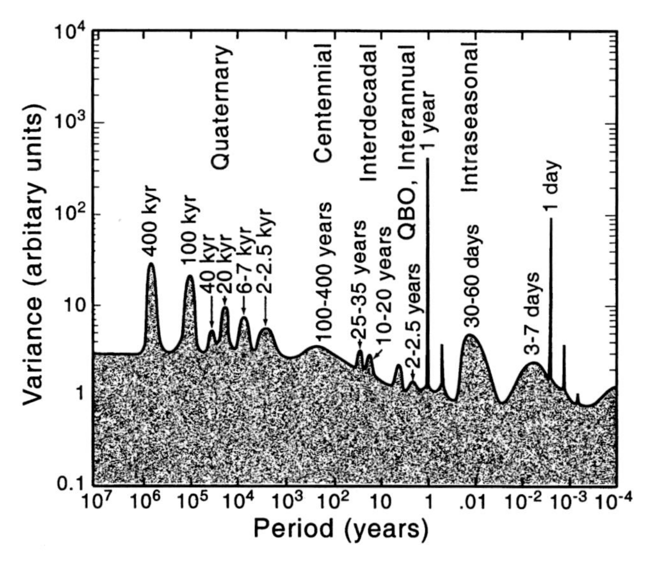

Figure 1.2 An artists' view on climate variability (Mitchell, 1976) displaying a 'hypothetical' spectrum based on information of different time series such as shown in Fig. 1.1 (from Dijkstra and Ghil, 2005).

The leftmost part of Fig. 1.2 represents palaeoclimatic variability. The information summarized here comes exclusively from proxy indicators of climate variables (Imbrie and Imbrie, 1986). These include coral records and tree rings for the historic past, as well as marine-sediment and ice-core records for the last two million years of Earth history, the Quaternary. Glaciation cycles, an alternation of warmer and colder climatic episodes, dominated the Quaternary era. The cyclicity is manifest in the broad peaks present in Fig. 1.2 between roughly 1 kyr and 1 Myr and can be seen in Fig. 1.1b. The two peaks at about 20 kyr and 40 kyr reflect variations in Earth's orbit, whereas the dominant peak at 100 kyr remains to be convincingly explained.

Within these glaciation cycles, there are higher-frequency oscillations prominent in the North Atlantic palaeoclimatic records, in particular in Greenland ice core records. These are the Heinrich events, with a near-periodicity of 6–7 kyr, and the Dansgaard-Oeschger cycles, which provide the peak at around 1–2.5 kyr in Fig. 1.2. Rapid changes in temperature, of up to one half of the amplitude of a typical glacial–interglacial temperature difference, occurred during Heinrich events, and somewhat smaller ones occurred over a Dansgaard-Oeschger cycle. None of these higher-frequency oscillations can be directly connected to orbital or other external forcings.

In summary, climate variations range from the large-amplitude climate excursions of the past millennia to smaller-amplitude fluctuations on shorter time scales. Fig. 1.2 reflects three types of variability: (i) sharp lines that correspond to periodically forced variations, at one day and one year; (ii) broader peaks such as that associated with

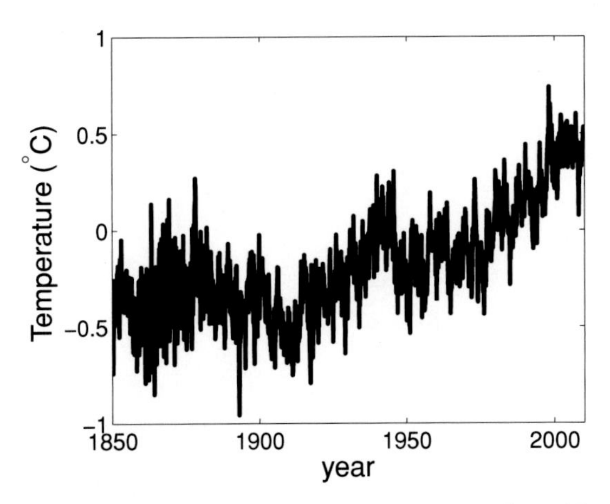

Figure 1.3 The global mean surface temperature anomaly with respect to the period 1961–1990 from the HadCRUT3 data set (http://www.cru.uea.ac.uk/cru/data/temperature/).

ENSO in the interannual frequency band; and (iii) a continuous background with a higher power in the lower frequencies. The understanding of these spectral properties of climate variability is crucial to interpret the time series of the global mean surface temperature from the instrumental record, as plotted in Fig. 1.3, and to assess its predictability.

1.3 The climate system

Although climate scientists' views on the climate system probably greatly differ, most would admit that it is a system displaying very complex spatio-temporal variability in many of its components, such as the atmosphere, the hydrosphere (including the oceans), the cryosphere, the biosphere and the lithosphere.

In a report to the NASA Advisory Council, Bretherton (1988) presented a sketch of the Earth System components and their interactions. The original figure (Bretherton, 1988), sometimes referred to as the 'horrendogram' of the climate system, and its simplification shown in Fig. 1.4 are certainly useful in recognizing many of the subcomponents of the climate system and identifying the important processes. The figure also provides a basis for understanding the transfer of properties (e.g., energy and mass) that are exchanged between these different subsystems. Examples of such interactions and associated fluxes are usually referred to as the energy cycle, the hydrological cycle and several biogeochemical cycles (e.g., the carbon, the sulphur and the nitrogen cycles).

To understand climate variability, it is important to realize that different characteristic time scales are introduced into the climate system by the different processes in the subsystems. One way of looking at these time scales is to perturb the specific subsystem out of an equilibrium and then monitor how long it takes for it to reach

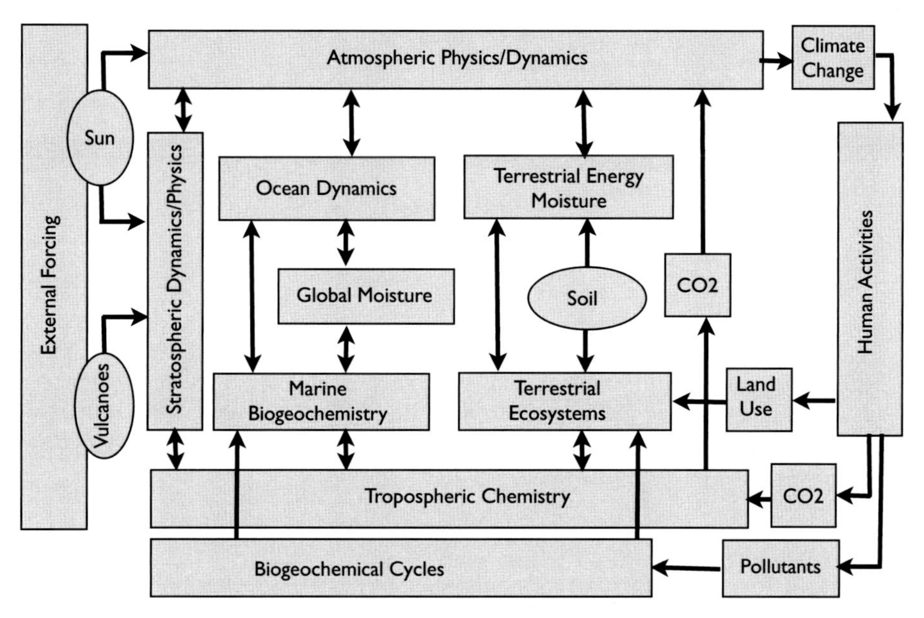

Figure 1.4 A schematic of the organization of the climate system, showing the different components and their connections (simplified from Bretherton [1988]).

equilibrium again. Such characteristic (or response) time scales of atmospheric processes range from a few seconds (e.g., formation of cloud droplets) to a few days (e.g., dissipation of midlatitude weather systems). For the ocean, these scales range from a few months (e.g., upper layer ocean mixing) to thousands of years (e.g., deep ocean circulation adjustment). The cryosphere has an even larger range because sea ice processes are much faster than those of ice on land. The time scales of the biosphere also have a very wide range, and those of the lithosphere (e.g., motion of continents) are up to millions of years.

In addition, feedbacks between the different components may also introduce new time scales of variability. One of the most important examples is the coupling between the equatorial ocean and the global atmosphere, which introduces the interannual time scale of variability associated with the ENSO (Fig. 1.1b). For feedbacks to occur, it is important to know the relevant strengths of the couplings between the different subsystems. The organizational structure of the climate system, as sketched in Fig. 1.4, together with the different response time scales and relatively coupling strength of the subsystems provides the basis for the stochastic dynamical systems framework, which is introduced in the next subsection.

1.4 The stochastic dynamical systems framework

To determine how the Earth system develops as a whole, humans included, appears an impossible task. Although there are basic mathematical equations available for the

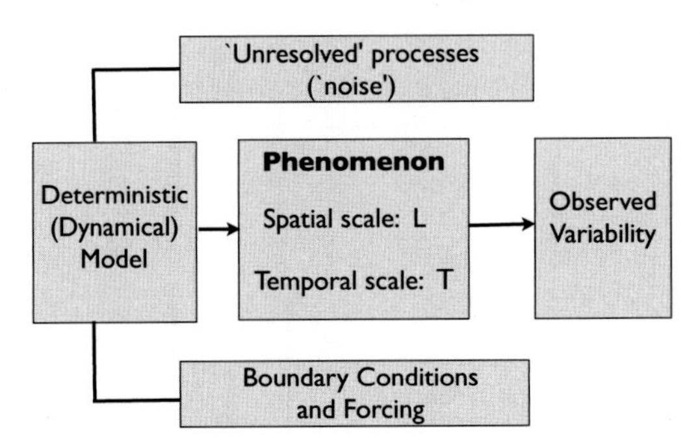

Figure 1.5 A schematic of the stochastic dynamical systems framework of climate variability. A deterministic dynamical system represents the particular phenomenon on a dominant spatial scale L and a temporal scale T. Smaller and faster scale processes (labelled here as 'unresolved' by the deterministic system) are represented as 'noise', whereas slower processes are fixed into the boundary conditions.

physical components of the system (i.e., conservation principles), such equations are (still) lacking for many other components, such as for the development of ecosystems. Missing such equations on highly relevant components and facing the problem of many spatial and temporal scales of the system, we necessarily must turn to more modest approaches.

The issue of trying to determine the development of the Earth System is closely related to the question one asks. Is the question related to the development of the El Niño over the next few months, is it to predict when the next ice age will occur, or is it about the changes of the temperature in Western Europe over the next 50 years due to the increase in atmospheric greenhouse gases? Once the question has been fixed, we can immediately take advantage of the fact that processes can be differentiated according to the spatial and temporal scales, and hence one can make adequate approximations. For example, we can make a very good approximation by assuming that the continents are at fixed positions when studying the development of the weather over the next few days.

Let us focus on a phenomenon with a characteristic time scale T occurring predominantly over a spatial scale L (Fig. 1.5). As an example, we can think of the interannual associated with the present-day El Niño with spatial patterns extending over a large part of the Pacific basin, that is, $T \approx 5$ years and $L \approx 10^7$ km. In a stochastic dynamical systems view of this phenomenon, all processes on time scales $\tau \gg T$ can be assumed to be fixed in time. For example, for present-day El Niño prediction, the ocean bathymetry can be fixed to present-day values, and orbital variations in insolation can be neglected.

How do we handle in this case the processes occurring on much smaller time and spatial scales, such as wind waves on the surface of the Pacific Ocean? This is in general a tricky issue, as collective behavior due to small-scale processes certainly can influence the large-scale behavior. Hence there is no general theory to cope with the small-scale processes apart from explicitly modelling them. When this is impossible, one can resort to several options, usually referred to as parameterisation of the small scales into large-scale descriptions. Parameterisations may be deterministic (i.e., given by explicit relationship between variables), or they may be stochastic (Majda and Wang, 2006). In the latter case, a stochastic model of the small scales has to be proposed. Both this stochastic model and the deterministic parameterisations are in most cases (at least partially) based on observations.

What results from this stochastic dynamical systems framework (Fig. 1.5) is a set of mathematical (in general, stochastic partial differential) equations for the description of the phenomena at the scales T and L. The equations for these large scales contain parameterisations in which the effects of the small scales are represented. Boundary conditions are formulated at areas where development of the system is much slower. Because of the cyclic nature of the insolation entering the climate system, the set of equations may contain a periodic forcing component. The response of this (in general, nonlinear) periodically forced stochastic dynamical system can then be compared with data sets from the instrumental record and from proxy records.

1.5 Overview of the book

The use of a stochastic dynamical systems framework for addressing problems in climate variability logically structures the book into three parts: a methods part (Chapters 2–5), a climate dynamics part (Chapters 6–11) and a climate predictability part (Chapter 12).

In Chapter 2, an introduction to the theory of deterministic dynamical systems is given. It is relatively short on bifurcation theory compared with material presented in Dijkstra (2005), but it provides more details on transient phenomena such as non-normal growth. In Chapter 3, methods of stochastic calculus are described starting from the very basics. The main aim of this chapter is to give a 'mild' mathematical description of stochastic methods as applied to climate dynamics. The general solution of the linear scalar stochastic differential equation and the basics of the Fokker-Planck equation are described. In Chapter 4, a short exposition of bifurcation theory in random dynamical systems is given. Chapter 5 focuses on data (either from models or observations) analysis techniques. We also start from the basis, discuss the traditional linear stationary statistical methodology (univariate and multivariate) and extend it slightly with modern nonstationary (e.g., wavelets) and nonlinear time series analysis (e.g., attractor embedding) techniques.

The climate dynamics part of the book starts with Chapter 6, where an introduction is given into the hierarchy of climate models that are used in subsequent chapters. In each of Chapters 7–11, an important problem of climate variability is discussed using the stochastic dynamical systems framework, that is, the North Atlantic Oscillation (Chapter 7), the El Niño/Southern Oscillation (Chapter 8), the Atlantic Multidecadal

Oscillation (Chapter 9), the Dansgaard-Oeschger events (Chapter 10) and the Pleistocene Ice Ages (Chapter 11). The choice of these problems is motivated by the following:

- The phenomena all occur on different time scales, and hence different processes affect the dynamics of each of the climate subsystems. Also the 'noise' has a different interpretation in each of the problems.
- For each of the phenomena, a reasonable amount of data (either from the instrumental record or from proxy archives) are available to provide model-data comparisons.
- The techniques from stochastic dynamical systems theory, as presented in Chapters 2–5, have been applied to these problems, so there is an extensive set of results from the literature.
- In each of the problems there are different stages of understanding of the phenomena based on the use of different levels of models (in the model hierarchy presented in Chapter 6).

The book concludes with Chapter 12, in which the application of stochastic dynamical systems techniques on predictability problems associated with present-day climate change is presented.

This book has been written with two types of readers in mind. Reader type A likes an introduction into the modern theory of stochastic dynamical systems using a less formal mathematical description than is given in most mathematics textbooks. Someone wanting to learn only about the methodology of random dynamical systems would focus on Chapters 3 and 4. Reader type B is curious about what the stochastic dynamical systems framework provides to the understanding of his/her topic of interest described in Chapters 7–11 or how these concepts are applied in climate predictability (Chapter 12). For example, someone interested in the Dansgaard-Oeschger events can turn directly to Chapter 10 and read background material from Chapters 2–6 when required. The material in each chapter is largely self-contained, although obviously in Chapters 7–11, extensive reference is made to material in Chapters 2–6.

2

Deterministic Dynamical Systems

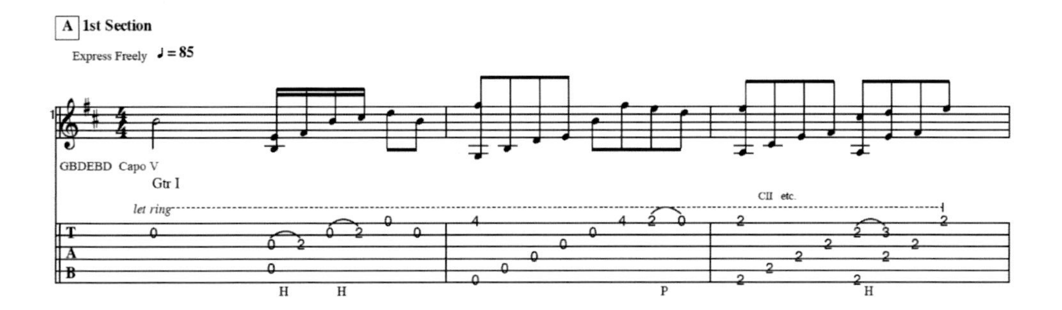

An elementary element of truth.
GBDEBD, Anne's Song, Will Ackerman

In this chapter, an overview is given of the theory and tools used to investigate transition behavior in deterministic dynamical systems.

2.1 Finite dimensional dynamical systems

A general first-order system of ordinary differential equations (ODEs) can be written as the continuous time dynamical system

$$\frac{d\mathbf{x}}{dt} = \mathbf{f}(\mathbf{x}, p, t), \tag{2.1}$$

where $\mathbf{x}$ is the state vector in the state space $\mathbb{R}^d$, $\mathbf{f}$ is a smooth (sufficiently differentiable) vector field, p is a parameter in the control or parameter space $\mathbb{R}^p$ and t denotes time. The number d is referred to as the dimension or the number of degrees of freedom of the dynamical system. A trajectory of the dynamical system, starting, for example, at $\mathbf{x}_0$, is a curve $\mathbf{x}(t)$ satisfying (2.1) (i.e., at each point, the vector field

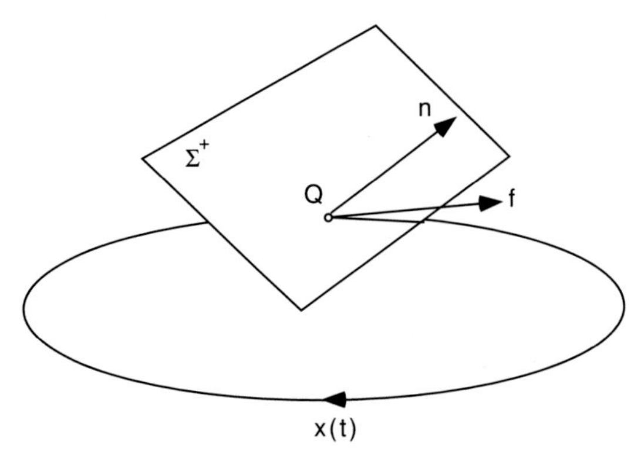

Figure 2.1 Sketch of a Poincaré section Σ^+. A periodic orbit is sketched, which intersects the Poincaré section at the point Q. The vector $\mathbf{f}$ is the tangent to the trajectory at Q, and $\mathbf{n}$ is the outward normal to Σ^+.

$\mathbf{f}$ is tangent to the curve). When the vector field $\mathbf{f}$ does not depend explicitly on time, that is,

$$\frac{d\mathbf{x}}{dt} = \mathbf{f}(\mathbf{x}, p), \tag{2.2}$$

the dynamical system is called autonomous; otherwise, it is called nonautonomous.

When time is discrete, with counter k, we obtain a discrete dynamical system of the form

$$\mathbf{x}_{k+1} = \mathbf{f}(\mathbf{x}_k, p, t_k). \tag{2.3}$$

A continuous time dynamical system can be analysed as a discrete time dynamical system through a so-called Poincaré map. To define a Poincaré map, a hypersurface Σ^+ in the state space $\mathbb{R}^d$, for example, a line segment in two-dimensional state space or a plane in three-dimensional state space, is chosen such that each trajectory is not tangent to it for all time t, that is, when

$$\mathbf{n} \cdot \mathbf{f} \neq 0. \tag{2.4}$$

Here, $\mathbf{n}$ is the normal to the hypersurface (Fig. 2.1) and $\mathbf{f}$ the right-hand side of (2.1); this hypersurface is called a Poincaré section. Let a trajectory intersect a Poincaré section at successive intersections indicated by $\{\mathbf{x}_1, \mathbf{x}_2, \mathbf{x}_3, \ldots\}$; then the Poincaré map $\mathcal{P} : \Sigma^+ \to \Sigma^+$ is defined as

$$\mathbf{x}_{k+1} = \mathcal{P}\mathbf{x}_k. \tag{2.5}$$

2.2 Stability theory and transient behaviour

A solution $\bar{\mathbf{x}}$ of an autonomous dynamical system at a parameter value p is a fixed point if

$$\mathbf{f}(\bar{\mathbf{x}}, p) = 0, \tag{2.6}$$

and hence any trajectory with initial conditions on the fixed point will remain there forever. In this section, we are interested in the transient behaviour of small perturbations on such a fixed point.

2.2.1 Linear stability

In the analysis of the linear stability of a particular fixed point $\bar{\mathbf{x}}$ at $\bar{p}$, small perturbations $\mathbf{y}$ are assumed to be present, that is,

$$\mathbf{x} = \bar{\mathbf{x}} + \mathbf{y}, \tag{2.7}$$

and linearisation of (2.2) around $\bar{\mathbf{x}}$ gives

$$\frac{d\mathbf{y}}{dt} = J(\bar{\mathbf{x}}, p)\mathbf{y}, \tag{2.8}$$

where J is the Jacobian matrix given by

$$J = \begin{bmatrix} \frac{\partial f_1}{\partial x_1} & \cdots & \frac{\partial f_1}{\partial x_n} \\ \cdots & \cdots & \cdots \\ \frac{\partial f_n}{\partial x_1} & \cdots & \frac{\partial f_n}{\partial x_n} \end{bmatrix}. \tag{2.9}$$

The solution of (2.8) with initial condition $\mathbf{y}_0$ is given by

$$\mathbf{y}(t) = e^{Jt}\mathbf{y}_0, \tag{2.10}$$

and hence the time behaviour depends on the eigenvalues of the Jacobian matrix J. The corresponding eigenvectors are usually referred to as the normal modes. If J is decomposed as $J = UDU^{-1}$, where D contains the eigenvalues of J, e^{Jt} is given by

$$e^{Jt} = \sum_{k=0}^{\infty} \frac{1}{k!}(Jt)^k = U\left[\sum_{k=0}^{\infty} \frac{1}{k!}(Dt)^k\right]U^{-1} = Ue^{Dt}U^{-1}. \tag{2.11}$$

If all eigenvalues of J, say, $\sigma_1, \ldots, \sigma_n$ have negative real parts, that is, $\mathcal{R}(\sigma_j) < 0$ for all j, then the fixed point is linearly stable. For this case, indeed all trajectories of (2.8) will approach $\mathbf{y} = 0$ asymptotically (i.e., for $t \to \infty$). When at least one of the eigenvalues σ_k has a positive real part, $\mathcal{R}(\sigma_k) > 0$, the fixed point is said to be unstable. The possible behavior of trajectories near a fixed point for a two-dimensional ($d = 2$) system is presented in Example 2.1.

Example 2.1 Linear stability of fixed points in $\mathbb{R}^2$ Consider the two-dimensional linear system $\dot{\mathbf{x}} = A\mathbf{x}$, written out as

$$\frac{dx}{dt} = ax + by, \tag{2.12a}$$

$$\frac{dy}{dt} = cx + dy, \tag{2.12b}$$

and assume that the coefficient determinant Det $A = ad - bc \neq 0$. The only fixed point of these equations is $x = y = 0$. Its stability is determined by the eigenvalues of A; indicate these with σ and τ. These can both be real or they can form a complex conjugate pair; in the latter case $\sigma = \alpha + i\beta$, $\tau = \alpha - i\beta$.

From linear algebra we use the fact that there exists a nonsingular matrix T such that with the transformation $\tilde{\mathbf{x}} = T\mathbf{x}$, the dynamical system is written as

$$\dot{\tilde{\mathbf{x}}} = TAT^{-1}\tilde{\mathbf{x}} = B\tilde{\mathbf{x}},$$

where B has the same eigenvalues as A.

The transformation is such that the matrix B has one of the following expressions, which are the Jordan forms of the matrix A:

(i) Saddle

$$B = \begin{pmatrix} \sigma & 0 \\ 0 & \tau \end{pmatrix}, \quad \sigma < 0 < \tau.$$

In this case, we have solutions $x(t) = C_1 e^{\sigma t}$, $y(t) = C_2 e^{\tau t}$, and hence the fixed point is always unstable.

(ii) Improper node I

$$B = \begin{pmatrix} \sigma & 0 \\ 0 & \tau \end{pmatrix}, \quad \sigma < \tau < 0 \text{ or } 0 < \sigma < \tau.$$

In this case, we have solutions $x(t) = C_1 e^{\sigma t}$, $y(t) = C_2 e^{\tau t}$. In case $\sigma < \tau < 0$, the fixed point is stable, and in case $0 < \sigma < \tau$, it is unstable.

(iii) Focus

$$B = \begin{pmatrix} \alpha & \beta \\ -\beta & \alpha \end{pmatrix}, \quad \alpha \neq 0, \beta \neq 0.$$

In this case, if we transform to polar coordinates $x = r\cos\theta$, $y = r\sin\theta$, we have solutions $r(t) = C_1 e^{\alpha t}$, $\theta(t) = -\beta t + C_2$. When $\alpha > 0$, the fixed point is unstable, and for $\alpha < 0$, it is stable.

(iv) Centre

$$B = \begin{pmatrix} 0 & \beta \\ -\beta & 0 \end{pmatrix}, \quad \beta \neq 0.$$

In this case, if we again transform to polar coordinates $x = r\cos\theta$, $y = r\sin\theta$, we have solutions $r(t) = C_1$, $\theta(t) = -\beta t + C_2$. The fixed point is neither stable nor unstable and hence it is neutrally stable.

(v) Improper node II

$$B = \begin{pmatrix} \sigma & 0 \\ 1 & \sigma \end{pmatrix}, \quad \sigma \neq 0, \sigma \in \mathbb{R}.$$

In this case, we have solutions $x(t) = C_1 e^{\sigma t}$, $y(t) = (C_2 + t)e^{\sigma t}$, and hence the fixed point is always unstable when $\sigma > 0$ and stable when $\sigma < 0$.

(vi) Proper node

$$B = \begin{pmatrix} \sigma & 0 \\ 0 & \sigma \end{pmatrix}, \quad \sigma \neq 0, \sigma \in \mathbb{R}.$$

In this case, we have solutions $x(t) = C_1 e^{\sigma t}$, $y(t) = C_2 e^{\sigma t}$, and hence $y = kx$; the fixed point is always unstable when $\sigma > 0$ and stable when $\sigma < 0$.

The different cases can be conveniently visualised by plotting the local solutions (also called phase portraits) in a diagram (Fig. 2.2) of the determinant of A versus the trace of A, where

$$\text{Det } A = ad - bc = \sigma\tau, \tag{2.13a}$$

$$\text{Trace } A = a + d = \sigma + \tau. \tag{2.13b}$$

The curve $(\text{Trace } A)^2 = 4 \text{ Det } A$ separates the nodes from the focus phase portraits. ■

2.2.2 Non-normal growth

In the previous section, stability was considered from the viewpoint of asymptotic behavior of the linearised system (2.8). Often, however, the transient behavior (short time) of perturbations of a fixed point of a system is also of interest. An important application comes from weather prediction: given a partially observed initial state, one wants to know the directions in phase space where over a short time the largest deviations from the initial state occur. A particular case of this is when the initial state is an equilibrium state (fixed point) of the system.

To analyse the short-time behavior of perturbations on a particular state, consider the problem

$$\frac{d\tilde{\mathbf{x}}}{dt} = J\tilde{\mathbf{x}}, \tag{2.14}$$

where J is a $d \times d$ Jacobian matrix and $\tilde{\mathbf{x}}$ is the d-dimensional state vector of the perturbation. Let the initial perturbation be given by $\tilde{\mathbf{x}}_0$, then the growth $G(t)$ of this

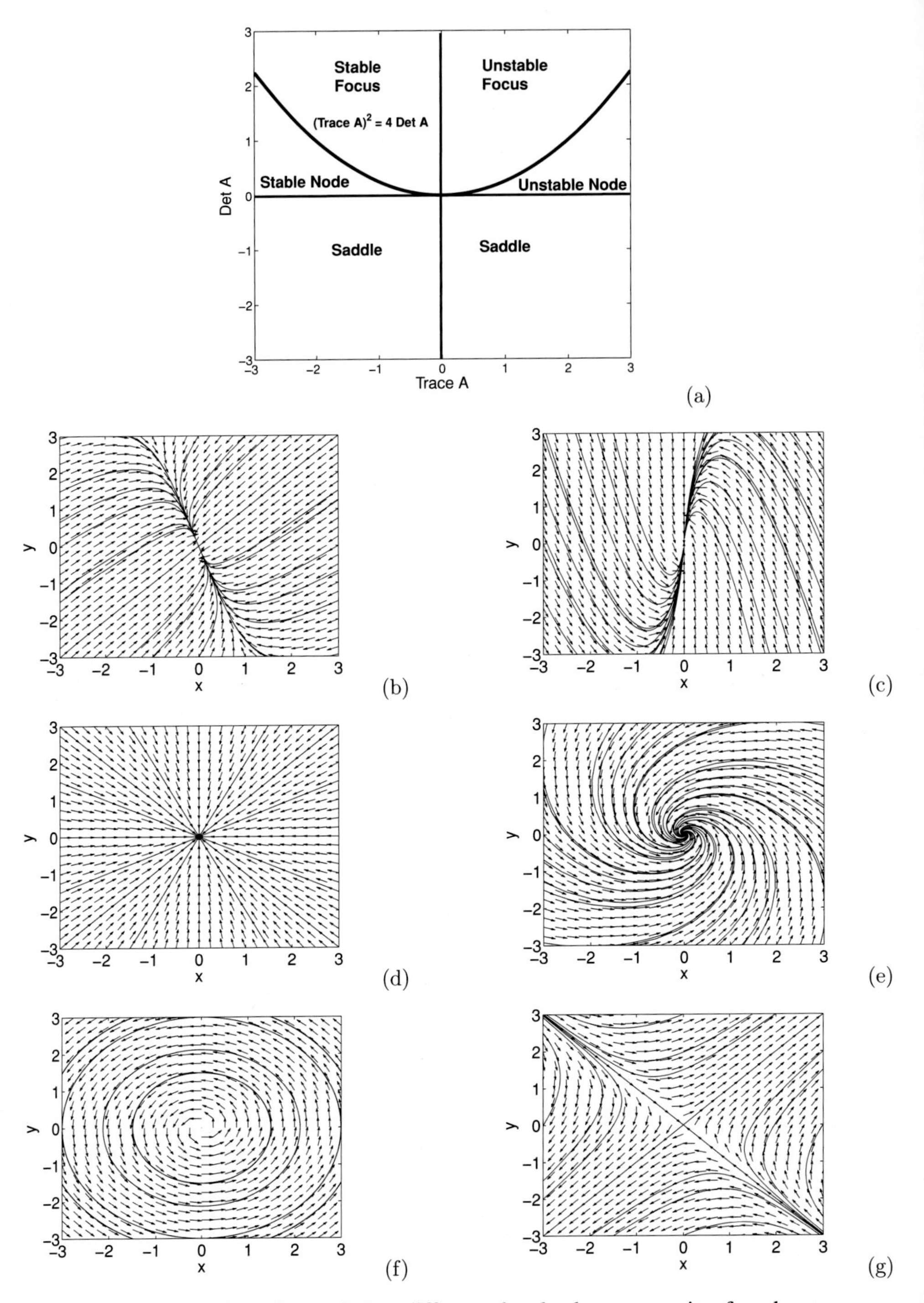

Figure 2.2 (a) Overview of the different local phase portraits for the two-dimensional dynamical system (2.12). The figures that follow were obtained through http://www.math.iupui.edu/~mtc/Chaos/phase.htm. (b) Stable improper node I; $a = -4,\ b = -1,\ c = -3,\ d = -2;\ \sigma = -1,\ \tau = -5$. (c) Stable improper node II; $a = -1, b = 0, c = 3, d = -1; \sigma = -1, \tau = -1$. (d) Stable proper node; $a = -2, b = 0, c = 0, d = -2; \sigma = -2, \tau = -2$. (e) Stable focus; $a = -2, b = -3, c = 3, d = -2;\ \sigma = -2 + 3i,\ \tau = -2 - 3i$. (f) Centre; $a = 0,\ b = -1,\ c = 1,\ d = 0; \sigma = i, \tau = -i$. (g) Saddle; $a = 1, b = 2, c = 2, d = 1; \sigma = -1, \tau = 3$.

perturbation at a time t is defined by

$$G(t) = \frac{\langle \tilde{\mathbf{x}}(t), \tilde{\mathbf{x}}(t) \rangle}{\langle \tilde{\mathbf{x}}_0, \tilde{\mathbf{x}}_0 \rangle}, \tag{2.15}$$

where the brackets represent the inner product on $\mathbb{R}^d$. As the solution of (2.14) is given by

$$\tilde{\mathbf{x}}(t) = e^{Jt} \tilde{\mathbf{x}}_0, \tag{2.16}$$

we find from (2.15) that

$$G(t) = \frac{\langle e^{Jt} \tilde{\mathbf{x}}_0, e^{Jt} \tilde{\mathbf{x}}_0 \rangle}{\langle \tilde{\mathbf{x}}_0, \tilde{\mathbf{x}}_0 \rangle}, \tag{2.17}$$

and because $\langle J\mathbf{x}, \mathbf{y} \rangle = \langle \mathbf{x}, J^\dagger \mathbf{y} \rangle$ where $J^\dagger$ is the adjoint matrix of J, we find

$$G(t) = \frac{\langle e^{J^\dagger t} e^{Jt} \tilde{\mathbf{x}}_0, \tilde{\mathbf{x}}_0 \rangle}{\langle \tilde{\mathbf{x}}_0, \tilde{\mathbf{x}}_0 \rangle}. \tag{2.18}$$

The asymptotic behavior of the system is controlled by the normal mode corresponding to the eigenvalue of J having the largest real part (Farrell and Ioannou, 1996). Hence, when all eigenvalues of J have negative real parts, $G(t) \to 0$ for $t \to \infty$, and hence the perturbation $\tilde{\mathbf{x}}_0$ will decay to zero. However, it is interesting that when J is a non-normal matrix ($JJ^\dagger \neq J^\dagger J$), initial growth may occur. Expanding $G(t)$ in (2.18) for $t \to 0$, that is,

$$e^{J^\dagger t} e^{Jt} \approx (I + J^\dagger t)(I + Jt) = I + (J + J^\dagger)t + \mathcal{O}(t^2), \tag{2.19}$$

where I is the identity matrix. Hence the initial growth is determined by the eigenvalues of the symmetric part of J (i.e., the matrix $\bar{J} = (J + J^\dagger)/2$). Eigenanalysis of $\bar{J}$ typically reveals that high growth rates over short times can be realized even though all normal modes of J are damped. This growth can only occur when the normal modes are nonorthogonal, which is connected to the non-normality of the matrix. An illustrative example (Farrell and Ioannou, 1996) follows.

Example 2.2 Non-normal growth behavior Consider a two-dimensional problem with a trivial equilibrium solution $(0, 0)$ for which the Jacobian matrix is given by

$$J = \begin{pmatrix} -1 & -\frac{\cos\theta}{\sin\theta} \\ 0 & -2 \end{pmatrix},$$

for $\theta \in (0, \pi)$. This matrix has eigenvalues $\sigma_1 = -1$ and $\sigma_2 = -2$, with corresponding eigenvectors $\mathbf{e}_1$ and $\mathbf{e}_2$ given by

$$\mathbf{e}_1 = \begin{pmatrix} 1 \\ 0 \end{pmatrix}; \ \mathbf{e}_2 = \begin{pmatrix} \cos\theta \\ \sin\theta \end{pmatrix},$$

and hence the equilibrium solution is linearly stable. The matrix $\bar{J} = (J + J^\dagger)/2$ is given by

$$\bar{J} = \begin{pmatrix} -1 & -\frac{1}{2}\frac{\cos\theta}{\sin\theta} \\ -\frac{1}{2}\frac{\cos\theta}{\sin\theta} & -2 \end{pmatrix},$$

and its eigenvalues are given by

$$\bar{\sigma}_1 = -\frac{3}{2} + \frac{1}{2\sin\theta}; \; \bar{\sigma}_2 = -\frac{3}{2} - \frac{1}{2\sin\theta}.$$

If $\sin\theta < 1/3$, then $\bar{\sigma}_1 > 0$, and hence there will be initial growth (along the first eigenvector $\bar{\mathbf{e}}_1$) corresponding to $\bar{\sigma}_1$.

This is a remarkable result! Although the fixed point (0, 0) is linearly stable, there is initial growth of some perturbations in a direction that is a linear combination of the eigenvectors $\mathbf{e}_1$ and $\mathbf{e}_2$. So what happens for intermediate time t? Let $E = (\mathbf{e}_1, \mathbf{e}_2)$; we then can write

$$J = E \begin{pmatrix} -1 & 0 \\ 0 & -2 \end{pmatrix} E^{-1},$$

and hence

$$e^{Jt} = E \begin{pmatrix} e^{-t} & 0 \\ 0 & e^{-2t} \end{pmatrix} E^{-1} = \begin{pmatrix} e^{-t} & \frac{\cos\theta}{\sin\theta}(e^{-2t} - e^{-t}) \\ 0 & e^{-2t} \end{pmatrix}.$$

If we take as initial condition a vector along the direction $(\sin\theta, -\cos\theta)$, then we find

$$e^{Jt}\begin{pmatrix} \sin\theta \\ -\cos\theta \end{pmatrix} = \begin{pmatrix} e^{-t}\sin\theta - \frac{\cos^2\theta}{\sin\theta}(e^{-2t} - e^{-t}) \\ -e^{-2t}\cos\theta \end{pmatrix},$$

which for intermediate time ($e^{-2t} \ll e^{-t}$) can be approximated by

$$e^{Jt}\begin{pmatrix} \sin\theta \\ -\cos\theta \end{pmatrix} \approx e^{-t}\frac{1}{\sin\theta}\begin{pmatrix} 1 \\ 0 \end{pmatrix},$$

and when $\sin\theta$ is small, this initial vector amplifies the least damped mode $\mathbf{e}_1$ with a factor $1/\sin\theta$. It is easy to check that the 'optimal' perturbation $(\sin\theta, -\cos\theta)$ is an eigenvector of the adjoint matrix $J^\dagger$ with eigenvalue $\sigma_1^\dagger = -1$.

From the example, we easily deduce that the transient amplification can occur only when both eigenvectors $\mathbf{e}_1$ and $\mathbf{e}_2$ are nonorthogonal. This is associated with the non-normal property of the matrix J; in this case,

$$JJ^\dagger - J^\dagger J = \frac{\cos\theta}{\sin\theta}\begin{pmatrix} \frac{\cos\theta}{\sin\theta} & 1 \\ 1 & -\frac{\cos\theta}{\sin\theta} \end{pmatrix},$$

and hence J is only a normal matrix for $\theta = \pi/2$. In the normal case, the eigenvalues are real and its eigenvectors orthogonal; indeed of $\theta = \pi/2$, then $\bar{\sigma}_1 < 0$, and there is no initial growth. ■

2.3 Bifurcation theory: fixed points

Bifurcation theory addresses changes in the qualitative behavior of a dynamical system as one or several of its parameters vary. The results of this theory permit one to follow systematically the behavior from the simplest kind of model solutions to the most complex, from single to multiple equilibria and from periodic, chaotic to fully turbulent solutions. In this section, we restrict discussion to the theory of bifurcations of fixed points.

2.3.1 Normal forms

Consider the autonomous dynamical system (2.2), for convenience again written here,

$$\frac{d\mathbf{x}}{dt} = \mathbf{f}(\mathbf{x}, p), \tag{2.20}$$

having a fixed point $\bar{\mathbf{x}}$ at a certain parameter value $\bar{p}$. For the linear stability of this fixed point, the transient development of small perturbations $\mathbf{y}$ is considered. Substituting the solutions $\mathbf{y}(t) = \mathrm{e}^{\sigma t}\hat{\mathbf{y}}$, an eigenvalue problem results:

$$J(\bar{\mathbf{x}}, \bar{p})\, \hat{\mathbf{y}} = \sigma \hat{\mathbf{y}}. \tag{2.21}$$

Here $J(\bar{\mathbf{x}}, \bar{p})$ is again the Jacobian matrix; the dependence on $\bar{\mathbf{x}}$ and $\bar{p}$ is now made explicit.

Fixed points for which there are eigenvalues $\sigma = \sigma_r \pm i\sigma_c$ of J with $\sigma_r > 0$ are unstable because the associated perturbations are exponentially growing, whereas fixed points for which $\sigma_r < 0$ are linearly stable. In the situation in which $\sigma_c \neq 0$, the associated eigenmodes will be oscillatory with frequency σ_c (i.e., with a characteristic period of $2\pi/\sigma_c$). The eigenspaces of J associated with eigenvalues $\sigma_r > 0$ (respectively $\sigma_r < 0$) are denoted E^u (respectively E^s), whereas the eigenspaces associated with eigenvalues $\sigma_r = 0$ are denoted E^c.

If J has no purely imaginary eigenvalues, $\bar{\mathbf{x}}$ is called a hyperbolic fixed point. Near a hyperbolic fixed point, the local solution structure of the linearised system is the same as that of the nonlinear system. This is a consequence of the so-called Hartman-Grobman theorem (Guckenheimer and Holmes, 1990). When qualitative changes occur in the fixed point solutions of the dynamical system, such as the changes in type or number of solutions, the dynamical system is said to have undergone a bifurcation. This can occur only at nonhyperbolic fixed points. In the state-parameter space formed by $(\mathbf{x}, p)$, locations at which bifurcations occur are called bifurcation points. A bifurcation that needs at least m parameters to occur is called a codimension-m bifurcation.

The stable (W^s) and unstable (W^u) manifolds of the fixed point $\bar{\mathbf{x}}$ are defined as

$$W^s(\bar{\mathbf{x}}) = \{\mathbf{x} \mid \lim_{t\to+\infty} \phi_t(\mathbf{x}) = \bar{\mathbf{x}}\}, \tag{2.22a}$$

$$W^u(\bar{\mathbf{x}}) = \{\mathbf{x} \mid \lim_{t\to-\infty} \phi_t(\mathbf{x}) = \bar{\mathbf{x}}\}, \tag{2.22b}$$

where $\phi_t(\mathbf{x})$ indicates the trajectory of the dynamical system starting at $\mathbf{x}$. In cases in which bifurcations occur, that is, there are n_0 eigenvalues with vanishing real part, $\sigma_r = 0$, the centre manifold theorem (Guckenheimer and Holmes, 1990) states that there exist unique stable and unstable manifolds W^u and W^s tangent to E^u and E^s at $\bar{\mathbf{x}}$ and a (nonunique) center manifold W^c tangent to E^c at $\bar{\mathbf{x}}$. The three manifolds are all invariant by the flow ϕ_t.

The centre manifold theorem also implies that it is possible to (locally) reduce the dynamics on the center manifold. Typically, taking $\bar{\mathbf{x}} = 0$ for simplicity, one has

$$\frac{d\mathbf{x}}{dt} = \mathbf{L}_0\mathbf{x} + \mathbf{N}(\mathbf{x}, p), \tag{2.23}$$

where $\mathbf{N}$, which depends on the parameter p, has a Taylor expansion starting with at least quadratic terms, $\mathbf{x} \in \mathbb{R}^{n_0}$ and $\mathbf{L}_0$ has n_0 eigenvalues with zero real part. Having reduced the system (2.20) into the system (2.23), it is possible to find a change of coordinates so that the system becomes 'as simple as possible'. The resulting vector field thus obtained is called the normal form. This procedure is an extension of the reduction to Jordan form for matrices to the nonlinear case. Normal form theory (Guckenheimer and Holmes, 1990) provides a way to classify the different kind of bifurcations that may occur with only knowledge of the eigenvalues that lie on the imaginary axis, that is, those of $\mathbf{L}_0$.

2.3.2 Local codimension-1 bifurcations of fixed points

We are now ready to present the most important bifurcation cases, starting with the situation of a single zero eigenvalue ($n_0 = 1$). In this case, there are three important normal forms:

1. Saddle-node bifurcation: it corresponds to the case in which the system (2.23), when reduced to its normal form, is

$$\frac{dx}{dt} = p \pm x^2. \tag{2.24}$$

The sign characterizes supercriticality ($p - x^2$) or subcriticality ($p + x^2$). In the supercritical case, it is straightforward to check that the branch of solutions $x = \sqrt{p}$ is linearly stable and the branch $x = -\sqrt{p}$ is unstable (see Fig. 2.3).

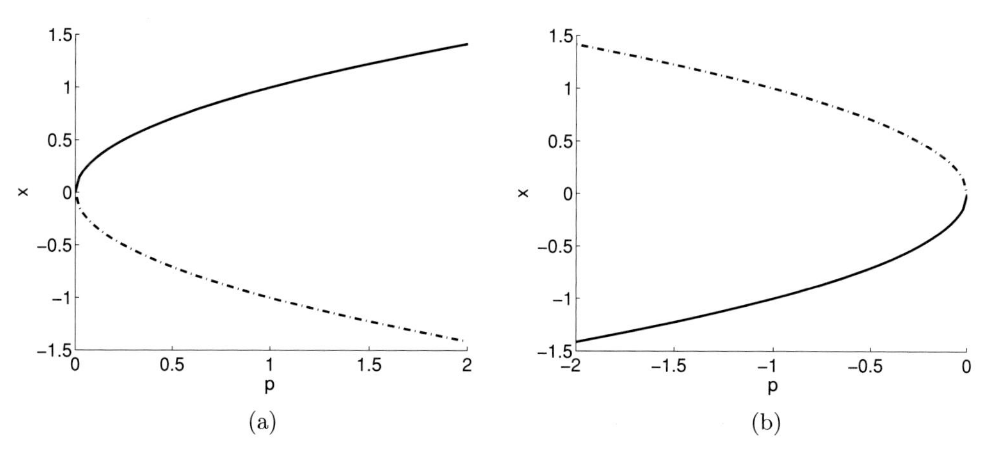

Figure 2.3 Supercritical (a) and subcritical (b) saddle-node bifurcation. The solid (*dash-dotted*) branches indicate stable (unstable) solutions.

2. Transcritical bifurcation: in this case, the normal form is given by

$$\frac{dx}{dt} = px \pm x^2. \tag{2.25}$$

In both subcritical and supercritical cases, there is an exchange of stability from stable to unstable fixed points and vice versa as the parameter p is varied through the bifurcation at $p = 0$ (Fig. 2.4).

3. Pitchfork bifurcation (symmetry breaking): the normal form is

$$\frac{dx}{dt} = px \pm x^3. \tag{2.26}$$

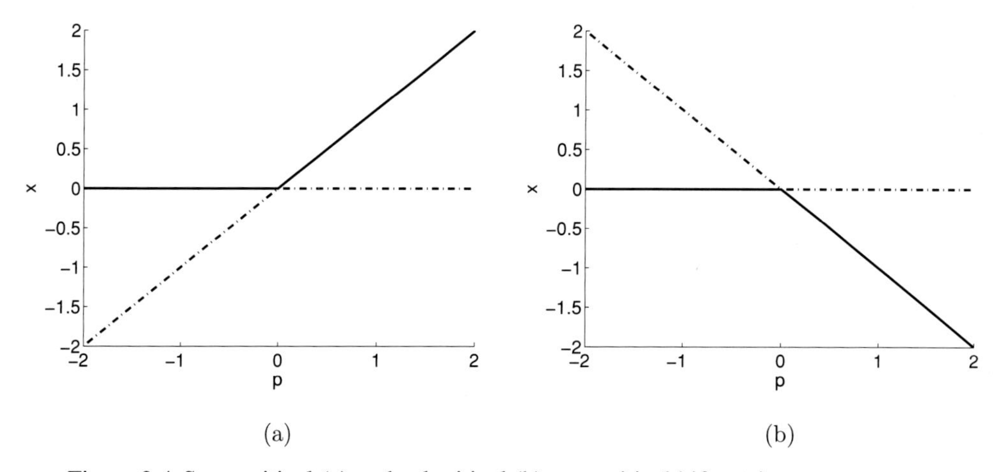

Figure 2.4 Supercritical (a) and subcritical (b) transcritical bifurcation.

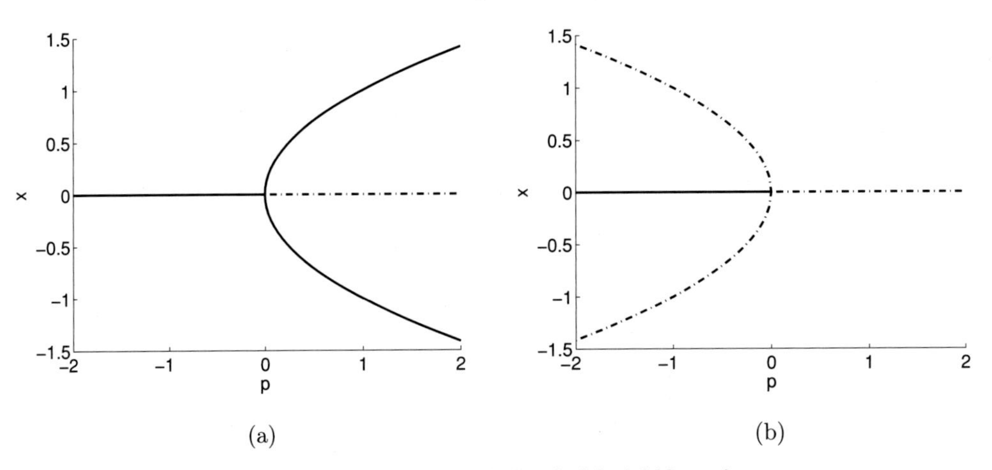

Figure 2.5 Supercritical (a) and subcritical (b) pitchfork bifurcation.

In the supercritical situation ($dx/dt = px - x^3$), there is a transfer of stability from the symmetric solution $x = 0$ to the pair of conjugated solutions $x = \pm\sqrt{p}$ (see Fig. 2.5a). In the supercritical case, it must be noted that the system remains in a neighbourhood of the equilibrium so that one observes a soft or noncatastrophic loss of stability. In the subcritical case ($dx/dt = px + x^3$), the situation differs markedly, as can been seen in Fig. 2.5b. The domain of attraction of the stable fixed point is bounded by the unstable fixed points and shrinks as the parameter p approaches zero to disappear. The system is thus pushed out from the neighbourhood of the now unstable fixed point, leading to a *sharp* or catastrophic loss of stability. Decreasing again the parameter to negative values will not necessarily return the system to the previously stable fixed point because it may have already left its domain of attraction.

Whereas, in the previous cases, the number of fixed points changed as the parameter was varied, it is also possible that a steady solution transfers its stability to a periodic orbit or limit cycle. This kind of transition is called a Hopf bifurcation. It corresponds to the special case of a simple conjugate pair of pure imaginary eigenvalues $\sigma = \pm i\omega$ ($n_0 = 2$) crossing the imaginary axis. The normal form can be written in polar coordinate as

$$\frac{dr}{dt} = pr \pm r^3, \tag{2.27a}$$

$$\dot{\theta} = \omega. \tag{2.27b}$$

Again, the sign determines whether the Hopf bifurcation is supercritical (Fig. 2.6) or subcritical, and the discussion is similar to the pitchfork bifurcation case.

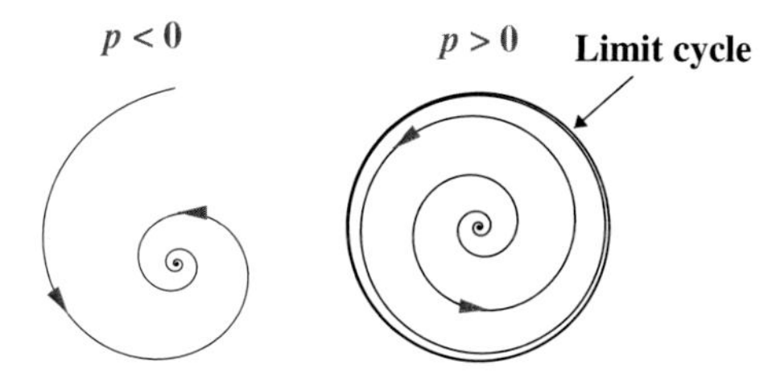

Figure 2.6 Phase-space trajectories associated with a supercritical Hopf bifurcation at $p = 0$. For $p < 0$, there is only one stable fixed point (*left panel*), whereas a stable limit cycle appears for $p > 0$ (*right panel*).

2.3.3 Fast-slow systems

In many models used to study climate variability, some of the parameters are slowly varying (compared with the relaxation time of the model system to perturbations) in time. Such a system can be represented by (Kuehn, 2011)

$$\frac{d\mathbf{x}}{dt} = \mathbf{f}(\mathbf{x}, \mathbf{y}), \tag{2.28a}$$

$$\frac{d\mathbf{y}}{dt} = \epsilon \mathbf{g}(\mathbf{x}, \mathbf{y}), \tag{2.28b}$$

where $\mathbf{x} \in \mathbb{R}^d$ is the state vector, $\mathbf{y} \in \mathbb{R}^p$ now indicates the parameter vector and $\epsilon \ll 1$ is a dimensionless positive real number.

When a slow time scale $\tau = \epsilon t$ is introduced, then with $d\mathbf{x}/dt = \epsilon d\mathbf{x}/d\tau$, we find the equivalent system to (2.28) as

$$\epsilon \frac{d\mathbf{x}}{d\tau} = \mathbf{f}(\mathbf{x}, \mathbf{y}), \tag{2.29a}$$

$$\frac{d\mathbf{y}}{d\tau} = \mathbf{g}(\mathbf{x}, \mathbf{y}), \tag{2.29b}$$

In the limit $\epsilon = 0$ these equations reduce to

$$0 = \mathbf{f}(\mathbf{x}, \mathbf{y}), \tag{2.30a}$$

$$\frac{d\mathbf{y}}{d\tau} = \mathbf{g}(\mathbf{x}, \mathbf{y}), \tag{2.30b}$$

which are referred to as the fast (2.30a) and the slow (2.30b) flow. The slow flow is hence constrained on the so-called critical manifold C with

$$C = \left\{(\mathbf{x}, \mathbf{y}) \in \mathbb{R}^{d+p} : \mathbf{f}(\mathbf{x}, \mathbf{y}) = 0\right\}, \tag{2.31}$$

which just contains the fixed points of the fast system. The eigenvalues of the Jacobian of $\mathbf{f}$ determine whether the critical manifold is hyperbolic (no eigenvalues on the imaginary axis).

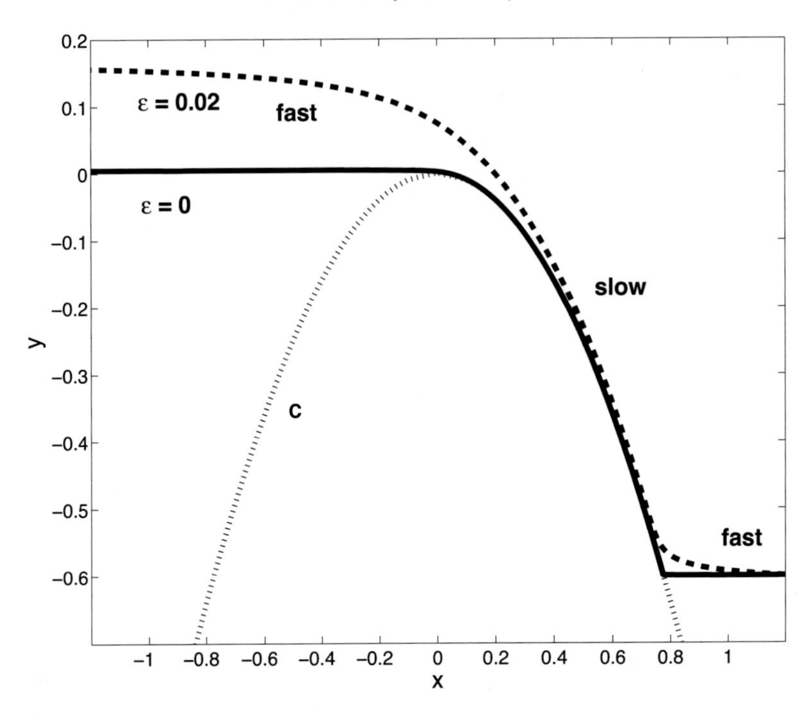

Figure 2.7 Trajectories for the slow-fast system (2.32) for $\epsilon = 0$ (drawn) with an initial condition $(1.2, -0.6)$ showing the slow and fast segments. The dashed curve is a trajectory with the same initial condition but with $\epsilon = 0.02$. The critical manifold is indicated by C (figure redrawn from Kuehn, 2011).

Example 2.3 The fast-slow saddle node Consider the one-dimensional system in the slow time scale τ given by

$$\epsilon \frac{dx}{d\tau} = -y - x^2, \tag{2.32a}$$

$$\frac{dy}{d\tau} = 1. \tag{2.32b}$$

The fixed points of the fast system exist only for $y \leq 0$ and are given by $x = \pm\sqrt{-y}$. As the Jacobian $J = -2x$, the point $(0, 0)$ is a nonhyperbolic point, and only the branch of fixed points $x = \sqrt{-y}$ is stable.

Suppose that the initial parameter value $y_0 < 0$; the parameter y is then slowly varying in time and will eventually become positive. A trajectory will (as long as $y < 0$) be attracted to the stable branch $x = \sqrt{-y}$ and then follow this branch slowly until $y = 0$. When $y > 0$, it escapes quickly to infinity. At $y = 0$, $x = 0$, it is said that a critical transition has occurred, as there is a change from a relatively slow flow to a relatively fast one (Fig. 2.7). For $\epsilon = 0.02$, a trajectory with the same initial condition is also plotted in Fig. 2.7 showing also a fast attraction to the critical manifold and a slow motion near this manifold followed by a fast motion from it later on. ■

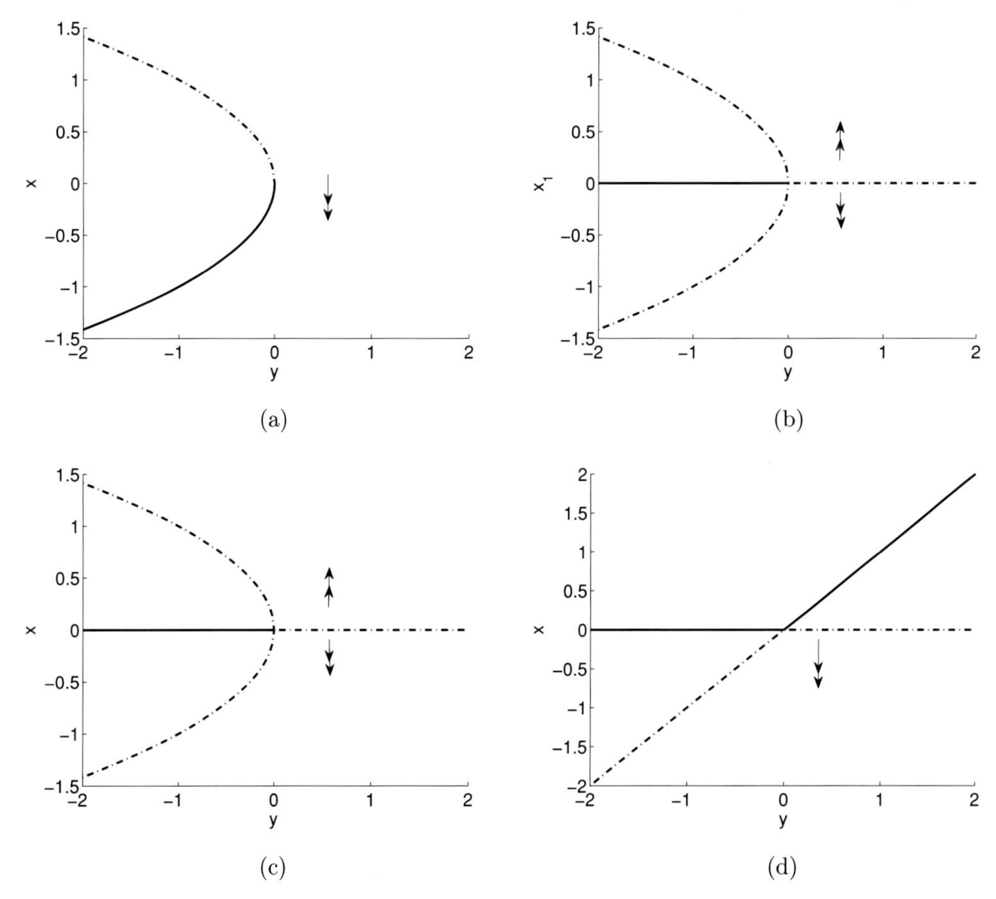

Figure 2.8 Fast subsystem bifurcation diagrams for four bifurcations that are critical transitions. Double arrows indicate the flow of the fast subsystem. (a) Saddle-node bifurcation, (b) the subcritical Hopf bifurcation (only one component x_1 is plotted), (c) the subcritical pitchfork bifurcation and the (d) transcritical bifurcation (redrawn from Kuehn, 2011).

The fast-slow systems are the prototype dynamical systems for studying the popular notion of a so-called tipping point, where small changes in parameters can lead to a large response over a relatively short time interval. Critical conditions can occur only for a saddle-node bifurcation (Fig 2.8a), the subcritical Hopf bifurcation (Fig 2.8b), the subcritical pitchfork bifurcation (Fig 2.8c) and the transcritical bifurcation (Fig 2.8d).

2.4 Bifurcation theory: periodic orbits

In this section, we present the theory of possible transitions in periodic orbits of autonomous dynamical systems when parameters are varied, mostly following Simonnet et al. (2009).

2.4.1 Local codimension-1 bifurcations of limit cycles

Assume that one has a limit cycle γ of the autonomous system (2.20) for a parameter $\bar{p}$ that we omit in the notations for simplicity and whose corresponding solution is $\bar{\mathbf{x}}(t) = \bar{\mathbf{x}}(t + T)$, where T is the period of the orbit. We consider an infinitesimal perturbation $\xi(t)$ of γ; that is, we let $\mathbf{x}(t) = \bar{\mathbf{x}}(t) + \xi(t)$ in (2.20), and, neglecting quadratic terms, one obtains

$$\dot{\xi} = J(\bar{\mathbf{x}}(t))\xi, \tag{2.33}$$

and $J(\bar{\mathbf{x}}(t))$ is now a T-periodic matrix.

It can be shown (Guckenheimer and Holmes, 1990) that the fundamental solution matrix X of the system (2.33) is written as

$$X(t + T) = MX(t). \tag{2.34}$$

The matrix M is called the monodromy matrix, and its eigenvalues $\rho_1, \ldots, \rho_d$ are called the Floquet multipliers. The monodromy matrix is not uniquely determined by the solutions of (2.33), but its eigenvalues are uniquely determined. Because the perturbation $\xi(t) = \bar{\mathbf{x}}(t + \epsilon) - \bar{\mathbf{x}}(t)$, where ϵ is small, is T periodic, it immediately implies that M has a unit eigenvalue (i.e., perturbations along γ neither diverge or converge). The linear stability of γ is thus determined by the remaining $d - 1$ eigenvalues.

Let Σ^+ be a (fixed) local cross-section of dimension $d - 1$ (cf. Fig. 2.1) of the limit cycle γ such that the periodic orbit is not tangent to this hypersurface and denote $\mathbf{x}^\star$ the intersection of Σ^+ with γ. There is a nice geometrical interpretation of the monodromy matrix in terms of the Poincaré map defined as $\mathcal{P}(\mathbf{x}) = \phi_\tau(\mathbf{x})$, where $\mathbf{x}$ is assumed to be in a neighbourhood of $\mathbf{x}^\star$, and τ is the time taken for the orbit $\phi_t(\mathbf{x})$ to first return to Σ^+ (as $\mathbf{x}$ approaches $\mathbf{x}^\star$, τ will tend to T). After a change of basis such that the matrix M has a column $(0, \ldots 0, 1)^T$ corresponding to the unit eigenvalue, the remaining block $(d - 1) \times (d - 1)$ matrix corresponds to the linearised Poincaré map.

These remarks show that the bifurcations of limit cycles are related to the behaviour of a discrete dynamical system (the Poincaré map) $\mathbf{x}_{n+1} = \mathcal{P}\mathbf{x}_n$ rather than a continuous dynamical system such as in the case of fixed points. The bifurcation theory for fixed points of the iterative map with eigenvalue having unit norm is completely analogous to the bifurcation theory for fixed points with an eigenvalue on the imaginary axis. Periodic orbits become unstable when Floquet multipliers ρ_i cross the unit circle as the parameter p is changed (remember that the Floquet multipliers depend on the parameter p). There are three important cases:

1. A real Floquet multiplier is crossing the unit circle $\rho(\bar{p}) = 1$ (saddle-node bifurcation of periodic orbit). This situation can be shown to be topologically equivalent to the one-dimensional discrete dynamical system,

$$u_{n+1} = \mathcal{P}(u_n), \text{ with } \mathcal{P}(u) = p + u \pm u^2. \tag{2.35}$$

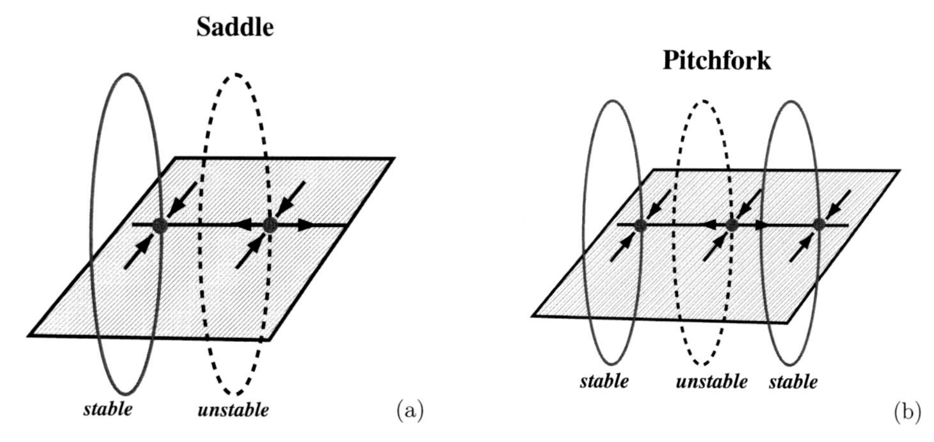

Figure 2.9 State space view associated with a saddle-node (a) and pitchfork (b) bifurcation of a periodic orbit.

Consider the supercritical case $\mathcal{P}(u) = p + u - u^2$ and assume that $\bar{p} = 0$ for simplicity. As p becomes positive, two fixed points $u_1^\star$ and $u_2^\star$ of the iterative map (2.35) appear, which are solutions of $\mathcal{P}(u) = u$. These two fixed points correspond to the appearance of two new families of periodic orbits. One family is stable ($\mathcal{P}'(u_1^\star) < 1$), whereas the other is unstable ($\mathcal{P}'(u_2^\star) > 1$). Like in the case of fixed points, particular constraints (such as symmetry) may lead to transcritical or pitchfork bifurcations (see Fig. 2.9).

2. A real Floquet multiplier is crossing the unit circle with $\rho(\bar{p}) = -1$. This situation is called flip or period-doubling bifurcation and has no equivalent for fixed points. The system is topologically equivalent to

$$u_{n+1} = \mathcal{P}(u_n), \text{ with } \mathcal{P}(u) = -(1+p)u \pm u^3. \tag{2.36}$$

This situation corresponds to the pitchfork case for the second iterate $\mathcal{P}^2$ map. Again, consider (with $\bar{p} = 0$) the supercritical case $\mathcal{P}(u) = -(1+p)u + u^3$. As p becomes positive, two fixed points of the second iterate $\mathcal{P}^2$ appear, which are not fixed points of the first iterate. This means that another stable periodic orbit of period $2T$ arises, whereas the original periodic orbit γ becomes unstable (see Fig. 2.10). The corresponding trajectories alternate from one side of γ to the other along the direction of the eigenvector associated with the Floquet multiplier $\rho = -1$.

3. The final example corresponds to the case of a pair of complex conjugate Floquet multipliers ρ crossing the unit circle such that $|\rho(\bar{p})| = |e^{i\phi}| = 1$. This bifurcation is called Neimark-Sacker or torus bifurcation. If one assumes after reduction on a two-dimensional invariant manifold that $d\rho(p)/dp \neq 0$ at $p = \bar{p}$, then there is a

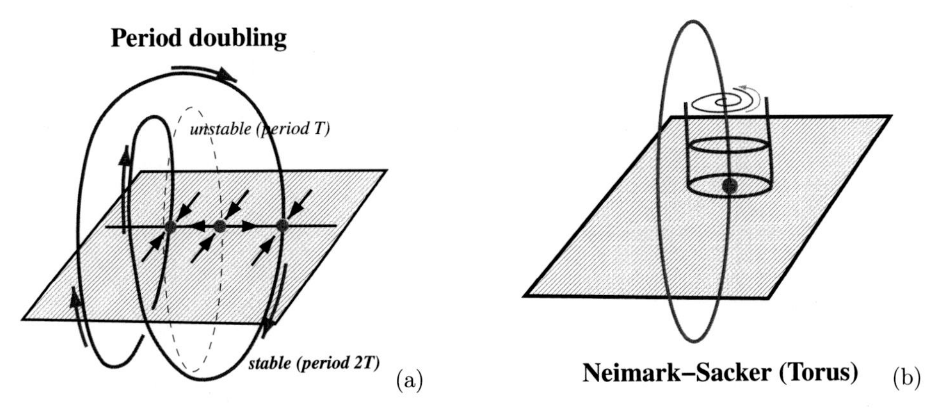

Figure 2.10 State space view of a period-doubling or flip bifurcation of a periodic orbit (a) and a Neimark-Sacker or torus bifurcation (b).

change of coordinates such that the Poincaré map takes the following form in polar coordinates (r, θ)

$$\begin{aligned} \mathcal{P}_r(r,\theta) &= r + c(p - \bar{p})r + ar^3 \\ \mathcal{P}_\theta(r,\theta) &= \theta + \varphi + br^2 \end{aligned}, \tag{2.37}$$

where a, b and c are parameters. Provided $a \neq 0$, this normal form indicates that a close curve generically bifurcates from the fixed point; this closed curve corresponds to a two-dimensional invariant torus (Kuznetsov, 1995).

2.5 Beyond simple behaviour: attractors

Although much information is available on the behaviour of dynamical systems beyond the stability boundaries of limit cycles, it is too diverse to be presented here. When we are interested in the long-term (near equilibrium) behaviour of dissipative dynamical systems (systems with friction and/or diffusion), trajectories always approach a so-called attractor, which is an attracting subset in state space with the following properties:

(i) It is an invariant set. If a trajectory in on the attractor, it will remain on this set.
(ii) Every point of the attractor will eventually be visited (recurrence).
(iii) There is no subset of the attractor with the same properties (i) and (ii) above (irreducability).

Of course, stable fixed points, limit cycles and tori are examples of such attractors. Before introducing specific properties of more complex attractors, an example is given of how these complicated objects arise in state-parameter space.

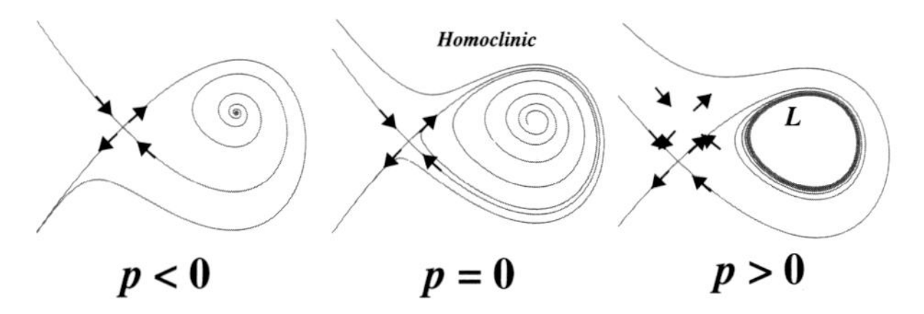

Figure 2.11 Phase trajectories of the system (2.38) in the (u, v) plane showing a homoclinic orbit (*middle panel*) arising as the interaction between a periodic orbit and a saddle fixed point.

2.5.1 Homoclinic and heteroclinic bifurcations of equilibria

One of the mechanisms of the emergence of complex behaviour in dynamical systems is when there is a global change of state space as one parameter is varied. This situation corresponds to a global bifurcation, and we focus here on the case of homoclinic orbits. This is a central issue in dynamical systems theory because global bifurcations in dynamical systems with $d > 2$ are in general responsible for the emergence of chaos and strange attractors. Ruelle and Takens (1970) introduced the notion of a strange attractor, in referring to a chaotic attractor characterised by sensitivity to the initial state and whose dimension is fractional, rather than integer.

Homoclinic orbits correspond to the interaction between an unstable fixed point and a periodic orbit. This is exemplified by Fig. 2.11, which corresponds to a planar homoclinic orbit to a saddle-node fixed point (middle panel of Fig. 2.11). The corresponding normal form is

$$\begin{aligned} \dot{u} &= -u + 2v + u^2 \\ \dot{v} &= (2 - p)u - v - 3u^2 + \tfrac{3}{2}uv \end{aligned} . \tag{2.38}$$

Indeed, it can be shown that homoclinic bifurcations on the plane are determined by the saddle quantity $\kappa = \sigma_1 + \sigma_2$, where σ_i are the eigenvalues at the (saddle-type) fixed point at the homoclinic bifurcation ($p = \bar{p}$). The system (2.38) corresponds to $\kappa = -2 < 0$, with $\bar{p} = 0$. As both objects connect to each other, the period of the orbit becomes longer as p approaches zero to become infinite at $p = 0$. The trajectory along the unstable manifold of the fixed point approaches in infinite time the same fixed point along its stable manifold. The homoclinic orbit is then attracting the trajectories inside. For p positive, a stable periodic orbit (indicated by L in Fig. 2.11) is created.

Similar to homoclinic orbits, which connect a fixed point to itself, a heteroclinic orbit (sometimes called a heteroclinic connection) is a path in phase space that joins two different fixed points. For a continuous dynamical system, suppose there are equilibria at $x = x_0$ and $x = x_1$; then a solution $x(t)$ is a heteroclinic orbit from x_0 to x_1 if

$$x(t) \to x_0 \quad \text{as} \quad t \to -\infty, \; x(t) \to x_1 \quad \text{as} \quad t \to +\infty, \tag{2.39}$$

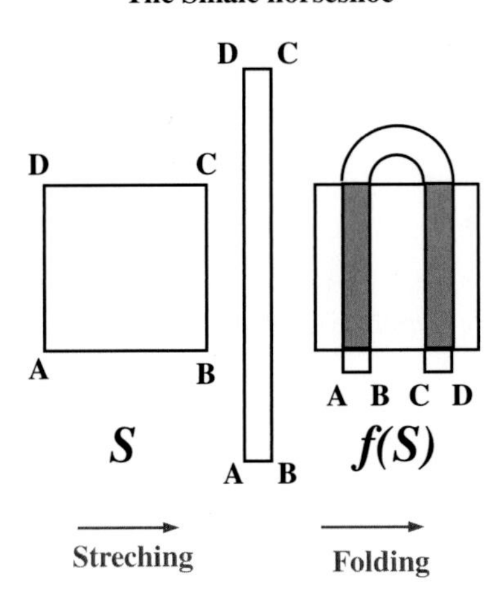

Figure 2.12 One iteration of the Smale horseshoe map f on the square S illustrating that complicated behaviour occurs through folding and stretching in state space.

which implies that the orbit is contained in the stable manifold of x_1 and in the unstable manifold of x_0.

2.5.2 *Chaos and fractal dimension*

Homoclinic bifurcations in systems with $d > 2$ are considerably more complicated. They most often induce stretching and folding in some bounded set of state space and lead to the appearance of (Smale) horseshoes and possibly chaos. Typical horseshoes are given by the celebrated map sketched in Fig. 2.12. In the limit where this map is iterated an infinite number of time, a fractal Cantor set is obtained (cf. Example 2.4). In the neighbourhood of the saddle fixed point where the homoclinic orbit is connected, the leading eigenvalues determine how much stretching and folding occurs in phase space. Consider a three-dimensional system with two eigenvalues $\sigma_1, \sigma_2 < 0$ and one with $\sigma_3 > 0$ at the hyperbolic fixed point (saddle) and assume that homoclinic orbits exist for $p = \bar{p}$. There are then two important cases: (i) $\sigma_i \in \mathbb{R}, i = 1, 2, 3$ and (ii) $\sigma_{1,2} = \rho \pm i\omega$ with $\rho < 0$ and $\sigma_3 > 0$.

In situations in which there is only one homoclinic orbit for case (i), the system behaves more or less similarly to the planar case (2.38). In particular, a periodic orbit is created for $p \neq \bar{p}$ (Wiggins, 1994). The interesting case corresponds to the situation in which two homoclinic orbits connect to the saddle point. In particular, when there is some reflection symmetry in the system, it amounts to having only one parameter controlling the existence of the two homoclinic orbits at the same time. The symmetric case is of particular historical interest, because this is precisely the situation that arises in the much-studied Lorenz system (Lorenz, 1963).

Under particular conditions on the eigenvalues (e.g., $-\sigma_2 < \sigma_3 < -\sigma_1$) and invariance of the system by symmetry ($(u_1, u_2, u_3) \to (-u_1, u_2, -u_3)$), one observes a one-sided homoclinic explosion. This means that for, say, $p \leq \bar{p}$, there is nothing spectacular associated with the dynamics near the broken homoclinic orbit, but for $p > \bar{p}$, horseshoes appear seemingly out of nowhere. There exists a value of p, say p_0, such that for all p such that $\bar{p} < p < p_0$, the Poincaré map exhibits an invariant Cantor set. An infinite number of unstable periodic orbits of all possible periods are created. Although no strange attractor actually exists, these horseshoes are considered as being the chaotic heart of numerically observed strange attractors.

We briefly illustrate the situation of the Lorenz system. This is a three-mode truncation of the Rayleigh-Bénard convection problem and corresponds to

$$\begin{aligned} \dot{x} &= s(y - x) \\ \dot{y} &= -xz + rx - y\,, \\ \dot{z} &= xy - bz \end{aligned} \tag{2.40}$$

with fixed parameters $s = 10$, $b = 10/3$ and control parameter r. As r increases, one observes the following bifurcations: for $0 < r < 1$, the origin is globally stable; at $r = 1$, there is a supercritical pitchfork bifurcation; at $r = 470/19$, there are subcritical asymmetric Hopf bifurcations; and the first homoclinic explosion is observed at $r \approx 13.93$. It must be noted that an important bifurcation occurs at $r \approx 24.06$, where the invariant Cantor set is destroyed through successive homoclinic explosions, which yields to the existence of a genuine strange attractor.

The case (ii), with $\sigma_{1,2} = \rho \pm i\omega$ with $\rho < 0$ and $\sigma_3 > 0$, was first studied by Shilnikov (1965), and the dynamics is known as the Shilnikov phenomenon; let $\delta = -\rho/\sigma_3$. It can be shown that, provided $\delta < 1$, there is a finite number of periodic orbits for $p < \bar{p}$ and $p > \bar{p}$ and a countable infinity of periodic orbits for $p = \bar{p}$. Thus, contrary to the Lorenz case (i), the Shilnikov case is two sided. We present in Fig. 2.13 the bifurcation behaviour of the periodic orbits for the principal homoclinic orbit; such behavior is often referred to as Shilnikov wiggles. For $\delta > 1$, the situation is simple because there are no horseshoes. Only one periodic orbit exists at one side of the critical parameter value (say $p > \bar{p}$), which becomes homoclinic at $p = \bar{p}$. When a reflection symmetry is present, this case is referred to as a gluing bifurcation.

Example 2.4 The Cantor set and fractal dimension The Cantor set arises when the interval [0, 1] is iteratively divided into smaller intervals. In the first iteration [0, 1] is cut into three equal intervals, and the middle interval [1/3, 2/3] is removed. This procedure is repeated with subsequent intervals: so in the second iteration, [0, 1/3] is cut into three intervals, and the interval [1/9, 2/9] is removed. One then asks: what set remains when the number of iterations $n \to \infty$? To answer this, we first look at

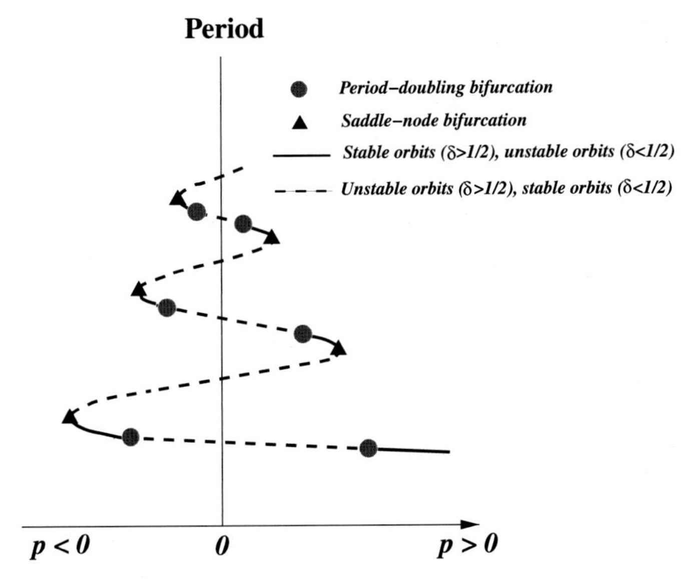

Figure 2.13 Dependence of the period and stability of the bifurcated periodic orbits for the case $\delta < 1$. The critical parameter value $\bar{p}$ is set to zero for convenience.

what is removed (Fig 2.14). At each iteration n, there are 2^n intervals, and the length of the intervals removed R_n is

$$R_n = \frac{1}{3} + 2\frac{1}{9} + 4\frac{1}{27} + \cdots + 2^{n-1}\left(\frac{1}{3}\right)^n, \tag{2.41}$$

and hence

$$\lim_{n\to\infty} R_n = \frac{1}{3}\sum_{i=0}^{\infty} 2^n \left(\frac{1}{3}\right)^n = \frac{1}{3}\frac{1}{1-\frac{2}{3}} = 1, \tag{2.42}$$

so the Cantor set has measure zero when using the Euclidian length measure.

The Cantor set also serves to illustrate the fractal properties of attractors. One of the fractal dimensions of a certain set is the so-called box-counting dimension D_0, which is defined as

$$D_0 = \lim_{r\to 0} \frac{\ln N(r)}{\ln(1/r)}, \tag{2.43}$$

where $N(r)$ is the number of intervals (boxes) of length r that are needed to completely cover the set. Examples are as follows:

(a) Consider the case of two isolated points. In this case, we can take $N(r) = 2$ for every r and hence

$$D_0 = \lim_{r\to 0} \frac{\ln 2}{\ln(1/r)} \to 0. \tag{2.44}$$

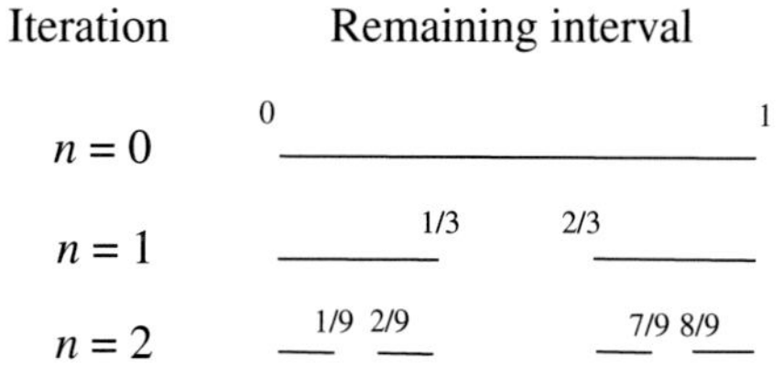

Iteration Remaining interval

$n = 0$ 0 ——— 1

$n = 1$ 1/3 2/3

$n = 2$ 1/9 2/9 7/9 8/9

Figure 2.14 Sketch of the iterative procedure, with iteration index n, leading to the Cantor set.

(b) Consider a line element of length L. In this case, we have $N(r) = L/r$ and

$$D_0 = \lim_{r\to 0} \frac{\ln(L/r)}{\ln(1/r)} \to 1. \tag{2.45}$$

(c) For the Cantor set, we need $N(r) = 2^n$ intervals of length $r = (1/3)^n$ to cover the set and hence

$$D_0 = \lim_{n\to\infty} \frac{\ln(2^n)}{\ln(3^n)} \to \frac{\ln 2}{\ln 3} \approx 0.631. \tag{2.46}$$

The examples show that the box-counting dimension reduces to the normal Euclidian measure for two points and a line element, but it gives noninteger dimensions for fractal sets such as the Cantor set. ■

2.5.3 Sensitivity to initial conditions

A final important notion is that of divergence of trajectories or sensitivity to initial conditions, which is characteristic of chaotic systems. Consider a trajectory $\phi_t(\mathbf{x}) \in \mathbb{R}^d$ of the dynamical system

$$\dot{\mathbf{x}} = \mathbf{f}(\mathbf{x}, t), \tag{2.47}$$

at some time t and study small perturbations $\mathbf{y}$ from this trajectory, that is,

$$\mathbf{x} = \phi_t(\mathbf{x}) + \mathbf{y}.$$

If we linearize around the trajectory, we find the linearized equations

$$\dot{\mathbf{y}} = J(\phi_t(\mathbf{x}))\mathbf{y}, \tag{2.48}$$

where J is the time-dependent Jacobian matrix.

The Lyapunov exponents λ_i are defined as

$$\lambda_i = \lim_{t\to\infty} \frac{1}{t} \ln |\mathbf{y}_i(t)| \tag{2.49}$$

for $i = 1, \ldots, d$, where the $\mathbf{y}_i$ are the solutions of (2.48). The largest Lyapunov exponent of a fixed point is smaller than zero, it is zero for a periodic orbit and a torus has two zero largest Lyapunov exponents. When the largest Lyapunov exponent is larger than zero, there is exponential divergence of trajectories, and the behavior of the dynamical system is called chaotic.

This completes a very brief description of the theory of deterministic dynamical systems, covering all concepts needed in Chapters 7–12 of this book. We have seen that when parameters are changed, the behaviour of a dynamical system can change qualitatively due to the occurrence of bifurcation points. When the ratio of forcing and dissipation is increased, in many cases the attractors of the system change from simple fixed points, through limit cycles or tori to (strange) attractors often associated with the appearance of sensitivity to initial conditions. The Lyapunov exponents previously introduced are the measures of the appearance of such complicated behaviour in a dynamical system.

3

Introduction to Stochastic Calculus

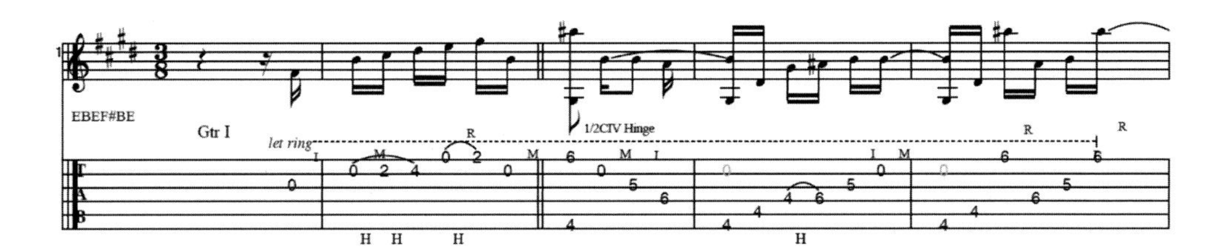

Ordering random elements.
EBEF♯BE, Causeway, Alex de Grassi

In this chapter, an introduction to the theory of stochastic calculus is given. Although there are many books on this topic (Van Kampen, 1981; Kloeden et al., 1994; Oksendal, 1995; Mikosch, 2000), many of these are quite mathematically oriented and hence less accessible for climate scientists. Here an attempt is made to present this material from a 'user' point of view (Higham, 2001), while being as rigorous as possible.

3.1 Random variables

The starting point is the concept of a random variable. We are all familiar with examples where random variables play a role, such as throwing a dice and tossing a coin. In this example, we have the only outcomes 'head' or 'tail'. If we attribute 0 to the outcome $\omega =$ 'head' and 1 to the outcome $\omega =$ 'tail', then we can define a random variable $X : \Omega \to \{0, 1\}$, where Ω is the outcome space. In general, a random variable $X(\omega)$ is a real-valued function defined on Ω.

To tackle questions related to the mean and spread of values of the random variable X, we first have to define an algebraic structure on the outcome space Ω. More specifically, a σ-algebra $\mathcal{F}$ is defined on Ω, which at least contains the empty set

$\emptyset$ and its complement Ω. Moreover, if $A \in \mathcal{F}$, then so is its complement A^c, and if $A, B \in \mathcal{F}$, then so are $A \cap B$, $A \cup B$, $A^c \cap B$, $A \cap B^c$ $A^c \cup B$ and $A \cup B^c$. The intuitive meaning of $\mathcal{F}$ is that if one applies the operations $\cap$, $\cup$ and ^{c} to the elements of $\mathcal{F}$ – the 'events' – the result is still an element of $\mathcal{F}$.

On a σ-algebra $\mathcal{F}$, a probability measure $P : \mathcal{F} \to [0, 1]$ is defined as (for $A, B \in \mathcal{F}$)

$$P(\emptyset) = 0 \; ; \; P(\Omega) = 1 \; ; \; P(A^c) = 1 - P(A), \tag{3.1a}$$

$$P(A \cup B) = P(A) + P(B) - P(A \cap B). \tag{3.1b}$$

For example, if a coin is 'fair', we assign a probability $1/2$ to both events 'head' and 'tail' and therefore $P(\{\omega : X(\omega) = 0\}) = P(\{\omega : X(\omega) = 1\}) = 1/2$.

The distribution function $F_X(x)$ of the random variable X provides a measure of the collection of the probabilities $P(X \leq x) = P(\{\omega : X(\omega) \leq x\})$ for certain $x \in \mathbb{R}$. A distribution function is either continuous or discrete. In the discrete case, with a discrete random variable X, the distribution function is given by

$$F_X(x) = \sum_{k : x_k \leq x} p_k, \tag{3.2}$$

where $p_k = P(X = x_k)$, $0 \leq p_k \leq 1$ for all k and $\sum_k p_k = 1$. Important discrete distributions are the binomial distribution $Bin(n, p)$ for $n \in \mathbb{N}$ and $p \in (0, 1)$ with

$$P(X = k) = \binom{n}{k} p^k (1 - p)^{n-k}, \tag{3.3}$$

for $k \in \mathbb{N}$ and the Poisson distribution $Poi(\lambda)$ containing a parameter $\lambda > 0$ with

$$P(X = k) = e^{-\lambda} \frac{\lambda^k}{k!}. \tag{3.4}$$

The Poisson distribution has a large applicability. The number of cars that pass a certain point over a given time period, the number of stars in a given volume of space, the spelling mistakes on a page of text all satisfy a Poisson distribution. Note that the Poisson distribution is the limit $n \to \infty$ of the binomial distribution with $p = \lambda/n$ because

$$\lim_{n \to \infty} \binom{n}{k} p^k (1 - p)^{n-k} = e^{-\lambda} \frac{\lambda^k}{k!}. \tag{3.5}$$

In contrast, a continuous random variable has a continuous distribution function F_X, which, in most cases, has a probability density function $f_X \geq 0$ such that

$$F_X(x) = \int_{-\infty}^{x} f_X(y)\, dy; \quad \int_{-\infty}^{\infty} f_X(y)\, dy = 1. \tag{3.6}$$

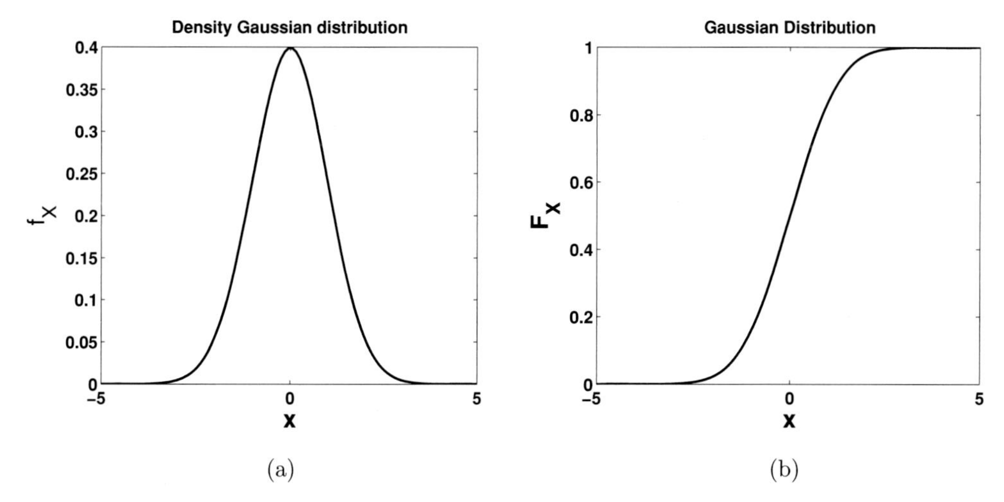

Figure 3.1 (a) Plot of the Gaussian probability density function (3.8) for $\sigma = 1$ and $\mu = 0$. (b) Plot of the Gaussian distribution function (3.6) for $\sigma = 1$ and $\mu = 0$.

An important example is the uniform distribution $U(a, b)$ on the interval $[a, b] \in \mathbb{R}$ with a probability density function

$$f_X(x) = \frac{1}{b-a};\; x \in (a, b) \tag{3.7}$$

and $f_X(x) = 0$ elsewhere. Another important example is the Gaussian (or normal) distribution $N(\mu, \sigma^2)$ with a density

$$f_X(x) = \frac{1}{\sqrt{2\pi}\sigma} e^{-\frac{(x-\mu)^2}{2\sigma^2}}, \tag{3.8}$$

with $\mu \in \mathbb{R}$ referred to as the mean and $\sigma \in \mathbb{R}$ as the standard deviation. As a special case, the distribution $N(0, 1)$ is called the standard normal distribution, and plots of f_X and F_X of this distribution are given in Fig. 3.1. One can easily show that when X has a distribution $N(\mu, \sigma^2)$, it follows that $Y = (X - \mu)/\sigma$ has a distribution $N(0, 1)$. About 68% of values drawn from a Gaussian distribution are within one standard deviation σ away from the mean, about 95% of the values lie within two standard deviations and about 99.7% are within three standard deviations.

To determine the statistical properties of continuous distributions, the first two moments (mean and variance) are defined as

$$\mu_X = E[X] = \int_{-\infty}^{\infty} x f_X(x)\, dx, \tag{3.9a}$$

$$\sigma_X^2 = Var[X] = E[(X - \mu_X)^2] = \int_{-\infty}^{\infty} (x - \mu_X)^2 f_X(x)\, dx. \tag{3.9b}$$

The quantity $E[X]$ is also called the expectation value. A relation exists between σ_X^2 and $E[X^2]$ through

$$\sigma_X^2 = E[(X - \mu_X)^2] = E[X^2] - 2\mu_X E[X] + (E[X])^2 = E[X^2] - \mu_X^2. \quad (3.10)$$

For the uniform distribution, it is easily calculated that

$$\mu_X = \int_a^b \frac{x}{b-a} dx = \frac{a+b}{2}, \quad (3.11a)$$

$$\sigma_X^2 = \int_a^b \frac{(x-\mu_X)^2}{b-a} dx = \frac{(b-a)^2}{12}, \quad (3.11b)$$

and for the Gaussian distribution (3.8), we find $\mu_X = \mu$ and $\sigma_X^2 = \sigma^2$.

3.2 Stochastic processes

In the previous section, we considered only a random variable X. This is easily generalised to a random vector $\mathbf{X} = (X_1, \ldots, X_n)$, where each X_i is a random variable. The distribution function $F_{\mathbf{X}}(\mathbf{x})$ is generalized as

$$F_{\mathbf{X}}(\mathbf{x}) = P(X_1 \leq x_1, \ldots, X_n \leq x_n) = \int_{-\infty}^{x_1} \cdots \int_{-\infty}^{x_n} f_{\mathbf{X}}(y_1, \ldots, y_n) dy_1 \ldots dy_n, \quad (3.12)$$

where again $f_{\mathbf{X}}$ is the corresponding probability density function. As an example, the multidimensional Gaussian distribution has a probability density function

$$f_{\mathbf{X}}(\mathbf{x}) = \frac{1}{(2\pi)^{n/2}(\det \Sigma)^{1/2}} e^{-\frac{1}{2}(\mathbf{x}-\mu_{\mathbf{X}})\Sigma^{-1}(\mathbf{x}-\mu_{\mathbf{X}})^T}, \quad (3.13)$$

where Σ is the covariance matrix with elements

$$\Sigma_{i,j} = \mathrm{Cov}[X_i, X_j] = E[(X_i - \mu_{X_i})(X_j - \mu_{X_j})]. \quad (3.14)$$

Two random variables X_1 and X_2 are independent if and only if

$$F_{X_1,X_2}(x_1, x_2) = F_{X_1}(x_1) F_{X_2}(x_2). \quad (3.15)$$

When two random variables X_i and X_j are independent, then $\mathrm{Cov}[X_i, X_j] = E[(X_i - \mu_{X_i})(X_j - \mu_{X_j})] = E[(X_i - \mu_{X_i})]E[(X_j - \mu_{X_j})] = 0$. If all components in a random vector are independent, then Σ is a diagonal matrix.

We are now ready to define a stochastic process X_t as a time series of random variables

$$(X_t, t \in T) = (X_t(\omega), t \in T, \omega \in \Omega), \quad (3.16)$$

where T denotes the time interval and Ω the outcome space. When t is fixed, then $X_t(\omega)$ is just a random variable. When ω is fixed, then $X_t(\omega)$ is a function of time,

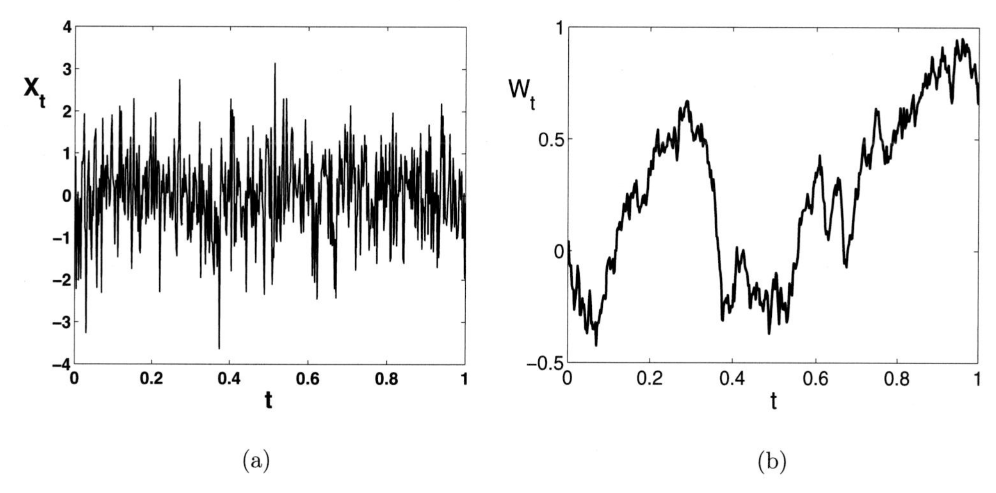

Figure 3.2 (a) A realization of a Gaussian process. (b) A realization of a Wiener process.

which is called a trajectory, a realization or a sample path. The expectation function of X_t is defined by $\mu_X(t) = E[X_t]$, and the covariance function $c_X(t, s)$ is defined as

$$c_X(t, s) = \mathrm{Cov}(X_t, X_s) = E[(X_t - \mu_X(t))(X_s - \mu_X(s))]. \tag{3.17}$$

In particular, the variance function is given by $\sigma_X^2(t) = c_X(t, t)$.

A Gaussian process (Fig. 3.2a) is defined over the interval $T = [0, 1]$ with $0 \leq t_1 \leq \ldots \leq t_n \leq 1$ such that all $X_{t_1}, \ldots, X_{t_n}$ are independent and standard normally distributed, that is, each X_{t_i} has a distribution function

$$\Phi(x) = \frac{1}{\sqrt{2\pi}} e^{-\frac{x^2}{2}}.$$

The multidimensional distribution function is then given by

$$F_{\mathbf{X}}(\mathbf{x}) = \Phi(x_1) \ldots \Phi(x_n), \tag{3.18}$$

and the expectation function and (co-)variance functions are given by

$$\mu_X(t) = 0\,;\; t \neq s : c_X(t, s) = 0\,;\; \sigma_X^2(t) = 1. \tag{3.19}$$

A stochastic process $(X_t, t \in T)$ is strictly stationary if for all choices of $t_1, \ldots, t_n$ and h such that $t_i + h \in T$ for all i, the finite dimensional distributions satisfy

$$(X_{t_1}, \ldots, X_{t_n}) \stackrel{d}{=} (X_{t_1+h}, \ldots, X_{t_n+h}), \tag{3.20}$$

where $\stackrel{d}{=}$ indicates equality in distribution sense. A stochastic process $(X_t, t \in T)$ has stationary increments if

$$X_t - X_s \stackrel{d}{=} X_{t+h} - X_{s+h}, \tag{3.21}$$

for each t, s. A stochastic process $(X_t, t \in T)$ has independent increments if the random variables $X_{t_2} - X_{t_1}, \ldots, X_{t_n} - X_{t_{n-1}}$ are independent for every $t_1 < \ldots < t_n$.

The Wiener process $W_t, t \in [0, \infty)$ is a stochastic process (Fig. 3.2b) with the following properties:

(i) $W_0 = 0$,
(ii) W_t has stationary, independent increments,
(iii) $\forall t > 0 : W_t$ has a $N(0, t)$ distribution, and
(iv) W_t has continuous sample paths.

From these properties, it follows immediately that with $0 \leq s < t \leq T$

$$W_t - W_s \stackrel{d}{=} N(0, t - s), \tag{3.22a}$$

$$\mu_W(t) = 0, \tag{3.22b}$$

$$c_W(t, s) = s. \tag{3.22c}$$

To obtain the result (3.22a), we note that from property (i) and (ii), it follows that $W_t - W_s \stackrel{d}{=} W_{t-s} - W_0 \stackrel{d}{=} W_{t-s}$, and (3.22a) follows then directly from property (iii). Equation (3.22b) follows directly from property (iii), and (3.22c) follows from $c_W(t, s) = E[W_t W_s] = E[(W_t - W_s + W_s)W_s] = E[(W_t - W_s)(W_s - W_0)] + E[W_s^2] = E[W_s^2] = s$, where the last two equalities follow from properties (ii) and (iii).

3.3 Stochastic calculus

In Subsections 3.1 and 3.2, the basic material on stochastic processes was introduced. In this subsection, we introduce basic material of stochastic calculus, in particular the stochastic integral and the Itô lemma.

3.3.1 *The stochastic integral*

Consider a smooth function $h : [0, T] \to \mathbb{R}$ for which the derivative h' is bounded on $[0, T]$. To define the Riemann integral of h, the interval $[0, T]$ is partitioned into subintervals $0 = t_0 < t_1 < \ldots < t_{N-1} < t_N = T$. The Riemann integral of h is then given by

$$\int_0^T h(t)dt = \lim_{N\to\infty} \sum_{j=0}^{N-1} h(t_j)(t_{j+1} - t_j). \tag{3.23}$$

The stochastic integral can be defined in a similar way, and two forms exist, the Itô form and the Stratonovich form. The Itô integral of h is

$$\int_0^T h(t)dW_t = \lim_{N\to\infty} \sum_{j=0}^{N-1} h(t_j)(W(t_{j+1}) - W(t_j)), \tag{3.24}$$

where $W(t_j)$ indicates the value of the Wiener process W_t at $t = t_j$. The Stratonovich integral of h is

$$\int_0^T h(t) \circ dW_t = \lim_{N\to\infty} \sum_{j=0}^{N-1} h\left(\frac{t_j + t_{j+1}}{2}\right)(W(t_{j+1}) - W(t_j)). \tag{3.25}$$

Note that the difference between the two forms of the stochastic integral is the time values of h considered with respect to the Wiener process values W. In the Itô integral, only h values at the left endpoint are considered, just as in the Riemann integral. In the Stratonovich integral, values of h at the midpoint of the interval are considered. The Itô and Stratonovich integral in general lead to different outcomes, but a relation exists between these results, and hence both definitions have their use.

Example 3.1 Itô integral of a Wiener process Consider $h(t) = W(t)$, where the function $W(t)$ indicates a Wiener process W_t. For this case, the Itô integral can be evaluated analytically because

$$\begin{aligned}
&\sum_{j=0}^{N-1} W(t_j)(W(t_{j+1}) - W(t_j)) \\
&= \sum_{j=0}^{N-1} \frac{1}{2}\left[W^2(t_{j+1}) - W^2(t_j) - (W(t_{j+1}) - W(t_j))^2\right] \\
&= \frac{1}{2}(W_T^2 - W_0^2) - \frac{1}{2}\sum_{j=0}^{N-1}(W(t_{j+1}) - W(t_j))^2.
\end{aligned}$$

With $dW_j = W(t_{j+1}) - W(t_j)$ and (from (3.22)) $E[(dW_j)^2] = t_{j+1} - t_j = dt$ we obtain

$$E\left[\sum_{j=0}^{N-1}(W(t_{j+1}) - W(t_j))^2\right] = \sum_{j=0}^{N-1} dt = T,$$

and, finally, we find

$$\int_0^T W(t)dW_t = \frac{1}{2}(W_T^2 - T), \tag{3.27}$$

where the equality is interpreted in the mean-square sense. ■

3.3.2 The Itô lemmas

We proceed with the Itô integral and consider the stochastic version of the main theorem of integral calculus,

$$f(b) - f(a) = \int_a^b f'(t)dt, \tag{3.28}$$

for a smooth function f on the interval $[a, b]$. To proceed, we use the notation $dW_t = W_{t+dt} - W_t$ and consider the Taylor-series expansion

$$f(W_x + dW_x) - f(W_x) = f'(W_x)dW_x + \frac{1}{2}f''(W_x)(dW_x)^2 + \cdots. \tag{3.29}$$

With $E[(dW_x)^2] = dx$, we obtain the first Itô lemma by integration of (3.29) over the interval $[s, t]$, that is,

$$f(W_t) - f(W_s) = \int_s^t f'(W_x)dW_x + \int_s^t \frac{1}{2}f''(W_x)dx, \tag{3.30}$$

again with equality in the mean-square sense.

We see that (3.30) is the generalization of (3.28) to the stochastic case. In addition to the first term on the right-hand side, there is now an additional Riemann integral involving the second derivative of f.

The first Itô lemma (3.30) is a powerful tool to compute stochastic integrals explicitly. Consider, for example, $f(t) = t^2$, with $f'(t) = 2t$ and $f''(t) = 2$; with (3.30) we then find

$$W_t^2 - W_s^2 = 2\int_s^t W_x dW_x + \int_s^t dx. \tag{3.31}$$

With $s = 0$ and $t = T$, this gives

$$W_T^2 - W_0^2 = 2\int_0^T W_x dW_x + \int_0^T dx \Rightarrow \int_0^T W_x dW_x = \frac{1}{2}(W_T^2 - T), \tag{3.32}$$

which is the same as (3.27).

There are two extensions of the first Itô lemma. Consider a stochastic process $f(t, W_t)$ for which the function $f(t, y)$ is smooth. Again by Taylor series expansion, we find

$$f(x+dx, W_{x+dx}) - f(x, W_x) = f_1(x, W_x)dx + f_2(x, W_x)dW_x$$
$$+\frac{1}{2}\left[f_{11}(x, W_x)(dx)^2 + 2f_{12}(x, W_x)dx dW_x + f_{22}(x, W_x)(dW_x)^2\right] + \cdots,$$

where $f_1 = \partial f/\partial t$, $f_2 = \partial f/\partial y$, $f_{11} = \partial^2 f/\partial t^2$, and so forth. We use again that $E[(dW_x)^2] = dx$, neglect higher-order terms dx^2 and $dx\, dW_x$ and integrate over the interval $[s, t]$ to obtain the second Itô lemma,

$$f(t, W_t) - f(s, W_s) = \int_s^t \left[f_1(x, W_x) + \frac{1}{2}f_{22}(x, W_x)\right]dx + \int_s^t f_2(x, W_x)dW_x. \tag{3.33}$$

If $f = f(W_t)$, then $f_1 = 0$, $f_2 = f'$, $f_{22} = f''$, and the second Itô lemma (3.33) reduces to the first Itô lemma (3.30).

Example 3.2 The Itô exponential An application of the Itô lemmas is the determination of the Itô exponential. In the deterministic case, we know that $f(t) = e^t$ has the special property that

$$f(t) - f(s) = \int_s^t f(x)dx.$$

The Itô exponential is a process X_t with the property

$$X_t - X_s = \int_s^t X_x\, dx.$$

To determine this process, we first try $f(t) = e^t$; using the first Itô lemma (3.30) then gives

$$e^{W_t} - e^{W_s} = \int_s^t e^{W_x}dW_x + \frac{1}{2}\int_s^t e^{W_x}dx,$$

and as the latter integral is always positive, the process $X_t = e^{W_t}$ is not the Itô exponential. The correct Itô exponential follows from the use of the second Itô lemma (3.33) with $f(t, y) = e^{y-t/2}$, such that with $f_1 = -\frac{1}{2}e^{(y-t/2)}$, $f_2 = f_{22} = f$, we get

$$e^{W_t - \frac{t}{2}} - e^{W_s - \frac{s}{2}} = \int_s^t e^{W_x - \frac{x}{2}}dW_x,$$

and hence $X_t = e^{W_t - t/2}$ is the desired Itô exponential. For illustration, both stochastic processes are plotted in Fig. 3.3. ■

A third extension of the main theorem of integral calculus is for stochastic processes of the form $f(t, X_t)$, where X_t is given by

$$X_t = X_0 + \int_0^t A^{(1)}(s, X_s)ds + \int_0^t A^{(2)}(s, X_s)dW_s. \tag{3.34}$$

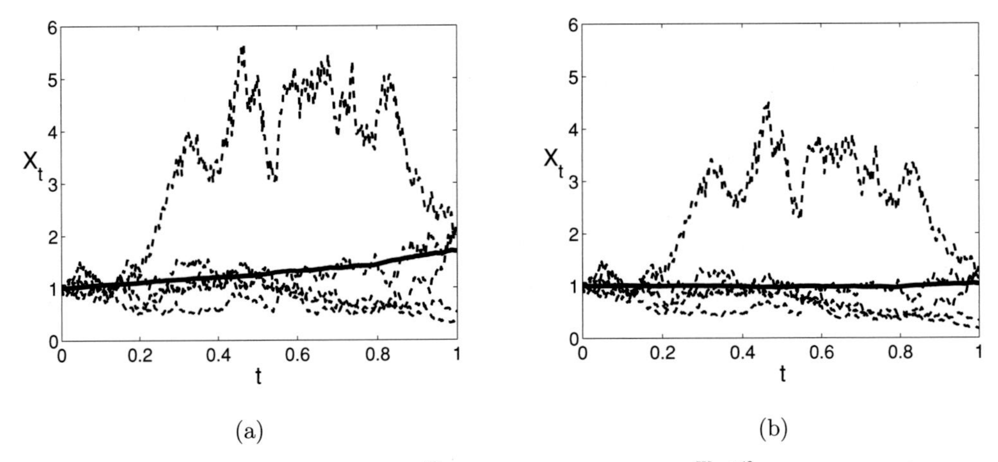

Figure 3.3 (a) The process $X_t = e^{W_t}$. (b) The process $X_t = e^{W_t - t/2}$. In each case, five sample paths are plotted together with the mean over 1,000 sample paths.

Here the $A^{(i)}$ are smooth functions of s and X_s. Using the same procedure as for the first two Itô lemmas (Taylor series, neglect higher-order terms and $E[(dW_x)^2] = dx$), leads to the third Itô lemma,

$$f(t, X_t) - f(s, X_s) = \int_s^t \left[f_1(x, X_x) + A_x^{(1)} f_2(x, X_x) + \frac{1}{2}(A_x^{(2)})^2 f_{22}(x, X_x) \right] dx + \int_s^t A_x^{(2)} f_2(x, X_x) dW_x, \tag{3.35}$$

where $A_x^{(i)} = A^{(i)}(x, X_x)$. This is the most general form of the Itô lemmas, and it is used in the study of stochastic differential equations in Section 3.4.

3.3.3 Itô versus Stratonovich

There is a direct relation between the Itô and Stratonovich integrals of a smooth function f, which is given by

$$\int_0^T f(W_t) \circ dW_t = \int_0^T f(W_t) dW_t + \frac{1}{2} \int_0^T f'(W_t) dt, \tag{3.36}$$

where the first integral on the right-hand side is the Itô integral. This can be shown by defining $y_j = (t_{j+1} + t_j)/2$ and then considering the sum

$$I_N = \sum_{j=0}^{N-1} f(W_{y_j})(W_{t_{j+1}} - W_{t_j}).$$

Using the Taylor expansion

$$f(W_{y_j}) = f(W_{t_j}) + f'(W_{t_j})(W_{y_j} - W_{t_j}) + \cdots,$$

gives

$$I_N \cong \sum_{j=0}^{N-1} \left[f(W_{t_j})(W_{t_{j+1}} - W_{t_j}) + f'(W_{t_j})(W_{y_j} - W_{t_j})(W_{t_{j+1}} - W_{t_j}) \right].$$

With $W_{t_{j+1}} - W_{t_j} = W_{t_{j+1}} - W_{y_j} + W_{y_j} - W_{t_j}$ in the second sum we obtain

$$\begin{aligned} I_N \cong & \sum_{j=0}^{N-1} f(W_{t_j})(W_{t_{j+1}} - W_{t_j}) \\ & + \sum_{j=0}^{N-1} f'(W_{t_j})(W_{y_j} - W_{t_j})^2 \\ & + \sum_{j=0}^{N-1} f'(W_{t_j})(W_{y_j} - W_{t_j})(W_{t_{j+1}} - W_{t_j}). \end{aligned}$$

In the limit $N \to \infty$, the first term converges (in the mean-square sense) to the Itô integral of f, the second term converges to the second term on the right-hand side of (3.36) and the third sum converges to zero (this requires some more detailed analysis; see Chapter 2 of Mikosch [2000]).

The Stratonovich calculus is more similar to the deterministic calculus. For example, if we take $f(t) = g'(t)$ in the first Itô lemma (3.30), we find

$$\begin{aligned} & \int_0^T g'(W_x) dW_x + \frac{1}{2} \int_0^T g''(W_x) dx = g(W_T) - g(W_0) \\ \Rightarrow & \int_0^T f(W_x) dW_x + \frac{1}{2} \int_0^T f'(W_x) dx = g(W_T) - g(W_0). \end{aligned}$$

With (3.36), we find

$$\int_0^T f(W_x) \circ dW_x = \int_0^T g'(W_x) \circ dW_x = g(W_T) - g(W_0), \tag{3.37}$$

which is similar to the classical main theorem of integral calculus. The Stratonovich exponential is, therefore, simply $X_t = e^{W_t}$ (Fig. 3.3a).

3.4 Stochastic differential equations

A general scalar ordinary differential equation (ODE) is written as

$$\frac{dx}{dt} = f(t, x) \to dx = f(t, x) dt. \tag{3.38}$$

With an initial condition $x(0) = x_0$, it has a formal solution,

$$x(t) = \int_0^t f(s, x)ds + x_0. \tag{3.39}$$

A general stochastic differential equation is written as

$$dX_t = a(t, X_t)dt + b(t, X_t)dW_t, \tag{3.40}$$

for smooth functions a, b and with initial condition X_0. The formal solution of (3.40) is given,

$$X_t - X_0 = \int_0^t a(s, X_s)ds + \int_0^t b(s, X_s)dW_s. \tag{3.41}$$

Either form (3.40) or (3.41) is referred to as the Itô stochastic differential equation (SDE). It can be deduced from (3.36) (see Section 3.2.3 of Mikosch [2000]) that, for each Itô SDE (3.40), there is an equivalent Stratonovich SDE of the form

$$dX_t = (a(t, X_t) - \frac{1}{2}b(t, X_t)\frac{\partial b}{\partial x}(t, X_t))dt + b(t, X_t) \circ dW_t. \tag{3.42}$$

A strong solution X_t of an Itô SDE (or equivalent Stratonovich SDE) has the following properties:

(i) X_t satisfies (3.41), and at time t, it is a function of W_s for $s \leq t$ and the coefficient functions a and b.
(ii) The integrals in (3.41) are well defined in terms of Riemann or Itô stochastic integrals.

The important notion here is that if we change the Wiener process W_t, then also the strong solution X_t changes, but the functional relation between X_t and W_t remains the same. For a weak solution X_t of (3.41), the sample path dependence is not needed; for given X_0, a and b, we just have to find a W_t for which (3.41) holds. There exist Itô SDEs that have only weak solutions!

Assume that the initial condition X_0 has a finite second moment ($E[X_0^2] < \infty$) and that for all $t \in [0, T]$ and $x, y \in \mathbb{R}$, the coefficient functions $a(t, x)$ and $b(t, x)$ are continuous and satisfy a Lipschitz condition in the second variable, that is,

$$|a(t, x) - a(t, y)| + |b(t, x) - b(t, y)| \leq K|x - y|, \tag{3.43}$$

for certain $K > 0$, then the Itô SDE (3.41) has a unique strong solution X_t on $[0, T]$. For the proof, see Oksendal (1995), Theorem 5.5. In most practical applications, such conditions are satisfied, and existence and uniqueness of strong solutions are guaranteed.

As an example, consider the simple case $a = 0$ and $b = 1$, which gives the SDE

$$X_t - X_0 = \int_0^t dW_s. \tag{3.44}$$

The solution is $X_t = W_t$ (just by evaluating the stochastic integral), and hence the Wiener process is itself a strong solution of this SDE.

The general linear scalar Itô SDE is defined as

$$X_t - X_0 = \int_0^t (c_1(s)X_s + c_2(s))ds + \int_0^t (\sigma_1(s)X_s + \sigma_2(s))dW_s, \tag{3.45}$$

with smooth and bounded functions c_1, c_2, σ_1 and σ_2 on the interval $[0, T]$. With $a(t, x) = c_1(t)x + c_2(t)$ and $b(t, x) = \sigma_1(t)x + \sigma_2(t)$, we find from the Lipschitz condition

$$\begin{aligned} |a(t,x) - a(t,y)| + |b(t,x) - b(t,y)| &= |c_1(t)(x-y)| + |\sigma_1(t)(x-y)| \\ &\leq K|(x-y)|, \end{aligned} \tag{3.46}$$

where $K = \max_{t\in[0,T]}(|c_1(t)| + |\sigma_1(t)|)$ and hence (3.45) has a unique strong solution on every interval $[0, T]$. In the following subsections, we consider some specific cases.

3.4.1 Pure additive noise

When $\sigma_1 = 0$, the stochastic integral in (3.45) does not depend on the solution X_t, and this case is referred to as 'pure additive' noise. The Itô SDE becomes

$$X_t - X_0 = \int_0^t (c_1(s)X_s + c_2(s))ds + \int_0^t \sigma_2(s)dW_s. \tag{3.47}$$

To solve this equation, we introduce the process $Y_t = f(t, X_t) = \alpha(t)X_t$ where

$$\alpha(t) = e^{-\int_0^t c_1(s)ds}.$$

With $A^{(1)} = c_1X + c_2$ and $A^{(2)} = \sigma_2$, the third Itô lemma (3.35) is applied to Y_t to give (with $f_1 = \alpha' x$, $f_2 = \alpha$ and $f_{22} = 0$)

$$\begin{aligned} \alpha(t)X_t - \alpha(0)X_0 = &\int_0^t \left[\alpha'(x)X_x + (c_1(x)X_x + c_2(x))\alpha(x)\right] dx \\ &+ \int_0^t \alpha(x)\sigma_2(x)dW_x. \end{aligned}$$

Because $\alpha' = -c_1\alpha$ and $\alpha(0) = 1$, we find

$$X_t = \frac{1}{\alpha(t)}\left[X_0 + \int_0^t \alpha(x)c_2(x)dx + \int_0^t \alpha(x)\sigma_2(x)dW_x\right]. \tag{3.48}$$

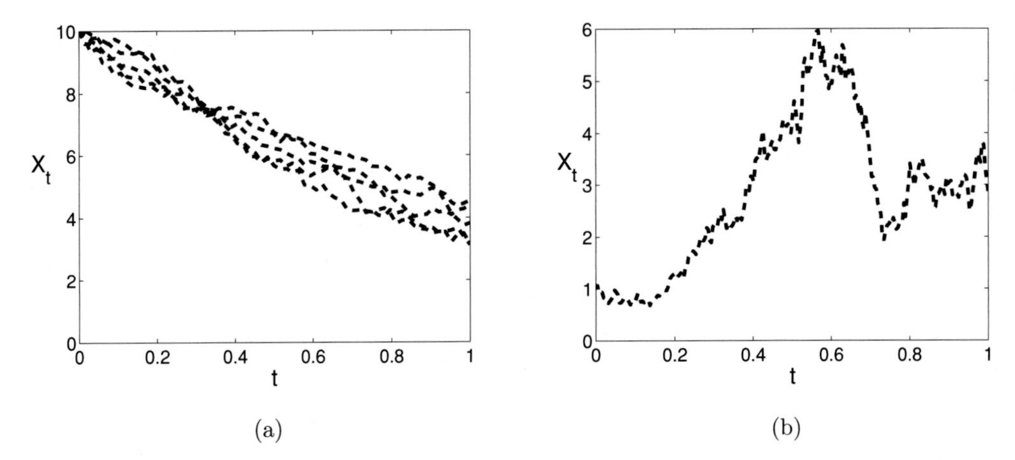

Figure 3.4 (a) Five sample paths of the Ornstein-Uhlenbeck process (3.49) with $\gamma = 1$, $\sigma = 1$ and $X_0 = 10$. (b) One sample path of the solution (3.53) for $\lambda = 2$, $\mu = 1$ and $X_0 = 1$.

A prominent example is the Langevin equation, for which $c_1(t) = -\gamma$, $c_2 = 0$ and $\sigma_2 = \sigma$, where γ and σ are constants. We find that in this case

$$\alpha(t) = e^{\int_0^t \gamma ds} = e^{\gamma t},$$

and the solution (3.48) becomes

$$X_t = e^{-\gamma t}\left[X_0 + \sigma \int_0^t e^{\gamma x}\, dW_x\right]. \tag{3.49}$$

The stochastic process associated with this solution is called the Ornstein-Uhlenbeck process (Fig. 3.4a).

3.4.2 Pure multiplicative noise

When $\sigma_2 = 0$, the stochastic integral in (3.45) depends only on the solution X_t, and this case is referred to as 'multiplicative' (or state-dependent) noise. With also $c_2 = 0$, the Itô SDE (3.45) becomes

$$X_t - X_0 = \int_0^t c_1(s) X_s ds + \int_0^t \sigma_1(s) X_s dW_s. \tag{3.50}$$

We apply again the third Itô lemma (3.35), but now on the process $Y_t = \ln X_t$. With $f(t, x) = \ln x$, $f_1 = 0$, $f_2 = 1/x$ and $f_{22} = -1/x^2$, we find

$$\begin{aligned} \ln X_t - \ln X_0 = &\int_0^t \left[c_1(x) X_x \frac{1}{X_x} + \frac{1}{2}(\sigma_1(x) X_x)^2 (-\frac{1}{X_x^2})\right] dx \\ &+ \int_0^t \sigma_1(x) X_x \frac{1}{X_x} dW_x, \end{aligned} \tag{3.51}$$

and hence the solution is given by

$$X_t = X_0 \exp\left\{\int_0^t \left[c_1(x) - \frac{1}{2}\sigma_1^2(x)\right] dx + \int_0^t \sigma_1(x) dW_x\right\}. \tag{3.52}$$

An example is given by the special case $c_1 = \lambda$ and $\sigma_1 = \mu$, where λ, μ are constants. In this case, the solution (Fig. 3.4b) follows immediately as

$$X_t = X_0 e^{(\lambda - \mu^2/2)t + \mu W(t)}. \tag{3.53}$$

This solution is often used to test the convergence of several numerical solution methods for SDEs (Section 3.6).

3.4.3 Complete solution of the linear scalar Itô SDE

The general linear one-dimensional Itô SDE (3.45), now written as

$$Y_t - Y_0 = \int_0^t (c_1(s)Y_s + c_2(s))ds + \int_0^t (\sigma_1(s)Y_s + \sigma_2(s))dW_s, \tag{3.54}$$

can be completely solved analytically. We refer to Section 3.3.3 of Mikosch (2000) for the derivation and present only the answer here for completeness as

$$Y_t = X_t(Y_0 + \int_0^t [c_2(x) - \sigma_1(x)\sigma_2(x)] X_x^{-1} dx + \int_0^t \sigma_2(x) X_x^{-1} dW_x), \tag{3.55}$$

where X_t is the solution (3.52).

3.4.4 Mean and variance

Once we have obtained solutions to SDEs, we want to know their statistical properties. Let $\mu_X(t)$ indicate the mean and $Var(X_t) = \sigma_X^2(t)$ the variance of the stochastic process X. Note that these in general depend on time, as for every fixed t, we have a distribution of X. There are basically two ways to obtain this information: (i) by direct evaluation using the discrete form of the stochastic integral and (ii) through formulation of ODEs for these statistical properties. We will illustrate both approaches with the Ornstein-Uhlenbeck process (3.49).

Following approach (i), we first determine

$$E[\int_0^t e^{\gamma x} dW_x] = E[\lim_{N\to\infty} \sum_{j=0}^{N-1} e^{\gamma t_j}(W(t_{j+1}) - W(t_j))]$$

$$= \lim_{N\to\infty} \sum_{j=0}^{N-1} e^{\gamma t_j} E[W(t_{j+1})] - E[W(t_j))] = 0,$$

and hence

$$\mu_X(t) = E[e^{-\gamma t}\left[X_0 + \int_0^t e^{\gamma x}\sigma\, dW_x\right]] = e^{-\gamma t}E[X_0]. \tag{3.56}$$

Next, we determine (using $Var(cX) = c^2 Var(X)$ and $Var(dW_j) = dt$)

$$\begin{aligned} Var\left(\int_0^t e^{\gamma x} dW_x\right) &= Var\left(\lim_{N\to\infty}\sum_{j=0}^{N-1} e^{\gamma t_j}(W(t_{j+1}) - W(t_j))\right) \\ &= \lim_{N\to\infty}\sum_{j=0}^{N-1} e^{2\gamma t_j} Var(W(t_{j+1}) - W(t_j)) \\ &= \lim_{N\to\infty}\sum_{j=0}^{N-1} e^{2\gamma t_j}(t_{j+1} - t_j) = \int_0^t e^{2\gamma x} dx = \frac{e^{2\gamma t} - 1}{2\gamma}, \end{aligned}$$

and hence for X_t in (3.49),

$$Var(X_t) = e^{-2\gamma t}\left(Var(X_0) + \sigma^2 \frac{e^{2\gamma t} - 1}{2\gamma}\right). \tag{3.57}$$

Following approach (ii), we consider the general linear scalar SDE given by (3.45) and apply directly the expectation operator, which gives (using $E[dW_s] = 0$),

$$\mu_X(t) = E[X_t] = E[X_0] + \int_0^t (c_1(s)E[X_s] + c_2(s))ds. \tag{3.58}$$

Differentiation of this relation to t gives

$$\mu_X'(t) = c_1(t)\mu_X(t) + c_2(t), \tag{3.59}$$

with initial conditions $\mu_X(0) = E[X_0]$. For the Ornstein-Uhlenbeck process (with $c_1 = -\gamma$, $c_2 = 0$), we find

$$\mu_X'(t) = -\gamma\mu_X(t) \Rightarrow \mu_X(t) = \mu_X(0)e^{-\gamma t}, \tag{3.60}$$

which is in correspondence with (3.56).

To obtain the variance using approach (ii), we define $q_X(t) = E[X_t^2]$. An expression for X_t^2 can be found using the third Itô lemma (3.35) with $f(t,x) = x^2$ (such that $f_1 = 0$, $f_2 = 2x$, $f_{22} = 2$). This gives for (3.45)

$$\begin{aligned} X_t^2 - X_0^2 = &\int_0^t \left[2X_s(c_1X_s + c_2) + (\sigma_1 X_s + \sigma_2)^2\right] ds \\ &+ \int_0^t 2X_s(\sigma_1 X_s + \sigma_2)dW_s. \end{aligned}$$

Taking the expectation operator and using the fact that X_s and dW_s are independent then gives

$$E[X_t^2] = E[X_0^2] + \int_0^t \left[2(c_1 E[X_s^2] + c_2 E[X_s]) + (\sigma_1^2 E[X_s^2] + 2\sigma_1\sigma_2 E[X_s] + \sigma_2^2\right] ds$$

or

$$q_X(t) = q_X(0) + \int_0^t \left[(2c_1 + \sigma_1^2) q_X(s) + (2c_2 + 2\sigma_1\sigma_2)\mu_X(s) + \sigma_2^2\right] ds.$$

Differentiating to time, we finally obtain

$$q_X'(t) = (2c_1 + \sigma_1^2) q_X(t) + (2c_2 + 2\sigma_1\sigma_2)\mu_X(t) + \sigma_2^2. \tag{3.61}$$

For the Ornstein-Uhlenbeck process (with $c_1 = -\gamma, c_2 = 0, \sigma_1 = 0, \sigma_2 = \sigma$), this reduces to

$$q_X'(t) = -2\gamma q_X(t) + \sigma^2,$$

with initial conditions $q_X(0) = E[X_0^2]$, having the solution

$$q_X(t) = e^{-2\gamma t}\left[\frac{\sigma^2}{2\gamma}(e^{2\gamma t} - 1) + q_X(0)\right].$$

Finally, the variance of the process follows from

$$\begin{aligned} Var(X_t) = E[X_t^2] - (E[X_t])^2 &= q_X(t) - \mu_X^2(t) \\ &= \frac{\sigma^2}{2\gamma}(1 - e^{-2\gamma t}) + e^{-2\gamma t}(q_X(0) - \mu_X^2(0)) \\ &= e^{-2\gamma t} Var(X_0) + \frac{\sigma^2}{2\gamma}(1 - e^{-2\gamma t}), \end{aligned} \tag{3.62}$$

which is in agreement with (3.57).

3.5 The Fokker-Planck equations

An alternative approach to computing statistical properties of a stochastic process associated with an SDE is the determination of its Fokker-Planck equation (FPE).

3.5.1 Markov processes

Recall from Section 3.2 that the distribution function $F_{\mathbf{X}}(\mathbf{x})$ for a random vector $\mathbf{X}$ was given by (3.12), for convenience here written again as

$$F_{\mathbf{X}}(\mathbf{x}) = P(X_1 \le x_1, \ldots, X_n \le x_n) = \int_{-\infty}^{x_1} \cdots \int_{-\infty}^{x_n} f_{\mathbf{X}}(y_1, \ldots, y_n) dy_1 \ldots dy_n, \tag{3.63}$$

where $f_{\mathbf{X}}$ is the (joint) probability density function. In the following, we associate the indices $1, \ldots, n$ with times $t_1, \ldots, t_n$ and write

$$f_{\mathbf{X}}(x_1, \ldots, x_n) = p(x_1, t_1; \ldots; x_n, t_n). \tag{3.64}$$

The conditional probability $P(A|B)$ of two events A and B is defined as

$$P(A|B) = \frac{P(A \cap B)}{P(B)}, \tag{3.65}$$

where $P(A)$ and $P(B)$ are the probabilities of the events A and B, respectively. Hence the probability $P(A|B)$ concerns events A, which are contained in the set B. We need this concept to define a Markov process for which the following property, the Markov property, holds,

$$p(x_1, t_1; \ldots; x_n, t_n | y_1, \tau_1; \ldots; y_n, \tau_n) = p(x_1, t_1; \ldots; x_n, t_n | y_1, \tau_1) \tag{3.66}$$

for $t_1 \geq \ldots \geq t_n \geq \tau_1 \geq \ldots \geq \tau_n$. Loosely speaking, for a Markov process, one can make future predictions based solely on its present state just as well as one could knowing the process's full history.

A well-known example of a Markov process is the following: suppose that someone is popping many kernels of popcorn, and each kernel will pop at an independent, uniformly random time within the next time interval. Let X_t denote the number of kernels that have popped up to time t. If after some amount of time, one wants to guess how many kernels will pop in the next second, one needs only know how many kernels have popped. It will not help to know when they popped, so knowing X_t for previous times t will not inform the guess any better.

For a Markov process, we find (using 3.65)

$$\begin{aligned}
p(x_1, t_1; \ldots; x_n, t_n) &= p(x_1, t_1; \ldots; x_{n-1}, t_{n-1} | x_n, t_n) p(x_n, t_n) \\
&= p(x_1, t_1; \ldots; x_{n-2}, t_{n-2} | x_{n-1}, t_{n-1}; x_n, t_n) p(x_{n-1}, t_{n-1} | x_n, t_n) p(x_n, t_n) \\
&= p(x_1, t_1; \ldots; x_{n-2}, t_{n-2} | x_{n-1}, t_{n-1}) p(x_{n-1}, t_{n-1} | x_n, t_n) p(x_n, t_n) \\
&= p(x_1, t_1 | x_2, t_2) \ldots p(x_{n-1}, t_{n-1} | x_n, t_n) p(x_n, t_n).
\end{aligned}$$

Hence only the so-called transition probability $p(x_{i-1}, t_{i-1} | x_i, t_i)$ is needed to describe the joint probability density function of a Markov process.

3.5.2 Forward Fokker-Planck equation

We now return to the general Itô SDE (3.40) given by

$$X_t = X_0 + \int_0^t a(X_s, s) ds + \int_0^t b(X_s, s) dW_s \tag{3.67}$$

for smooth functions a and b. With X_0 given, the future time development is uniquely determined by $W_t, t > 0$. As W_t for $t > 0$ is independent of X_t for $t < 0$, we conclude that X_t for $t > 0$ is independent of X_t for $t < 0$, provided X_0 is known, and hence X_t is a Markov process (for a more extensive discussion, see Gardiner [2002], Section 4.3.2). Let the transition probability be indicated by $p(x, t) = p(x, t|x_0, t_0)$.

Now by definition of the expectation operator, for any smooth function $f : \mathbb{R} \to \mathbb{R}$, we find

$$E[f(X_t)] = \int_{-\infty}^{\infty} f(x)p(x, t)dx \tag{3.68}$$

and hence

$$\frac{d}{dt}E[f(X_t)] = \int_{-\infty}^{\infty} f(x)\frac{\partial p}{\partial t}(x, t)dx. \tag{3.69}$$

However, when we use the third Itô lemma (3.35) for f, we find

$$\begin{aligned} f(X_t) - f(X_0) = &\int_0^t \left[a(X_s, s)\frac{\partial f}{\partial x}(X_s) + \frac{1}{2}b^2(X_s, s)\frac{\partial^2 f}{\partial x^2}(X_s)\right] ds \\ &+ \int_0^t b(X_s, s)\frac{\partial f}{\partial x}(X_s)dW_s. \end{aligned} \tag{3.70}$$

Taking the expectation operator of (3.70), using (3.68) and differentiating the result to t then gives

$$\frac{d}{dt}E[f(X_t)] = \int_{-\infty}^{\infty} \left[a(x, t)\frac{\partial f}{\partial x}(x) + \frac{1}{2}b^2(x, t)\frac{\partial^2 f}{\partial x^2}(x)\right] p(x, t)dx. \tag{3.71}$$

Combining (3.71) and (3.69), we find

$$\int_{-\infty}^{\infty} \left[(a(x, t)\frac{\partial f}{\partial x}(x) + \frac{1}{2}b^2(x, t)\frac{\partial^2 f}{\partial x^2}(x))p(x, t) - f(x)\frac{\partial p}{\partial t}(x, t)\right] dx = 0.$$

When furthermore it is assumed that $p, \partial p/\partial x \to 0$ for $x \to \pm\infty$, then partial integration of the terms with a and b gives

$$\int_{-\infty}^{\infty} f\left(\frac{\partial p}{\partial t} + \frac{\partial(ap)}{\partial x} - \frac{1}{2}\frac{\partial^2(pb^2)}{\partial x^2}\right) dx = 0, \tag{3.72}$$

and as f is arbitrary, we find for p the forward Fokker-Planck equation

$$\frac{\partial p}{\partial t} + \frac{\partial(ap)}{\partial x} - \frac{1}{2}\frac{\partial^2(pb^2)}{\partial x^2} = 0. \tag{3.73}$$

Once this Fokker-Planck equation is solved, the probability distribution of the stochastic process X_t is totally determined.

Example 3.3 Forward FPE of the Ornstein-Uhlenbeck process For the Ornstein-Uhlenbeck process we have $a(x,t) = -\gamma x$ and $b(x,t) = \sigma$. The Fokker-Planck equation (3.73), then becomes

$$\frac{\partial p}{\partial t} = \frac{\partial(\gamma x p)}{\partial x} + \frac{\sigma^2}{2}\frac{\partial^2 p}{\partial x^2},$$

with $p, \partial p/\partial x \to 0$ for $x \to \pm\infty$. In many cases, only the stationary distribution of p is desired. When putting the time derivative to zero, the resulting equation can be integrated to x to give

$$\gamma x p + \frac{\sigma^2}{2}\frac{\partial p}{\partial x} = C_1,$$

and $C_1 = 0$ through the boundary conditions. Integrating once more, we find

$$p(x) = C_2 e^{-\frac{\gamma}{\sigma^2}x^2}; \quad \int_{-\infty}^{\infty} p(x)dx = 1 \Rightarrow C_2 = \sqrt{\frac{\gamma}{\pi}\frac{1}{\sigma}}.$$

We immediately conclude from this normal distribution that $\mu[X] = 0$ and that $Var[X] = \sigma^2/(2\gamma)$, which is exactly the limit $t \to \infty$ of the expression (3.62). ■

3.5.3 Backward Fokker-Planck equation

In a forward problem, we have information about the state X_t of a particular process at time t (i.e., a probability distribution $p(x,t)$), and we want to know the probability distribution at a later time $s > t$. Hence $p(x,t)$ serves as an initial condition for the forward Fokker-Planck equation. In many problems, however, we want to know for every state at time t the probability of ending up at a future time $s > t$ in a target state X_s. A typical example is a so-called exit time problem, where the target state is outside a specific domain, as discussed in the next subsection. To solve this problem, the probability distribution $p(x,s)$ serves as a final condition, and we have to integrate the adjoint equation of the forward Fokker-Planck equation back in time. This adjoint equation is called the backward Fokker-Planck equation. With the inner product

$$< f, g > = \int_{-\infty}^{\infty} f(x)g(x)dx, \tag{3.74}$$

the adjoint operator $L^\dagger$ is defined as $< Lf, g > = < f, L^\dagger g >$. Writing (3.73) as $\partial p/\partial t = Lp$, we can write

$$< Lf, g > = \int_{-\infty}^{\infty} g\left(-\frac{\partial(af)}{\partial x} + \frac{1}{2}\frac{\partial^2(fb^2)}{\partial x^2}\right)dx. \tag{3.75}$$

Using partial integration (and with vanishing boundary conditions for f and g at $x = \pm\infty$) leads to

$$< f, L^\dagger g > \int_{-\infty}^{\infty} \left(a \frac{\partial g}{\partial x} + \frac{b^2}{2} \frac{\partial^2 g}{\partial x^2} \right) f dx \tag{3.76}$$

and hence the backward Fokker-Planck equation is given by (see, e.g., Risken, 1989)

$$-\frac{\partial p}{\partial t} = a \frac{\partial p}{\partial x} + \frac{b^2}{2} \frac{\partial^2 p}{\partial x^2}. \tag{3.77}$$

3.5.4 Exit-time problems

The backward Fokker-Planck equation is used to solve so-called exit-time problems. Let a particle be at a position x at $t = 0$ and assume that its position for later times is described by a stochastic process X_t satisfying the SDE (3.40). The problem is to estimate the average time it will take to leave a certain interval $x \in (x_0, x_1)$. When the particle reaches either x_0 or x_1, it is removed from the system; hence when it is still in (x_0, x_1), it has never left the interval.

To solve this exit-time problem, let the time when the particle leaves the interval be indicated by $T(x)$. For the probability $P(T(x) \geq t)$ we can write

$$P(T(x) \geq t) = \int_{x_1}^{x_0} p(y, t|x, 0) dy \equiv G(x, t), \tag{3.78}$$

because the right-hand side is precisely the probability that the particle is still in the interval (x_0, x_1).

As argued in the previous subsection, the probability $p(y, t|x, 0)$ satisfies a backward Fokker-Planck equation (3.77). Because $p(y, t|x, 0) = p(y, 0|x, -t)$ (time-shift), we find that $G(x, t)$ satisfies

$$\frac{\partial G}{\partial t} = a \frac{\partial G}{\partial x} + \frac{b^2}{2} \frac{\partial^2 G}{\partial x^2}. \tag{3.79}$$

Because $p(y, 0|x, 0) = \delta(x - y)$, the initial conditions are $G(x, 0) = 1$ for $x \in (x_0, x_1)$ and $G(x, 0) = 0$ elsewhere. If $x = x_0$ or $x = x_1$, the particle is absorbed immediately; hence $P(T(x) \geq t) = 0$. The boundary conditions for G hence become

$$G(x_0, t) = G(x_1, t) = 0. \tag{3.80}$$

Because G is the probability that $T(x) \geq t$, the mean first passage time $\overline{T}(x)$ can be written as the expectation value

$$\overline{T}(x) = \int_0^{\infty} t \, d(1 - G(x, t)) = -\int_0^{\infty} t \frac{\partial G}{\partial t} \, dt = \int_0^{\infty} G(x, t) \, dt. \tag{3.81}$$

When (3.79) is integrated in time over the interval $(0, \infty)$ we find

$$G(x,\infty) - G(x,0) = -1 = a\frac{\partial \overline{T}}{\partial x} + \frac{b^2}{2}\frac{\partial^2 \overline{T}}{\partial x^2}, \tag{3.82}$$

with boundary conditions $\overline{T}(x_0) = \overline{T}(x_1) = 0$; the general solution for $\overline{T}(x)$ can be found in Gardiner (2002), Section 5.2.7.

We now consider the special case that $x_0 \to -\infty$, and the particle is only removed at $x = x_1$. Because at $x = x_0$ the probability will not depend on x anymore, we now require boundary conditions

$$\frac{\partial G}{\partial x}(x_0, t) = G(x_1, t) = 0, \tag{3.83}$$

instead of (3.80) and, consequently, $\overline{T}'(x_0) = \overline{T}(x_1) = 0$. To solve (3.82), we put $S(x) = \overline{T}'(x)$, solve for the homogeneous problem first and then determine a particular solution by variation of constants. The result (with the boundary condition at $x = x_0$ included) is

$$S(x) = \int_{-\infty}^{x} \frac{-2}{b^2(s)}\psi(s)ds, \tag{3.84}$$

where $\psi(x)$ is defined as

$$\psi(x) = e^{\int_{-\infty}^{x} 2\frac{a(s)}{b^2(s)}ds}. \tag{3.85}$$

By integrating $S(x)$, the solution for the mean first passage time is given by

$$\overline{T}(x) = 2\int_{x}^{x_1}\left[\frac{1}{\psi(y)}\int_{-\infty}^{y}\frac{\psi(z)}{b^2(z)}dz\right]dy. \tag{3.86}$$

Example 3.4 Exit times for a double-well potential Consider the Itô stochastic differential equation

$$dX_t = -V'(x)dt + \sigma dW_t,$$

for which the stationary probability density function can be explicitly solved from the forward Fokker-Planck equation (just as in Example 3.3) as

$$p(x) = Ce^{-2\frac{V(x)}{\sigma^2}}, \quad \int_{-\infty}^{\infty} p(x)\,dx = 1.$$

The function $V(x)$ can be viewed as a potential, and let us assume that it has a minimum at $x = x_0$ and a maximum at $x = x_1$. In Fig. 3.5, the potential $V(x) = x^4/4 - x^2/2$ is plotted, for which $x_0 = -1$ and $x_1 = 0$.

We want to determine the mean first passage time of a particle in the potential well near $x = x_0$ over the potential barrier at $x = x_1$ and use the notation $T(x_0 \to x_1)$. In the solution (3.86), the function ψ is given by

$$\psi(x) = e^{-\frac{2V(x)}{\sigma^2}},$$

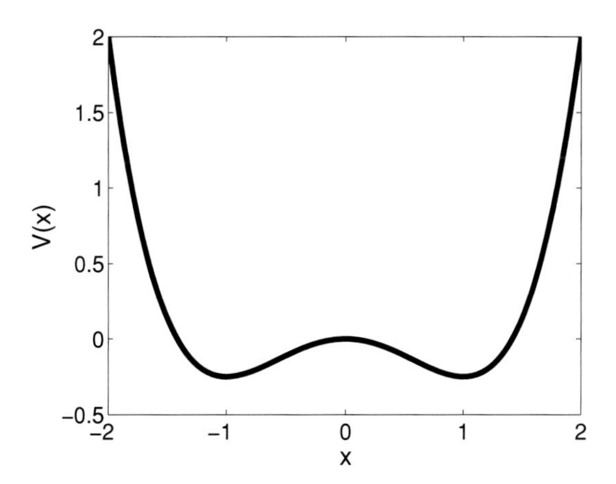

Figure 3.5 Plot of the potential function $V(x) = x^4/4 - x^2/2$ with $x_0 = -1$ (minimum) and $x_1 = 0$ (maximum).

and hence $T(x_0 \to x_1)$ is given by

$$T(x_0 \to x_1) = \frac{2}{\sigma^2} \int_{x_0}^{x_1} \left[e^{\frac{2V(y)}{\sigma^2}} \int_{-\infty}^{y} e^{-\frac{2V(z)}{\sigma^2}} dz \right] dy.$$

When σ is small, the integral over z will only be relevant for values of y near x_1 because, otherwise, the integral values of the first integrand will be small. Hence we can approximate the integral by taking $y = x_1$ in the inner integral, giving

$$T(x_0 \to x_1) \approx \frac{2}{\sigma^2} \int_{x_0}^{x_1} e^{\frac{2V(y)}{\sigma^2}} dy \int_{-\infty}^{x_1} e^{-\frac{2V(z)}{\sigma^2}} dz. \tag{3.87}$$

■

3.5.5 *Variance in fast-slow systems*

The Fokker-Planck approach can also be applied efficiently to study the change in variance in fast-slow systems (as discussed in Section 2.3.3) when a bifurcation point is approached. Consider, for example, the two-dimensional Itô SDE (Kuehn, 2011)

$$dX_t = f(X_t, Y_t)dt + \sigma dW_t, \tag{3.88a}$$

$$dY_t = \epsilon \, dt, \tag{3.88b}$$

where X_t, Y_t are both one-dimensional stochastic variables and σ is constant.

For $\epsilon = 0$, the forward Fokker-Planck for the probability density function $p^y(x, t)$ is

$$\frac{\partial p^y}{\partial t} = -\frac{\partial (f(x, y)p^y)}{\partial x} + \frac{\sigma^2}{2} \frac{\partial^2 p^y}{\partial x^2}, \tag{3.89}$$

and the stationary distribution $\bar{p}^y$ is given by

$$\bar{p}^y(x) = \frac{1}{N} e^{\int_a^x \frac{2}{\sigma^2} f(s,y)\, ds}, \tag{3.90}$$

where N is again a normalization factor.

Consider now, for example, the normal form of the saddle node, given by $f_1(x, y) = -y - x^2$ (cf. Example 2.3). Here, the interval (a, b) consists of the points that do not escape to infinity, that is, $a = -\sqrt{-y}$, $b = \infty$. Evaluating the integral in (3.90) then gives

$$\int_{-\sqrt{-y}}^{x} (-y - s^2) ds = -yx - \frac{1}{3}x^3 + \frac{2}{3}(-y)^{3/2},$$

and hence the equilibrium probability density function is given by

$$\bar{p}_1^y(x) = \frac{1}{N_1} e^{\frac{2}{\sigma^2}(-yx - \frac{1}{3}x^3 + \frac{2}{3}(-y)^{3/2})}. \tag{3.91}$$

Similar expressions can be derived for the transcritical bifurcation ($f_2(x, y) = yx - x^2$, $a = y$, $b = \infty$) and the subcritical pitchfork bifurcation ($f_3(x, y) = yx + x^3$, $a = -\sqrt{-y}$, $b = \sqrt{-y}$), giving expressions (Kuehn, 2011)

$$\bar{p}_2^y(x) = \frac{1}{N_2} e^{\frac{2}{\sigma^2}(\frac{1}{2}yx^2 - \frac{1}{3}x^3 - \frac{1}{6}y^3)}, \tag{3.92a}$$

$$\bar{p}_3^y(x) = \frac{1}{N_3} e^{\frac{2}{\sigma^2}(\frac{1}{2}yx^2 + \frac{1}{4}x^4 + \frac{1}{4}y^2)}. \tag{3.92b}$$

The variance of the system can be directly determined from the probability density function and is plotted for the three different bifurcations as a function of y in Fig. 3.6, for $\sigma = 0.1$. For each case, the variance increases sharply when the critical transition at $y = 0$ is approached. The distance of the local maximum to the critical transition decreases with decreasing noise level σ. This suggests that increased variance may be used as indicator of the approach to the critical transition.

3.6 Numerical solutions of SDEs

The numerical solution of SDEs is more involved than the solution of the deterministic counterparts. Consider an Itô SDE of the form

$$X(t) = X(0) + \int_0^t f(X(s))ds + \int_0^t g(X(s))dW(s), \tag{3.93}$$

where the notation of the stochastic integral is slightly changed. Let us define a partition $\tau_j = j\Delta t$, $j = 0, \ldots, n$ on $[0, T]$ with $\Delta t = T/n$ and indicate the numerical solution at τ_j with X_j (which is the reason for changing the notation) and the analytical solution with $X(\tau_j)$.

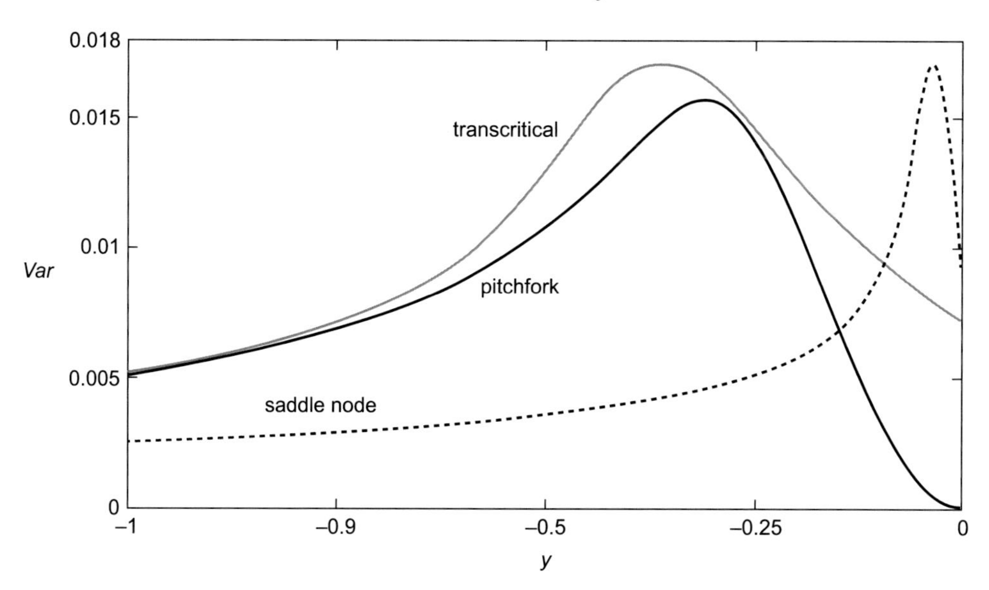

Figure 3.6 Variance for the stochastic system (3.88) for normal forms of three bifurcations and $\sigma = 0.1$ (figure slightly modified from Kuehn [2011]).

The order η of strong convergence for fixed k is such that

$$E[|X_k - X(\tau_k)|] \leq (\Delta t)^\eta. \tag{3.94}$$

Strong convergence therefore implies that the mean of the error converges to zero. On the contrary, weak convergence indicates only convergence of the expectation (error in the mean), and its order η is determined by

$$|E[X_k] - E[X(\tau_k)]| \leq (\Delta t)^\eta. \tag{3.95}$$

In the following subsections, two much-used schemes and their convergence behaviour are presented.

3.6.1 The Euler-Maruyama scheme

The Euler-Maruyama scheme for (3.93) is

$$X_j - X_{j-1} = f(X_{j-1})\Delta t + g(X_{j-1})(W(\tau_j) - W(\tau_{j-1})). \tag{3.96}$$

As an example, we consider $f(X) = \lambda X$ and $g(X) = \mu X$ for which we derived the analytical solution (3.53) as

$$X(t) = X(0)e^{(\lambda - \frac{\mu^2}{2})t + \mu W(t)}, \tag{3.97}$$

and we take $\lambda = 2$, $\mu = 1$ on the domain [0, 1]. First, a sample path of a Wiener process with a time step $\delta t = 2^{-8}$ and the exact solution (3.97) is shown as the drawn

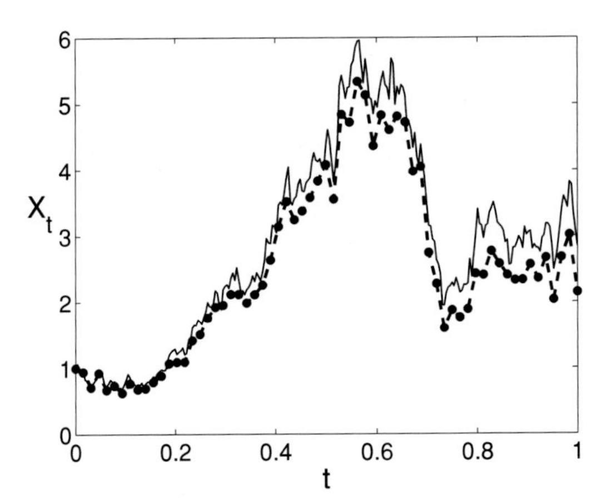

Figure 3.7 Application of the Euler scheme to the Itô SDE (3.50) for $X_0 = 1, \lambda = 2$ and $\mu = 1$. The drawn curve is the exact solution, and the labelled curve is the numerical solution with $\Delta t = 2^6$.

curve in Fig. 3.7. The numerical solution is computed with values of $\Delta t = R\delta t$ with $R = 4, 2$ and 1, and the solution for $R = 4$ is plotted in Fig. 3.7. For this case, the endpoint errors for $R = 4, 2, 1$ are 0.6907, 0.1595 and 0.0821, respectively.

To consider the strong and weak convergence of the Euler-Maruyama scheme, we choose $\delta t = 2^{-9}$ and compute 1,000 different sample paths of the Wiener process. For each path, the SDE is integrated with five different step sizes ($\Delta t = 2^{p-1}\delta t$, $p = 1, \ldots, 5$. The endpoint errors are computed for each of these paths, and the sample mean is computed over the 1,000 sample paths for each Δt. The result is shown in Fig. 3.8a, from which we see that the order of strong convergence is indeed near to a 1/2. A power law fit to the four points gives an exponent $\eta = 0.5384$. The same is done to study weak convergence of the Euler-Maruyama method. In Fig. 3.8b, the error in the expectation value is plotted versus Δt. A power-law fit to the four points gives an exponent $\eta = 0.9858$. These results confirm that the Euler-Maruyama scheme has a strong convergence with order $\eta = 1/2$ and a weak convergence with order $\eta = 1$ (Kloeden and Platen, 1999).

3.6.2 The Milstein scheme

To improve the order of strong convergence, we need higher-order terms to be included into the discretization scheme. One of these schemes is the Milstein scheme, which we present now for the Itô SDE (3.93). We first write the discretization as

$$X_{t_j} - X_{t_{j-1}} = \int_{t_{j-1}}^{t_j} f(X_s)ds + \int_{t_{j-1}}^{t_j} g(X_s)dW_s, \tag{3.98}$$

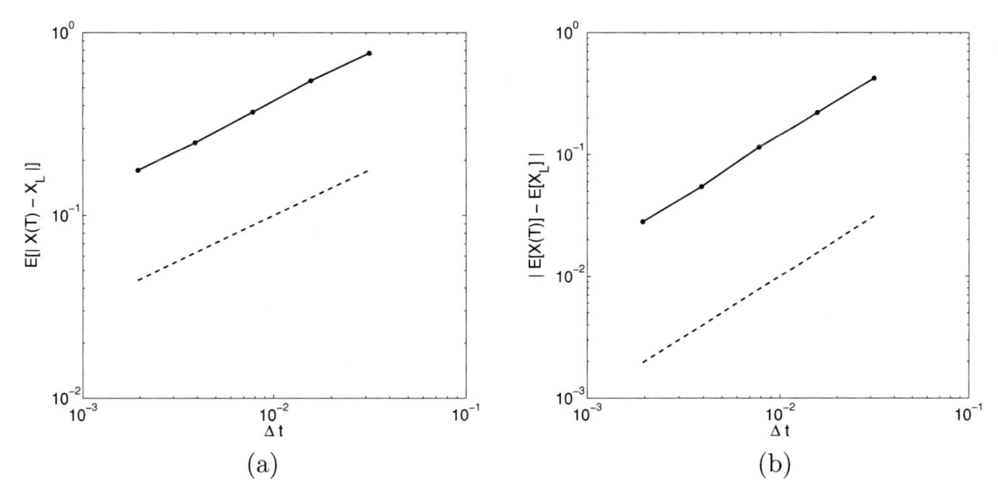

Figure 3.8 (a) Strong convergence of the Euler-Maruyama scheme to the Itô SDE (3.50) with $c_1 = \lambda$ and $\sigma_1 = \mu$ for $X_0 = 1$, $\lambda = 2$ and $\mu = 1$. The dotted curve is a line with slope $1/2$. (b) Weak convergence for the same case. The dotted curve is a line with slope 1.

and recover the Euler-Maruyama scheme, with $\Delta_j = t_j - t_{j-1}$ and $\Delta_j W = W_{t_j} - W_{t_{j-1}}$, as

$$\int_{t_{j-1}}^{t_j} f(X_s)ds = f(X_{t_{j-1}})\Delta_j, \tag{3.99a}$$

$$\int_{t_{j-1}}^{t_j} g(X_s)dW_s = g(X_{t_{j-1}})\Delta_j W. \tag{3.99b}$$

The crucial step in the derivation of higher-order schemes is the application of the third Itô lemma (3.35) for a function $f(x)$, with $f_1 = 0$, $f_2 = f'$, $f_{22} = f''$, whereas $A^{(1)} = f$ and $A^{(2)} = g$, according to (3.98). We then find

$$f(X_s) - f(X_{t_{j-1}}) = \int_{t_{j-1}}^{s} \left[ff' + \frac{1}{2}g^2 f''\right] dy + \int_{t_{j-1}}^{s} gf'dW_y, \tag{3.100}$$

where the integration argument, y, has been suppressed for clarity. We do the same for the function g to obtain

$$g(X_s) - g(X_{t_{j-1}}) = \int_{t_{j-1}}^{s} \left[fg' + \frac{1}{2}g^2 g''\right] dy + \int_{t_{j-1}}^{s} gg'dW_y. \tag{3.101}$$

Next we substitute the last two expressions into (3.98) and obtain

$$X_{t_j} - X_{t_{j-1}} = f(X_{t_{j-1}})\Delta_j + g(X_{t_{j-1}})\Delta_j W + R_j^1 + R_j^2, \tag{3.102}$$

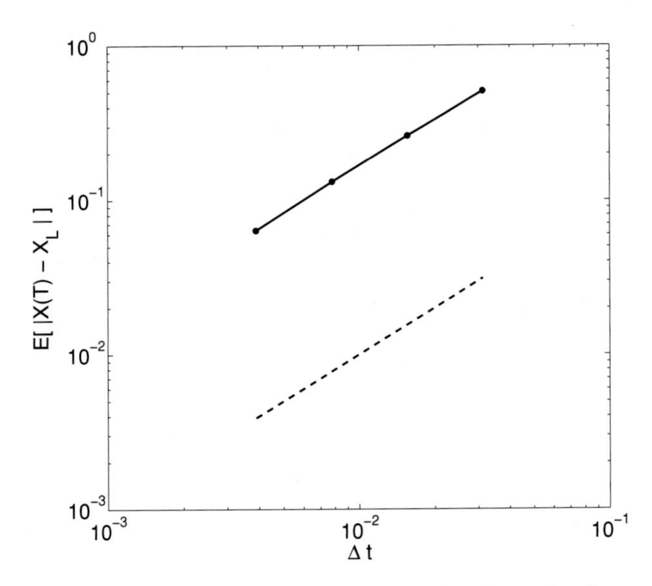

Figure 3.9 Test of strong convergence for the Milstein scheme for the same problem as in Fig 3.8. The dotted line has a slope 1.

with

$$R_j^1 = \int_{t_{j-1}}^{t_j} \int_{t_{j-1}}^{s} gg' dW_y dW_s. \tag{3.103}$$

As R_j^2 can be shown to be of smaller magnitude than R_j^1 (Kloeden and Platen, 1999), what remains is to evaluate R_j^1 as

$$R_j^1 = \frac{1}{2} g(X_{t_{j-1}}) g'(X_{t_{j-1}}) ((\Delta_j W)^2 - \Delta_j) \tag{3.104}$$

to finally give the Milstein scheme

$$X_{t_j} - X_{t_{j-1}} = f(X_{t_{j-1}})\Delta_j + g(X_{t_{j-1}})\Delta_j W + \frac{1}{2} g(X_{t_{j-1}}) g'(X_{t_{j-1}}) ((\Delta_j W)^2 - \Delta_j). \tag{3.105}$$

Again, we consider the Itô SDE with $f(X) = \lambda X$ and $g(X) = \mu X$ for which we derived the analytical solution (3.97), and we take $\lambda = 2$, $\mu = 1$ on the domain [0, 1]. To consider the strong and weak convergence of the Milstein scheme, we choose $\delta t = 2^{-12}$ and compute 1,000 different sample paths of the Wiener process. For each path, the SDE is integrated with four different step sizes $\Delta t = R\delta t$ with $R = 16, 32, 64$ and 128. The endpoint errors are computed for each of these paths, and the sample mean is computed over the 1,000 sample paths for each Δt. The results are shown in Fig. 3.9, from which we see that the order of convergence is indeed near to 1. A power-law fit to the four points gives an exponent $\eta = 0.9992$, and hence the Milstein scheme has a faster strong convergence than the Euler-Maruyama scheme.

By repeatedly using the Itô lemma, even higher-order schemes can be used. A summary of these schemes is given in Kloeden and Platen (1999). A Fortran 95 version of some of these schemes is available at http://steck.us/computer.html.

4

Stochastic Dynamical Systems

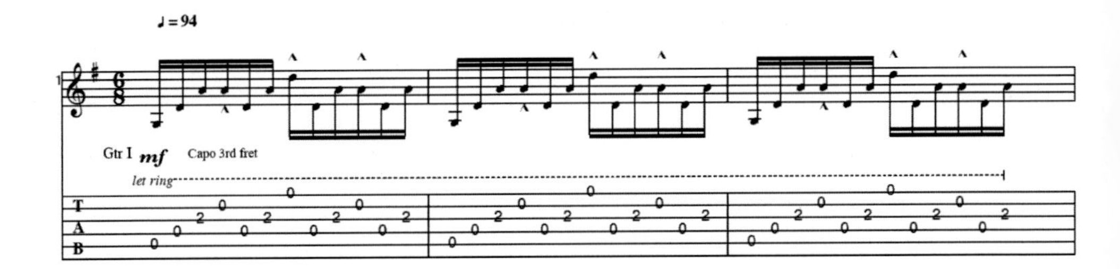

Invariants are everywhere
DGDGAD, Order of Magnitude, Dough Smith

When considering dynamical systems with additive or state-dependent noise, the immediate question that comes to mind is: how are bifurcation points defined in such systems? Although many results from the theory of random dynamical systems are available and ready to be applied (Arnold, 1998), there is quite a barrier in terms of the mathematics. Hence, in this chapter, the aim is to provide an introduction for this subject, with focus on the concepts and applications. Enough details are provided for the chapter to remain self-contained. We illustrate much of the material using the example of the stochastic pitchfork bifurcation and use this to present the more general theory.

4.1 Nonautonomous dynamical systems

When noise is included in a dynamical system as a representation of nonresolved processes, the dynamical system immediately becomes nonautonomous. The essential differences between autonomous and nonautonomous systems are discussed in the first subsection that follows while simultaneously introducing the mathematical concept of a flow. In the next subsection, an example is used to introduce the notion of a pullback attractor, which is the relevant dynamical object in the nonautonomous (and random) case.

4.1.1 Dynamical systems and flows

Let us first go back to the one-dimensional autonomous case in which the dynamical system is defined as

$$\frac{dx}{dt} = f(x), \tag{4.1a}$$

$$x(s) = x_0, \tag{4.1b}$$

where the initial time is indicated by $t = s$; let the state space be indicated by X. Instead of studying one trajectory of this dynamical system, we now want to study families of trajectories for different s. In this case, the flow $\phi(t, s) : X \to X$ is defined with $\phi(s, s) = Id$ (the identity flow) such that $\phi(t, s)[x]$ is a solution of (4.1). For example, when f is linear, say, of the form $-ax$, the solution is easily given as

$$x(t) = x_0 e^{-a(t-s)} \to \phi(t, s)[x] = e^{-a(t-s)} x. \tag{4.2}$$

For autonomous systems, these flows satisfy the group property (with composite operator $\circ$)

$$\phi(t_1 + t_2, s) = \phi(t_2, 0) \circ \phi(t_1, s), \tag{4.3}$$

which can be easily verified for the preceding example as

$$\begin{aligned} \phi(t_1 + t_2, s)[x] &= e^{-a(t_1 - s + t_2)} x \\ &= e^{-at_2} e^{-a(t_1 - s)} x = \phi(t_2, 0)[\phi(t_1, s)[x]]. \end{aligned} \tag{4.4}$$

An important consequence of the group property is that solutions of autonomous equations cannot intersect in isolated points without coinciding on the whole time interval.

For nonautonomous systems, defined by

$$\frac{dx}{dt} = f(x, t), \tag{4.5a}$$

$$x(s) = x_0, \tag{4.5b}$$

this group property no longer holds, as can be easily seen by considering the example $f(x, t) = -tx$ for which the solution is given by

$$\phi(t, s)[x] = x e^{-\frac{t^2 - s^2}{2}}, \tag{4.6}$$

and hence

$$\phi(t_1 + t_2, s)[x] = xe^{-\left(\frac{t_1+t_2}{2}\right)^2 + \frac{s^2}{2}}$$
$$\neq \phi(t_2, 0)[\phi(t_1, s)[x]] = e^{-\frac{t_2^2}{2}} e^{-\frac{(t_1^2 - s^2)}{2}} x. \tag{4.7}$$

One consequence is that solutions of nonautonomous systems can intersect themselves, which can also be seen by writing (4.5) as the autonomous system

$$\frac{dx}{dt} = f(x, y), \tag{4.8a}$$

$$\frac{dy}{dt} = 1. \tag{4.8b}$$

The solutions of (4.5) can be seen as a projection, and hence intersections are possible.

4.1.2 Pullback attractors

For autonomous systems, the asymptotic properties of the flow are discussed in terms of the global attractor of the system. For many of these systems, there exists a bounded set $A \in \mathcal{X}$, where $\mathcal{X}$ is the total space, into which all solutions enter whatever initial condition at $t = s$; hence

$$\phi(t, s)\mathcal{X} = A, \text{ as } \quad t \to \infty. \tag{4.9}$$

When one considers nonautonomous systems, then the form (4.8) already shows that the variable y becomes unbounded, and the contracting properties of the system will be lost. Hence a bounded set A with the property (4.9) in general does not exist.

For the nonautonomous case, the concept of pullback attractor is used. Instead of letting $t \to \infty$ (and s fixed), one considers $s \to -\infty$ and t fixed. For nonautonomous systems, an invariant set $A(t)$ may then exist such that

$$\lim_{s \to -\infty} |\phi(t, s)[x] - A(t)| = 0. \tag{4.10}$$

Example 4.1 A one-dimensional pullback attractor Consider the nonautonomous system

$$\frac{dx}{dt} = -x + t,$$

with the solution (which can be determined by variation of constants) for which $x(s) = x_0$ given by

$$x(t) = x_0 e^{s-t} + (t - s e^{s-t}) - (1 - e^{s-t}).$$

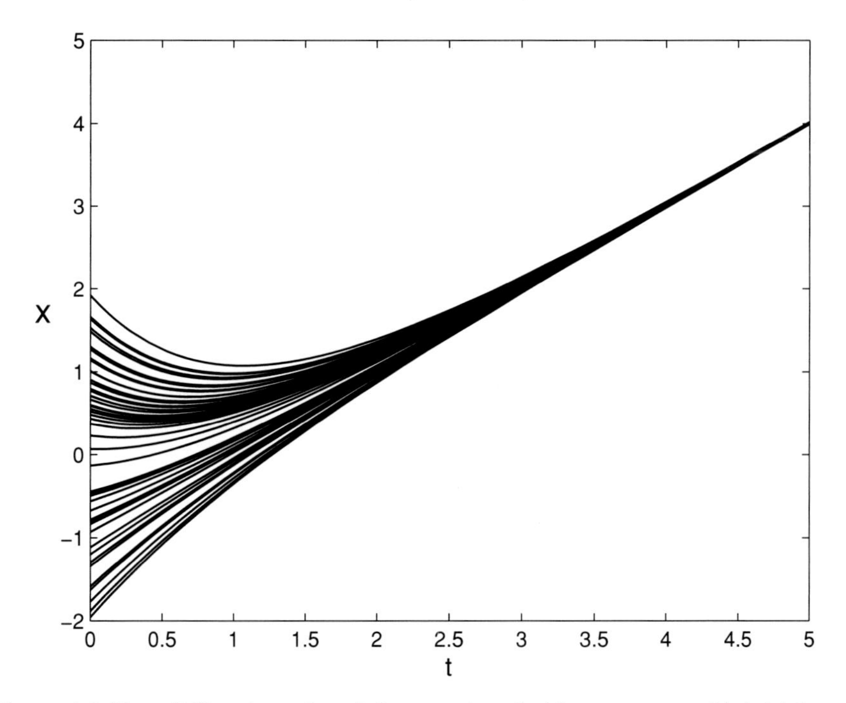

Figure 4.1 Plot of 50 trajectories of the equation $dx/dt = -x + t$ with initial conditions chosen from a uniform distribution on the interval $[-2, 2]$. The pullback attractor is given by $A(t) = t - 1$.

If we fix s and let $t \to \infty$, all trajectories become unbounded through the second term. The flow of the system is given by (just substitute $x_0 \to x$)

$$\phi(t, s)[x] = e^{(s-t)}(x - s + 1) + t - 1.$$

When letting $s \to -\infty$ while t is fixed, there is a set $A(t) = t - 1$ for which

$$\lim_{s \to -\infty} |\phi(t, s)[x] - (t - 1)| = \lim_{s \to -\infty} |e^{(s-t)}(x - s + 1)| = 0,$$

and hence the set $A(t)$ is the pullback attractor of the system (Fig. 4.1). ■

4.2 Random dynamical systems

In this section, a very brief introduction into the theory of random dynamical systems is given in which many details on the calculations are provided to make it accessible to less mathematically trained readers.

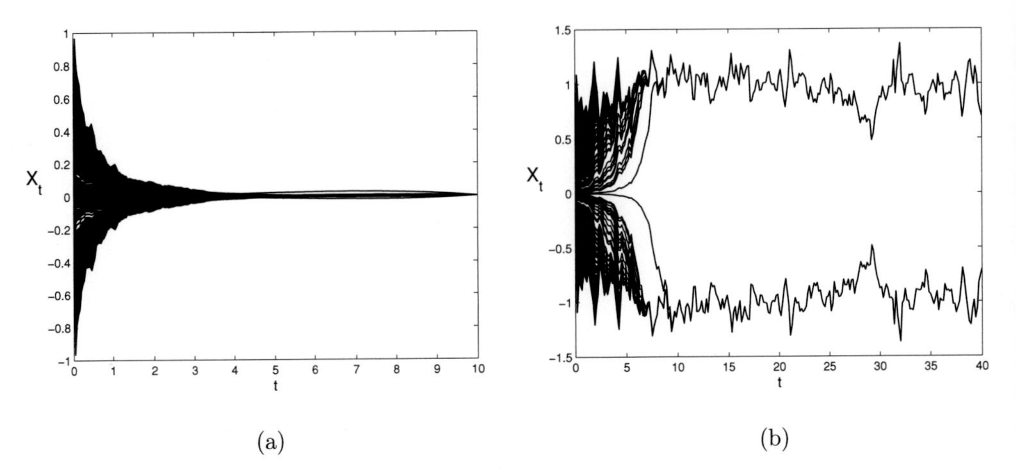

Figure 4.2 Plot of 500 sample paths for the SDE given by (4.11) for $\beta = 1$ and $\sigma = 0.25$. The initial conditions are uniformly distributed on the interval $[-1, 1]$. (a) $\alpha = -1.0$, (b) $\alpha = 1.0$.

4.2.1 An illustrative problem

We start with an example of the following stochastic Stratonovich differential equation:

$$dX_t = (\alpha X_t - \beta X_t^3)dt + \sigma X_t \circ dW_t, \tag{4.11}$$

with α, β being the parameters and σ the noise amplitude. The deterministic part can be recognized as the normal form of the pitchfork bifurcation (2.26), and the noise represented is state-dependent. Using the Euler-Maruyama scheme described in Section 3.5, we compute 500 trajectories of (4.11) for two different values of α for fixed $\beta = 1$ and $\sigma = 0.25$; the results are plotted in Fig. 4.2. These sample paths already show several important features, such as the clustering of the trajectories, which leads to changes of the density distribution of the trajectories. Whereas for $\alpha = -1$ (Fig. 4.2a) the sample paths cluster around zero, for $\alpha = 1$ (Fig. 4.2b) they cluster near $+1$ and near -1. There is a clear qualitative change in the behaviour of the solutions (4.11) when α changes, which is characteristic of a (in this case stochastic) bifurcation.

We can solve (4.11) analytically by first writing it as an Itô SDE, using (3.42),

$$dX_t = \left(\left(\alpha + \frac{\sigma^2}{2}\right) X_t - \beta X_t^3\right) dt + \sigma X_t dW_t. \tag{4.12}$$

Next we use the transformation $Y_t = 1/X_t^2$ and use the second Itô lemma (3.33), with $f(t, x) = 1/x^2$. We then find for Y_t the linear scalar Itô SDE

$$Y_t - Y_0 = 2\left[\int_0^t ((\sigma^2 - \alpha)Y_s + \beta)ds - \int_0^t \sigma Y_s dW_s\right]. \tag{4.13}$$

The solution was discussed in Section 3.4.3, and (with $c_1 = 2(\sigma^2 - \alpha)$, $c_2 = 2\beta$, $\sigma_1 = -2\sigma$ and $\sigma_2 = 0$) we find that the general solution Y_t to (4.13) is given by

$$Y_t = e^{-2(\alpha t + \sigma W_t)}(Y_0 + 2\beta \int_0^t e^{2(\alpha s + \sigma W_s)} ds), \tag{4.14}$$

and hence $X_t = 1/\sqrt{Y_t}$ is finally given by

$$X_t = \frac{X_0 e^{\alpha t + \sigma W_t}}{(1 + 2X_0^2 \beta \int_0^t e^{2\alpha u + 2\sigma W_u} du)^{1/2}}. \tag{4.15}$$

As can be seen from this solution, the sample paths depends on the particular Wiener path W_t chosen, the initial condition X_0 and time t. Hence a flow ϕ can be defined that takes the initial condition X_0 to the point X_t over a time t using the Wiener path W_t. For the case of (4.12), this mapping becomes

$$\phi(t, \omega)[x] = \frac{x e^{\alpha t + \sigma W_t(\omega)}}{(1 + 2x^2 \beta \int_0^t e^{2\alpha u + 2\sigma W_u(\omega)} du)^{\frac{1}{2}}}, \tag{4.16}$$

where the explicit dependence of the path W_t on ω stresses the dependence on the particular noise realization. So it is this mapping that takes a 'density' of initial conditions at $t = 0$ to a 'density' of sample paths at time t.

4.2.2 General theory

To be able to understand the asymptotic behaviour of nonlinear dynamical systems with noise (in the pullback sense), a precise concept of a random dynamical system is needed (Arnold, 1998).

The first ingredient is to reformulate the noise in terms of flows. Therefore, a probability space $(\Omega, \mathcal{F}, \mathbb{P})$, as defined in Section 3.1, is needed. Here, $\mathbb{P}$ is called the probability measure on $\mathcal{F}$, and it is a mapping $\mathcal{F} \to \mathbb{R}$ with a meaning of the probability of a certain event. On this probability space, a driving system (or base flow) is defined as a dynamical system θ_t (a mapping $\Omega \to \Omega$) such that (i) $\theta_0 = Id_\Omega$ and (ii) $\theta_{t+s} = \theta_t \circ \theta_s$. There are other more technical conditions on measurability properties, but we omit these here.

For an Itô SDE of the form

$$dX_t = a(t, X_t)dt + b(t, X_t)dW_t, \tag{4.17}$$

the Wiener process W_t for a given fixed realization is a path $t \mapsto W_t(\omega), t \in \mathbb{R}$, whereas for fixed t, it represents a random variable $\omega \mapsto W_t(\omega) = \omega(t), \omega \in \Omega, \omega(t) \in \mathbb{R}$ with $\omega(0) = 0$. The driving system then becomes

$$\theta_t[\omega(s)] = \omega(t + s) - \omega(t). \tag{4.18}$$

Indeed, $\theta_0[\omega(s)] = \omega(s) - \omega(0) = \omega(s)$ and hence $\theta_0 = Id_\Omega$. To show that property (ii) above holds, we compute

$$\theta_s[\omega(x)] = \omega(s + x) - \omega(s), \tag{4.19}$$

and hence

$$\begin{aligned}\theta_t[\theta_s[\omega(x)]] &= \theta_t[\omega(s + x) - \omega(s)] \\ &= \omega(t + s + x) - \omega(t) - (\omega(t + s) - \omega(t)) \\ &= \omega(t + s + x) - \omega(t + s) = \theta_{t+s}[\omega(x)].\end{aligned} \tag{4.20}$$

For later reference, from the driving system also a relation for the Wiener process $W_t(\omega) = \omega(t)$ follows as

$$W_{t+s}(\omega) = \omega(t + s) = \omega(t) + \theta_t[\omega(s)] = W_t(\omega) + W_s(\theta_t\omega). \tag{4.21}$$

Having formulated the driving system as a flow, a random dynamical system is defined as a mapping,

$$(t, \omega, x) \mapsto \phi(t, \omega)[x], \tag{4.22}$$

where $x \in X$, t indicates time and $\omega \in \Omega$. This mapping has the property $\phi(0, \omega) = Id_X$ and the co-cycle property holds, that is,

$$\phi(t + s, \omega) = \phi(t, \theta_s\omega) \circ \phi(s, \omega). \tag{4.23}$$

When one considers the space $\Omega \times X$, the random dynamical system can be viewed as a flow,

$$\Theta(t) : (\omega, x) \mapsto (\theta_t\omega, \phi(t, \omega)[x]), \tag{4.24}$$

over this space (Fig. 4.3) with driving function θ. Note that this is very similar to how one would view a nonautonomous system, where the driving system is just $\theta_t = t$, in an autonomous way (e.g., through the extended system [4.8]).

It may be instructive to see how one technically handles random dynamical systems by investigating the co-cycle property for the analytical solution (4.16). Therefore, we use the notation

$$G(t, \omega) = 2(\alpha t + \sigma W_t(\omega)), \tag{4.25}$$

such that the solution (4.16) can be written as

$$\phi(t, \omega)[x] = \frac{x e^{G(t,\omega)/2}}{(1 + 2x^2\beta \int_0^t e^{G(u,\omega)} du)^{\frac{1}{2}}}. \tag{4.26}$$

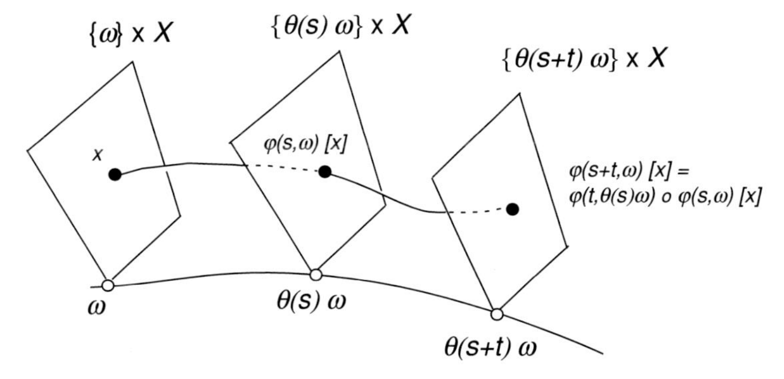

Figure 4.3 View of a random dynamical system (RDS) as a flow on the 'probability times dynamical space' $\Omega \times X$, where we indicate $\theta_s = \theta(s)$. For a given state x and a realization ω, the RDS is such that $\Theta(t)(x, \omega) = (\theta_t \omega, \phi(t, \omega)[x])$ is the flow on this composite space (figure based on Ghil et al. [2008]).

As a first step to determining $\phi(t + s, \omega)[x]$, we evaluate

$$\begin{aligned} G(t+s, \omega) &= 2(\alpha(t+s) + \sigma W_{t+s}(\omega)) \\ &= 2(\alpha(t+s) + \sigma(W_t(\omega) + W_s(\theta_t \omega))) \\ &= G(t, \omega) + G(s, \theta_t \omega), \end{aligned} \tag{4.27}$$

where (4.21) is used. As the second step, we evaluate

$$\begin{aligned} \int_0^{t+s} e^{G(u,\omega)} du &= \int_0^s e^{G(u,\omega)} du + \int_s^{t+s} e^{G(u,\omega)} du \\ &= \int_0^s e^{G(u,\omega)} du + \int_0^t e^{G(s+u',\omega)} du' \\ &= \int_0^s e^{G(u,\omega)} du + \int_0^t e^{G(s,\omega)+G(u',\theta_s\omega)} du' \\ &= \int_0^s e^{G(u,\omega)} du + e^{G(s,\omega)} \int_0^t e^{G(u',\theta_s\omega)} du'. \end{aligned} \tag{4.28}$$

This gives, for $\phi(t + s, \omega)[x]$, the expression

$$\phi(t+s, \omega)[x] = \frac{x e^{(G(t,\omega)+G(s,\theta_t\omega))/2}}{(1 + 2x^2 \beta (\int_0^s e^{G(u,\omega)} du + e^{G(s,\omega)} \int_0^t e^{G(u',\theta_s\omega)} du'))^{\frac{1}{2}}}. \tag{4.29}$$

From the analytic solution (4.15), we find that

$$\phi(s, \omega)[x] = \frac{x e^{G(s,\omega)/2}}{(1 + 2x^2 \beta \int_0^s e^{G(u,\omega)} du)^{\frac{1}{2}}} \equiv y, \tag{4.30}$$

and now $\phi(t, \theta_s\omega) \circ \phi(s, \omega)[x]$ can be written as

$$\begin{aligned}
&\phi(t, \theta_s\omega)[y] \\
&= \frac{y e^{G(t,\theta_s\omega)/2}}{(1 + 2y^2\beta \int_0^t e^{G(u,\theta_s\omega)} du)^{\frac{1}{2}}} \\
&= \frac{x e^{G(s,\omega)/2}}{(1 + 2x^2\beta \int_0^s e^{G(u,\omega)} du)^{\frac{1}{2}}} \frac{e^{G(t,\theta_s\omega)/2}}{\left(1 + 2\left(\frac{x^2 e^{G(s,\omega)}}{(1+2x^2\beta \int_0^s e^{G(u,\omega)} du)}\right) \beta \int_0^t e^{G(u,\theta_s\omega)} du\right)^{\frac{1}{2}}} \\
&= \frac{x e^{(G(s,\omega)+G(t,\theta_s\omega))/2}}{(1 + 2x^2\beta \int_0^s e^{G(u,\omega)} du + 2x^2\beta e^{G(s,\omega)} \int_0^t e^{G(u,\theta_s\omega)} du)^{\frac{1}{2}}} \\
&= \phi(t + s, \omega)[x],
\end{aligned} \tag{4.31}$$

and hence the co-cycle property is satisfied.

4.2.3 Invariant measures

For a certain random dynamical system, we can ask what sets remain invariant under the flow. Examples of such sets for the SDE (4.12) were already shown in Fig. 4.2. The concept of an invariant measure in random dynamical systems is a direct extension of that of a fixed point for autonomous deterministic dynamical systems. To define the invariant measures, let μ be a probability measure on $\Omega \times X$ and Θ defined by (4.24); μ reduces to the probability measure $\mathbb{P}$ when projected on Ω. An invariant measure μ satisfies

$$\Theta(t)\mu = \mu. \tag{4.32}$$

Note that this is quite an intuitive notation as μ operates on sets, and hence a more precise notation would be $\Theta(t)\mu(A) = \mu(\Theta^{-1}(A))$ for every measurable set $A \in \Omega \times X$. The reader is referred to Arnold (1998) for the mathematical justification of the preceding 'sloppy' notation.

By factorization of the measure μ into $\mathbb{P}$ on Ω and the measure μ_ω on X, it can be shown that in terms of the flow ϕ, the invariant measure μ_ω satisfies

$$\phi(t, \omega)[\mu_\omega] = \mu_{\theta_t\omega}. \tag{4.33}$$

An invariant measure is called a random Dirac measure if there exists a random variable $\chi : \Omega \to X$ with $\mu_\omega = \delta_{\chi(\omega)}$ and

$$\phi(t, \omega)[\chi(\omega)] = \chi(\theta_t\omega). \tag{4.34}$$

For a Dirac measure, the probability measure on X is a delta function. In Fig. 4.2, we already saw that (for a single noise realization) trajectories clustered onto a 'single trajectory' in time, which illustrates this delta function type invariant measure.

For the example system (4.11), the invariant measures are given by

$$\mu_\omega^0 = \delta_0, \quad \alpha \in \mathbb{R}, \tag{4.35a}$$

$$\mu_\omega^\pm = \delta_{\pm\chi(\omega)}, \quad \alpha > 0, \tag{4.35b}$$

with

$$\chi(\omega) = \frac{1}{(2\beta \int_{-\infty}^0 e^{G(u,\omega)} du)^{\frac{1}{2}}} = \frac{1}{(2\beta \int_{-\infty}^0 e^{2(\alpha u + \sigma W_u(\omega))} du)^{\frac{1}{2}}}. \tag{4.36}$$

Again, it is instructive to verify by direct computation that the $\mu_\omega^\pm$ are invariant measures.

From the definition of the flow $\phi(t, \omega)$ in (4.26), it follows directly that

$$\begin{aligned}
\phi(t,\omega)[\chi(\omega)] &= \frac{\chi(\omega) e^{G(t,\omega)/2}}{(1 + 2\chi^2(\omega)\beta \int_0^t e^{G(u,\omega)} du)^{\frac{1}{2}}} \\
&= \frac{1}{(2\beta \int_{-\infty}^0 e^{G(u,\omega)} du)^{\frac{1}{2}}} \frac{e^{G(t,\omega)/2}}{(1 + \frac{\int_0^t e^{G(u,\omega)} du}{\int_{-\infty}^0 e^{G(u,\omega)} du})^{\frac{1}{2}}} \\
&= \frac{1}{(2\beta)^{\frac{1}{2}}} \frac{e^{G(t,\omega)/2}}{(\int_{-\infty}^0 e^{G(u,\omega)} du + \int_0^t e^{G(u,\omega)} du)^{\frac{1}{2}}} \\
&= \frac{1}{(2\beta)^{\frac{1}{2}}} \frac{1}{(\int_{-\infty}^t e^{G(u,\omega) - G(t,\omega)} du)^{\frac{1}{2}}} \\
&= \frac{1}{(2\beta)^{\frac{1}{2}}} \frac{1}{(\int_{-\infty}^0 e^{G(u'+t,\omega) - G(t,\omega)} du')^{\frac{1}{2}}},
\end{aligned} \tag{4.37}$$

where $u' = u - t$ was used in the last equality. From the definition of $G(t, \omega)$, in (4.25) we derive

$$\begin{aligned}
G(u' + t, \omega) - G(t, \omega) &= 2(\alpha(u' + t) + \sigma W_{u'+t}(\omega)) - 2(\alpha t + \sigma W_t(\omega)) \\
&= 2\alpha u' + 2\sigma(W_{u'+t}(\omega) - W_t(\omega)) \\
&= 2\alpha u' + 2\sigma W_{u'}(\theta_t \omega),
\end{aligned}$$

where (4.21) was used in the last equality. With this expression, it follows that

$$\phi(t,\omega)[\chi(\omega)] = \frac{1}{(2\beta)^{\frac{1}{2}}} \frac{1}{(\int_{-\infty}^0 e^{2\alpha u' + 2\sigma W_{u'}(\theta_t \omega)} du')^{\frac{1}{2}}} = \chi(\theta_t \omega), \tag{4.38}$$

which shows that $\mu_\omega^\pm$ are invariant measures.

To demonstrate that the invariant measure concept extends the fixed point concept for the deterministic case, let $\sigma = 0$. In that case,

$$\chi(\omega) = \frac{1}{(2\beta \int_{-\infty}^{0} e^{2\alpha u} du)^{\frac{1}{2}}} = \sqrt{\frac{\alpha}{\beta}}, \tag{4.39}$$

and hence the Dirac measures $\mu^{\pm}$ become the fixed points $\pm\sqrt{\alpha/\beta}$.

To determine the invariant measures (4.36) directly from the flow, the pullback approach is followed. The variable s is now written as $\theta_{-t}\omega$, and hence to determine the limiting set in the pullback sense, we need to evaluate the limit

$$\lim_{t\to\infty} \phi(t, \theta_{-t}\omega)[x] = \lim_{t\to\infty} \frac{x e^{G(t,\theta_{-t}\omega)/2}}{(1 + 2x^2\beta \int_0^t e^{G(u,\theta_{-t}\omega)} du)^{\frac{1}{2}}}. \tag{4.40}$$

To evaluate the limit, we use (4.25) and (4.21) to give

$$G(u, \theta_{-t}\omega) = 2(\alpha u + \sigma W_u(\theta_{-t}\omega) = 2(\alpha u + \sigma(W_{u-t}(\omega) - W_{-t}(\omega))),$$
$$G(t, \theta_{-t}\omega) = 2(\alpha t + \sigma W_t(\theta_{-t}\omega) = 2(\alpha t + \sigma(W_0(\omega) - W_{-t}(\omega)),$$

and hence (with $W_0(\omega) = 0$ and $u' = u - t$),

$$\begin{aligned}
\phi(t, \theta_{-t}\omega)[x] &= \frac{x e^{\alpha t - \sigma W_{-t}(\omega)}}{(1 + 2x^2\beta \int_0^t e^{2(\alpha u + \sigma(W_{u-t}(\omega) - W_{-t}(\omega)))} du)^{\frac{1}{2}}} \\
&= \frac{x e^{\alpha t - \sigma W_{-t}(\omega)}}{(1 + 2x^2\beta \int_{-t}^0 e^{2(\alpha(u'+t) + \sigma(W_{u'}(\omega) - W_{-t}(\omega)))} du')^{\frac{1}{2}}} \\
&= \frac{x}{(e^{-2\alpha t + 2\sigma W_{-t}(\omega)} + 2x^2\beta \int_{-t}^0 e^{2\alpha u' + 2\sigma W_{u'}(\omega)} du')^{\frac{1}{2}}},
\end{aligned} \tag{4.42}$$

and, finally,

$$\begin{aligned}
\lim_{t\to\infty} \phi(t, \theta_{-t}\omega)[x] &= \frac{x}{(2x^2\beta \int_{-\infty}^0 e^{2\alpha u' + 2\sigma W_{u'}(\omega)} du')^{\frac{1}{2}}} \\
&= \frac{\pm 1}{(2\beta \int_{-\infty}^0 e^{G(u',\omega)} du')^{\frac{1}{2}}} = \pm\chi(\omega),
\end{aligned} \tag{4.43}$$

the latter equality holding for $x \neq 0$; for $x = 0$, the limit is trivial.

Example 4.2 The stochastic Lorenz (1963) system As an illustration of the use of the theory of random dynamical systems in analyzing the behavior of a nonlinear stochastic system, we consider the stochastic Lorenz (1963) system with multiplicative

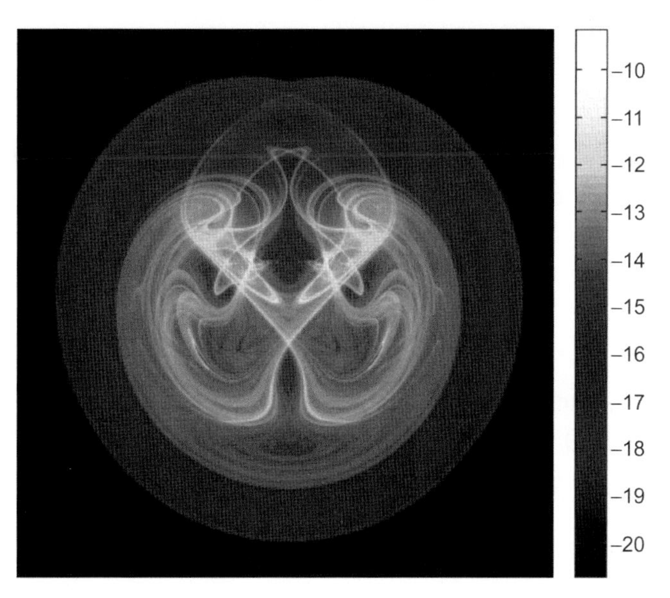

Figure 4.4 Density of trajectories of the random Lorenz system (4.44), projected on the (Y_t, Z_t) plane. The colour scale is logarithmic, with yellow indicating a high density of sample paths; see text for further description (figure from Chekroun et al., 2011). (See Colour Plate.)

noise, given by the set of Itô SDEs,

$$dX_t = s(Y_t - X_t)dt + \sigma X_t dW_t, \tag{4.44a}$$

$$dY_t = (rX_t - Y_t - X_t Z_t)dt + \sigma Y_t dW_t, \tag{4.44b}$$

$$dZ_t = (-bZ_t + X_t Y_t)dt + \sigma Z_t dW_t, \tag{4.44c}$$

with parameters $s = 10$, $r = 28$ and $b = 8/3$. It was shown in Dorfle and Graham (1983) that the stationary solution of the Fokker-Planck equation for the Lorenz model (4.44) looks very much like that of the unperturbed system. It is only a bit slightly fuzzier as the noise smoothes the small-scale structures of the attractor.

Snapshots of the density of sample paths of (4.44) at different times were presented in Chekroun et al. (2011). An example obtained at $t = 40$ with 10^9 initial conditions, $\sigma = 0.3$ and a time step of $\Delta t = 5 \times 10^{-3}$ is shown in Fig. 4.4. This a projection of the density of the trajectories onto the (Y_t, Z_t) plane. The colour bar to the right is on a log-scale and quantifies the probability of ending up in a particular region of phase space. Notice the interlaced filament structures between highly (yellow) and moderately (red) populated regions. There is no fuzziness whatsoever in the topological structure of this filamentation, which evokes the Cantor-set foliation of the deterministic attractor. ■

4.2.4 Stability of invariant measures

Lyapunov exponents were introduced in Section 2.5.3 in relation to divergence of nearby trajectories on an attractor. If the largest Lyapunov exponent on an attractor is positive, there is sensitive dependence on initial conditions. The definition of the Lyapunov exponent can, however, also be used to determine the stability of fixed points. To illustrate this, consider the normal form of the deterministic pitchfork bifurcation,

$$\frac{dx}{dt} = \alpha x - \beta x^3, \tag{4.45}$$

and suppose we want to study the stability of the fixed point $x = \bar{x}$. We put $x = \bar{x} + v(t)$, where $v(t)$ is a small perturbation and linearise the system to obtain the equation for v as

$$\frac{dv}{dt} = (\alpha - 3\beta\bar{x}^2)v \to v(t) = e^{(\alpha - 3\beta\bar{x}^2)t} v_0 \equiv \phi(t)[v_0], \tag{4.46}$$

where $\phi(t)$ is the flow associated with the linear equation for v. The Lyapunov exponent $\lambda(\bar{x})$ of the fixed point $x = \bar{x}$ is then defined as (cf. (2.49)),

$$\lambda(\bar{x}) = \lim_{t\to\infty} \frac{1}{t} \ln |v(t)| = \lim_{t\to\infty} \frac{1}{t} \ln |\phi(t)[v_0]| = \alpha - 3\beta\bar{x}^2, \tag{4.47}$$

independent of v_0. For the fixed points $\bar{x} = \pm\sqrt{\alpha/\beta}$, we find $\lambda = -2\alpha$, and for the fixed point $\bar{x} = 0$, we find $\lambda = \alpha$. Hence the Lyapunov exponent reduces to the eigenvalue determining the stability of the fixed point, and the pitchfork bifurcation on $\bar{x} = 0$ is characterized by a change in sign of the Lyapunov exponent at $\alpha = 0$. Note that a positive Lyapunov exponent (e.g., for $\alpha > 0$) does not indicate chaos here, as the fixed point $\bar{x} = 0$ for $\alpha > 0$ is not an attractor.

The Lyapunov exponent, generalized to the stochastic case, can be used to determine the stability of invariant measures. We illustrate this using the example problem (4.11), written here again as

$$dX_t = (\alpha X_t - \beta X_t^3)dt + \sigma X_t \circ dW_t. \tag{4.48}$$

For this equation, the invariant measures μ_ω were obtained in the previous section and satisfied

$$\phi(t, \omega)[\mu_\omega] = \mu_{\theta_t \omega}. \tag{4.49}$$

In correspondence with the fixed point notation in autonomous systems, we denote the flow of these invariant measures by $\phi(t, \omega)[\bar{\mu}]$.

Similarly to the autonomous case, we put $X_t = \phi(t, \omega)[\bar{\mu}] + V_t$ and linearize (4.48) in V_t to give

$$dV_t = (\alpha - 3(\phi(t, \omega)[\bar{\mu}])^2)V_t\, dt + \sigma V_t \circ dW_t, \tag{4.50}$$

which has the solution

$$V_t = V_0 e^{\alpha t + \sigma W_t(\omega) - 3\beta \int_0^t (\phi(s,\omega)[\bar{\mu}])^2 ds} \equiv D\phi(t, \omega, \bar{\mu})[V_0]. \tag{4.51}$$

The Lyapunov exponent of the invariant measure $\bar{\mu}$ is then defined as

$$\begin{aligned}\lambda(\bar{\mu}) &= \lim_{t\to\infty} \frac{1}{t} \ln |D\phi(t, \omega, \bar{\mu})[v]| \\ &= \alpha - 3\beta \lim_{t\to\infty} \frac{1}{t} \int_0^t (\phi(s, \omega)[\bar{\mu}])^2 ds.\end{aligned} \tag{4.52}$$

For the equation (4.48), we found the invariant measures (4.35). For $\bar{\mu}(\omega) = \delta_0(\omega)$, we have $\phi(t, \omega)[\bar{\mu}] = 0$, and hence $\lambda(\delta_0(\omega)) = \alpha$. For $\bar{\mu}(\omega) = \delta_{\pm\chi(\omega)}$, we obtain from the results in the previous section that

$$(\phi(s, \omega)[\chi(\omega)])^2 = \chi^2(\theta_s\omega) = \frac{1}{2\beta} \frac{1}{\int_{-\infty}^0 e^{2\alpha u' + 2\sigma W_{u'}(\theta_s\omega)} du'}. \tag{4.53}$$

We show in the next subsection that the integral in (4.52) can be evaluated using the Fokker-Planck equation, eventually, with (4.53), giving $\lambda(\delta_{\pm\chi(\omega)}) = -2\alpha$.

4.3 Stochastic bifurcations

Having described the extension of fixed points (invariant measures) and linear stability (Lyapunov exponents) to random dynamical systems in the previous section, we next present elementary bifurcation theory for random dynamical systems (Chapter 9 in Arnold, 1998).

Bifurcation means that a 'qualitative change' in the behavior of the system occurs when a parameter is varied. The notion of change, however, assumes that there is a rigorous notion of similarity between the solutions of two dynamical systems. Random dynamical systems can be compared at different levels, for example, at the statistical level and at the topological level. If the comparison involves too-fine details, then the systems always differ, and if the comparison is at too coarse a level, then all systems look the same.

This is the reason that for random dynamical systems, the concept of bifurcation is more complicated, and there are P-bifurcations (statistical measure of change) and D-bifurcations (topological measure of change). A necessary condition for a D-bifurcation to occur is that one Lyapunov exponent vanishes, but this is, however, no sufficient condition. A P-bifurcation occurs at a parameter value where the stationary probability density function (as determined through the Fokker-Planck equation) changes shape.

As we have worked through the example of the pitchfork bifurcation system (4.11) in great detail throughout the last subsections, this case is discussed first, and the stochastic Hopf bifurcation is briefly mentioned in the next subsection. The results for the stochastic transcritical and saddle-node bifurcations can be found in Chapter 9 of Arnold (1998).

4.3.1 Pitchfork: multiplicative noise

We first consider the case with multiplicative noise for which the equation (4.11), in which we choose $\beta = 1$, becomes

$$dX_t = (\alpha X_t - X_t^3)dt + \sigma X_t \circ dW_t, \tag{4.54}$$

and we take α as the main control parameter.

To detect a P-bifurcation, we have to consider the Fokker-Planck equation for (4.54) and determine the stationary probability density function, written as $p_\alpha(x)$, as α is changed. The Itô form of (4.54) is given by

$$dX_t = \left(\left(\alpha + \frac{\sigma^2}{2}\right) X_t - X_t^3\right) dt + \sigma X_t dW_t, \tag{4.55}$$

and hence, according to (3.73), p_α is determined from

$$-\left[\left(\left(\alpha + \frac{\sigma^2}{2}\right) x - x^3\right) p_\alpha\right]' + \frac{1}{2}\left[\sigma^2 x^2 p_\alpha\right]'' = 0, \tag{4.56}$$

where the $'$ indicates differentiation to x. This equation has a solution $p_\alpha = \delta_0$ for all values of α, which corresponds to the invariant measure μ_ω^0.

With the boundary condition that p_α and its first derivative vanish for $x \to \pm\infty$, we solve for p_α through

$$\frac{p'_\alpha}{p_\alpha} = \left(\frac{2\alpha}{\sigma^2} - 1\right)\frac{1}{x} - \frac{2x}{\sigma^2},$$

which can be integrated once to give

$$\ln p_\alpha = \left(\frac{2\alpha}{\sigma^2} - 1\right) \ln|x| - \frac{x^2}{\sigma^2} + C_\alpha.$$

For $x > 0$ (corresponding to the invariant measure μ_ω^+) the solution, indicated by $p_\alpha^+(x)$, is given by

$$p_\alpha^+(x) = N_\alpha^+ \, x^{\frac{2\alpha}{\sigma^2} - 1} \, e^{-\frac{x^2}{\sigma^2}}, \tag{4.57}$$

and $p_\alpha^+(x) = 0$ for $x \leq 0$. Similarly, for the invariant measure μ_ω^-, for $x < 0$, the solution, indicated by $p_\alpha^-(x)$, is given by

$$p_\alpha^-(x) = N_\alpha^- \, (-x)^{\frac{2\alpha}{\sigma^2} - 1} \, e^{-\frac{x^2}{\sigma^2}}, \tag{4.58}$$

and $p_\alpha^-(x) = 0$ for $x \geq 0$. The normalization constants are determined by requiring

$$\int_0^\infty p_\alpha^+(x)dx = \int_{-\infty}^0 p_\alpha^-(x)dx = 1 \rightarrow N_\alpha^\pm = \frac{e^{\frac{2\alpha}{\sigma^2}}}{\Gamma(\frac{\alpha}{\sigma^2})}, \tag{4.59}$$

where Γ is the Gamma function, defined by

$$\Gamma(x) = \int_0^\infty e^{-t} t^{x-1} dt.$$

Both probability density functions $p_\alpha^\pm$ change shape when

$$0 = \frac{2\alpha}{\sigma^2} - 1 \rightarrow \alpha = \frac{\sigma^2}{2}, \tag{4.60}$$

as the behaviour is of the form $x^{-1}\, e^{-x^2/2}$ for $2\alpha/\sigma^2 - 1 < 0$ and of the form $x\, e^{-x^2/2}$ for $2\alpha/\sigma^2 - 1 > 0$. Hence a P-bifurcation occurs at $\alpha_P = \sigma^2/2$, as shown in Fig. 4.5.

To detect a D-bifurcation, we need the Lyapunov exponents of all invariant (Dirac) measures for this case, and these were discussed in Section 4.2.4. The Lyapunov exponent of the invariant measure $\mu_\omega^0 = \delta_0$ was determined as $\lambda(\delta_0(\omega)) = \alpha$. The Lyapunov exponent for the invariant measures $\mu_\omega^\pm$, which exist for $\alpha > 0$, can be evaluated from (4.52) and (4.53) using the following identity for ergodic systems (Gardiner, 2002):

$$\lim_{t\to\infty} \frac{1}{t} \int_0^t (\phi(s,\omega)[\bar{\mu}])^2 ds = \int_{-\infty}^\infty x^2 p_\alpha(x) dx. \tag{4.61}$$

For $\bar{\mu} = \delta_{\pm\chi(\omega)}$, we then find

$$\lambda = \alpha - 3 \int_{-\infty}^\infty x^2 p_\alpha(x) dx = -2\alpha, \tag{4.62}$$

and hence $\lambda(\delta_{\pm\chi(\omega)}) = -2\alpha$. As a consequence, there is a D-bifurcation at $\alpha_D = 0$ (Fig. 4.5). This provides the details for the bifurcation diagram of the stochastic pitchfork bifurcation with multiplicative noise, as plotted in Fig. 4.5.

4.3.2 Pitchfork: additive noise

The situation differs markedly for the stochastic pitchfork bifurcation under additive noise (for which the Stratonovich form is the same as the Itô form), and the SDE (again for $\beta = 1$) is given by

$$dX_t = (\alpha X_t - X_t^3)dt + \sigma dW_t. \tag{4.63}$$

For this case, there is no explicit expression for the flow of the invariant measures, but let us indicate them here with $\bar{X}_t$.

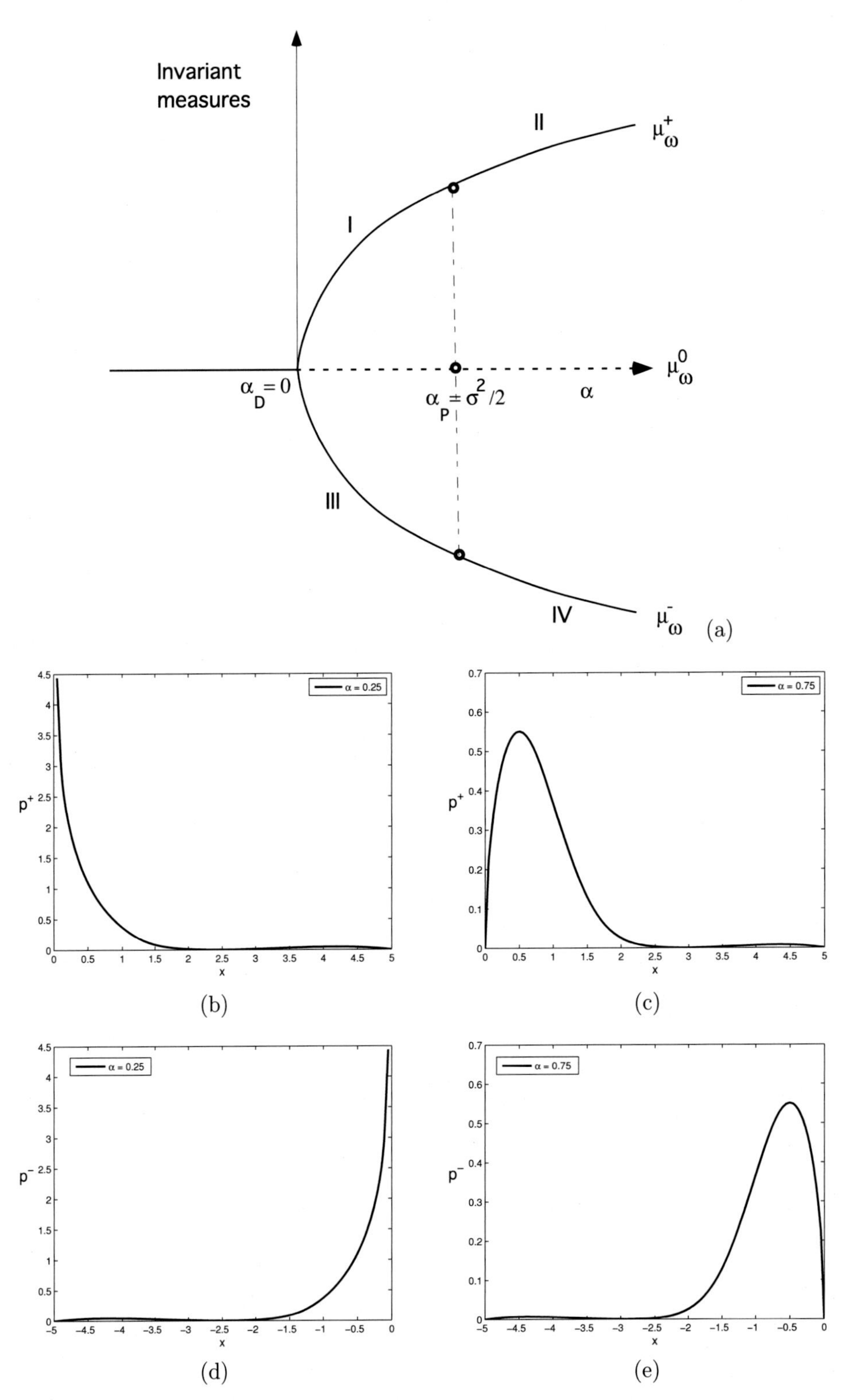

Figure 4.5 (a) Bifurcation diagram of the stochastic pitchfork bifurcation, from equation (4.11), with $\beta = 1$. (b–e) The probability density functions $p^{\pm}$ are plotted for $\sigma = 1$ and different values of α and illustrate the different statistical regimes I–IV in (a); figures based on Arnold (1998).

The Fokker-Planck equation determines the equilibrium probability density function p_α, which follows from

$$-\left[(\alpha x - x^3)p_\alpha\right]' + \frac{\sigma^2}{2}p_\alpha'' = 0 \qquad (4.64)$$

and has the solution

$$p_\alpha(x) = N_\alpha\, e^{\frac{\alpha x^2 - x^4/2}{\sigma^2}}, \qquad (4.65)$$

where N_α is again a normalization constant. For $\alpha < 0$, $p_\alpha(x)$ is unimodal, whereas for $\alpha > 0$, it is bimodal. Hence a P-bifurcation occurs at $\alpha = 0$.

The Lyapunov exponent λ follows from the linearized equation

$$dV_t = (\alpha - 3\bar{X}_t^2)V_t\, dt, \qquad (4.66)$$

where the stochastic term is now absent, as it does not depend on X_t. Hence λ is determined by

$$\lambda = \alpha - \lim_{t\to\infty} \frac{3}{t}\int_0^t (\bar{X}_s)^2 ds, \qquad (4.67)$$

which can again be evaluated through the identity (4.61) as

$$\lambda = \alpha - 3\int_{-\infty}^{\infty} x^2 p_\alpha(x)dx = \int_{-\infty}^{\infty} (\alpha - 3x^2)p_\alpha(x)dx$$
$$= -\int_{-\infty}^{\infty} (\alpha x - x^3)p_\alpha'(x)dx = -\frac{2}{\sigma^2}\int_{-\infty}^{\infty} (\alpha x - x^3)^2 p_\alpha(x)dx, \qquad (4.68)$$

where in the last step (4.64) is used. This shows that $\lambda < 0$ for all values of α, and hence no D-bifurcation occurs. The bifurcation behaviour of the additive noise case hence substantially differs from that of the multiplicative noise case.

4.3.3 The stochastic Hopf bifurcation

The theory of the stochastic Hopf bifurcation is presented in Chapter 9 of Arnold (1998), but, as it is quite complicated, it is not discussed here in detail. We only present one important phenomenon due to the presence of such a bifurcation, that is, the excitation of oscillatory behavior.

As we have seen in Section 2.3.2, the simplest dynamical system exhibiting a supercritical Hopf bifurcation is the system (cf. (2.27))

$$\frac{dr}{dt} = \lambda r - r^3, \qquad (4.69a)$$

$$\frac{d\theta}{dt} = \omega, \qquad (4.69b)$$

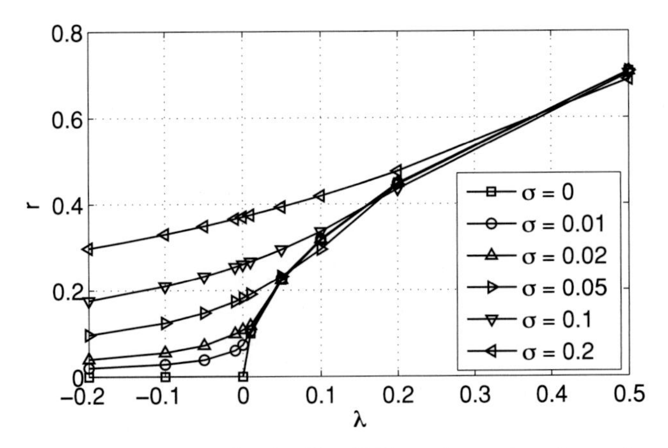

Figure 4.6 Response near a stochastic Hopf bifurcation at $\lambda = 0$ monitored through solutions of (4.71). In the deterministic case ($\sigma = 0$), $r = 0$ for $\lambda < 0$. When noise is included, the expectation value $r = E[R_t]$, where $R_t^2 = X_t^2 + Y_t^2$, increases with increasing σ for any value of λ. In the cases with stochastic forcing, r is determined over a long time interval integration.

in polar coordinates (r, θ). For $\lambda < 0$, there is only one steady state $r = 0$. For $\lambda > 0$, however, there are two solutions of the steady equation (4.69a), that is, $r = 0$ and $r = \sqrt{\lambda}$. The latter corresponds through equation (4.69b) with a periodic orbit with an angular frequency ω and period $2\pi/\omega$. Hence at the Hopf bifurcation ($\lambda = 0$), periodic behavior with a frequency ω is spontaneously generated through an instability of the trivial solution $r = 0$.

When we change from polar coordinates (r, θ) to Cartesian coordinates, (x, y), with $x = r\cos\theta$ and $y = r\sin\theta$, (4.69) becomes

$$\frac{dx}{dt} = \lambda x - \omega y - x(x^2 + y^2), \tag{4.70a}$$

$$\frac{dy}{dt} = \lambda y + \omega x - y(x^2 + y^2), \tag{4.70b}$$

and hence the stochastic Itô SDE with additive noise is formulated as

$$dX_t = (\lambda X_t - \omega Y_t - X_t(X_t^2 + Y_t^2))dt + \sigma\, dW_t, \tag{4.71a}$$

$$dY_t = (\lambda Y_t + \omega X_t - Y_t(X_t^2 + Y_t^2))dt + \sigma\, dW_t, \tag{4.71b}$$

where σ is the amplitude of the additive noise and W_t is a Wiener process with increment dW_t. The expectation value $r = E[R_t]$, where $R_t = \sqrt{X_t^2 + Y_t^2}$, resulting from the stochastic integration of (4.71) is shown in Fig. 4.6 for several values of σ; the deterministic case is shown for $\sigma = 0$. Clearly, there is a response for values $\lambda < 0$ that increases with increasing noise level σ. This indicates that oscillatory behavior can be excited by additive noise below $\lambda = 0$.

5

Analysing Data from Stochastic Dynamical Systems

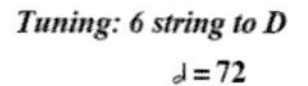

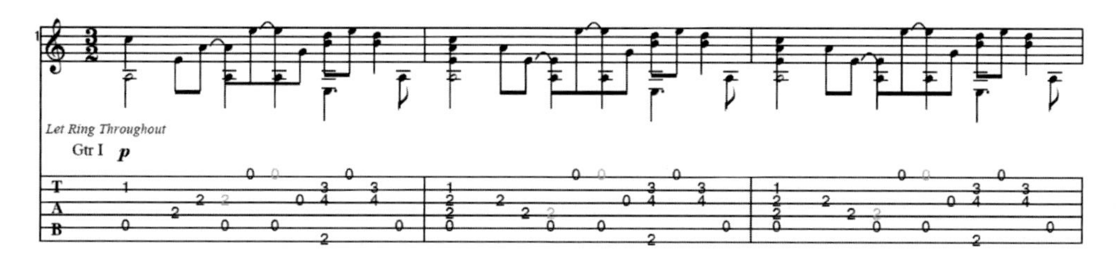

Let the data speak and conquer the noise.
DADGBE, Andecy, Andrew York

In the previous chapters, we were given a stochastic model for a random variable X_t and computed the statistics of the solutions from the properties of the equations. However, often only a time series (e.g., from observations) is given, and we a priori may not know the underlying equations. Hence we have to estimate the statistical quantities we are interested in from the time series itself.

We start with a brief overview of classical methods of univariate, bivariate and multivariate time series analysis. The literature on this subject is very broad, and for climate research, one of the main references is Von Storch and Zwiers (1999). It is useful here, however, to repeat some of the main results, as they are needed to understand modern nonlinear techniques, of which some are presented later in this chapter.

5.1 Classical univariate methods

In general, we have a time series at discrete time intervals Δt, the sampling interval. If we denote the variable by X, then we have a discrete time series X_t at points $t = k\Delta t$.

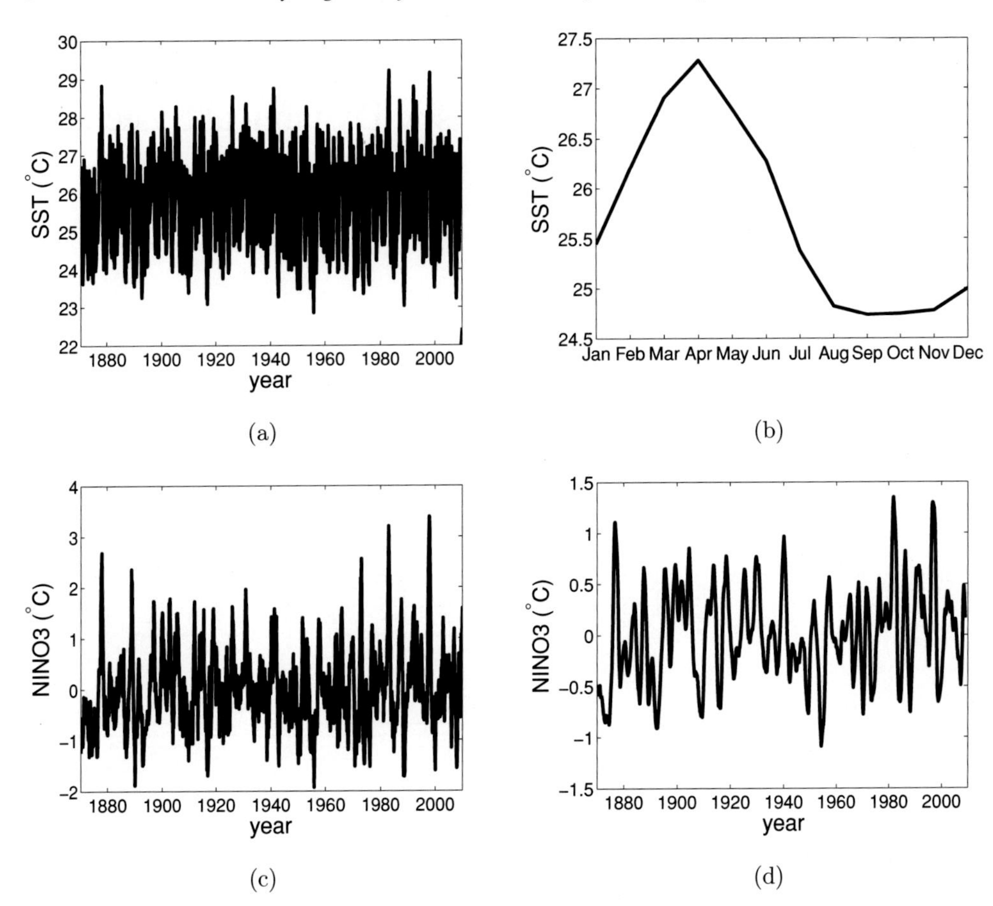

Figure 5.1 (a) Time series over the period 1870–2010 of the sea surface temperature (SST) in the Eastern Equatorial Pacific averaged over the domain [150°W–90°W] × [5°S–5°N] as provided in the HadISST1 data set (Rayner et al., 2003). (b) The mean seasonal cycle obtained from the time series in (a) by averaging over all January months, February, . . . , December. (c) Difference of the time series in (a) and (b) giving the NINO3 time series. (d) Low-pass-filtered time series of (c) with a cut-off value of 24 months.

We illustrate the methodology of analysing such a time series with help of a specific time series (Fig. 5.1a): the sea surface temperature in the Eastern Equatorial Pacific averaged over the domain [150°W–90°W] × [5°S–5°N].

The process of analysis of the time series consists of the following four steps (Chatfield, 2004):

(i) Plot and inspect the time series. This may seem trivial, but it is important to get an impression of trends, periodicities and rapid changes.
(ii) Remove any undesired known periodic cycles (diurnal, seasonal) and long-term trends. Filter the time series, if needed.

(iii) Analyse the result after (ii) in the time domain (correlation analysis) and frequency domain (spectral analysis).
(iv) Obtain the 'signal' from the 'noise'. Formulate a null hypothesis and try to reject each null hypothesis based on a test statistic and a stochastic process representing the noise.

5.1.1 Trend removal and filtering

In many cases, the trend may be just the quantity one is after. Think of the warming trend of surface temperatures due to the increase of greenhouse gases, such as in Fig. 1.3. Or one may just be interested in the seasonal cycle of a particular quantity, say, of the surface pressure in station De Bilt in the Netherlands.

Often, however, climate scientists are interested in the processes causing variability in a particular frequency band, for example, the interannual variability associate with the El Niño/Southern Oscillation (ENSO) phenomenon. A question here may be what the dominant frequency of the ENSO variability is in a particular time series such as that of the SST in Fig. 5.1a. In this case, one wants to remove the long-term trend (maybe caused by global warming) and the seasonal cycle.

There are many methods of linear trend removal. The general model is that the time series can be represented by

$$X_t = X_t^d + \alpha + \beta t + \epsilon_t, \tag{5.1}$$

with constants α and β, a noise term ϵ_t and X_t^d as the detrended time series. There are now two cases:

(i) The observations have a seasonal signal. In this case, one computes annual averages of the data as a new time series Y_t, fits a line to these annual averages to determine α and β in (5.1) and subtracts this line from the original time series.
(ii) When the data have no seasonal signal, one computes the difference time series

$$Y_t = X_{t+1} - 2X_t + X_{t-1}$$

to remove the linear trend and then puts $X_t^d = Y_t$.

To remove a seasonal signal S_t, one first needs to know whether the seasonal signal is additive or multiplicative. In the (most common) additive case, the general model is

$$X_t = X_t^d + S_t + \epsilon_t. \tag{5.2}$$

One can test whether this model is appropriate by checking whether

$$\sum_{t=0}^{n\Delta t} S_t \approx 0, \tag{5.3}$$

where n is the number of observations within a year. In the multiplicative case, the model is $X_t = X_t^d S_t + \epsilon_t$, and the sum (5.3) is approximately unity. For the time series in Fig. 5.1a, the trend is very small, and the seasonal cycle is shown in Fig. 5.1b. The SST anomaly time series with respect to this seasonal cycle is plotted in Fig. 5.1c; this time series is also known as the NINO3 time series. When one has monthly data, one can also remove the seasonal cycle (in the additive case) by computing

$$X_t^d = \frac{1}{12}\left(\frac{1}{2}X_{t-6} + X_{t-5} + \cdots + X_{t+5} + \frac{1}{2}X_{t+6}\right), \tag{5.4}$$

where the coefficients are chosen in such a way as to add up to 12.

The procedure (5.4) is an example of a so-called running mean or moving average filter. A much-used expression is

$$Sm(X_t) = \frac{1}{2q+1}\sum_{r=-q}^{q} X_{t+r}, \tag{5.5}$$

for $t = q + 1$ up to $t = N - q$. In a low-pass filter, only the low-frequency components are retained. We can, for example, study the low-frequency variability of the NINO3 monthly mean time series (Fig. 5.1d) by computing a running mean filter with a time scale of two years ($q = 24$). In a high-pass filter, we perform the same operation but then consider the time series $X_t - Sm(X_t)$.

5.1.2 Elementary statistical properties

The detrended (and maybe filtered) time series is the starting point for further statistical analysis. A first issue to check is whether the time series is stationary, that is, for which the statistical quantities such as mean and variance do not explicitly depend on time. One method to test this is to divide the time series in several blocks and perform the analysis of mean and variance on each block separately. In the following, we assume that we have a stationary time series.

Given a time series $X_t, t = \Delta t, \ldots, i\Delta t, \ldots, N\Delta t$, or for convenience $X_i, i = 1, \ldots, N$, the so-called estimators (denoted by a quantity with a hat) for the mean and variance are given by

$$\hat{\mu}_X = \frac{1}{N}\sum_{i=1}^{N} X_i, \tag{5.6a}$$

$$\hat{\sigma}_X^2 = \frac{1}{N-1}\sum_{i=1}^{N}(X_i - \hat{\mu}_X)^2. \tag{5.6b}$$

We can immediately compute the mean and variance of these estimators when it is assumed that every X_i is from a Gaussian distribution $N(\mu, \sigma^2)$ and that they are independent. In this case, we find for the mean, using the expectation E and variance

Var operators,

$$E[\hat{\mu}_X] = \mu, \tag{5.7a}$$

$$Var[\hat{\mu}_X] = \frac{\sigma^2}{N}. \tag{5.7b}$$

To determine $E[\hat{\sigma}_X^2]$, we first consider

$$\begin{aligned}\sum_{i=1}^{N}(X_i - \hat{\mu}_X)^2 &= \sum_{i=1}^{N}(X_i - \mu + \mu - \hat{\mu}_X)^2 \\ &= \sum_{i=1}^{N}(X_i - \mu)^2 + 2\sum_{i=1}^{N}(X_i - \mu)(\mu - \hat{\mu}_X) + N(\mu - \hat{\mu}_X)^2 \\ &= \sum_{i=1}^{N}(X_i - \mu)^2 - N(\hat{\mu}_X - \mu)^2,\end{aligned}$$

and hence

$$\begin{aligned}E\left[\frac{1}{N-1}\sum_{i=1}^{N}(X_i - \hat{\mu}_X)^2\right] &= \frac{1}{N-1}\sum_{i=1}^{N} E[(X_i - \hat{\mu}_X)^2] \\ &= \frac{1}{N-1}\sum_{i=1}^{N} E[(X_i - \mu)^2] - \frac{N}{N-1}E[(\hat{\mu}_X - \mu)^2],\end{aligned}$$

and using (5.7b), we find

$$E[\hat{\sigma}_X^2] = \frac{N}{N-1}\sigma^2 - \frac{1}{N-1}\sigma^2 = \sigma^2.$$

From a similar calculation for the variance of the estimator $\hat{\sigma}_X^2$, it is found that

$$E[\hat{\sigma}_X^2] = \sigma^2, \tag{5.8a}$$

$$Var[\hat{\sigma}_X^2] = \frac{2}{N-1}\sigma^4. \tag{5.8b}$$

Estimators preferably should be unbiased for a Gaussian time series. If $\hat{\alpha}$ is an estimator of a quantity α, then the bias $B(\hat{\alpha})$ of $\hat{\alpha}$ is defined as

$$B(\hat{\alpha}) = E[\hat{\alpha}] - \alpha. \tag{5.9}$$

Suppose that every X_i is from a Gaussian distribution $N(\mu, \sigma^2)$ and that they are independent; we then compute the bias of the estimators (5.6) as

$$B(\hat{\mu}_X) = E[\hat{\mu}_X] - \mu = 0, \tag{5.10a}$$

$$B(\hat{\sigma}_X^2) = E[\hat{\sigma}_X^2] - \sigma^2 = 0, \tag{5.10b}$$

and hence both are unbiased.

Once the estimator of the mean $\hat{\mu}_X$ is determined, one usually proceeds to subtract this mean from the signal (if only interested in the variability). The next phase is to analyse the degree of correlation in the time resulting series. When we anticipate a periodic signal, the time series will repeat itself after some lag k (which corresponds to a period $k\Delta t$). To determine this correlation, one determines estimators of the covariance coefficients $\hat{c}_k$ with

$$\hat{c}_k = \frac{1}{N}\sum_{i=1}^{N-k}(X_i - \hat{\mu}_X)(X_{i+k} - \hat{\mu}_X), \tag{5.11}$$

and correlation coefficients $\hat{r}_k$ defined by

$$\hat{r}_k = \frac{\hat{c}_k}{\hat{c}_0}. \tag{5.12}$$

It is instructive to inspect the plot of $\hat{r}_k$ versus k. For a random Gaussian process, we expect $\hat{r}_k = 0$ for every $k \neq 0$, whereas $\hat{r}_k$ will oscillate in k when the time series has an oscillatory component. A plot of the function $\hat{r}_k$ versus k for the NINO3 time series in Fig. 5.1c (restricted to the years 1950–2000) is plotted in Fig. 5.2a. Here one can see that there is a dominant periodic signal (with a period of about 50 months) in this time series.

5.1.3 Spectral analysis

Spectral analysis is used to determine dominant frequencies in time series. A most obvious method would be to compute the discrete Fourier transform of the covariance coefficients $\hat{c}_k$. However, it turns out that this is no unbiased estimator of the spectrum and that windowing methods have to be used. The Fourier transform actually used is

$$\hat{f}(\omega) = \frac{1}{\pi}\left[\lambda_0\hat{c}_0 + 2\sum_{k=1}^{M}\lambda_k\hat{c}_k\cos\omega k\right], \tag{5.13}$$

where ω is the frequency and the λ_k are windowing coefficients. If the sampling time is Δt, the largest frequency that can be resolved is the Nyquist frequency $\omega = \pi/\Delta t$ (or the period $\tau = 2\Delta t$), and hence we can take $\omega_j = (j\pi/M\Delta t)$, $j = 1, \ldots, M$ in (5.13).

Several windows are in use and give biased estimators of the amplitude of the spectral coefficients $|\hat{f}(\omega)|$. In the so-called Tukey window, the coefficients λ_k are taken as

$$\lambda_k = \frac{1}{2}\left(1 + \cos\frac{\pi k}{M}\right), \quad k = 0, \ldots, M. \tag{5.14}$$

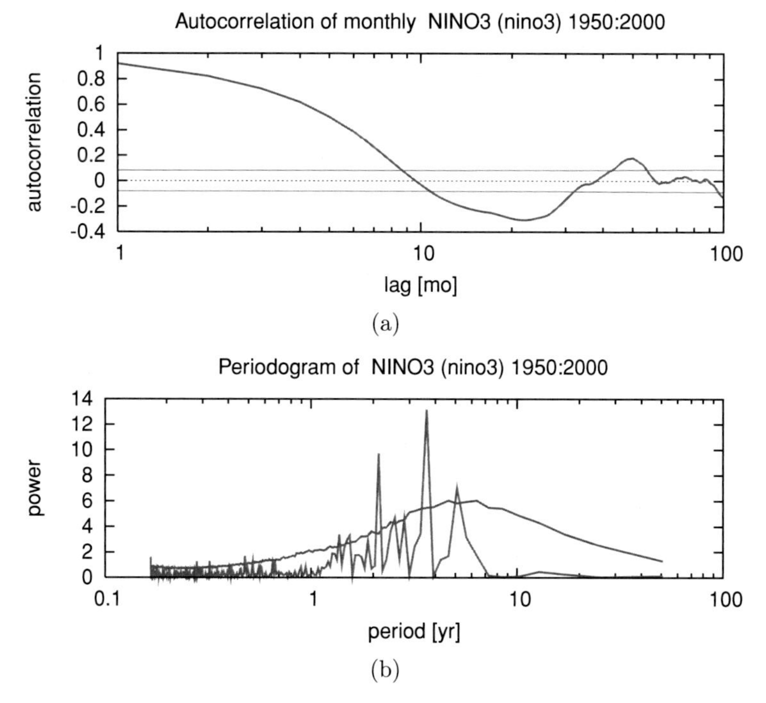

Figure 5.2 (a) Autocorrelation function of the NINO3 time series over the period 1950–2000. The horizontal lines give the 95% significance for a single point in the case of white noise, assuming all measurements are independent. (b) Fourier spectrum (periodogram, light curve) of the NINO3 time series in Fig. 5.1c over the period 1950–2000. The dark curve denotes the 95% highest spectrum of 2669 AR(1) processes with the same autocorrelation (0.939). The most significant peaks are at 2.12 and 3.64 years.

In the Parzen window, the coefficients are chosen as

$$\lambda_k = 1 - 6\left(\frac{k}{M}\right)^2 + 6\left(\frac{k}{M}\right)^3, 0 \le k \le M/2, \tag{5.15a}$$

$$\lambda_k = 2 - 6\left(1 - \frac{k}{M}\right)^3, \quad M/2 \le k \le M. \tag{5.15b}$$

A 'rule of thumb' is the choice $M \approx 2\sqrt{N}$. A spectrum (without window, this is a periodogram) for the NINO3 time series is shown in Fig. 5.2b. The question now arises: are the peaks in this spectrum statistically significant? In other words, can these peaks also be produced by just some noise process?

5.2 Hypothesis testing and significance

The general idea behind hypothesis testing is that we pose some null hypothesis that we try to reject using either a statistical test variable or a computed distribution obtained through (e.g., Monte Carlo) simulation. We illustrate these procedures with two examples.

5.2.1 Test variables

Suppose we have two time series $X_i, i = 1, \ldots, N_x$ and $Y_j, j = 1, \ldots, N_y$, and we want to deduce the relation between the mean μ_X and μ_Y of these two time series. The null hypothesis H_0 is

$$H_0 : \mu_X = \mu_Y. \tag{5.16}$$

The optimal test variable for this null hypothesis is given by (Von Storch and Zwiers, 1999)

$$\hat{T} = \frac{\hat{\mu}_X - \hat{\mu}_Y}{\hat{S}_P\sqrt{\frac{1}{N_x} + \frac{1}{N_y}}}, \tag{5.17}$$

where

$$\hat{S}_p^2 = \frac{\sum_{i=1}^{N_x}(X_i - \hat{\mu}_X)^2 + \sum_{j=1}^{N_y}(Y_j - \hat{\mu}_Y)^2}{N_x + N_y - 2}$$

and $\hat{\mu}_X$ and $\hat{\mu}_Y$ are determined from (5.6a).

Consider the following assumptions:

(i) Every realization of X is independent of every other realization. Same for Y.
(ii) The distribution that generates X_i is the same for every i. Same for Y.
(iii) The distributions of X and Y are both normal.

Under these assumptions, the test variable $\hat{T}$ in (5.17) has a so-called t-distribution with $k = N_x + N_y - 2$ degrees of freedom, given by

$$f_t(x) = \frac{1}{\sqrt{k\pi}} \frac{\Gamma\left(\frac{k+1}{2}\right)}{\Gamma\left(\frac{k}{2}\right)} \left(1 + \frac{x^2}{k}\right)^{-\left(\frac{k+1}{2}\right)}, \tag{5.18}$$

where Γ, the Gamma function, is defined by

$$\Gamma(x) = \int_0^\infty e^{-t} t^{x-1} dt. \tag{5.19}$$

The probability density function $f_t(x)$ is plotted for different k in Fig. 5.3a. In Fig. 5.3b, df indicates the number of degrees of freedom, and the values in the table indicate

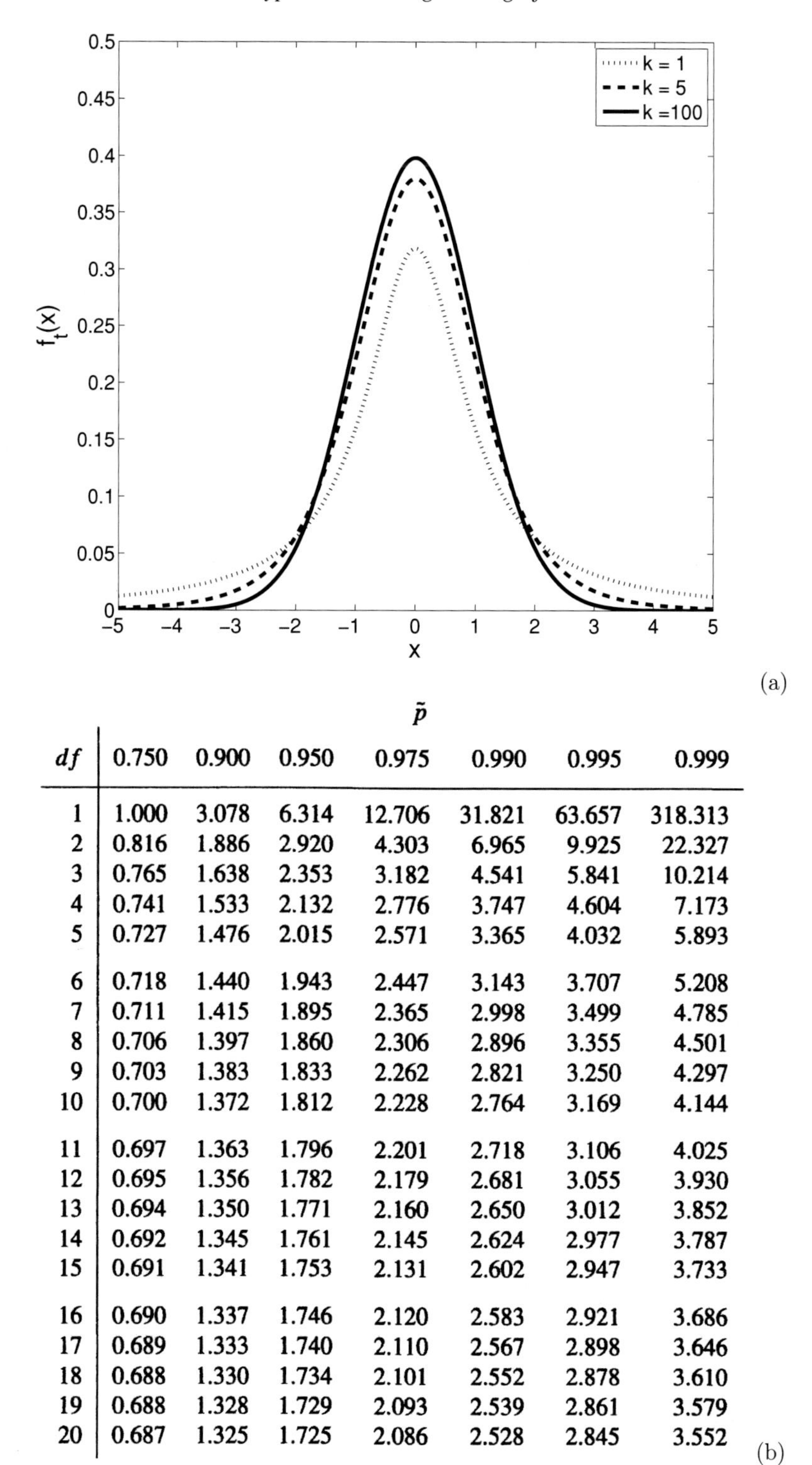

df	$\tilde{p}$ 0.750	0.900	0.950	0.975	0.990	0.995	0.999
1	1.000	3.078	6.314	12.706	31.821	63.657	318.313
2	0.816	1.886	2.920	4.303	6.965	9.925	22.327
3	0.765	1.638	2.353	3.182	4.541	5.841	10.214
4	0.741	1.533	2.132	2.776	3.747	4.604	7.173
5	0.727	1.476	2.015	2.571	3.365	4.032	5.893
6	0.718	1.440	1.943	2.447	3.143	3.707	5.208
7	0.711	1.415	1.895	2.365	2.998	3.499	4.785
8	0.706	1.397	1.860	2.306	2.896	3.355	4.501
9	0.703	1.383	1.833	2.262	2.821	3.250	4.297
10	0.700	1.372	1.812	2.228	2.764	3.169	4.144
11	0.697	1.363	1.796	2.201	2.718	3.106	4.025
12	0.695	1.356	1.782	2.179	2.681	3.055	3.930
13	0.694	1.350	1.771	2.160	2.650	3.012	3.852
14	0.692	1.345	1.761	2.145	2.624	2.977	3.787
15	0.691	1.341	1.753	2.131	2.602	2.947	3.733
16	0.690	1.337	1.746	2.120	2.583	2.921	3.686
17	0.689	1.333	1.740	2.110	2.567	2.898	3.646
18	0.688	1.330	1.734	2.101	2.552	2.878	3.610
19	0.688	1.328	1.729	2.093	2.539	2.861	3.579
20	0.687	1.325	1.725	2.086	2.528	2.845	3.552

Figure 5.3 (a) Probability density of the t-distribution, with k being the number of degrees of freedom. (b) Values of T_0 for different degrees of freedom ($df = k$) and significance levels ($\tilde{p}$).

the value of $\hat{T}$, say T_0, for which $P(\hat{T} > T_0) = 1 - \tilde{p}$, for a specific significance level $\tilde{p}$. The test procedure is now as follows:

(i) Choose a significance level $\tilde{p}$.
(ii) Compute the number of degrees of freedom $N_x + N_y - 2$ and the test variable $\hat{T}$ from (5.17).
(iii) Determine from the table the value of T_0 at the particular value of $\tilde{p}$.
(iv) If $\hat{T} \geq T_0$, then the null hypothesis can be rejected at the $100\tilde{p}\%$ significance level. If $\hat{T} < T_0$, then the null hypothesis cannot be rejected at that significance level.

5.2.2 Stochastic simulation

In many situations, it is difficult to determine test variables such as $\hat{T}$ in (5.17). In that case, we can sometimes choose a stochastic model, representing our knowledge of the noise, and use the sample paths of this model to test the null hypothesis H_0. As an example, consider the significance of a peak of the spectrum from a certain time series X. We want to be sure that this peak cannot be caused by a certain stochastic process, which we think is a good model for the noise. There are many discrete stochastic models around, for example, white noise, random walk (Wiener process), red noise, autoregressive (AR) processes, moving average (MA) processes and combinations of the latter two processes (ARMA). For easy reference, we list two of these processes and their properties.

(i) White noise. In this case,

$$X_t = Z_t, \tag{5.20}$$

where Z_t are independent identically distributed Gaussian $N(0, \sigma)$ stochastic variables. For the autocovariance, we find directly from (5.11) that

$$\hat{c}_k = 0, k = \pm 1, \pm 2, \ldots, \tag{5.21}$$

and the autocorrelation $\rho_k = 1$ for $k = 0$ and zero otherwise.

(ii) Red noise. This is a so-called autoregressive process of order one (also abbreviated as AR(1)), defined by

$$X_t = \alpha X_{t-1} + Z_t, \tag{5.22}$$

where $\alpha \in (0, 1)$ and Z_t has a $N(0, \sigma^2)$ distribution. To compute the autocovariance and autocorrelation functions, we realise that

$$X_t = \alpha(\alpha X_{t-2} + Z_{t-1}) + Z_t = Z_t + \alpha Z_{t-1} + \alpha^2 Z_{t-2} + \cdots ,$$

and hence

$$E[X_t] = 0\ ;\ Var[X_t] = \sigma^2(1 + \alpha^2 + \alpha^4 + \cdots) = \frac{\sigma^2}{1 - \alpha^2}, \tag{5.23}$$

and, finally,

$$\hat{c}_k = \sigma^2 \alpha^k \sum_{i=0}^{\infty} (\alpha^i)^2 = \frac{\alpha^k \sigma^2}{1 - \alpha^2}, \tag{5.24a}$$

$$\rho_k = \alpha^k. \tag{5.24b}$$

Once the discrete stochastic model is chosen, say as AR(1), the procedure to compute significance levels is as follows:

(i) Run many realizations of the AR(1) process with the same time step as that in the data. Choose α such that the autocorrelation function ρ_k has the same decay as that of the data.
(ii) Determine spectra for each realization and compute the distribution function of the spectra.
(iii) Choose a significance level $\tilde{p}$ and plot the spectrum that has a probability $1 - \tilde{p}$. The peaks that extend above the latter spectrum are significant at the $100\tilde{p}\%$ level.

The spectrum at the 95% ($\tilde{p} = 0.95$) significance level of an AR(1) process is plotted for the NINO3 time series as the dark curve in Fig. 5.2b. Here 2,669 AR(1) sample paths were computed.

5.3 Bivariate and multivariate analysis

In this section, the results of the univariate analysis are extended to study statistical relations between two time series or between time series of fields, such as spatial fields of climate variables.

5.3.1 Cross-covariance and co-spectra

Suppose we have two time series, for example, of NINO3 (Fig. 5.1) and of the global mean temperature (Fig. 1.3), and ask the question whether there is any causal relation between the data in these series. In the time domain, we are hence interested in the cross-covariance of both time series. When the time series are denoted by $X_i, i = 1, \ldots, N_x$ and $Y_j, j = 1, \ldots, N_y$, the cross-covariance coefficients with lag k then are defined as (for convenience, we assume that $N_x = N_y = N$)

$$k = 0, \ldots, N-1 : \hat{c}_{xy}(k) = \frac{1}{N} \sum_{i=1}^{N-k} (X_i - \hat{\mu}_X)(Y_{i+k} - \hat{\mu}_Y), \tag{5.25a}$$

$$k = -(N-1), \ldots, -1 : \hat{c}_{xy}(k) = \frac{1}{N} \sum_{i=1-k}^{N} (X_i - \hat{\mu}_X)(Y_{i+k} - \hat{\mu}_Y), \tag{5.25b}$$

and the cross-correlation coefficients are defined by

$$\hat{r}_{xy}(k) = \frac{\hat{c}_{xy}(k)}{\sqrt{\hat{c}_{xx}(0)\hat{c}_{yy}(0)}}. \tag{5.26}$$

In addition, the co-spectrum $\hat{c}$ and the quadrature spectrum $\hat{q}$ are defined as

$$\hat{c}(\omega) = \frac{1}{\pi}\left[\sum_{k=-M}^{M} \lambda_k \hat{c}_{xy}(k)\cos\omega k\right], \tag{5.27a}$$

$$\hat{q}(\omega) = \frac{1}{\pi}\left[\sum_{k=-M}^{M} \lambda_k \hat{c}_{xy}(k)\sin\omega k\right], \tag{5.27b}$$

and the cross-amplitude, phase and coherency spectra follow from

$$\hat{\alpha}(\omega) = \sqrt{\hat{c}^2(\omega) + \hat{q}^2(\omega)}, \tag{5.28a}$$

$$\tan\hat{\phi}(\omega) = -\frac{\hat{q}(\omega)}{\hat{c}(\omega)}, \tag{5.28b}$$

$$\hat{C}(\omega) = \frac{\hat{\alpha}^2(\omega)}{\hat{f}_x(\omega)\hat{f}_y(\omega)}, \tag{5.28c}$$

where $\hat{f}_x$ and $\hat{f}_y$ are estimates of the individual power spectra of X and Y.

For fields of variables, such relations can be determined between quantities of two different locations. The analysis can be easily extended to deal with correlations in the whole field simultaneously, as we see in the next section.

5.3.2 Empirical orthogonal functions

Suppose we now have a vector time series $\mathbf{X}^k, k = 1, \ldots, N$ where each $\mathbf{X}^k$ has a dimension d. Such vector time series arise naturally from grid point values of a climate quantity, such as the monthly mean sea-surface temperature T. For example, if we sample T at a $I \times J$ grid for N years, then $d = I \times J$, and each vector $\mathbf{X}^k = (T^k_{1,1}, \ldots, T^k_{I,J})$ for $k = 1, \ldots, N$. Just as in the bivariate case, we would like to analyse the covariance between the time series at each grid point.

The bivariate approach to calculate cross-covariance coefficients is easily generalised. We first estimate elements of the covariance matrix $\hat{\Sigma}$, with

$$\hat{\Sigma} = \frac{1}{N}\sum_{k=1}^{N}(\mathbf{X}^k - \hat{\mu})(\mathbf{X}^k - \hat{\mu})^T, \tag{5.29}$$

where the vector mean $\hat{\mu}$ is determined from

$$\hat{\mu} = \frac{1}{N}\sum_{k=1}^{N}\mathbf{X}^k, \tag{5.30}$$

and hence an element of $\hat{\Sigma}$, say $\hat{\Sigma}_{ij}$, is given by

$$\hat{\Sigma}_{ij} = \frac{1}{N}\sum_{k=1}^{N}(\mathbf{X}_i^k - \hat{\mu}_i)(\mathbf{X}_j^k - \hat{\mu}_j), \tag{5.31}$$

which is a similar expression as for the cross-covariance coefficients (5.25).

As the $d \times d$ real matrix $\hat{\Sigma}$ is symmetric and positive definite, it has real positive eigenvalues $\hat{\lambda}_i$, $i = 1, \ldots, d$. Let these eigenvalues be ordered by $\hat{\lambda}_1 \geq \hat{\lambda}_2 \geq \ldots \geq \hat{\lambda}_d$ and the corresponding eigenvectors be indicated by $\hat{\mathbf{e}}_1, \ldots, \hat{\mathbf{e}}_d$. These eigenvectors are called the empirical orthogonal functions (EOFs). As the EOFs form an orthonormal basis in $\mathbb{R}^d$, every vector $\mathbf{X}^k$ can be represented as

$$\mathbf{X}^k = \sum_{i=1}^{d} \hat{\alpha}_i^k \hat{\mathbf{e}}_i. \tag{5.32}$$

For fixed i, the coefficient $\hat{\alpha}_i^k$ is called the principal component (PC) corresponding to the ith EOF. It can be shown (Von Storch and Zwiers, 1999) that

$$\hat{\alpha}_i^k = < \mathbf{X}^k, \hat{\mathbf{e}}_i >, \tag{5.33}$$

with $<, >$ the inner product in $\mathbb{R}^d$ minimizes the error

$$\hat{\epsilon}_M = \sum_{k=1}^{N} |\mathbf{X}^k - \sum_{i=1}^{M} \hat{\alpha}_i^k \hat{\mathbf{e}}_i|^2, \tag{5.34}$$

for every M. This means that the EOFs are optimal in representing the variance in the stochastic process $\mathbf{X}^1, \ldots, \mathbf{X}^N$. Note that each EOF represents a spatial pattern on the grid $I \times J$ if each $\mathbf{X}^k$ is derived from a grid-point representation of a continuous variable.

If for fixed i, $\hat{\alpha}_i$ indicates the stochastic variable associated with the time series $\hat{\alpha}_i^k$, $k = 1, \ldots, N$, then it can also be shown that for each i,

$$Var[\hat{\alpha}_i] = \hat{\lambda}_i. \tag{5.35}$$

The variance $\hat{\sigma}_i^2$ accounted for by the ith EOF when a total of M EOFs are used to represent the signal is usually written as

$$\hat{\sigma}_i^2 = \frac{\hat{\lambda}_i}{\sum_{j=1}^{M} \hat{\lambda}_j} \times 100\%. \tag{5.36}$$

In many practical problems, a large part of the variance can already be accounted for by only a few EOFs, which makes this method an attractive approach to data reduction.

Table 5.1. *First column: first five eigenvalues $\hat{\lambda}_i$ of the covariance matrix for the SST data over the period January 1950 to December 2000 from the HadISST1 data set. Second column: explained variance of the EOF. Third column: cumulative explained variance*

# EOF	Eigenvalue	Explained variance (%)	Cumulative (%)
1	1078.8	55.44	55.44
2	228.91	11.76	67.20
3	98.110	5.04	72.24
4	69.068	3.55	75.79
5	55.494	2.85	78.65

Measures of uncertainty in the EOFs $\hat{\mathbf{e}}_i$ and eigenvalues $\hat{\lambda}_i$ are given (North et al., 1982) by

$$\Delta\hat{\lambda}_i = \sqrt{\frac{2}{N^*}}\hat{\lambda}_i \ ; \ \Delta\hat{\mathbf{e}}_i = \frac{\Delta\hat{\lambda}_i}{\hat{\lambda}_j - \hat{\lambda}_i}\hat{\mathbf{e}}_i, \tag{5.37}$$

where $\hat{\lambda}_j$ is the closest eigenvalue to $\hat{\lambda}_i$ and N^* is the number of independent observations in the sample. This indicates that one has to be careful if the variance explained by two EOFs is nearly equal.

Example 5.1 EOFs of Pacific SST Consider monthly mean data of the sea-surface temperature (SST) in the equatorial Pacific ([120°E–70°W] × [20°S–20°N]) over the period January 1950 to December 2000 from the HadISST1 data set (Rayner et al., 2003). These are available on a 1° × 1° grid, but we will compute the EOFs for data on a 2° × 2° grid. In this case, $N = 12 \times 51 = 612$ and $d = 170/2 \times 40/2 = 85 \times 20 = 1{,}700$. The seasonal cycle is calculated by averaging over all January months, February months, and so forth, and then subtracting from the original data; the resulting vector time series consists only of SST anomalies.

The eigenvalues of the covariance matrix are shown in Table 5.1. The first two EOFs already explain about 67% of the variance in the data and are very dominant compared with the rest of the EOFs. The uncertainty in the eigenvalues can be computed from (5.37) with $N^* \approx N$. The pattern of the first EOF (Fig. 5.4a) is the well-known ENSO pattern with large amplitudes of the SST variability in the eastern Pacific. The SST variability associated with the second EOF is mainly confined to the coast of South America. The time series of the first two principal components (Fig. 5.5) show the characteristic interannual variability of SST in the Eastern Equatorial Pacific; PC1 is well correlated with NINO3. ■

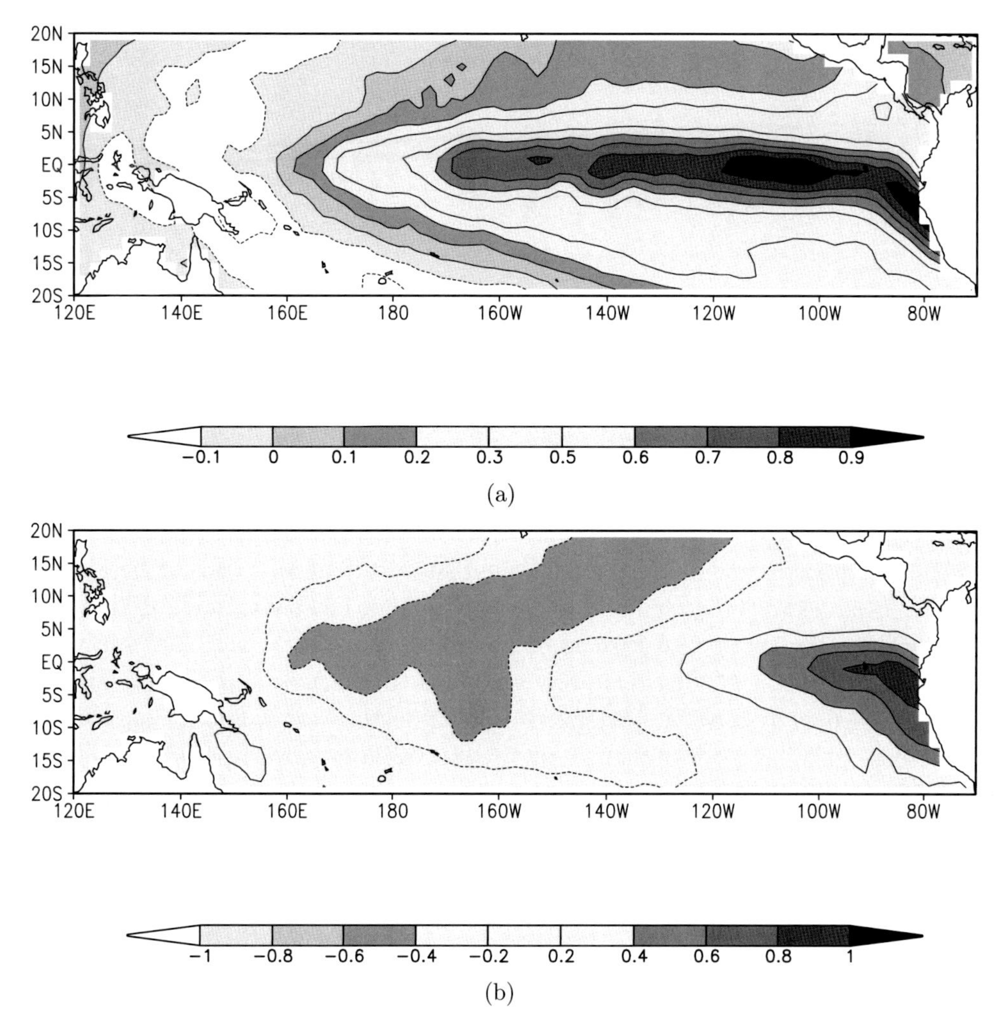

Figure 5.4 The first (a) and the second (b) EOF of equatorial SST from the HadISST1 data set over the period 1950–2000.

5.3.3 Singular spectrum analysis

In recent years, a combination of time series analysis and dynamical systems ideas has led to spectral techniques by which important information (such as dominant frequencies and patterns of variability) can be extracted from short and noisy time series (Ghil et al., 2002). Singular spectrum analysis (SSA) allows one to unravel the information embedded in a delay-coordinate state space (as described next) by decomposing the sequence of augmented vectors thus obtained into elementary patterns of behaviour. It does so by providing data-adaptive filters that help separate the time series into components that are statistically independent, at zero lag, in the augmented vector space of interest. These components can be classified essentially into trends,

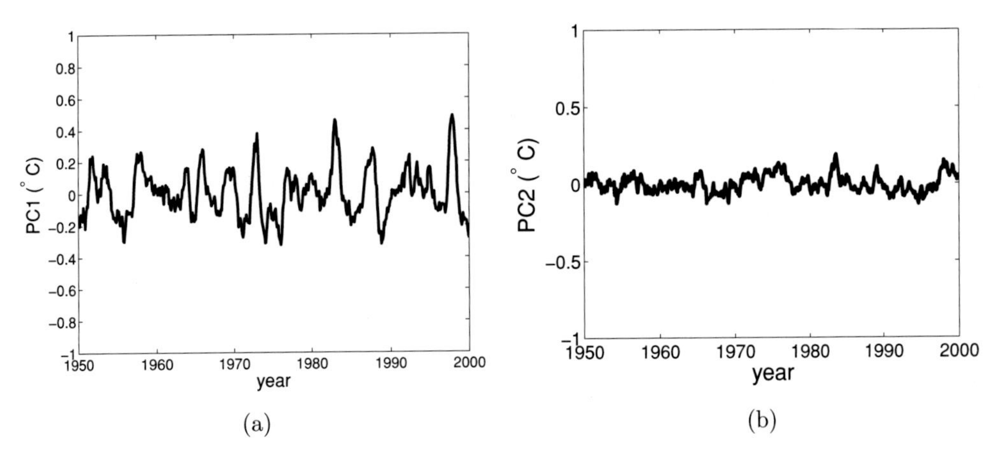

Figure 5.5 Time series of the leading two PCs corresponding to the leading two EOFs, (a) PC1 and (b) PC2.

oscillatory patterns and noise. It is an important feature of SSA that the trends need not be linear and that the oscillations can be amplitude and phase modulated.

The starting point of SSA is to embed a time series X^k, $k = 1, \ldots, N$ in a vector space of dimension M, that is, to represent it as a trajectory in the phase space of the hypothetical system that generated X^k. In concrete terms, this is equivalent to representing the behavior of the system by a succession of overlapping views of the series through a sliding M point window.

The embedding procedure applied so constructs a sequence $\tilde{\mathbf{X}}^k$ of M-dimensional vectors from the original time series, by using lagged copies of the scalar data X^k, as follows:

$$\tilde{\mathbf{X}}^l = (X^l, X^{l+1}, \ldots, X^{l+M-1}), \tag{5.38}$$

where the vectors $\tilde{\mathbf{X}}^l$ are indexed by $l = 1, \ldots, N'$, where $N' = N - M + 1$. The $M \times M$ covariance matrix C_X of the vector time series (5.38) is a so-called Toeplitz matrix with elements

$$\hat{c}_{ij} = \frac{1}{N - |i - j|} \sum_{l=1}^{N-|i-j|} X^l X^{l+|i-j|}, \tag{5.39}$$

for $i = 1, \ldots, M$ and $j = 1, \ldots, M$.

In SSA, the eigenvalues λ_m and eigenvectors ρ_m, $m = 1, \ldots, M$ of the matrix C_X are determined by solving the problem

$$C_X \rho_m = \lambda_m \rho_m. \tag{5.40}$$

Because the ρ_m are also eigenvectors of a covariance matrix, they are indicated as t-EOFs and the corresponding PCs as t-PCs.

Example 5.2 SSA of NINO3 For the NINO3 (1950–2000) time series in Fig. 5.1c, as derived from the HadISST1 data set (Rayner et al., 2003), the eigenvalues λ_m are plotted in Fig. 5.6a; here $N = 612$ and $M = 80$. The first two eigenvalues are paired and indicate an oscillatory signal in the time series. Statistical tests are available (Ghil et al., 2012) to determine the significance of this pair (e.g., under a red-noise hypothesis). The error bars on the eigenvalues are computed here by $\sqrt{2/N}\lambda_m$.

The first two t-EOFs and t-PCs are shown in Fig. 5.6b–e, where the $N - M$ number t-PCs A_m are calculated from

$$A_m^l = \sum_{j=1}^{M} X^{l+j-1} \rho_m^j \tag{5.41}$$

and clearly indicate that t-PC1 and t-PC2 are in quadrature (they are shifted by a quarter period, here about 10 months).

As the dominant signal seems to be represented by the first few t-EOFs, we can restrict the signal to a small set of t-EOFs and obtain the reconstructed time series (with a length N). Using six tEOFs, the reconstructed component of the NINO3 time series is plotted in Fig. 5.7a, together with the original time series. A spectrum of the reconstructed component (Fig. 5.7b) now shows two spectral peaks at frequencies 0.019/month (period 52 months) and 0.034/month (period 29 months). ■

Multichannel singular spectral analysis (M-SSA) (Plaut and Vautard, 1994) is a direct extension of the SSA technique, which aims to extract propagating patterns that are optimal in representing variance. M-SSA is mathematically equivalent to extended EOF analysis (EEOF) (Weare and Nasstrom, 1982), but in M-SSA, focus is on the temporal structure of the variability, whereas in EEOF, the spatial variability is emphasized.

Let a data set $\mathbf{X}$ consist of a multichannel time series X_l^i, $i = 1, \ldots, N$; $l = 1, \ldots, L$, where i represents time and l the channel number. Index l may represent a point number on a specific grid or a principal component (PC) if the data are prefiltered with principal component analysis (Preisendorfer, 1988). We assume that $\mathbf{X}$ has zero mean and is stationary. By making M lagged copies of $\mathbf{X}$, the state vector at time i is given by

$$(X_1^{i+1}, X_1^{i+2}, \ldots, X_1^{i+M}, \ldots, X_L^{i+1}, \ldots, X_L^{i+M})^T, \tag{5.42}$$

where M is the window length. The covariance matrix T for a chosen window length M has a general block-Toeplitz form in which each block $\mathsf{T}_{ll'}$ is the lag covariance matrix (with maximum lag M) between channel l and channel l'. The $L \times M$ real eigenvalues λ_k of the symmetric matrix T are sorted in decreasing order where an eigenvector (referred to as a ST-EOF) $\mathbf{E}^k$ is associated with the k-th eigenvalue λ_k. The $\mathbf{E}^k$ are M-long time sequences of vectors, describing space-time patterns of decreasing importance as their order k increases. A space-time principal component (referred to as a ST-PC) α^k can be computed by projecting $\mathbf{X}$ onto $\mathbf{E}^k$; λ_k is the

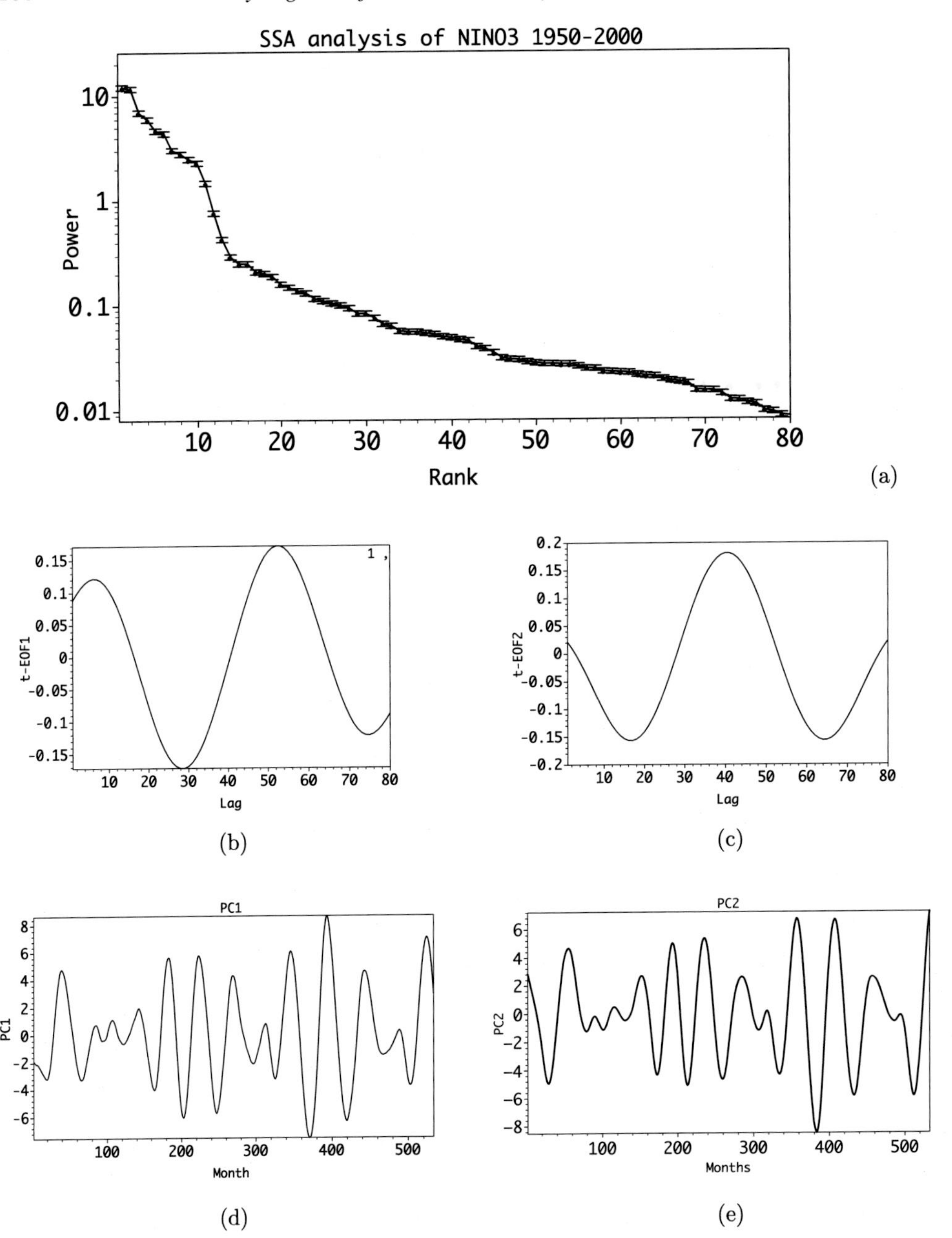

Figure 5.6 (a) The eigenvalues of the problem (5.40) for the NINO3 time series over the period 1950–2000 with a window $M = 80$. (b–c) The two leading t-EOFs and (d–e) the corresponding t-PCs.

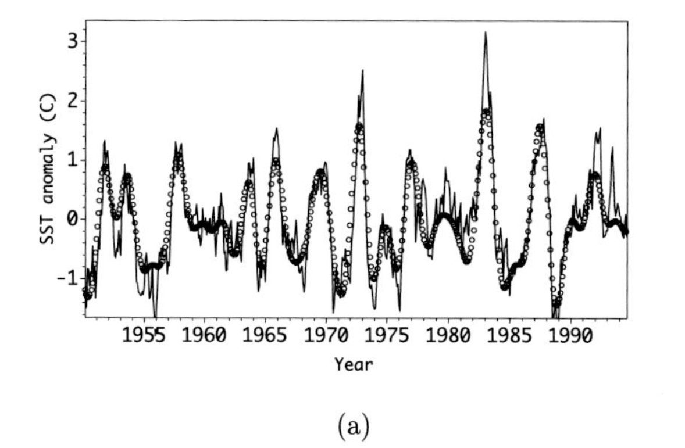

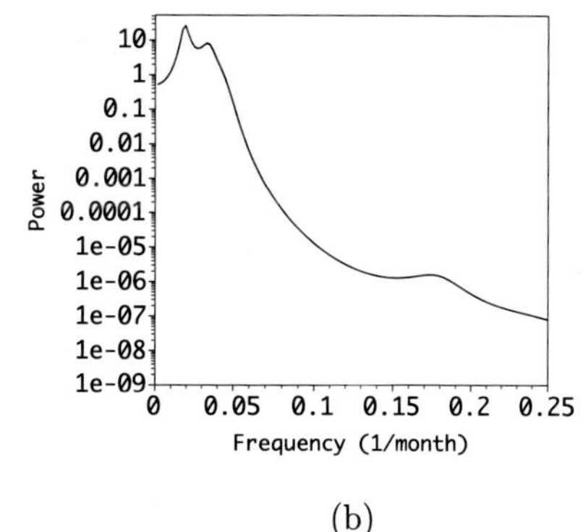

(a) (b)

Figure 5.7 (a) Reconstructed component (using the first six t-EOFs) for the NINO3 time series over the period 1950–2000 with a window $M = 80$. (b) Maximum entropy spectrum of the time series in (a) with order 20; this spectrum was made using the software package kSpectra (http://www.spectraworks.com/).

variance in α^k. In this way, the M-SSA expansion of the original data series is given by

$$X_l^{i+j} = \sum_{k=1}^{L \times M} \alpha_i^k \mathbf{E}_{lj}^k, \quad j = 1, \ldots, M. \tag{5.43}$$

Principal component analysis and SSA (Vautard and Ghil, 1989; Vautard et al., 1992) are particular cases of M-SSA: PCA can be derived from M-SSA with $M = 1$ and from SSA with $L = 1$.

When two consecutive eigenvalues are nearly equal and the two corresponding $\mathbf{E}^k$ as well as the associated α^k are in quadrature, then the data possess an oscillation whose period is given by that of α^k and whose spatial pattern is that of $\mathbf{E}^k$ (Plaut and Vautard, 1994). The sum on the right-hand side of (5.43), restricted to one or several terms, describes the part of the signal behaving as the corresponding $\mathbf{E}^k$. The components constructed in this way are called reconstructed components (RCs) and the part of the signal involved with an oscillation can be isolated. The original signal is exactly the sum of all the RCs.

As Allen and Robertson (1996) have pointed out, the presence of an eigenvalue pair in M-SSA is not sufficient grounds to conclude that the data exhibit an oscillation. Moreover, low-frequency eigenvalue pairs, which are entirely due to red noise, will appear high in the eigenvalue rank order. A Monte Carlo red-noise significance test for M-SSA was therefore constructed (Allen and Robertson, 1996). This is an objective hypothesis test for the presence of oscillations at low signal-to-noise ratios in multivariate data. Rejection of the red-noise null hypothesis using the test should be considered a necessary condition for M-SSA to have detected an oscillation, although in certain situations, non-oscillatory processes might also lead to rejection.

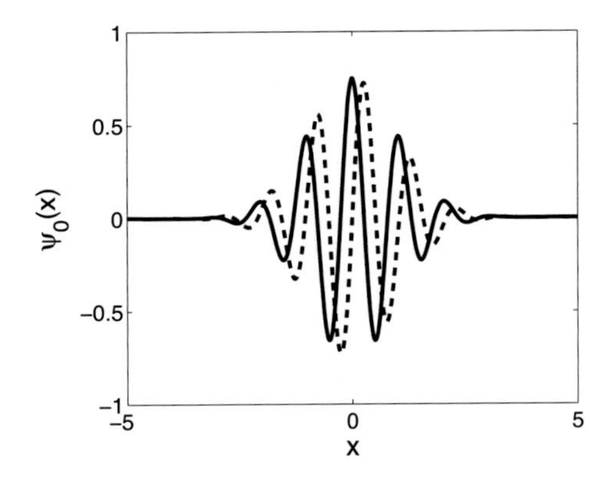

Figure 5.8 Plot of the real (drawn) and imaginary (dashed) part of the Morlet wavelet (5.44) with $\omega_0 = 6$.

5.4 Nonstationary methods: wavelets

Many time series are nonstationary, with, for example, the mean and variance changing in time. For the analysis of these time series, so-called wavelet techniques have been developed; with these techniques, the changes in the amplitude and period of an oscillation with time can be determined. The wavelet basis differs from the Fourier basis in that the functions have limited extent and can be stretched and translated. As an example, consider the much used Morlet wavelet, where the mother wavelet is given by

$$\psi_0(\eta) = \pi^{-\frac{1}{4}} e^{i\omega_0\eta} e^{-\frac{\eta^2}{2}}. \tag{5.44}$$

The function ψ_0 is a complex valued function of which the real and imaginary parts are plotted in Fig. 5.8 for $\omega_0 = 6$. The independent quantity η will be used to stretch and translate the waveform as plotted in Fig. 5.8.

After choosing a certain mother wavelet, a range of scales must be selected to adequately sample the time series. For example, if the data are sampled with a time interval (Δt), then the smallest period that can be resolved (the Nyquist period) is $2\Delta t$, which is usually chosen as the scale s_0. The larger scales (longer periods) are then chosen as

$$s_j = 2^j s_0, \quad j = 1, \ldots, J, \tag{5.45}$$

where J is determined such that the largest period is less than half the length of the time series.

Once these scales are chosen, the wavelet transform $W_n(s)$ of the time series $X_k, k = 1, \ldots, N$ is calculated (for each $s = s_j$) from

$$W_n(s) = \sqrt{\frac{\Delta t}{s}} \sum_{l=1}^{N} X_l \, \psi_0^\dagger \left(\frac{l-n}{s} \Delta t \right), \tag{5.46}$$

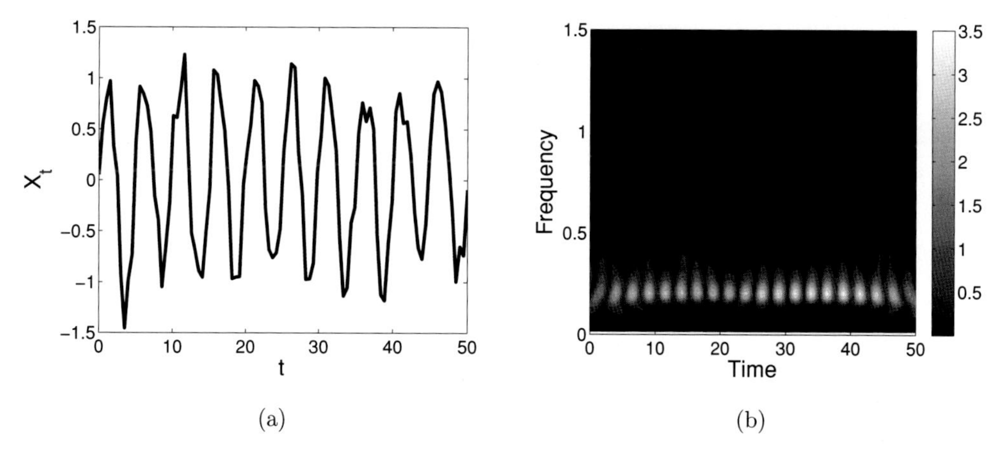

Figure 5.9 (a) Plot of the simple time series (5.47). (b) Wavelet spectrum of the time series in (a).

where the † superscript indicates the complex conjugate. Hence $n = 1, \ldots, N$ serves as a translation and s as the stretching (determining the scale) parameter. While 'moving over' the time series, the wavelet 'senses' the power $|W_n(s)|^2$ at the scales s.

Example 5.3 Examples of wavelet transforms As an example (Trauth, 2007), consider the pure sine wave with a period 5 with added Gaussian noise, defined by

$$X(t) = \sin\left(\frac{2\pi t}{5}\right) + \epsilon\zeta(t), \tag{5.47}$$

where ζ has the standard normal distribution $N(0, 1)$. The (stationary) time series (for $N = 100$, so with $\Delta t = 1/2$ and $s_0 = 1$) is plotted in Fig. 5.9a for $\epsilon = 0.2$. We use the Morlet wavelet over scales $s_j = 2^j$ and compute the wavelet spectrum (using MATLAB) as shown in Fig. 5.9b. Indeed, over the whole time period, the frequency 0.2 is the dominant one.

An example from climate data is that of the NINO3 time series as plotted in Fig. 5.1c. Here the number of data points $N = 612$ and $\Delta t = 1/12$ years. Hence $s_0 = 1/6$ years, and scales up to 26 years are considered. The wavelet spectrum is plotted in Fig. 5.10. The drawn contour in Fig. 5.10 indicates the region of frequencies where end effects (due to the finite length of the time series) are deteriorating the spectral analysis; this contour itself is the so-called cone of influence. ■

5.5 Nonlinear methods

Typical time series from observations of climate variables result from nonlinear deterministic processes (usually at large scales) combined with noise (usually from the small scales). Nonlinear deterministic processes may give rise to chaotic dynamics

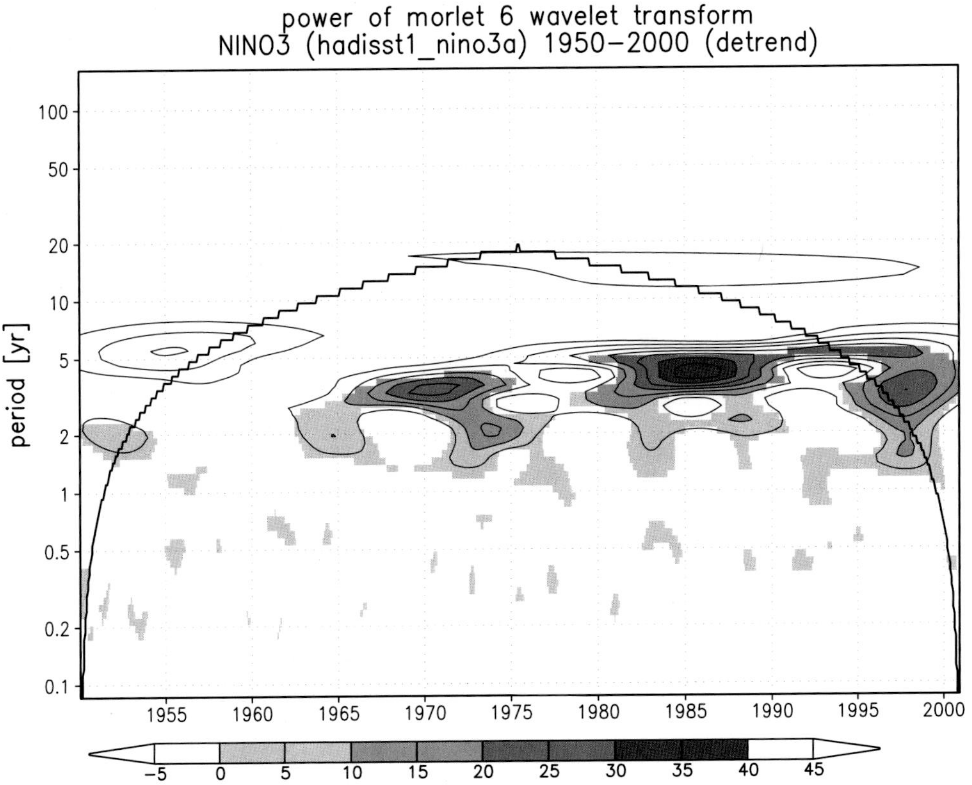

power of morlet 6 wavelet transform
NINO3 (hadisst1_nino3a) 1950–2000 (detrend)

Figure 5.10 Wavelet spectrum of the NINO3 time series in Fig. 5.1c showing contour values of the power $|W_n(s)|^2$ (figure made with the Climate Explorer, http://climexp.knmi.nl).

with extremely complicated time series. In this section, an introduction will be given into techniques to analyze such time series, following Abarbanel (1996). Computations can be done, for example, by using the software package TISEAN, which can be downloaded at http://www.mpipks-dresden.mpg.de/~tisean/.

5.5.1 Attractor embedding

Let $\mathbf{X}_k$, $k = 1, \ldots, n$ indicate a vector of climate variables with sampling times indicated by $t_k = k\tau_s$. Assume that we observe only a single time series from this climate variable, that is,

$$s_k = s(t_k) = h(\mathbf{X}_k), \tag{5.48}$$

where h acts as an observation mapping. Although this seems rather limited information about the full state vector $\mathbf{X}_k$, in the early 1980s, it was realised that any subsequent element in the series s_k is related to all the other variables in the system (i.e., also those not observed) through nonlinear couplings.

All time derivatives of the underlying observable $s(t)$, for example, $\frac{ds}{dt}$, $\frac{d^2s}{dt^2}$, are also represented by the series $s(t)$ as

$$\frac{ds}{dt}(t) \approx \frac{s(t+\tau_s) - s(t)}{\tau_s}, \tag{5.49a}$$

$$\frac{d^2s}{dt^2}(t) \approx \frac{s(t+\tau_s) - 2s(t) + s(t-\tau_s)}{\tau_s^2}. \tag{5.49b}$$

A famous embedding theorem (Abarbanel, 1996) states that for a certain delay time $T = l\tau_s$ and a dimension d, the trajectories of the d-dimensional vector,

$$\mathbf{y}_k = (s(t_k), s(t_k + l\tau_s), s(t_k + 2l\tau_s), \ldots, s(t_k + l\tau_s(d-1)))^T, \tag{5.50}$$

have the 'same' behavior as that of the original state vector $\mathbf{X}_k$. The theorem does not provide a recipe for choosing the so-called embedding dimension d and the time delay T; the data have to be analysed to provide good choices.

An attractor of an autonomous dynamical system will be a geometrical set $\mathcal{A}$ (possibly with a complicated fractal structure); let a proper fractal dimension measure (such as the box-counting dimension D_0, defined in Example 2.3) indicate that the set $\mathcal{A}$ has dimension d_A. Orbits on this attractor do not have self-intersections, and hence in our reconstructed state space of dimension d, the trajectories should also not self-intersect; obviously $d > d_A$. Consider the subspace of dimension d_A within d. When two subspaces of dimension d_1 and d_2 intersect, the dimension of the intersection has a dimension $d_1 + d_2 - d$ (e.g., the intersection of two planes $[d_1 = d_2 = 2]$ in three-dimensional space $[d = 3]$ is in general a line $d_1 + d_2 - d = 1$). Hence a sufficient condition that the subset of dimension d_A does not self-intersect is that

$$d_A + d_A - d < 0 \rightarrow d > 2d_A. \tag{5.51}$$

However, this is an upper boundary for d, as we may embed the attractor without self intersections into a smaller space than that determined by $d > 2d_A$. In addition, this estimate is only useful if we have an estimate of the attractor dimension d_A.

To determine a minimal value for d, the method of false nearest neighbours (FNN) is useful. Here the distances between nearby points on the reconstructed trajectories are compared in dimension d and $d + 1$. Let these points be indicated by $\mathbf{y}_k$ and $\mathbf{y}_k^{NN}$, where the subscript NN indicates nearest neighbour. We then have for their distance R^2 in dimensions d and $d + 1$,

$$R_d^2(k) = \sum_{m=1}^{d} (s(t_k + (m-1)T) - s^{NN}(t_k + (m-1)T))^2, \tag{5.52a}$$

$$R_{d+1}^2(k) = R_d^2(k) + (s(t_k + dT) - s^{NN}(t_k + dT))^2, \tag{5.52b}$$

where T is again the time delay. If the relative distance error F, given by

$$F = \sqrt{\frac{R_{d+1}^2(k) - R_d^2(k)}{R_d^2(k)}} = \frac{|(s(t_k + dT) - s^{NN}(t_k + dT))|}{R_d(k)}, \tag{5.53}$$

is small, then the points are 'true' neighbours in going from dimension d to $d+1$. However, when F is large apparently close points in dimension d are not close in dimension $d+1$.

Example 5.4 The method of false nearest neighbours Consider the time series given by (with an amplitude $A > 0$)

$$s(t_k) = A \sin t_k, \tag{5.54}$$

for $t_k = k\tau_s \in (0, 2\pi]$ observed by monitoring the displacement component $s(t) = x(t)$ of the harmonic oscillator, which is a solution to the differential equation

$$\frac{d^2x}{dt^2} + x = 0. \tag{5.55}$$

Clearly the phase space of this deterministic model is two-dimensional, and trajectories are circles.

With embedding dimensions $d = 1$ and $d = 2$, we have the reconstructions

$$y_k = A \sin t_k, \tag{5.56a}$$

$$\mathbf{y}_k = (A \sin t_k, A \sin(t_k + T))^T, \tag{5.56b}$$

for a certain time delay T. For the case $d = 1$, the 'orbits' move in an interval, whereas in the $d = 2$ case, the orbits move on closed curves (Fig. 5.11). An example of a pair of such FNN for $d = 1$ is shown in Fig. 5.11; in this case R_1 is small, whereas R_2 is large and hence F is large. When we consider $d = 3$ for this simple example, we notice that all points are real neighbours, and hence $R_2 \approx R_3$ and F is small. The necessary embedding dimension d is hence obtained from the data at the dimension where F rapidly decreases. ■

At this point, the delay time T is still unknown. Based on classical analysis, one could base a choice of T on the autocorrelation function (e.g., take T to be the lag where the autocorrelation has the first zero), but in a chaotic system, this is usually not very revealing. Intuitively T should not be too small, because otherwise $s(t_k)$ and $s(t_k + T)$ are too much related, and the attractor is not sampled well. However, T should not be too large, because otherwise $s(t_k)$ and $s(t_k + T)$ are totally uncorrelated. To determine T, a nonlinear extension of the autocorrelation function, called the mutual information $I(T)$, is borrowed from information theory.

Consider realizations $A_i, i = 1, \ldots, N_A$ and $B_j, j = 1, \ldots, N_B$ of event A and event B, and we ask the question: what amount of bits is learned from the measurement

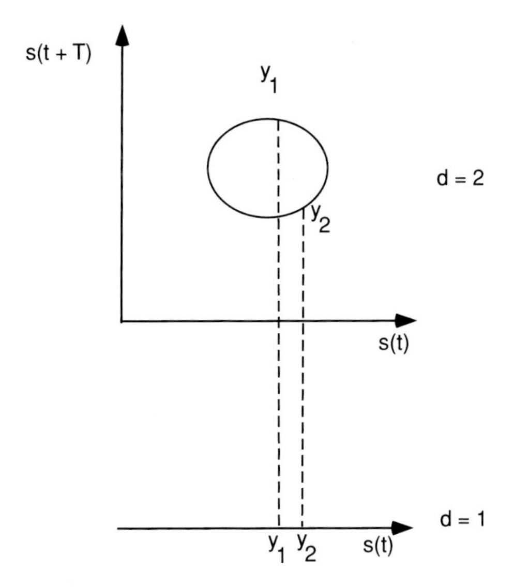

Figure 5.11 Sketch of the nearest neighbours in the reconstruction of the phase space of the harmonic oscillator.

of A_i about the measurement of B_j? From information theory, a measure of this amount of information is defined as

$$I_{AB} = \sum_{A_i, B_j} \log_2 \frac{P_{AB}(A_i, B_j)}{P_A(A_i) P_B(B_j)}, \tag{5.57}$$

where P_{AB} is the joint probability for the measurements of A and B and P_A (P_B) is the individual probability of A (B). If the A_i and B_j are uncorrelated, then $P_{AB}(A_i, B_j) = P_A(A_i) P_B(B_j)$ and $I_{AB} = 0$. The average mutual information is given by

$$\bar{I}_{AB} = \sum_{A_i, B_j} P_{AB}(A_i, B_j)\, I_{AB}. \tag{5.58}$$

If we now take for event A the measurements $s(t_k)$ and for B the events $s(t_k + T)$, then the average mutual information is defined as

$$I(T) = \sum_{s(t_k), s(t_k+T)} P(s(t_k), s(t_k + T)) \log_2 \frac{P(s(t_k), s(t_k + T))}{P(s(t_k)) P(s(t_k + T))}. \tag{5.59}$$

To determine $P(s(t_k))$, we make a histogram of the values of $s(t_k)$ versus k; same for $P(s(t_k + T))$. For the joint probability, we compute a two-dimensional histogram in the same way. The first minimum of the function $I(T)$ versus T will give a proper estimate of the time lag T.

5.5.2 Estimation of attractor properties

Once we have found an appropriate T and an embedding dimension d and the corresponding reconstruction of the dynamical system, that is,

$$\mathbf{y}_k = (s(t_k), s(t_k + T), \ldots, s(t_k + T(d-1)))^T, \tag{5.60}$$

for $k = 1, \ldots, N$, we are able to determine further properties of the attractor. A generalised view of the geometric structure of the attractor can be obtained by determining the number of data points $n(\mathbf{x}, r)$ within a radius r of a certain chosen point $\mathbf{x}$ in the d-dimensional space. With $\mathcal{H}$ being the Heaviside function, we have

$$n(\mathbf{x}, r) = \frac{1}{N} \sum_{k=1}^{N} \mathcal{H}(r - |\mathbf{y}_k - \mathbf{x}|), \tag{5.61}$$

where the norm is taken in $\mathbb{R}^d$. When we take $\mathbf{x}$ equal to all the $\mathbf{y}_k$, an average density is given by the correlation function

$$C(q, r) = \frac{1}{N} \sum_{k=1}^{N} \left[\frac{1}{K} \sum_{l=1}^{K} \mathcal{H}(r - |\mathbf{y}_k - \mathbf{y}_l|) \right]^{q-1}. \tag{5.62}$$

If the number of points $n(\mathbf{x}, r) \approx r^{D_q}$, then we expect an average density

$$C(q, r) \approx r^{(q-1)D_q} = e^{(q-1)D_q \ln r} \to \ln C(q, r) \approx (q-1)D_q \ln r. \tag{5.63}$$

Hence by plotting $\ln C(q, r)$ versus $\ln r$, we may find a slope $(q-1)D_q$. As we have chosen q, we find an estimate of the so-called correlation dimension of the attractor by

$$D_q = \lim_{r \to 0} \frac{\ln C(q, r)}{(q-1) \ln r}. \tag{5.64}$$

The box-counting dimension D_0 discussed in (2.43) is a special case. It is defined as

$$D_0 = \lim_{r \to 0} \frac{\ln C(0, r)}{-\ln r}; \quad C(0, r) \approx N(r). \tag{5.65}$$

The dimension D_2 is usually referred to as the Grassberger-Procaccia dimension.

As we saw in Section 2.4.3, an important property of an attractor is whether there is sensitivity to initial conditions; if so, there should be a positive Lyapunov exponent. The issue here is that now we have to determine these exponents from the data using the reconstructed attractor. This is not an easy issue in practice, but the procedure is as follows. The Lyapunov exponents require an estimate of the Jacobian matrix. In the reconstructed space, the different points on the trajectory are related through

$$\mathbf{y}_{k+1} = \mathbf{F}(\mathbf{y}_k), \tag{5.66}$$

for some mapping $\mathbf{F}$. When a perturbation Δ_k is assumed on a certain $\bar{\mathbf{y}}_k$, that is, $\mathbf{y}_k = \bar{\mathbf{y}}_k + \Delta_k$, then linearization leads to

$$\Delta_{k+1} = J(\bar{\mathbf{y}}_k)\Delta_k, \tag{5.67}$$

where J is the local Jacobian matrix. We hence find through iteration that

$$\Delta_{k+n} = J^n(\bar{\mathbf{y}}_k)\Delta_k. \tag{5.68}$$

At this point a famous theorem from ergodic theory due to Oseledec (Abarbanel, 1996) can be used, which states that:

(i) The length of the vector $|\Delta_{k+n}|$ is given by

$$|\Delta_{k+n}|^2 = \Delta_k^T (J^n(\bar{\mathbf{y}}_k))^T J^n(\bar{\mathbf{y}}_k)\Delta_k. \tag{5.69}$$

(ii) The Lyapunov exponents λ_i are given by

$$\lambda_i = \ln \sigma_i, \tag{5.70}$$

where σ_i are the eigenvalues of the Oseledec matrix

$$\mathcal{S} = \lim_{n\to\infty} \left[(J^n)^T J^n\right]^{\frac{1}{2n}}. \tag{5.71}$$

(iii) When we diagonalise the Oseledec matrix $\mathcal{S}$ into the form

$$\mathcal{S} = P^T \Sigma P,$$

where Σ is the diagonal matrix with the eigenvalues σ_i, then from (i) we find for large n

$$|\Delta_{k+n}|^2 \approx \Delta_k^T P^T (\Sigma)^{2n} P \Delta_k$$

and hence with $\tilde{\Delta}_k = P\Delta_k$, eventually

$$|\tilde{\Delta}_{k+n}|^2 \approx \sigma_1^{2n} |\tilde{\Delta}_k|^2 = e^{2n\lambda_1} |\tilde{\Delta}_k|^2, \tag{5.72}$$

where λ_1 is the largest Lyapunov exponent. This expression shows that the norm of the perturbations exponentially diverges if $\lambda_1 > 0$.

To compute the Oseledec matrix, we must determine an approximation of the Jacobian matrix from the data. Thereto we make use of the fact that every point on the attractor will be visited in time (because of the recurrence property). Choosing $\mathbf{y}_k$ and its N_B nearest neighbours $\mathbf{y}_k^r$, $r = 1, \ldots, N_B$, we next follow all the orbits of these nearest neighbours and choose a polynomial basis $\phi_m(\mathbf{x})$ to represent every vector into the form

$$\mathbf{y}_k = \sum_{m=1}^{M} c_{m,k}\phi_m(\mathbf{y}_k), \tag{5.73}$$

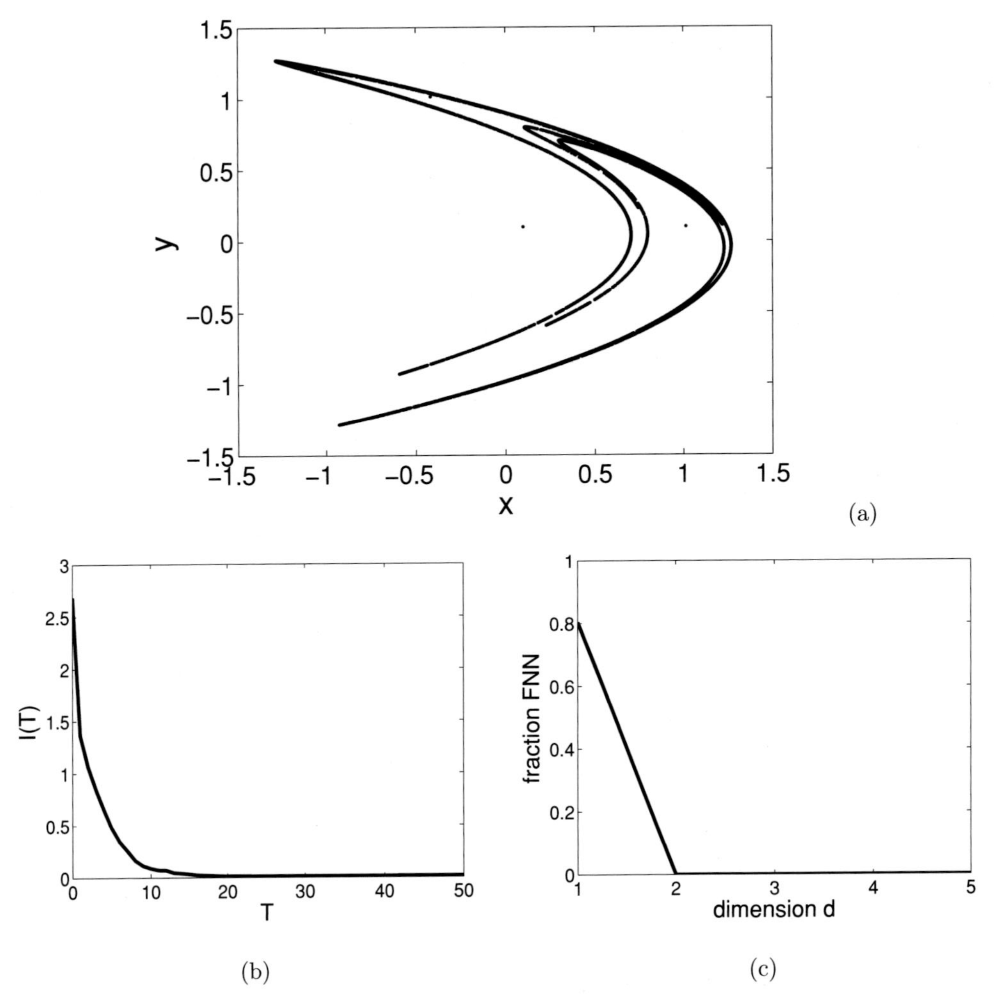

Figure 5.12 (a) Plot of the Hénon attractor obtained by iteration of the map (5.74). (b) Average mutual information for the Hénon map. (c) False nearest neighbours versus the embedding dimension d for the Hénon map (with a delay $T = 1$).

where the coefficients $c_{m,k}$ are determined through least squares. In this way we can compute the derivatives and hence an approximation of the Jacobian matrix.

Example 5.5 The Hénon map The equations for the Hénon map are given by

$$x_{k+1} = 1 - ax_k^2 + y_k, \tag{5.74a}$$

$$y_{k+1} = bx_k, \tag{5.74b}$$

with two parameters $a = 1.4$ and $b = 0.3$. By applying the iteration from any starting value, one eventually obtains the picture Fig. 5.12a, which is the famous Hénon attractor.

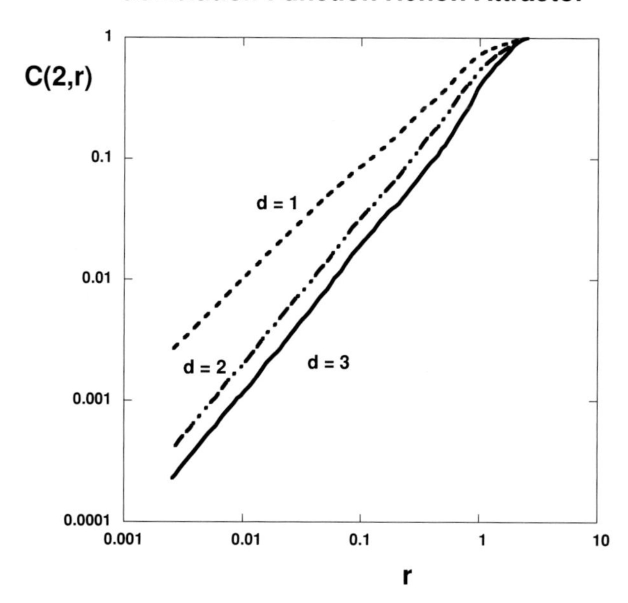

Figure 5.13 Correlation dimension of the Hénon map for different embedding dimensions d.

Here the natural delay time is the iteration step and hence $T = 1$. When we plot the average mutual information (Fig. 5.12b), we see that $I(T)$ is monotonically decreasing, and hence there is indeed no better choice than $T = 1$. With this delay, the FNN method is applied, and the percentage of FNN versus the embedding dimension (Fig. 5.12c) indicates that the Hénon attractor can be embedded into two-dimensional space, and hence $d = 2$ is a good choice.

In a logarithmic plot of $C(2, r)$ (calculated with TISEAN) for different embedding dimensions (Fig. 5.13), we obtain from the slope of $\log C(2, r)$ versus $\log r$ (and from a power-law fit for both $d = 2$ and $d = 3$) an estimate $D_2 = 1.23$. This is close to the standard value of 1.21 ± 0.01; the Hénon attractor is hence a fractal set. For the Hénon attractor, the TISEAN program provides values $\lambda_1 = 0.42$ and $\lambda_2 = -1.58$ for the Lyapunov exponents and hence the Hénon dynamical system is a chaotic system. ■

Of course, these approaches have their pitfalls. One of them is that one generally needs an enormous amount of data to accurately determine the desired properties of the attractors. Climate time series from observations are often too short, and the false nearest neighbour method may indicate a finite dimension of the attractor just because of lack of data.

5.5.3 Detection of critical transitions

In Subsection 3.5.5, we have seen that the approach to critical conditions in fast-slow systems is associated with an increase of variance in the time series. In this section, methodology is presented to detect the occurrence of such critical conditions from time series.

Suppose we have a time series of a certain quantity $x_n = x(t_n)$, where n indicates the time index. Near critical conditions, we know that one eigenvalue of the Jacobian matrix of the dynamical system that has generated x_n moves through the imaginary axis, and hence one normal mode switches from being damped to being amplified. To mimic this change in damping, suppose x is generated by the stochastic Itô SDE of the form (3.47) (Section 3.4.1),

$$X_t = X_0 + \int_0^t \lambda X_s ds + \sigma \int_0^t dW_s, \tag{5.75}$$

describing an Ornstein-Uhlenbeck process with damping coefficient λ. The solution to this equation is (3.49) with $\gamma = -\lambda$,

$$X_t = e^{\lambda t}\left(X_0 + \sigma \int_0^t e^{-\lambda s} dW_s\right). \tag{5.76}$$

The discrete version of (5.76) is (at t_{n+1} and t_n),

$$x_{n+1} = e^{\lambda t_{n+1}} X_0 + \sigma e^{\lambda t_{n+1}} \int_0^{t_{n+1}} e^{-\lambda s} dW_s, \tag{5.77a}$$

$$x_n = e^{\lambda t_n} X_0 + \sigma e^{\lambda t_n} \int_0^{t_n} e^{-\lambda s} dW_s \tag{5.77b}$$

and subtracting both equations gives

$$x_{n+1} - e^{\lambda \Delta t} x_n = \sigma \epsilon_n, \tag{5.78}$$

where ϵ_n represents the stochastic part and $\Delta t = t_{n+1} - t_n$. This is an AR(1) process (Section 5.2.2) of which the variance and autocorrelation follow from (5.24) as

$$Var[X] = \frac{\sigma^2}{1 - e^{2\lambda \Delta t}}, \tag{5.79a}$$

$$\rho_n = e^{n\lambda \Delta t}. \tag{5.79b}$$

This shows that the variance of the time series increases as $\lambda \to 0$ and also that long-range correlations do appear.

There are several techniques to detect long-range correlations and variance increase in time series, of which we discuss two: degenerate fingerprinting (Held and Kleinen, 2004) and detrented fluctuation analysis (Kantelhardt et al., 2001; Livina and Lenton, 2007). We see applications of these techniques in Section 10.3 when we discuss the Dansgaard-Oeschger oscillations.

Degenerate fingerprinting

The basis for this method is the equation (5.78), written in the form

$$x_{n+1} = c x_n + \sigma \epsilon_n, \tag{5.80}$$

with $c = e^{\lambda \Delta t}$. After the linear trend of the time series is removed, the quantity c is obtained by fitting. The error bars on c are given by (Held and Kleinen, 2004)

$$\Delta c^2 \approx (1 - c^2)\frac{\Delta t}{T} \rightarrow \Delta\lambda \cong \sqrt{\frac{2\lambda}{T}}, \tag{5.81}$$

where T is the total length of the time series. From this equation, one can also deduce the minimum length of the time series to obtain a specific accuracy in λ.

Detrended fluctuation analysis

This method is designed to determine long-range correlations in time series. In general, we have a time series $X(i)$, $i = 1, \ldots, N$ and determine the autocorrelation as

$$C(s) = \frac{1}{N - s} \sum_{i=1}^{N-s} X(i) X(i + s), \tag{5.82}$$

For processes with short-range correlations, we expect $C(s) \approx e^{-s/\tau}$, where τ is the decay time scale. However, for long-range correlations, an autocorrelation of the form $C(s) = s^{-\gamma}$, with $0 < \gamma < 1$ can be expected.

The detrended fluctuation analysis essentially consists of three steps:

1. Let $Y(i) = \sum_{k=1}^{i} X(k)$. The profile of $Y(i)$ is cut into precisely $N_s = N/s$ nonoverlapping segments of equal length s.
2. Determine the local trend P_ν of each segment ν by a least-squares polynomial fit of order n to the data. The detrended time series for a segment of duration s, denoted by $Y_s(i)$, is then given by

$$Y_s(i) = Y(i) - P_\nu(i). \tag{5.83}$$

   An example of the construction of the $Y_s(i)$ for $s = 100$ and $s = 200$ is shown in Fig. 5.14.
3. Calculate for each of the segments the quantity

$$F_s^2(\nu) = \frac{1}{s} \sum_{i=1}^{s} (Y_s((\nu - 1)s + i))^2, \tag{5.84}$$

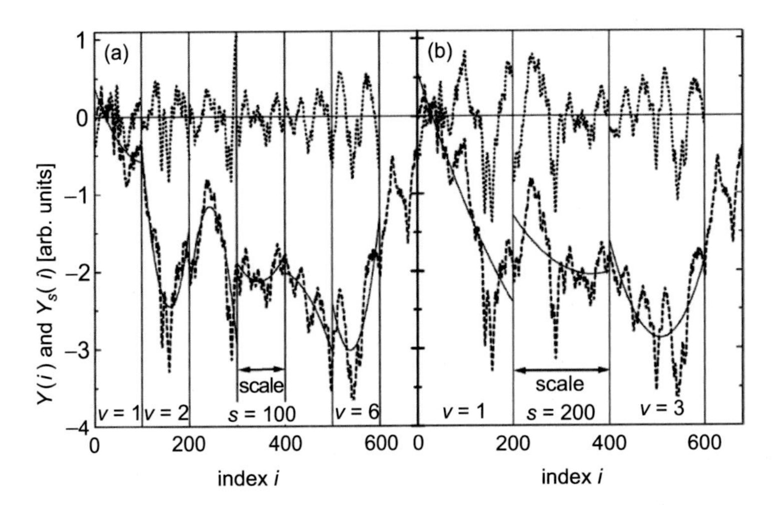

Figure 5.14 Illustration of the detrending procedure in the detrended fluctuation analysis for (a) $s = 100$ and (b) $s = 200$. The time series $Y(i)$ is the dashed curve, the drawn curves are the least-squares quadratic fits $P_2(i)$ and the dotted curve is the time series $Y_s(i)$ (figure from Kantelhardt et al., 2001).

and then from this the fluctuation function

$$F(s) = \left[\frac{1}{N_s} \sum_{\nu=1}^{N_s} F_s^2(\nu) \right]^{1/2}, \tag{5.85}$$

for several different polynomial orders n.

From Fig. 5.14, it is seen that the variance in Y_s, as measured by $F(s)$ increases with s. Suppose the original time series has zero mean and no trend; then $Y_s(i) = Y(i)$ and hence

$$Y^2(i) = \left(\sum_{k=1}^{i} X(k) \right)^2 = \sum_{k=1}^{i} (X(k))^2 + \sum_{k \neq j}^{j,k \leq i} X(j)X(k). \tag{5.86}$$

The second term can be written as

$$\sum_{k \neq j}^{j,k \leq i} X(j)X(k) = 2 \sum_{k=1}^{i-1} (i-k)C(k),$$

with $C(k)$ defined in (5.82). For long-range correlations, we have $C(k) \sim k^{-\gamma}$, for $0 < \gamma < 1$, and hence

$$\sum_{k=1}^{i-1} kC(k) \sim \sum_{k=1}^{i} k^{1-\gamma} \sim \int_1^i k^{1-\gamma} dk \sim i^{2-\gamma}, \tag{5.87}$$

which will dominate the variance in Y for large i.

Substituting this result into (5.85), the dependence of $F(s)$ on s is given by

$$F(s) \sim \langle Y_s^2 \rangle^{1/2} \sim s^{1-\gamma/2} = s^{\alpha}, \tag{5.88}$$

where $\alpha = 1 - \gamma/2$. For short-range correlations (or uncorrelated) data, $C(s)$ decays exponentially, and the first term on the right-hand side of (5.86) dominates. As $Y^2(i) \sim i$, this gives $F(s) \sim s^{1/2}$, and hence $\alpha = 1/2$. Indications for long-range correlations in the time series are hence found when $\alpha > 1/2$.

6
The Climate Modelling Hierarchy

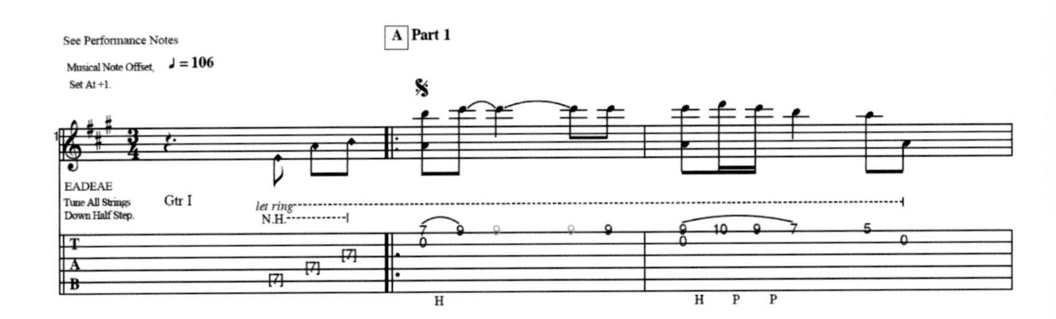

One half step at a time.
EADEAE, The Handing Down, Edward Gerhard

Climate phenomena are studied using observations and climate models. Because we cannot investigate most of these phenomena in the laboratory, climate models are a central component of climate research. A wide range of models is in use with on the one hand 'very simple' conceptual climate models and on the other hand 'very complex' state-of-the-art global climate models (GCMs). It would be impossible (and also useless) to try to provide an overview of all the models around. However, general notions on the use and importance of a climate modelling hierarchy can be given and are the main focus of this chapter.

In Section 6.1, the main model 'traits', that is, scales and processes, are described, and we argue that climate models can be roughly distinguished using these two attributes. In Section 6.2, we then address the intricate coupling between model choice and scientific question. Deterministic versus stochastic modelling approaches are described in Section 6.3, and in the last section (Section 6.4), sources of modelling errors are discussed.

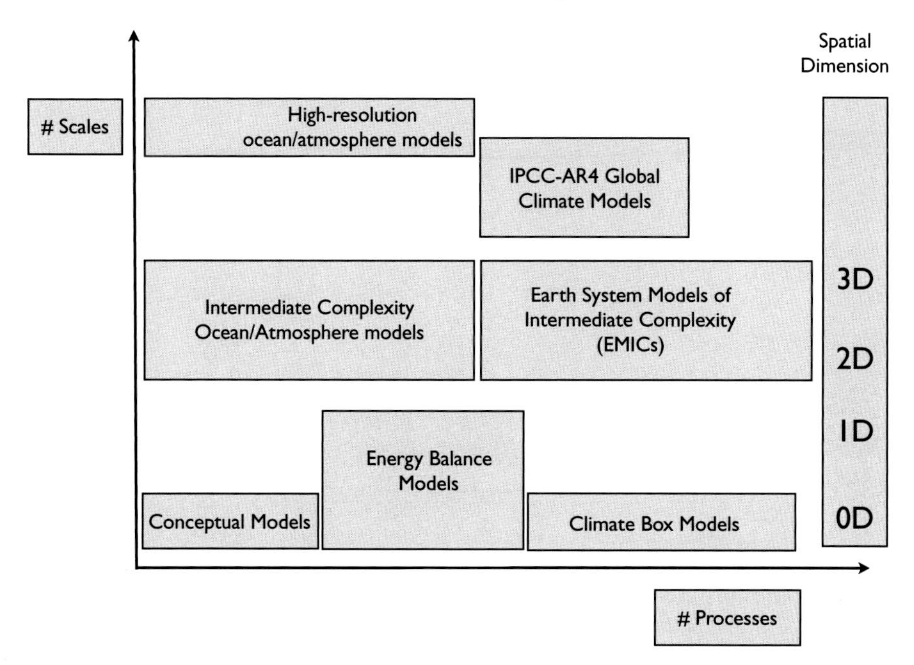

Figure 6.1 'Classification' of climate models according to the two model traits: number of processes and number of scales. There is, of course, overlapping between the different model types, but for simplicity they are sketched here as non-overlapping. Here 0D (zero dimensional), 1D (one dimensional), 2D (two dimensional) and 3D (three dimensional) indicate the spatial dimension of the model.

6.1 Model traits: scales and processes

As we have seen in Chapter 1, climate phenomena occur on different spatial and temporal scales. Climate variations associated with the Pleistocene Ice Ages have typical time scales of 100 kyr and a global spatial pattern. In contrast, thunderstorms affect only a local area and have typical time scales of hours. Based on observations, it is known that distinct processes are often involved in these different phenomena. The processes driving El Niño variability differ substantially from those that are involved in glacial-interglacial cycles.

As scales and processes are such important properties of phenomena, it is important to classify climate models using these two traits (Fig. 6.1). Here the trait 'scales' refers to both spatial and temporal scales, as there exists a relation between both: on smaller spatial scales, usually faster processes take place. 'Processes' refers to physical, chemical or biological processes taking place in the different climate compartments (atmosphere, ocean, cryosphere, biosphere, lithosphere). Both traits affect the dimension of the state vector of the dynamical system in a different way. An increase in scales increases the state vector dimension, with a fixed relation between the state vector variables due to the spatial coupling. For example, when the resolution is doubled of a one-dimensional scalar variable, the dimension increase due to that

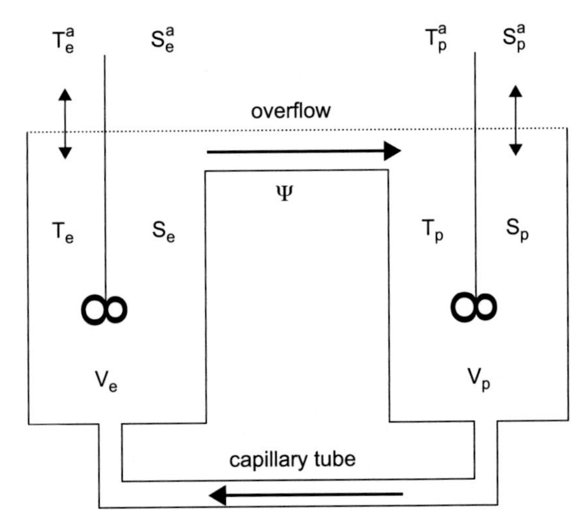

Figure 6.2 Sketch of the two-box model as in Stommel (1961). Two reservoirs contain well-mixed water and are connected through an overflow and a capillary tube. The circulation is driven by density gradients between the boxes, which are set up by the exchange of heat and freshwater at the surface.

variable is doubled. Such a relation is absent when the state vector is increased due to the increase in the number of processes.

Models with a limited number of processes and scales are usually referred to as conceptual climate models. In these models, only very specific interactions in the climate system are described. A typical example is a box model; one of the simplest ones was used by Stommel (1961) to study the stability of the ocean's thermohaline circulation (Fig. 6.2). As we use this model in Chapter 10, the equations are provided in Example 6.1.

Example 6.1 The Stommel two-box model In the model proposed by Stommel (1961), two boxes, having volumes V_p and V_e, contain well-mixed water of temperature and salinity (T_e, S_e) and (T_p, S_p), with the subscripts e and p indicating the equatorial and polar box, respectively. The boxes are connected at the surface by an overflow region and at the bottom by a capillary tube. The flow rate Ψ is directed from high to low pressure and is assumed linearly related to the density difference of the liquid between the boxes, that is,

$$\Psi = \gamma \, \frac{\rho_p - \rho_e}{\rho_0}, \tag{6.1}$$

where ρ_0 is a reference density and γ a hydraulic constant. Hence the flow rate is taken positive if the liquid is heavier in the polar box. The exchange of properties does not depend on the sign of Ψ, because it only matters that properties from one box are transported to the other box. Because mass is conserved, the pathway (either through the overflow or through the capillary) is unimportant. A linear equation of state is assumed of the form

$$\rho = \rho_0(1 - \alpha_T(T - T_0) + \alpha_S(S - S_0)), \tag{6.2}$$

where T_0 and S_0 are reference values for temperature and salinity and α_T and α_S are thermal expansion and saline contraction coefficients, respectively.

Exchange of heat and salt in each box due the surface forcing is modelled through a relaxation to a prescribed surface temperature and salinity (T^a, S^a) with relaxation coefficients C^T and C^S. These coefficients differ for each box and for each quantity considered (heat or salt). In this way, the balances of heat and salt in each box are given by

$$V_p \frac{dT_p}{dt} = C_p^T(T_p^a - T_p) + \mid \Psi \mid (T_e - T_p), \tag{6.3a}$$

$$V_e \frac{dT_e}{dt} = C_e^T(T_e^a - T_e) + \mid \Psi \mid (T_p - T_e), \tag{6.3b}$$

$$V_p \frac{dS_p}{dt} = C_p^S(S_p^a - S_p) + \mid \Psi \mid (S_e - S_p), \tag{6.3c}$$

$$V_e \frac{dS_e}{dt} = C_e^S(S_e^a - S_e) + \mid \Psi \mid (S_p - S_e). \tag{6.3d}$$

In the following, we restrict consideration to the case of realistic forcing, for which $T_e^a - T_p^a > 0$ and $S_e^a - S_p^a > 0$. For simplicity, it is assumed that the relaxation times for temperature to the surface forcing in both boxes is proportional to their volume, and hence $C_p^T / V_p = C_e^T / V_e \equiv R_T$ is constant. The same simplification is made for salinity with $R_S = C_p^S / V_p = C_e^S / V_e$.

When time, temperature, salinity and flow rate are scaled with $1/R_T$, $V_e V_p R_T / (\gamma \alpha_T (V_e + V_p))$, $V_e V_p R_T / (\gamma \alpha_S (V_e + V_p))$ and $V_e V_p R_T / ((V_e + V_p)$, respectively, the dimensionless equations become

$$\frac{d\bar{T}}{dt} = \eta_1 - \bar{T}(1 + \mid \bar{T} - S \mid), \tag{6.4a}$$

$$\frac{d\bar{S}}{dt} = \eta_2 - \bar{S}(\eta_3 + \mid \bar{T} - \bar{S} \mid), \tag{6.4b}$$

where $\bar{T} = T_e - T_p$, $\bar{S} = S_e - S_p$ and $\bar{\Psi} = T - S$ is the dimensionless flow rate. Three parameters appear in the equations (6.4), which are given by $\eta_1 = (T_e^a - T_p^a)\ \gamma \alpha_T (V_e + V_p) / (V_e V_p R_T)$, $\eta_2 = (R_S / R_T)(S_e^a - S_p^a)\ \gamma \alpha_S (V_e + V_p) / (V_e V_p R_T)$ and $\eta_3 = R_S / R_T$.

The explicit equations (6.4) show that the dimension of the state vector d is very small ($d = 2$), and, consequently, the model is a very limited representation of scales (box dimension, overturning time scale). In addition, it is clear that the model contains several (here, three) parameters that have to be estimated ad hoc. The ad hoc parameters arise through simplifying assumptions on the form of the different fluxes (mass, momentum, heat, salt) in the ocean system. ■

The number of processes between the boxes can be extended, for example, by including sea ice, land ice processes and/or biogeochemical processes. In this way, one ends up in the right lower part of the diagram in Fig. 6.1 because the number of scales is still relatively small. Models of this type are abundant in the area of biogeochemical climate modelling, and a prominent example is the model of Gildor et al. (2002), which we use in Chapter 11. An increase in the number of boxes will increase the number of scales represented, but the model will still consist of a system of ordinary differential equations (ODEs). The dimension of the state vector in these models is mainly increased because of an increase in the number of dependent quantities. Another example is the integrated models (Kemfert, 2005), where, apart from physical, chemical and biological processes, societal aspects are also included. A third example is the low-dimensional ODE-type models of glacial-interglacial cycles (Saltzmann, 2001).

Limiting the number of processes, scales can be added by discretising the governing partial differential equations spatially up to three dimensions. A prominent class of early climate models of this type are so-called energy balance models (EBMs) described in North et al. (1981). As we use these models in Chapter 11, the equations of a typical EBM are provided in Example 6.2.

Example 6.2 Energy balance models In the most simple EBMs, the state vector is composed only of the atmospheric surface temperature T_a, and a heat balance gives an equation of the form

$$\rho_a H_a C_{pa} \frac{\partial T}{\partial t} = \nabla \cdot \mathbf{F} + Q_S - Q_L - Q_{oa} - Q_{la} - Q_{ia}, \tag{6.5}$$

where H_a is a typical atmospheric vertical (depth) scale. In addition, ρ_a and C_{pa} are the density and the heat capacity of air, respectively. The quantity Q_S is the incoming shortwave radiation, Q_L is the outgoing longwave radiation and Q_{oa}, Q_{la} and Q_{ia} are the heat fluxes into the ocean, land and ice, respectively. The flux $\mathbf{F}$ is the meridional heat transport, which is mostly accomplished by atmospheric eddy processes.

The various energy balance models differ in the representation of physical processes through the choices of $\mathbf{F}$, the fluxes Q and the spatial dimension of the problem (0D, 1D or 2D). One of the earliest EBMs was the 0D model presented by Budyko (1969). The equation is

$$\rho_a H_a C_{pa} \frac{\partial T}{\partial t} = \frac{\Sigma_0}{4}(1 - \alpha) - \sigma T^4,$$

where σ is the Stefan-Boltzmann constant. In this EBM, only a simplified globally integrated radiation balance is represented.

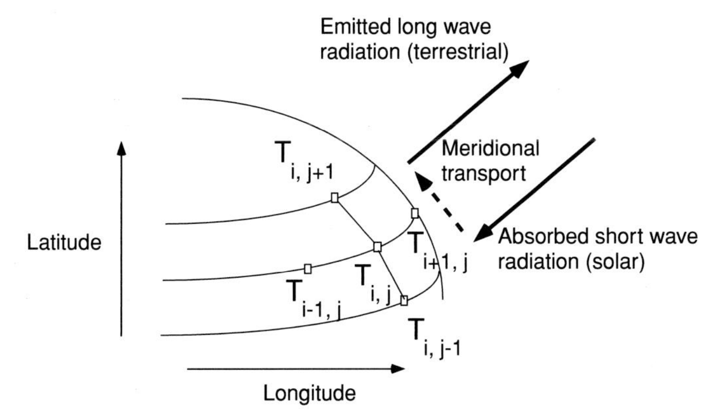

Figure 6.3 Sketch of a two-dimensional EBM, modelling the surface temperature T of the Earth.

For such a 0D EBM, the number of degrees of freedom $d = 1$. However, when a spatially one-dimensional version is considered, with latitude as spatial dimension, and the latitudinal grid has M points, the number of degrees of freedom will scale with M, that is, the values of T on the grid points. Similarly, discretising the equation (6.5) for the two-dimensional temperature field T (Fig. 6.3) using a finite difference, finite element or spectral method will result in a discrete state vector $\mathbf{x}$. The dimension of $\mathbf{x}$ will increase due to an increase in resolution to present T over the sphere. For example, using a finite difference method on a grid with longitudes $\phi_i = i\Delta\phi, i = 1, \ldots, N$ and latitudes $\theta_j = j\Delta\theta, j = 1, \ldots, M$ will result in a state vector $\mathbf{x}$ of dimension $N \times M$. ■

A higher spatial resolution and inclusion of more processes will give models located in the right upper part of the diagram. In a GCM, we divide the atmosphere, ocean, ice and land components into grid boxes (Fig. 6.4). Over such a three-dimensional grid box, we consider the budgets of momentum, mass and, for example, heat. Momentum budgets basically follow from Navier-Stokes equations formulated for air and water. The difference on what goes into a box minus what goes out of that box leads to an increase/decrease of a particular quantity, say temperature. Once the distribution of a quantity is known at a certain time, then these budgets provide an evolution equation to determine the quantity some time later.

The advantage of more boxes is that we resolve the temperature better (more points in a certain area). With an increasing number of grid boxes, however, the time development of an increasing number of quantities (at each grid box) has to be calculated. The same holds for the number of processes included in a GCM: more processes simply mean more calculations. Also, the longer time period over which we want to compute the development of each quantity, the longer it takes to do the calculation on a computer. An overview of the IPCC-AR4 models is provided in

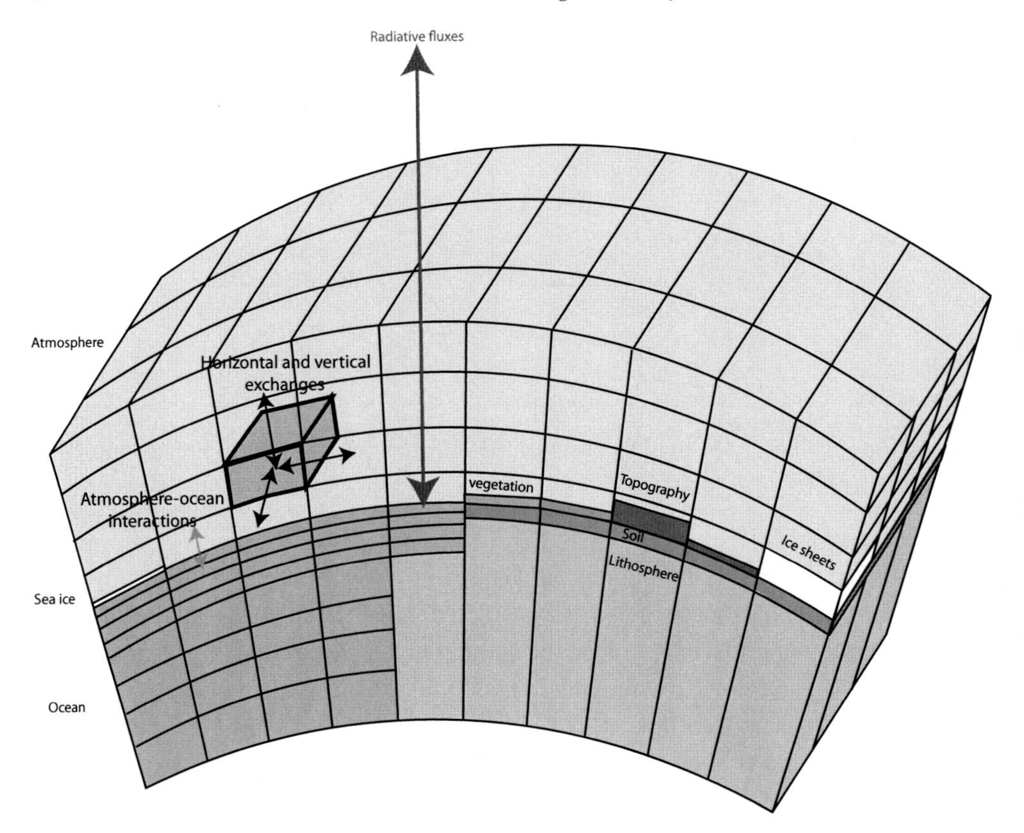

Figure 6.4 A typical structure of a GCM; the number of grid boxes in each of the components determines the spatial resolution of the model (figure from http://stratus.astr.ucl.ac.be/textbook/ and courtesy of Hugues Goosse). (See Colour Plate.)

Chapter 8 (pp. 597–600) of WGI (Meehl et al., 2007). A data archive of the results of these model simulations is at PCMDI (http://www-pcmdi.llnl.gov/).

The state-of-the-art GCMs are located above the Earth System Models of Intermediate Complexity (EMICs) because they represent a larger number of scales (Claussen et al., 2002). Compared with GCMs, the ocean and atmosphere models in EMICs are strongly reduced in the number of scales. For example, the atmospheric model may consist of a quasi-geostrophic or shallow-water model, and the ocean component may be a zonally averaged model. The advantage of EMICs is therefore that they are computationally less demanding than GCMs, and hence many more long-time scale processes, such as land-ice and carbon cycle processes, can be included. Each of the individual component models of EMICs may also be used to study the interaction of a limited number of processes. Such models are usually referred to as Intermediate Complexity Models (ICMs). A prominent example is the Zebiak-Cane model of the El Niño/Southern Oscillation phenomenon (Zebiak and Cane, 1987). In time, the GCMs of today will be the EMICs of the future, and the state-of-the-art GCMs will shift towards the upper right corner in Fig. 6.1.

6.2 Scientific questions and model choice

There is no 'unified' climate model with which one is able to tackle any scientific question on the climate system, and there will never be. Model choice is therefore tightly coupled to (i) the specific question one wants to answer and (ii) practical limitations, such as the computational platform that is available.

For example, regarding the El Niño/Southern Oscillation (ENSO) phenomenon, important scientific questions include the following. Which processes are responsible for the dominant ENSO time scale (about 4 years)? Can we predict the development of an ENSO over a certain time period, say a few months? How does ENSO affect the weather over the globe? Will the dominant period and amplitude of ENSO change under an increase of the global mean surface temperature?

How do climate scientists, motivated by one or more of these questions, make a model choice? For example, a conceptual model (implemented on a PC) may be very successfully used to understand particular aspects of ENSO (i.e., the dominant time scale), but it may be useless for an adequate quantitative prediction. However, a GCM (implemented on a supercomputer) may be fully capable of simulating ENSO under changing external conditions, but the analysis of the output may not lead to a detailed understanding of the dominant time scale of variability.

Let us restrict the discussion here to questions related to the understanding of processes (physical mechanisms) giving rise to a particular phenomenon. Such questions are usually about consistent relations between dependent quantities that eventually lead to a theory of the phenomenon, for example, a theory of ENSO. By choosing a certain model out of the hierarchy presented in the previous subsection, one is limited to the range of questions that can be 'usefully' posed. Or the other way around: with a specific question as target, not all models will be suited to answer it. Of course, one is tempted to answer questions with results of models that are not suited for their task, but this usually leads to confusion rather than understanding.

Obviously, observations are crucial for theory development, as otherwise there is no way to falsify the proposed theories. However, theories based only on observations do not have to be consistent with the underlying conservation laws that must be satisfied in the system. Models are therefore always needed to interpret the observations, and the whole model hierarchy (from conceptual models to GCMs) contributes to this interpretation. With the conceptual models, where the causal chains are precisely known, specific relations between quantities can be discovered, whereas with the GCMs, the degree of realism (the quality of model-observation agreement) can be improved.

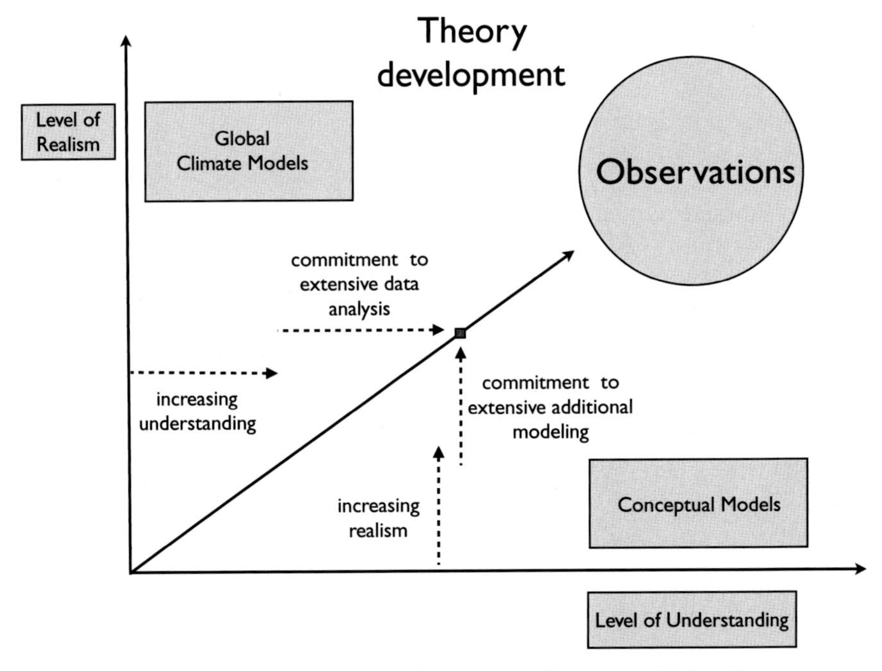

Figure 6.5 Model usage for theory development on phenomena in the climate system.

When one starts with a model high up in the hierarchy, it may be difficult to connect the description of causal chains to well-understood 'building blocks' as present in conceptual models. In this case, there must be a commitment to in-depth analysis of the model results. When starting with a model low in the hierarchy, the connection to observations will be difficult. Hence there must be a commitment to demonstrate that the processes identified in the conceptual model are also dominant in models higher up in the hierarchy. Theories then develop in stages and towards the interpretation of the observations, using successively better models of which the behavior is better understood (Fig. 6.5).

6.3 Deterministic versus stochastic models

In principle, we know the equations governing the climate system's development of the physical and chemical processes. In other words, no new physical 'laws' have to be discovered. However, because of the large-scale differences between processes in the climate system, we are unfortunately not able to use these equations. Consider, for example, the flow of salt water in an ocean basin such as the Atlantic. The flow is governed by the Navier-Stokes equations. The horizontal spatial scale is typically $L = 1{,}000$ km and the vertical dimension $D = 1$ km, whereas the smallest (Kolmogorov) scale is typically $l = 0.1$ m. The ratio of scales $L/l = 10^7$ in only one horizontal dimension, 10^{14} in the horizontal plane and the ratio of scales in the vertical $D/l = 10^4$. Hence one needs at least 10^{18} grid points to resolve the smallest scale processes in the ocean. At the moment there is no computer that can even store one state vector of a solution to these equations.

This effectively means that the governing equations of the physical processes are not known and have to be developed for each problem. An additional complication is that when including biological processes in the marine and/or terrestrial biosphere in climate models, we lack a first-principles description guideline (such as we have for the physical processes), and many of the governing equations are semiempirical. Again, the questions asked and problems targeted strongly determine which model is useful. In addition, should we opt for a stochastic description or for a deterministic approach?

6.3.1 Deterministic modelling approach

When we are restricted to questions about understanding, a deterministic approach will be appropriate if the (parameterized) small-scale processes are not crucially involved in the phenomenon. In many cases later in this book, small-case processes are, for example, represented as friction or diffusion, which are needed for consistency (closing balances), but these are not the crucial processes controlling the particular climate phenomenon.

One way to understand the processes driving a climate phenomenon using this approach is to start low in the hierarchy of deterministic models, say with a conceptual model. One identifies a basic mechanism consisting of the interaction of certain processes giving rise to the phenomenon. The next step is then to identify characteristics of this mechanism, so-called mechanistic indicators, which can be used to monitor whether this mechanism is present in models up in the model hierarchy. It is then crucial that the studies up to a model in the top of the hierarchy are really carried out and the properties of the mechanistic indicators investigated. When indeed the mechanism proposed from the conceptual model can be traced by its characteristics in a GCM and the latter provides an adequate simulation of observed phenomena, then we proceed in the right direction towards a theory of the phenomenon (Fig. 6.5). In the following chapters of the book, examples of this approach are given.

6.3.2 Stochastic modelling approach

When a stochastic approach is followed, it is assumed that the small-scale processes (at least the statistics of these processes) are crucial for the understanding of the phenomenon. In this case, a statistical model of the small-scale processes has to be explicitly provided.

The approach followed by Hasselmann (1976) starts in abstract form from the following set of equations:

$$\frac{d\mathbf{x}}{dt} = \mathbf{f}(\mathbf{x}, \mathbf{y}), \tag{6.6a}$$

$$\frac{d\mathbf{y}}{dt} = \mathbf{g}(\mathbf{x}, \mathbf{y}), \tag{6.6b}$$

where **f** and **g** are given vector functions of the 'slow' (climate) variable **y** and the 'fast' (weather) variable **x**. For example, the variable **x** can be associated with baroclinic instability in the atmosphere with a characteristic response time scale τ_x of less than a few days. In Example 6.3, these would be processes affecting T_a. The slow variable **y** is that associated with the long-term development of climate, say on a typical time scale τ_y; in Example 6.3, this is the sea-surface temperature T.

We want to know the statistics of the variable **y** under the condition of a time scale separation, that is, $\tau_x \ll \tau_y$. Let $\tilde{\mathbf{y}} = \mathbf{y} - \mathbf{y}_0$, where $\mathbf{y}_0$ is the initial condition for **y**. If $t \ll \tau_y$, then **y** can be assumed constant in the right-hand side of (6.6b) and hence

$$\frac{d\tilde{\mathbf{y}}}{dt} \approx \mathbf{g}(\mathbf{x}). \tag{6.7}$$

Let us now consider an ensemble of realizations over the short time scale τ_x, then by taking the ensemble average $\langle\ \rangle$ of (6.7), we find

$$\langle\tilde{\mathbf{y}}\rangle \approx \langle\mathbf{g}(\mathbf{x})\rangle t, \tag{6.8}$$

and hence for the fluctuations $\mathbf{y}' = \tilde{\mathbf{y}} - \langle\tilde{\mathbf{y}}\rangle$, it follows that

$$\frac{d\mathbf{y}'}{dt} \approx \mathbf{g}(\mathbf{x}) - \langle\mathbf{g}(\mathbf{x})\rangle = \mathbf{g}'(\mathbf{x}), \tag{6.9}$$

with $\langle\mathbf{g}'(\mathbf{x})\rangle = 0$. Hence on the short time scales $\mathbf{y}'$ will fluctuate rapidly compared to **y** and can be represented by a stochastic process.

Example 6.3 Red noise spectrum of SST A prominent example of stochastic modelling is the explanation of the red-noise background in the spectrum of sea-surface temperature (SST) anomalies in observations (Hasselmann, 1976). To explain the spectrum of SST, the response of the oceanic mixed layer temperature T to an atmospheric heat flux Q_{oa} is considered. If we assume that the mixed layer depth h is fixed, the governing equation for T is

$$\rho C_p h \frac{dT}{dt} = Q_{oa}, \tag{6.10}$$

where ρ and C_p are the constant density and heat capacity of the ocean water, respectively (Fig. 6.6).

The heat flux can be approximated as

$$Q_{oa} = \mu(T_a - T), \tag{6.11}$$

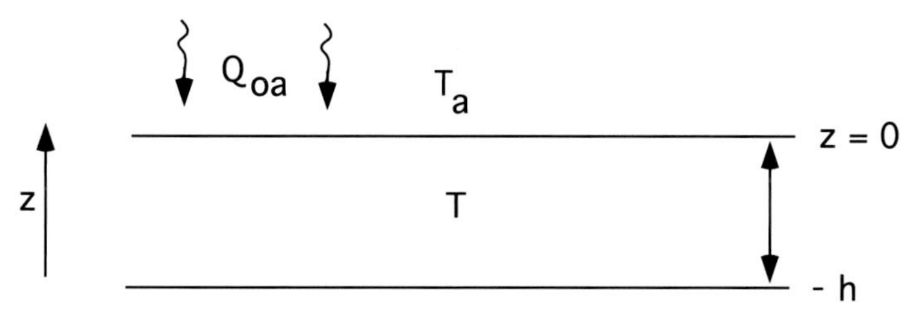

Figure 6.6 Sketch of an upper ocean mixed layer forced by a heat flux Q_{oa}. The mixed-layer depth is indicated by h.

where μ ($Wm^{-2}K^{-1}$) is a heat exchange coefficient and T_a the atmospheric temperature just above the ocean surface. Combining (6.10) and (6.11), we obtain

$$\frac{dT}{dt} = -\gamma T + \gamma T_a \; ; \; \gamma = \frac{\mu}{\rho C_p h}. \tag{6.12}$$

In general, the forcing part of this equation (here represented by T_a) will be very irregular (have energy in a wide range of frequencies), which can be represented by white additive noise.

This implies that we can write the temperature equation as the Itô SDE of the form

$$dT_t = -\gamma T \; dt + \sigma \; dW_t, \tag{6.13}$$

where σ represents the amplitude of the noise.

The spectrum of T follows directly from the solution of (6.13), which was given in (3.49) as (with initial condition T_0)

$$T_t = e^{-\gamma t} \left[T_0 + \sigma \int_0^t e^{\gamma x} \; dW_x \right], \tag{6.14}$$

with variance given by (3.57) as

$$Var(T_t) = e^{-2\gamma t} \left(Var(T_0) + \sigma^2 \frac{e^{2\gamma t} - 1}{2\gamma} \right) \tag{6.15}$$

and a spectrum

$$\hat{f}(\omega) = \frac{\sigma^2}{\gamma^2 + \omega^2}. \tag{6.16}$$

For high frequencies, $\omega \to \infty$, this spectrum decays as ω^{-2}, but for low frequencies $\omega \to 0$, it approaches a constant. These features of the spectrum are also seen in observations of SST anomalies (Dommenget and Latif, 2002). Hence when the 'noise' is represented as being temporally uncorrelated ('white'), the response shows higher energy in the low frequencies than in the high frequencies (hence it is 'red').

In a way, the slow dynamical (ocean) component is integrating the high-frequency (atmospheric) noise. ■

From Example 6.3, it is clear that it is important to have an appropriate statistical model of the small-scale processes available. Whether to use an Itô or Stratonovich formulation is less crucial. As the representations of small-scale processes come from physics for which classical rules of calculus should be valid, a Stratonovich formulation is a priori preferred. However, as shown in Chapter 3, an equivalent Itô formulation can always be found from this Stratonovich formulation.

6.4 Model error

Apart from many conceptual models, a deterministic climate model (ICM, EMIC, GCM) consists of a set of partial differential equations (representing the basic balances of momentum, energy, etc.) that can always be formulated into the operator form

$$\mathcal{M}\frac{\partial \mathbf{u}}{\partial t} + \mathcal{L}\mathbf{u} + \mathcal{N}(\mathbf{u}) = \mathcal{F}, \tag{6.17}$$

where $\mathcal{L}, \mathcal{M}$ are linear operators, $\mathcal{N}$ is a nonlinear operator, $\mathbf{u}$ is the vector of dependent quantities and $\mathcal{F}$ contains the forcing of the system. To get a well-posed problem, appropriate boundary and initial conditions have to be added to this set of equations.

Under any numerical method, the discretised equations can be written as a nonlinear system of ODEs (in many cases with algebraic constraints), which has the form

$$\mathcal{M}_N\frac{\partial \mathbf{x}}{\partial t} + \mathcal{L}_N\mathbf{x} + \mathcal{N}_N(\mathbf{x}) = \mathcal{F}_N, \tag{6.18}$$

where $\mathbf{x}$ indicates the total N-dimensional vector of unknowns. The operators depend on parameters, and their subscript N indicates that they are discrete equivalents of the continuous operators.

To systematically discuss the errors of such a numerical climate model, it is instructive to start with the situation that the discretized equations (6.18) model a laboratory fluid experiment. The properties of the fluid such a viscosity and thermal diffusivity are accurately known, and we can use Navier-Stokes and/or Boussinesq equations to model the flow. The discretised equations describing the evolution of the state vector $\mathbf{x}$ are then known. In this case, the model errors will consist of two components:

(i) *Discretisation error*. Suppose we use a uniform grid size Δx and a time step Δt in a finite-difference method using direct numerical simulation. When the method is, for example, second order in time and space (such as a central difference Adams-Bashforth method), then there is a discretisation error $\mathcal{O}(\Delta x^2, \Delta t^2)$, as we are basically solving slightly different equations than the continuous equations. We can choose Δx and Δt to achieve a desired accuracy of the solutions.

(ii) *Context error.* There are inaccuracies in representing the physical situation such as the dimensions of the problem, the forcing functions, the liquid properties, the influence of background laboratory conditions and so forth. These context errors enter the equations through uncertainties in the parameters occurring in the operators and the forcing $\mathcal{F}_N$.

When all the scales of the flow can be captured by the numerical method (i.e., Δx and Δt can be chosen sufficiently small), very accurate solutions can be obtained, and hence a close correspondence with observed phenomena can be achieved. In this case, one can develop benchmark problems where different numerical approaches can be compared and their convergence to the 'true' solution can be tested. In many fluid laboratory experiments, however, the flow is turbulent, and not all relevant scales can be captured. There are then two options: the scales are resolved by choosing a certain Δx (e.g., the direct numerical simulation [DNS] method) or a separate model of the behavior of the small scales is used (e.g., by using a large eddy simulation [LES] method). In either case, an additional error appears, which may be called a

(iii) *Scale representation error.* Part of the turbulent cascading process in fluids is not represented, which may be crucial for certain phenomena. We may also misrepresent specific instability phenomena as the particular scales are not resolved. This scale representation essentially differs from the discretisation error, as the latter error is only connected to the numerical representation of the actual equations used.

With a scale-representation error, not all equations for the complete state vector $\mathbf{x}$ in (6.18) are known. Hence $\mathbf{x}$ is split into $\mathbf{x} = \mathbf{x}_r + \mathbf{x}_u$, where the subscripts r and u indicate resolved and unresolved, respectively. In DNS the equations for $\mathbf{x}_u$ are simply omitted, whereas in LES, a separate model (a parameterization) is used, which results in $\mathbf{x}_u = P(\mathbf{x}_r)$.

In climate models the three error types (i)–(iii) above are all present. The context error is, for example, associated with uncertainties in the forcing of the model, which may arise from uncertain observations. The scale representation error appears in many ocean and atmosphere models due to a lack of sufficient horizontal and vertical resolution. In many GCMs and EMICs, for example, the effects of ocean eddies on the transport of heat and salt are represented by the Gent-McWilliams parameterization (Gent and McWilliams, 1990).

However, in climate models there is an additional error related to the representation of processes, that is,

(iv) *Process representation error.* Relevant processes for the particular phenomenon are neglected or have an inadequate representation; for example, there is no representation based on first principles or not even based on semiempirical studies.

In this case either the state vector **x** may be too limited or the operators contain errors. An example is the omission of processes controlling the carbon cycle in GCMs used in IPCC-AR4, such as those in the marine and terrestrial biosphere.

Suppose we have decided on our model type, with appropriate parameterizations, fixed parameters and chosen forcing functions. Even when realising that such a model contains severe idealisations and has all types [(i)–(iv)] of errors, there are still important a priori checks that can be done before using such a model.

(1) *Consistency*. When the continuous formulation is chosen (with all its idealizations and errors) and the discretisation method is chosen, it is very important to check that the grid and time step used provide sufficiently accurate results to these equations. Hence for fixed parameters, it is always necessary to perform calculations with different grid sizes and time steps to investigate whether one solves the 'correct' equations (the equations one intends to solve).
(2) *Integral balances*. For every model, one can integrate the continuous equations used over the domain and obtain usually much more simplified integral equations, representing overall conservation of mass/energy and heat/salt components. For a solution of the discrete model, the discrete integral balances should be satisfied up to a certain accuracy, say a relative error (with respect to the largest term) of a few percentages. Numerical solutions where mass, energy and so forth have large artificial sources or sinks in the system are meaningless.

With this overview of the hierarchy of models and their error sources, we are now ready to tackle several important problems in climate physics using dynamical systems theory as will be done in the next chapters.

7
The North Atlantic Oscillation

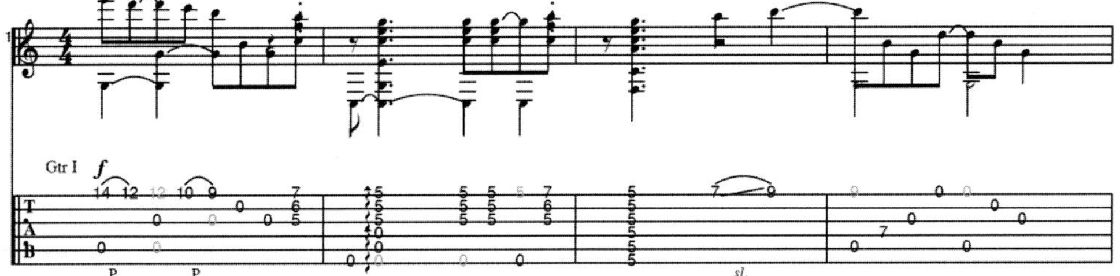

Timing is strange, but the pattern is clear.
CGEGBD, The Unexpected Visitor, Michael Hedges

As a first application of stochastic dynamical systems theory to understand phenomena of climate variability, we consider midlatitude atmospheric variability with a focus on the North Atlantic.

7.1 Midlatitude atmospheric variability

The midlatitude atmosphere varies on many time scales and spatial scales, and from daily weather maps we see the high and low pressure systems moving seemingly irregularly from west to east. When longer time scales than the synoptic weather time scale (1–10 days) are considered, it appears that some patterns of variability appear more frequently than others. Variability on time scales larger than 10 days is referred to as low-frequency variability (LFV). The patterns of this LFV exert an organizing tendency on the weather and hence may be important for weather predictability on time scales longer than 10 days.

7.1.1 Low-frequency variability

The total root mean square (rms) Northern Hemispheric (NH) 500 hPa geopotential height variance (in m) during the winter season over the years 1958–1999 is plotted in Fig. 7.1a. The LFV component of this variability obtained by a low-pass (>10 days) filter (Fig. 7.1b) represents a substantial component that is about a factor 3 larger (Fig. 7.1c) than the variability on the synoptic scale (2.5–6 days). The maxima in the LFV component (Fig. 7.1b) are almost equal in amplitude to those in total variance (Fig. 7.1a) and almost precisely collocated with them. Two of these winter maxima occur over the northeastern part of the North Pacific and North Atlantic, the third over the Siberian Arctic. The maxima in the 'weather band' (Fig. 7.1c) occur slightly upstream of the LFV maxima, mainly over the storm tracks off the east coasts of North America and Asia.

There are a large number of studies on the patterns of LFV using many different methods and data sets. These patterns are determined to be related to two features: (i) intraseasonal oscillations and (ii) planetary flow (or weather) regimes (Ghil and Robertson, 2002). Intraseasonal oscillations are specific propagating patterns of variability with time scales in the range of 30–70 days. The planetary flow regimes are patterns that explain high variance and are classified according to hemispheric regimes (Section 7.1.3) or regional regimes (Section 7.1.4).

7.1.2 Intraseasonal oscillations

Signatures of intraseasonal variability were found (Dickey et al., 1991) in data of the global atmospheric angular momentum (AAM). Variations in global AAM and in the length of day are highly correlated with each other on intraseasonal time scales. Essentially, the Earth-atmosphere system is closed with respect to angular momentum exchanges on this time scale, except for the well-known tidal effects of the Sun and Moon, which can be easily computed and eliminated. When the strength of the midlatitude westerly winds increases or that of the tropical trade winds decreases, the solid earth slows down in its rotation, and the length of day increases. A power spectrum of the observed AAM variance for the NH extratropics (26°N–90°N), see, for example, Fig. 7a in Dickey et al. (1991), indicates a peak at about 40 days, which is largely associated with fluctuations in the strength of the midlatitude westerlies.

Patterns of a 30- to 35-day oscillation in the NH winter have been found in 700 hPa geopotential height data using M-SSA (cf. Section 5.3.3) analysis. Phases of the oscillation are plotted in Fig. 7.2, where it can be seen that the dipole-type pressure anomalies with a dominant amplitude in the Atlantic region propagate westwards.

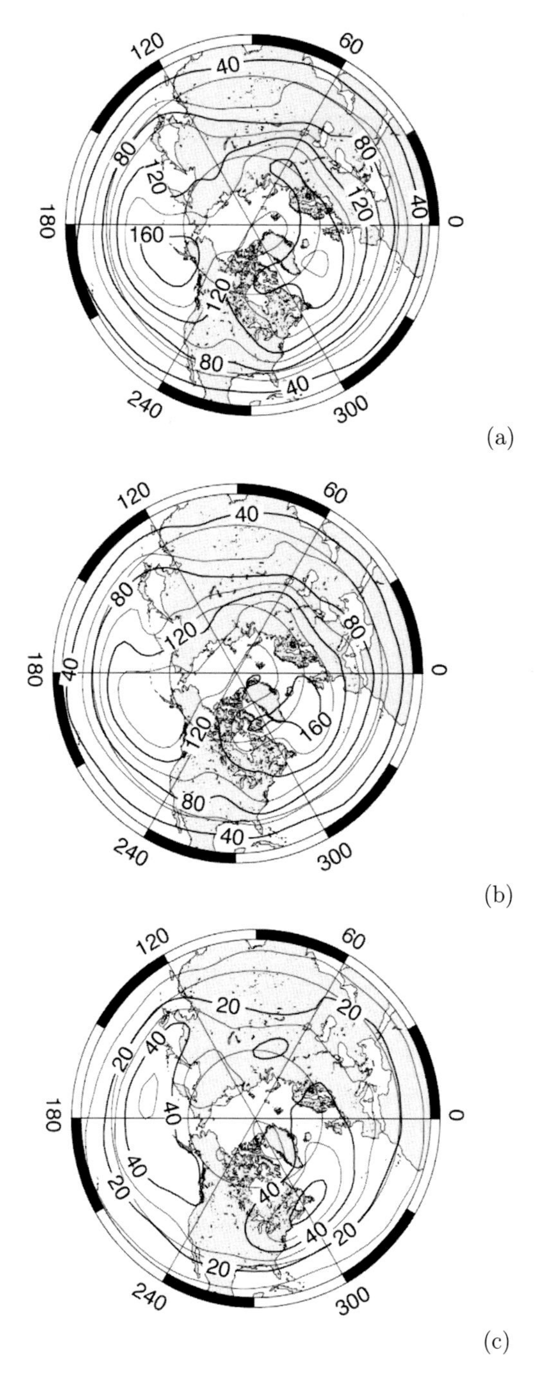

Figure 7.1 Maps of Northern Hemisphere 500-hPa geopotential height variance (shown as standard deviation (rms) at each point) for the winter months (December, January, and February) and the years 1958–1999, computed from NCEP/NCAR reanalysis data (Kalnay et al., 1996). (a) Total, contour interval 20 m; (b) low-pass (>10 days) filtered, contour interval 20 m, and (c) band-pass (2.5–6 days) filtered, contour interval 10 m (figure taken from Ghil and Robertson, 2002).

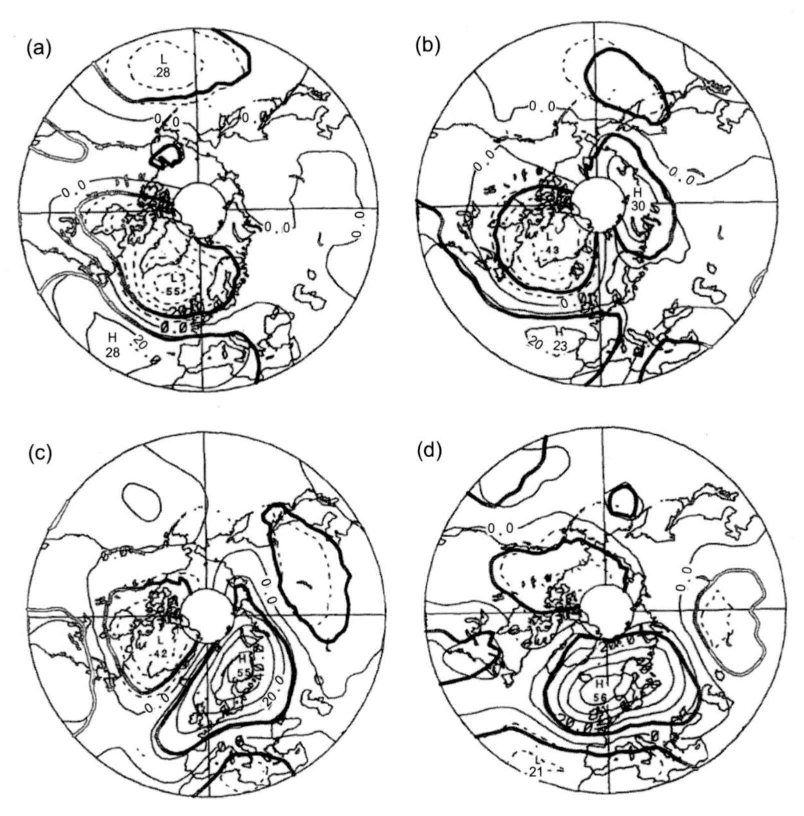

Figure 7.2 Different phases [time increases from (a) to (d)] of the pattern of 700 hPa geopotential height variability (dominant M-SSA mode, units in m) of the 30- to 35-day oscillation (figure from Plaut and Vautard [1994]).

In the same analysis (Plaut and Vautard, 1994), also intraseasonal oscillations with dominant periods of 40–45 days and 70 days are found. The 70-day oscillation consists of fluctuations in both position and amplitude of the Atlantic jet stream, with poleward-propagating anomaly patterns. The 40- to 45-day oscillation is specific to the Pacific sector.

7.1.3 Hemispheric weather regimes

The existence of hemispheric regimes is motivated by results (Wallace and Gutzler, 1981; Cheng and Wallace, 1993; Smyth et al., 1999) that suggest that circulation patterns with hemispheric coherence do exist or that regionally confined ones can be identified from hemispheric data. The hemispheric coherence may reflect a fundamental dynamical mode of the atmosphere, the Northern Annular Mode (NAM) (Thompson and Wallace, 2000; Thompson et al., 2000), also called the Arctic Oscillation, as shown in Fig. 7.3.

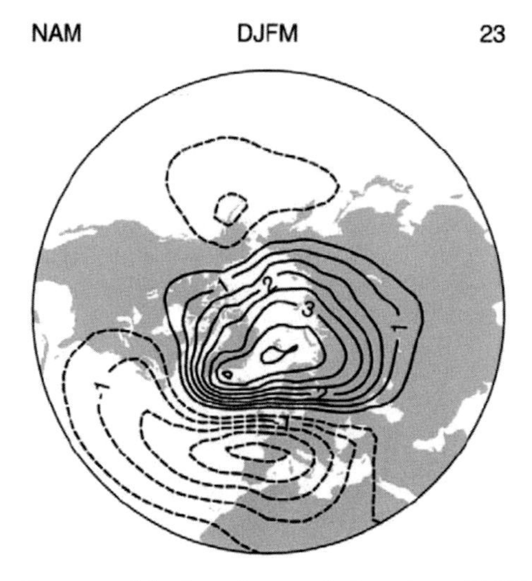

Figure 7.3 (a) Leading empirical orthogonal function (EOF 1) of the winter (December–March) mean sea level pressure (1899–2006) anomalies over the Northern Hemisphere (20°N–90°N) and the percentage of the total variance it explains. The pattern is displayed in terms of amplitude (hPa, contour increment 0.5 hPa), obtained by regressing the hemispheric sea level pressure anomalies on the leading principal component time series, plotted in Fig. 7.4b (figure from Hurrell et al., 2003).

The NAM pattern is the leading EOF of nonseasonal sea level pressure variations north of 20°N latitude, and it is characterized by pressure anomalies of one sign in the Arctic with the opposite anomalies centered about 37°N–45°N. The NAM is related to the degree to which Arctic air masses penetrate into midlatitudes. When the surface pressure anomaly is low in the polar region (positive NAM index), the midlatitude zonal winds are strong and cold Arctic air is locked in the polar region. When the NAM index is negative, there is relatively high pressure in the polar region with weaker zonal winds at midlatitudes and more movement of Arctic air southward. However, the NAM may also be a mere coincidence of separate regional patterns (Ambaum et al., 2001) that are discussed in the next section.

7.1.4 Regional weather regimes: Atlantic LFV

The regional classifications are motivated by evidence that the strongest patterns of Northern Hemispheric LFV are confined to either the Pacific/North American or the Atlantic-Eurasian sectors. When monthly mean anomalies of sea level pressure data over the North Atlantic region (20°N–70°N; 90°W–40°E) are analysed for dominant patterns of variability, the first EOF (explaining 36% of the variance) for the DJFM season shows a pattern as in Fig. 7.4a. This is the pattern of the North Atlantic Oscillation (NAO), displayed here in terms of amplitude, obtained by regressing the hemispheric sea level pressure anomalies on the leading principal component (PC)

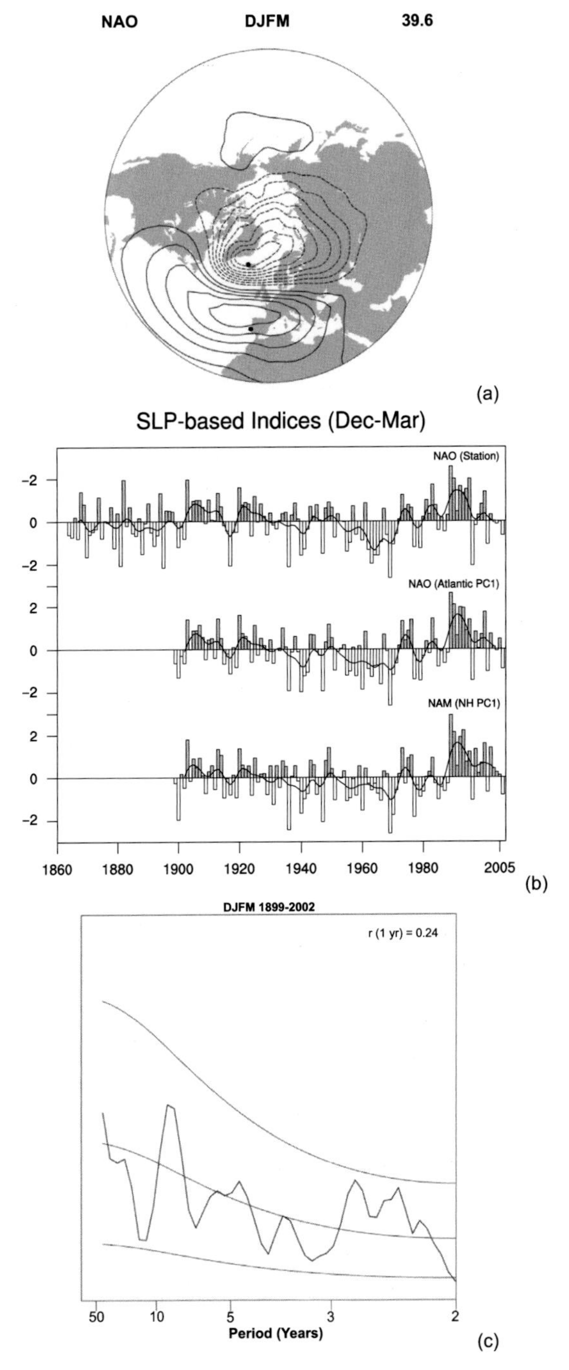

Figure 7.4 (a) Leading EOF of the DJFM mean sea level pressure anomalies (hPa, contour increment 0.5 hPa) in the North Atlantic sector (20°N–70°N; 90°W–40°E) over the years 1899–2001. (b) Normalised indices of the mean winter (December–March) NAO constructed from SLP data. Top panel: the difference of normalised sea level pressure between Lisbon and Stykkisholmur [dots in (a)]. Middle panel: the PC time series of the leading EOF as presented in (a). Lower panel: PC time series of the NAM. (c) Power spectrum of time series in the middle panel of (b) with the 5% and 95% confidence limits of the corresponding red noise spectrum; the lag-one autocorrelation coefficient is 0.24 (figure from Hurrell et al., 2003).

time series from the Atlantic domain, the NAO index. The centres with highest amplitude are located near Stykkisholmur (Iceland) and Lisbon (Portugal).

The winter-mean NAO index (Hurrell, 1995) based on station data (Lisbon-Stykkisholmur) is shown in the upper curve in Fig. 7.4b. The average winter sea level pressure data at each station were normalised by division of each seasonal pressure by the long-term mean (1864–1983) standard deviation. The station-based index for the winter season agrees well with PC1 of Atlantic-sector sea level pressure anomalies (middle curve). The correlation coefficient between the two is 0.92 over the common period 1899–2002, indicating that the station-based index adequately represents the time variability of the winter-mean NAO spatial pattern. Moreover, it correlates well ($r = 0.95$) with PC1 of the NAM (Thompson and Wallace, 2000). These results emphasise that the NAO and NAM reflect essentially the same mode of tropospheric variability. The spectrum in (c) of the winter-mean NAO index (middle panel of [b]) is slightly red, and the major conclusion is that there is no preferred time scale of NAO variability. However, the spectral peak between 6 and 9 years may be important, as time series of climate variables related to the NAO do show peaks in the decadal range of the spectrum (Da Costa and Colin de Verdiere, 2004).

During the winter season, the NAO accounts for more than one-third of the total variance in sea level pressure over the North Atlantic. In the so-called positive phase, higher-than-normal surface pressures south of 55°N combine with a broad region of anomalously low pressure throughout the Arctic to enhance the climatological meridional pressure gradient. The largest amplitude anomalies occur in the vicinity of Iceland and across the Iberian Peninsula. The positive phase of the NAO is associated with stronger-than-average surface westerlies across the middle latitudes of the Atlantic onto Europe, with anomalous southerly flow over the eastern United States and anomalous northerly flow across the Canadian Arctic and the Mediterranean (Hurrell and Deser, 2010).

In summary, the NAO is an important phenomenon in the NH winter climate. Although there appears to be no preferred time scale of variability and the time series in Fig. 7.4b are too short to demonstrate significant decadal-interdecadal variability, there is definitely a preferred pattern in the Atlantic SLP anomalies, which also shows up at synoptical time scale, from a few days to several weeks. One of the most intriguing questions arising from the results is which physical processes are responsible for the NAO and hence control the low-frequency variability in the North Atlantic midlatitude winter atmosphere.

7.2 Minimal model

It is remarkable that a barotropic quasi-geostrophic model, with appropriate boundary conditions (orography) when properly forced, is able to display one of the main properties of the NAO variability, that is, its spatial pattern. The dimensionless equations for this model can be written in streamfunction ψ and relative vorticity ζ (vertical component of the vorticity vector) formulation as (Selten, 1995)

$$\frac{\partial \zeta}{\partial t} + \mathcal{J}(\psi, \zeta + f + h) = F + D(\zeta), \tag{7.1a}$$

$$\zeta = \nabla^2 \psi. \tag{7.1b}$$

The equations have been nondimensionalised using the radius of the Earth as unit of length and the inverse of its angular velocity as unit of time. The quantity $\mathcal{J}(a, b)$ is the Jacobi operator, which, with longitude ϕ and latitude θ, is given by (with $\mu = \sin\theta$)

$$\mathcal{J}(a, b) = \frac{\partial a}{\partial \phi}\frac{\partial b}{\partial \mu} - \frac{\partial a}{\partial \mu}\frac{\partial b}{\partial \phi}. \tag{7.2}$$

Furthermore, F represents the forcing of the flow, D represents the friction (or damping) of the flow, $f = \sin\theta$ is the Coriolis parameter and h represents a prescribed orography.

In Selten (1995), realistic orography h is used and the damping is formulated as

$$D(\zeta) = k_1\zeta + k_2\nabla^3\zeta, \tag{7.3}$$

where k_1 measures the strength of the Ekman damping and k_2 that of a scale-selective damping. The forcing is derived from the 500-hPa winter mean values of streamfunction ψ_{cl} and vorticity ζ_{cl} and a component of the transient eddy forcing based on 10-day running mean (winter) anomalies ψ' and ζ', that is,

$$F = \mathcal{J}(\psi_{cl}, \zeta_{cl} + f + h) - D(\zeta_{cl}) + \overline{\mathcal{J}(\psi', \zeta')}, \tag{7.4}$$

where the overbar indicates a time average; hence the forcing is time-independent.

Spectral truncations of this model are often used (Selten, 1995) where the equation (7.1) in spherical coordinates is projected onto spherical harmonics and triangularly truncated at wave number 21. More specifically, ψ is expanded as

$$\psi(\phi, \mu, t) = \sum_{n=1}^{N} \sum_{m=-n}^{n} \psi_{m,n}(t) Y_{n,m}(\phi, \mu), \tag{7.5}$$

where $m + n$ is odd in the inner summation and the $Y_{n,m}$ are the spherical harmonics. The latter are meridionally bounded eigenfunctions of the problem

$$\nabla^2 Y_{n,m} = -n(n+1)\, Y_{n,m}, \tag{7.6}$$

with periodic boundary conditions in zonal direction. The expansion (7.5) is substituted into the equation (7.1), and the residual is projected on the functions $Y_{n,m}$. This gives a set of ordinary differential equations for the coefficients $\psi_{n,m}$.

In Legras and Ghil (1985), a spectrally truncated model with $N = 9$ in (7.5) of the spherical barotropic vorticity equation is studied. The results of this model are

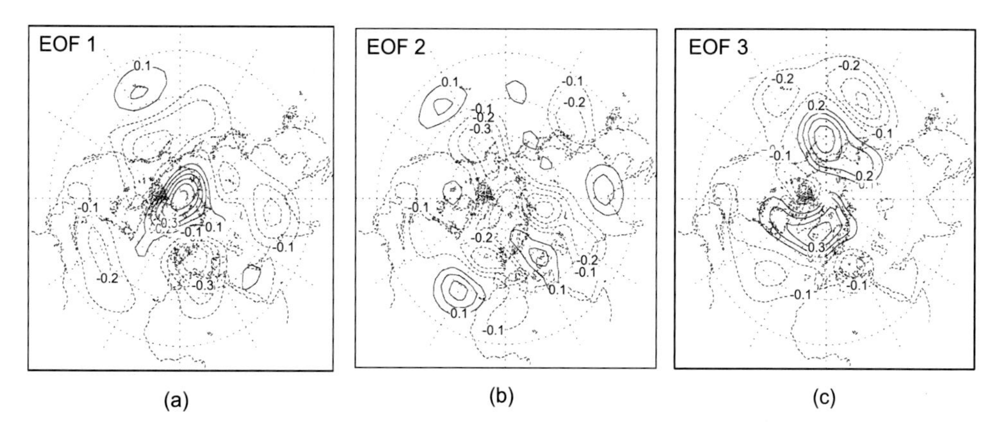

Figure 7.5 (a–c) EOF1–EOF3 calculated from a 200-yr data set produced by the T21 barotropic model. The kinetic energy norm was used for the calculation of the EOF as a measure of the variance of ψ (figure from Crommelin, 2003).

extensively discussed in Sections 6.4 and 6.5 of Ghil and Childress (1987), and particular aspects of the behavior are further investigated in Jin and Ghil (1990) and Strong et al. (1995).

In Crommelin (2003), several 200-year simulations were performed with the so-called T21 ($N = 21$ in [7.5]) version of the model of Selten (1995). The damping coefficient k_1 in (7.3) corresponds to a damping time scale of 15 days, and the scale selective damping k_2 is such that largest wave numbers are damped on a time scale of 3 days. The patterns of the first three EOFs of ψ are shown in Fig. 7.5, which explain 18.3%, 9.8% and 6.8% of the total kinetic energy in the flow. The first EOF has a similar pattern as that of the Northern Annular Mode (or Arctic Oscillation) associated with the strength of the polar vortex. The second EOF shows successive positive and negative anomalies over Western Europe. The third EOF shows a strong dipole character over the North Atlantic and resembles the pattern of the NAO.

A limited view of the probability density function (PDF) was obtained by projecting the full data set onto the PC1-PC2 plane. The result is shown as the contour plot in Fig. 7.6a, which reveals a bimodal structure with two maxima MAX1 and MAX2. The 500-hPa geopotential height anomaly patterns Z_{500} are shown in Fig. 7.6c–d. The flow pattern of MAX1 represents a situation of intensified zonal flow, and the pattern of MAX2 shows a blocked flow over Europe. The PDF-maxima MAX1 and MAX2 can be interpreted as flow regimes and the transitions between them (Fig. 7.6b) as so-called vacillatory behavior.

7.3 Variability in conceptual models

To answer the question of why the PDF in the previous section is bimodal and what processes drive the transitions between the states MAX1 and MAX2, so-called

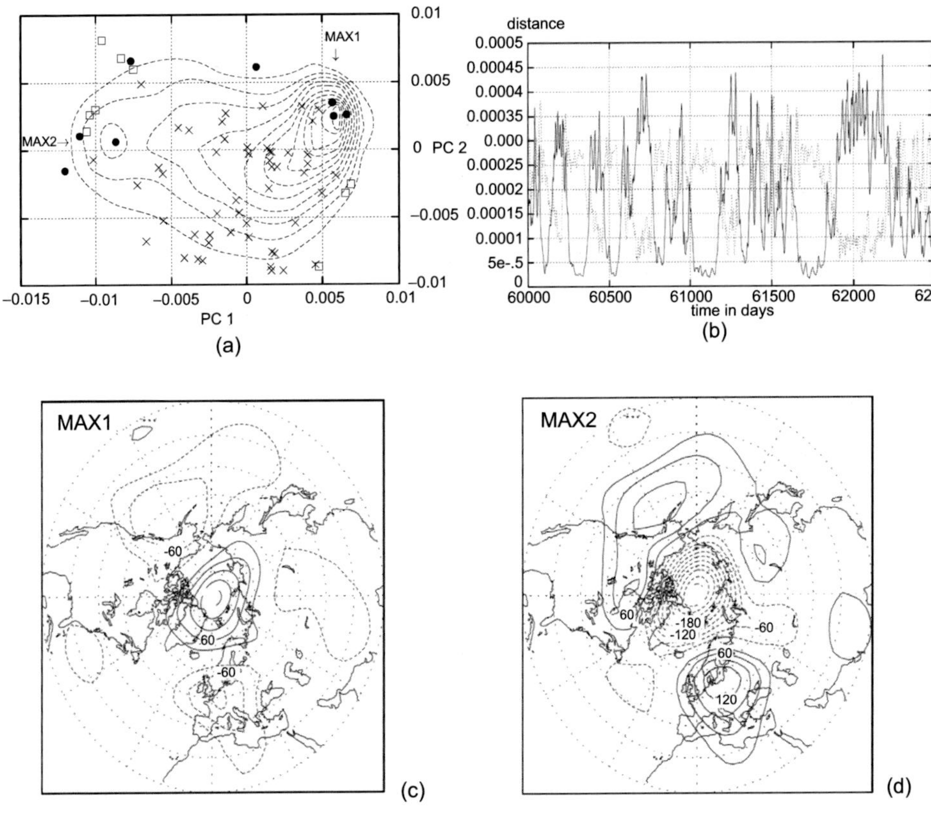

Figure 7.6 (a) Contour plot of the probability density function in the PC1-PC2 space. (b) Distances to two maxima MAX1 (solid) and MAX2 (dotted) during a data segment of 2,500 days. Distances are given as differences in turbulent kinetic energy; horizontal axis denotes time in days. (c–d) Patterns of the Z_{500} anomalies (contour interval 30 m) corresponding to the maxima of the PDF (figure from Crommelin, 2003).

reduced models have been used. Although the number of degrees of freedom d of the T21 truncated model in the previous section is not large ($d = 231$), an even further reduction allows for an easier mechanistic analysis (De Swart, 1989).

7.3.1 A six-mode model

In many studies with reduced models, the geometry is idealized to a zonal channel of width πbL and length $2\pi L$ on a midlatitude β plane with Coriolis parameter $f = f_0 + \beta_0 y$. In Cartesian coordinates, the dimensional equations are

$$\frac{\partial \zeta}{\partial t} + \mathcal{J}\left(\psi, \zeta + \beta_0 y + \frac{f_0}{H}h\right) = F + D(\zeta), \tag{7.7a}$$

$$\zeta = \nabla^2 \psi, \tag{7.7b}$$

where H is the thickness of the layer and

$$\mathcal{J}(a,b) = \frac{\partial a}{\partial x}\frac{\partial b}{\partial y} - \frac{\partial b}{\partial x}\frac{\partial a}{\partial y}.$$

In Crommelin et al. (2004), the dissipation and forcing are taken as

$$F = k\zeta^* \; ; \; D = -k\zeta, \tag{7.8}$$

where ζ^* is a prescribed function and k represents Ekman damping. When (7.7) is nondimensionalised using scales $1/f_0$ for time, L for length x and y, and $L^2 f_0$ for the streamfunction, f_0 for vorticity and H for topography, the dimensionless equation for ψ becomes

$$N(\psi, h, \psi^*) = \frac{\partial \nabla^2 \psi}{\partial t} + \mathcal{J}(\psi, \nabla^2\psi + \beta y + \gamma h) + C\nabla^2(\psi - \psi^*) = 0 \tag{7.9}$$

with $\beta = \beta_0 L/f_0$ and $C = k/f_0$, and the dimensionless parameter γ is introduced as a scaling factor of the orography. From now on, all dependent quantities are dimensionless in this subsection.

The set of basis functions used to derive the low-order model is chosen as the eigenfunctions of the Laplace equation on the domain $[0, 2\pi] \times [0, \pi b]$, that is,

$$\nabla^2 \phi_{n,m} = -\lambda_{n,m}\, \phi_{n,m}, \tag{7.10}$$

which satisfy periodic boundary conditions on the x direction and $\partial\phi_{n,m}/\partial x = 0$ at the channel walls $y = 0$ and $y = \pi b$. These eigenfunctions, when chosen to be orthonormal, are given by

$$\phi_{0,m}(y) = \sqrt{2}\cos\frac{my}{b}, \tag{7.11a}$$

$$\phi_{n,m}(x,y) = \sqrt{2}e^{inx}\sin\frac{my}{b}. \tag{7.11b}$$

The streamfunction ψ, the topography h and the forcing field ψ^* are now expanded as

$$(\psi, h, \psi^*) = \sum_{n,m}(\psi_{n,m}(t), h_{n,m}(t), \psi^*_{n,m}(t))\phi_{n,m}. \tag{7.12}$$

These expansions are substituted into the equation (7.9) and projected using a Galerkin method, that is, for every n, m

$$\langle N(\psi, h, \psi_*), \psi_{n,m}\rangle = \frac{1}{2\pi^2 b}\int_0^{2\pi}\int_0^{\pi b} N(\psi, h, \psi_*)\psi_{n,m}\, dxdy = 0, \tag{7.13}$$

where $\langle ., .\rangle$ is the inner product and N given by (7.9).

Next, the topography and forcing are chosen as

$$h(x,y) = \cos x\ \sin\frac{y}{b} \; ; \; \psi^*(x,y) = \Phi(y), \tag{7.14}$$

with $\Phi(y)$ a given function. This implies that $h_{1,1} = h_{-1,1} = 1/(2\sqrt{2})$ and all other $h_{n,m}$ are zero. Furthermore, only modes $\phi_{0,1}, \phi_{0,2}, \phi_{1,1}, \phi_{-1,1}, \phi_{1,2}$ and $\phi_{-1,2}$ are considered. With

$$x_1 = \frac{1}{b}\psi_{0,1} \; ; \; x_2 = \frac{1}{\sqrt{2}b}(\psi_{1,1} + \psi_{-1,1}) \; ; \; x_3 = \frac{i}{\sqrt{2}b}(\psi_{1,1} - \psi_{-1,1})$$

$$x_4 = \frac{1}{b}\psi_{0,2} \; ; \; x_5 = \frac{1}{\sqrt{2}b}(\psi_{1,2} + \psi_{-1,2}) \; ; \; x_6 = \frac{i}{\sqrt{2}b}(\psi_{1,2} - \psi_{-1,2}),$$

the projected equations (for $n = -1, 0, 1$, $m = 0, 2$) become

$$\frac{dx_1}{dt} = \tilde{\gamma}_1 x_3 - C(x_1 - x_1^*), \tag{7.15a}$$

$$\frac{dx_2}{dt} = -(\alpha_1 x_1 - \beta_1)x_3 - Cx_2 - \delta_1 x_4 x_6, \tag{7.15b}$$

$$\frac{dx_3}{dt} = (\alpha_1 x_1 - \beta_1)x_2 - \gamma_1 x_1 - Cx_3 + \delta_1 x_4 x_5, \tag{7.15c}$$

$$\frac{dx_4}{dt} = \tilde{\gamma}_2 x_6 - C(x_4 - x_4^*) + \epsilon(x_2 x_6 - x_3 x_5), \tag{7.15d}$$

$$\frac{dx_5}{dt} = -(\alpha_2 x_1 - \beta_2)x_6 - Cx_5 - \delta_2 x_4 x_3, \tag{7.15e}$$

$$\frac{dx_6}{dt} = (\alpha_2 x_1 - \beta_2)x_5 - \gamma_2 x_4 - Cx_6 + \delta_2 x_4 x_3, \tag{7.15f}$$

with chosen coefficients $x_4^* = rx_1^*$ (note that x_1^* and x_4^* are the only two nonzero coefficients when ψ^* is only a function of y) and

$$\alpha_m = \frac{8\sqrt{2}}{\pi}\frac{m^2}{4m^2-1}\frac{b^2+m^2-1}{b^2+m^2} \; ; \; \beta_m = \frac{\beta b^2}{b^2+m^2},$$

$$\gamma_m = \gamma\frac{4m^3}{4m^2-1}\frac{\sqrt{2}b}{\pi(b^2+m^2)} \; ; \; \delta_m = \frac{64\sqrt{2}}{15\pi}\frac{b^2-m^2+1}{b^2+m^2},$$

$$\tilde{\gamma}_m = \gamma\frac{4m}{4m^2-1}\frac{\sqrt{2}b}{\pi} \; ; \; \epsilon = \frac{16\sqrt{2}}{5\pi},$$

for $m = 0, 2$. In these equations, the terms with the δ_m and ϵ describe the nonlinear triad interactions between $\phi_{0,2}$ and the $\phi_{1,1}$ and $\phi_{2,2}$ modes. The terms with β represent the advection of planetary vorticity, the ones with α represent the advection of the modes by the zonal flow and the terms with γ represent the effect of topography.

In Crommelin (2004), the parameters $\beta = 1.25$, $b = 0.5$, $C = 0.1$ and $\gamma = 1.0$ are taken fixed, and either the parameter r or x_1^* is used as a control parameter. For $r = -0.4$, the bifurcation diagram of the model (showing x_1 versus x_1^*) is shown in Fig. 7.7. There are two saddle-node bifurcations (labelled as sn1 and sn2) that induce a regime of multiple steady states. The three equilibria at $x_1^* = 6.0$, for which

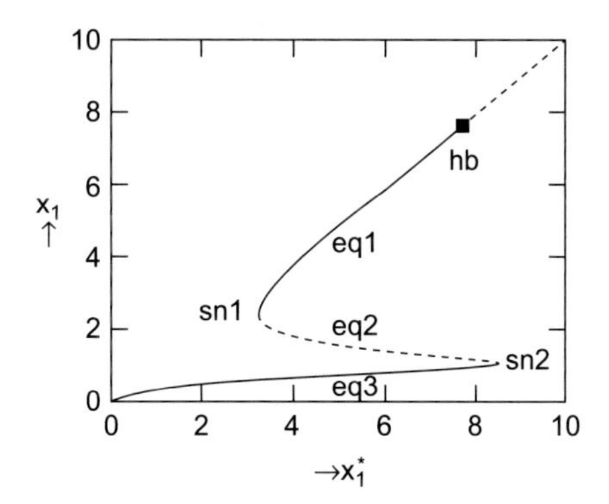

Figure 7.7 Bifurcation diagram showing x_1 versus x_1^* of the reduced atmospheric model (7.15) with $r = -0.4$. The drawn (dashed) curve represents stable (unstable) steady solutions (figure from Crommelin et al., 2004).

streamfunction patterns are shown in Fig. 7.8, are associated with three different ways in which the advection of potential vorticity and the vortex stretching caused by flow over topography can balance.

The stable flow eq1 in Fig. 7.8 is the so-called zonal flow, and the stable equilibrium eq3 in Fig. 7.8 is the blocked flow. The third equilibrium flow (eq2 in Fig. 7.8) is

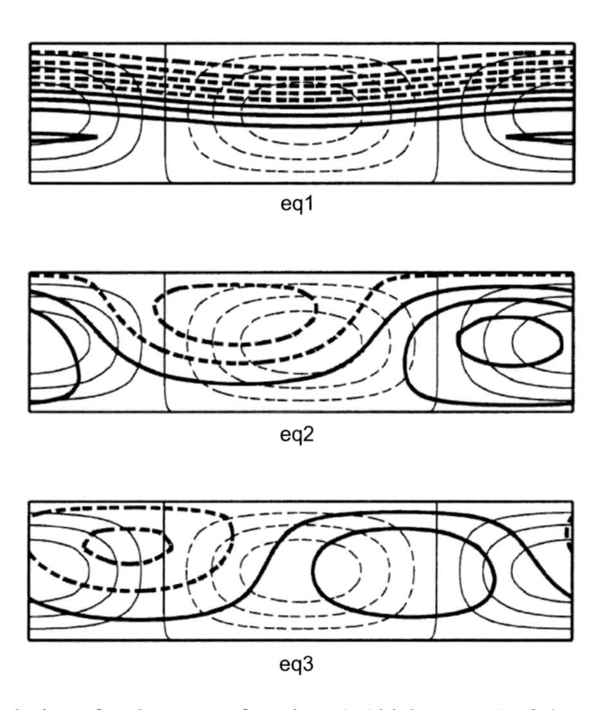

Figure 7.8 Solutions for the streamfunction ψ (thick curves) of the reduced atmospheric model (7.15) with $r = -0.4$ and $x_1^* = 6.0$. The contour levels of the topography are shown as the thinner curves (figure from Crommelin et al., 2004).

unstable (cf. the dashed curve in Fig. 7.7). For larger values of x_1^*, the zonal flow state becomes unstable through a Hopf bifurcation, which, as we learned in Chapter 2, is an oscillatory instability. We now turn to the physical mechanisms of these basic instabilities.

7.3.2 Instability mechanisms

The first issue to explain is the occurrence of the saddle-node bifurcations and hence the multiple equilibrium regime. When the initial conditions are such that $x_4 = x_5 = x_6 = 0$, we see from (7.15) that these amplitudes remain zero if $x_4^* = 0$ $(r = 0)$. The system of equations (7.15) then reduces to

$$\frac{dx_1}{dt} = \tilde{\gamma}_1 x_3 - C(x_1 - x_1^*), \tag{7.16a}$$

$$\frac{dx_2}{dt} = -(\alpha_1 x_1 - \beta_1)x_3 - Cx_2, \tag{7.16b}$$

$$\frac{dx_3}{dt} = (\alpha_1 x_1 - \beta_1)x_2 - \gamma_1 x_1 - Cx_3, \tag{7.16c}$$

which is often referred to as the Charney-DeVore model (Charney and DeVore, 1979).

The steady states of the Charney-deVore model are easily solved by introducing $\omega = \alpha_1 x_1 - \beta_1$ and solving the system (7.16b,c) for x_2 and x_3 in terms of x_1, which gives

$$x_2 = \frac{\omega\gamma_1}{\omega^2 + C^2} x_1 \; ; \; x_3 = -\frac{C\gamma_1}{\omega^2 + C^2} x_1, \tag{7.17}$$

Substituting the result into (7.16a) gives a third-order equation for x_1 as

$$F(x_1) = \left(\frac{\tilde{\gamma}_1 \gamma_1}{\omega^2 + C^2} + 1 \right) x_1 = x_1^*. \tag{7.18}$$

The function $F(x_1)$ is shown for several values of γ in Fig. 7.9. For small x_1^*, there is only one solution of (7.18), but as the topography height γ increases, eventually three solutions appear. From this analysis, it is clear that the modes x_4, x_5 and x_6 in (7.15) are not needed to explain the existence of the multiple equilibria.

The linear stability analysis of a steady state $(\bar{x}_1, \bar{x}_2, \bar{x}_3)$ leads to an eigenvalue problem for the Jacobian matrix J of (7.16), which is

$$J = \begin{pmatrix} -C & 0 & \tilde{\gamma}_1 \\ -\alpha_1 \bar{x}_3 & -C & \beta_1 - \alpha_1 \bar{x}_1 \\ -\gamma_1 + \alpha_1 \bar{x}_2 & -\beta_1 + \alpha_1 \bar{x}_1 & -C \end{pmatrix}. \tag{7.19}$$

The value of $\omega(\bar{x}_1)$ is plotted versus x_1^* for $\gamma = 1$ in Fig. 7.10a, and the eigenvalues σ of (7.19) along the branches are plotted in Fig. 7.10b–c. At small $\bar{x}_1$, there is one real

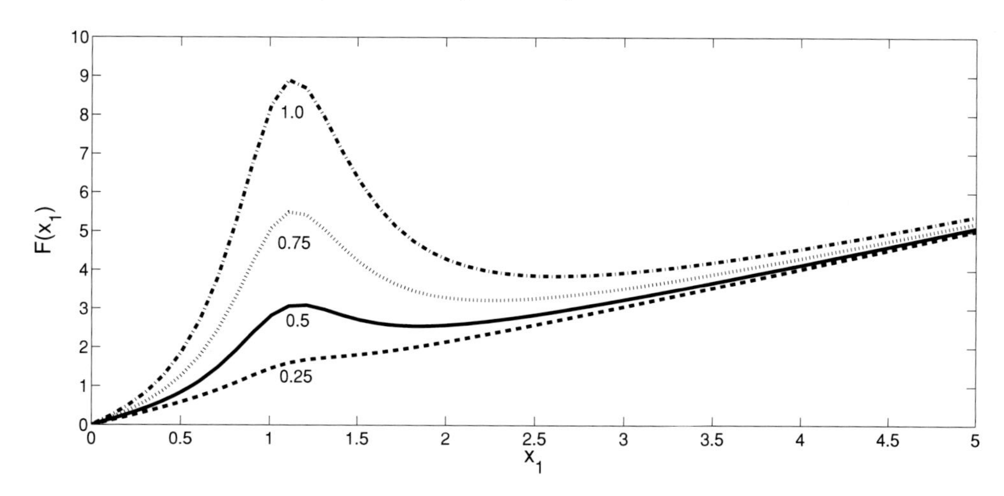

Figure 7.9 The function $F(x_1)$ in (7.18) for several values of γ.

eigenvalue σ_1 and a complex conjugate pair $\sigma_{2,3}$ (indicated by the circles and crosses in Fig. 7.10b,c), all with a negative real part. When the first saddle-node bifurcation point is approached, the value of $\omega \approx 0$ and $\sigma_1 \approx 0$, and while crossing the saddle node, the value of $\sigma_1 > 0$, whereas the real part of the pair $\sigma_{2,3} < 0$. On the unstable branch, between the saddle nodes, only $\sigma_1 > 0$, and it passes through zero again at the second saddle-node bifurcation. Hence the instability is associated with only one real eigenvalue (σ_1), whereas the oscillatory pair plays no role.

The instability mechanism associated with the real eigenvalue is called topographic instability (Charney and DeVore, 1979). The mechanism is most easily illustrated by considering the nondissipative case $C = 0$, for which the equations (7.16) reduce to

$$\frac{dx_1}{dt} = \tilde{\gamma}_1 x_3, \tag{7.20a}$$

$$\frac{dx_2}{dt} = -(\alpha_1 x_1 - \beta_1) x_3, \tag{7.20b}$$

$$\frac{dx_3}{dt} = (\alpha_1 x_1 - \beta_1) x_2 - \gamma_1 x_1. \tag{7.20c}$$

Here the steady states are given by $\bar{x}_3 = 0$ and $\bar{x}_2 = \gamma_1 \bar{x}_1 / \omega$, with $\omega = (\alpha_1 \bar{x}_1 - \beta_1)$ and $\bar{x}_1$ arbitrary (for example, $\bar{x}_1 = x_1^*$). Perturbations $(\tilde{x}_1, \tilde{x}_2, \tilde{x}_3)$ from these steady states satisfy the linearised equations

$$\frac{d\tilde{x}_1}{dt} = \tilde{\gamma}_1 \tilde{x}_3, \tag{7.21a}$$

$$\frac{d\tilde{x}_2}{dt} = -\omega \tilde{x}_3, \tag{7.21b}$$

$$\frac{d\tilde{x}_3}{dt} = \omega \tilde{x}_2 - (\gamma_1 - \alpha_1 \bar{x}_2) \tilde{x}_1 = \omega \tilde{x}_2 + \frac{\beta_1 \gamma_1}{\omega} \tilde{x}_1. \tag{7.21c}$$

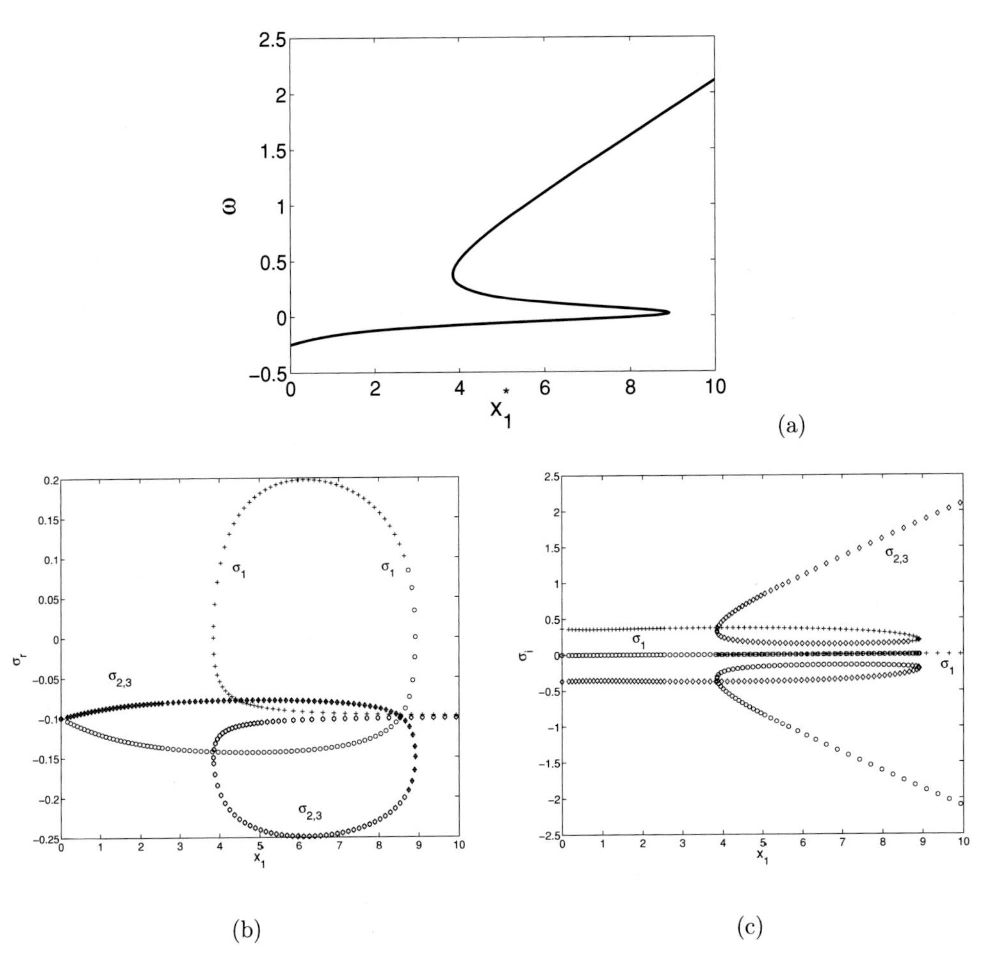

Figure 7.10 (a) Value of the parameter $\omega = \alpha x_1 - \beta_1$ along the branches of steady states when x_1^* is changed. (b) Real part σ_r of the eigenvalues of J in (7.19). (c) Imaginary part σ_i of the eigenvalues of J.

Differentiating (7.21c) to time and using (7.21a) and (7.21b) leads to

$$\frac{d^2\tilde{x}_3}{dt^2} + \left(\omega^2 - \frac{\beta_1\tilde{\gamma}_1\gamma_1}{\omega}\right)\tilde{x}_3 = 0, \tag{7.22}$$

and hence exponential growth will occur when $\omega^2 - \beta_1\tilde{\gamma}_1\gamma_1/\omega < 0$. The presence of the topography is essential because when $\gamma = 0$, then $\gamma_1 = \tilde{\gamma}_1 = 0$ and there is only oscillatory behaviour.

The instability mechanism is sometimes also referred to as topographic resonance because it involves a special case of parametric resonance that occurs in oscillators where the frequency varies with time according to $A\cos\omega_1 t$, giving an equation of the form

$$\frac{d^2x}{dt^2} + \omega_0^2(1 + A\cos\omega_1 t)x = 0. \tag{7.23}$$

In this case, resonance occurs for small A and $\omega_1 = 2\omega_0/n$ for integer n. In topographic instability, $\omega_1 \approx 0$, and resonance occurs, due to the topography, with a near stationary change in frequency (Ghil and Childress, 1987).

The second instability mechanism is related to the occurrence of the Hopf bifurcation on the zonal flow solution branch in Fig. 7.7. As can be concluded from the previous analysis, such a Hopf bifurcation does not occur in the three-mode (Charney-deVore) model, and hence patterns of $\phi_{0,2}$, $\phi_{1,2}$ and $\phi_{-1,2}$ are necessary to represent the instability. This is a so-called barotropic instability, where the energy of the perturbations is extracted from the horizontal shear of the background steady state through the Reynolds' stresses. As the instability mechanism is well described in other textbooks on dynamical meteorology and oceanography (Holton, 1992; Pedlosky, 1987), we do not expand on this here in more detail.

7.3.3 Low-frequency variability

When a trajectory is computed in the original six-mode model (7.15) with $\gamma = 0.2$, $x_1^* = 0.95$ and $r = -0.801$, the result of Fig. 7.11a is found, with the spatial patterns of the flows labelled eq1 and eq2,3 shown in Fig. 7.11b. Clearly, there is low-frequency variability, where the pattern seems to make excursions between a pattern of a zonal flow and a blocked flow.

The origin of the low-frequency variability in this model is due to homoclinic and heteroclinic bifurcations (cf. Section 2.5.1). These arise here through a so-called fold-Hopf bifurcation. The value of x_1^* at the Hopf bifurcation and the saddle-node bifurcations in Fig. 7.7 depend on other parameter values, for example, r and γ. The position of the saddle-node sn1 does not depend much on r (which is the forcing of the ϕ_{02} pattern), whereas the Hopf bifurcation will shift to smaller values of x_1^* when r becomes more negative.

For the case $\gamma = 0.2$, this can be seen in three bifurcation diagrams (where x_1 is plotted versus x_1^*) that are plotted for different values of r in Fig. 7.12. The bifurcation diagram for $r = -0.7$ is qualitatively similar to that in Fig. 7.7. Near $r = -0.8$, which is the case considered in Fig. 7.11, the saddle-node sn1 and the Hopf bifurcation H have nearly merged. For $r = -0.9$, the Hopf bifurcation is already located on the unstable branch of steady solutions.

From Fig. 7.12, it can be concluded that there exists a value of r such that the saddle-node sn1 and the Hopf bifurcation coalesce (occur at the same value of x_1^*). This is the location of the fold-Hopf bifurcation. The different possibilities of the behaviour of trajectories near the fold-Hopf bifurcation can be found, for example, in

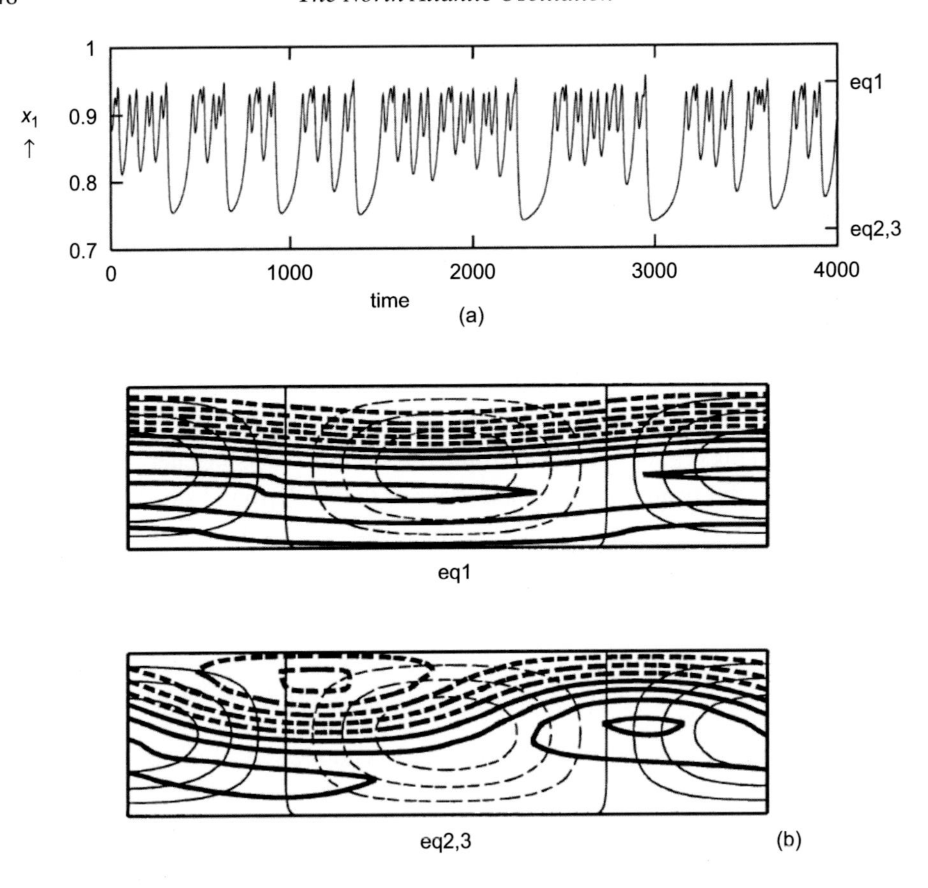

Figure 7.11 (a) Trajectory of the six-mode model with $\gamma = 0.2$, $x_1^* = 0.95$ and $r = -0.801$. (b) Streamfunction patterns of the states eq1 and eq2,3 in (a); figure from Crommelin et al. (2004).

detail in Kuznetsov (1995). As the Hopf bifurcation is supercritical, there is only one possibility, and the phase portraits are shown in Fig. 7.13. Here ρ_1 and ρ_2 are the control parameters, which can be thought of as r and x_1^*, respectively. In the middle figure, the path of the Hopf bifurcations (hb) intersects the path of the saddle node (sn).

To interpret the diagrams in Fig. 7.13, consider a fixed value of $\rho_2 < 0$ (fixed r) and a decreasing ρ_1 (increasing x_1^*). This is similar to investigating the behaviour locally near the $sn1 - H$ branch for $r = -0.7$ in Fig. 7.12 with increasing x_1^*. For x_1^* smaller than the value of sn1, locally no steady states exist, and there are no periodic orbits (of course, there is a steady state far away, but note that this is a local analysis), and hence this is regime a in Fig. 7.12. When the saddle node (sn) is passed (at $\rho_1 = 0$), two new equilibria are created, of which one is stable and one is unstable. These equilibria are sketched at the dots in regime f, where the upper one is attracting all trajectories (and hence this is the stable one) and the lower one is unstable. At slightly more negative ρ_1 (slightly larger x_1^*), the Hopf bifurcation H (hb in Fig. 7.12)

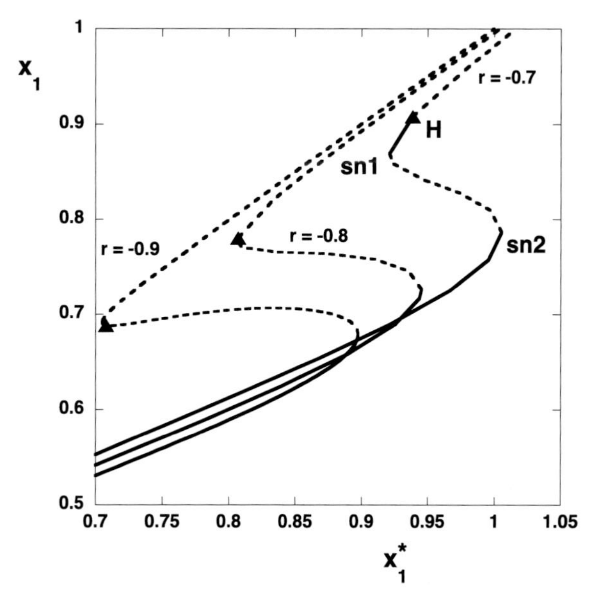

Figure 7.12 Bifurcation diagrams of the six-mode model with $\gamma = 0.2$ and different values of r. Solid (dashed) branches indicate stable (unstable) solutions, and the Hopf bifurcation (H) is indicated by a triangle.

is found, and the upper fixed point in regime f becomes unstable and a periodic orbit exists, which is case d. The amplitude of the periodic orbit becomes eventually so large that it touches both equilibria, and hence a heteroclinic connection (hc) is established. The case $r = -0.9$ corresponds to the case $\rho_2 > 0$, where all steady states and periodic orbits are unstable (cases b and c). As we have seen in Section 2.5.1, homoclinic and heteroclinic connections are associated with irregular dynamics and with trajectories displaying variability at low frequency. The existence of these heteroclinic connections explains the behaviour (low-frequency switching between two equilibria) displayed in Fig. 7.11.

7.4 Beyond conceptual models

Although the reduced (low-order) models have been very instructive for illuminating mechanisms of variability in atmospheric midlatitude flows, it is not at all clear whether the mechanisms identified are responsible for low-frequency behaviour in the minimal model (Section 7.2), for models higher up in the hierarchy as described in Chapter 6, and observations.

7.4.1 Barotropic quasi-geostrophic models

In Fig. 7.6a, the crosses, dots and squares indicate steady states of the minimal model (7.1), where the markers distinguish the degree of projection of these steady states

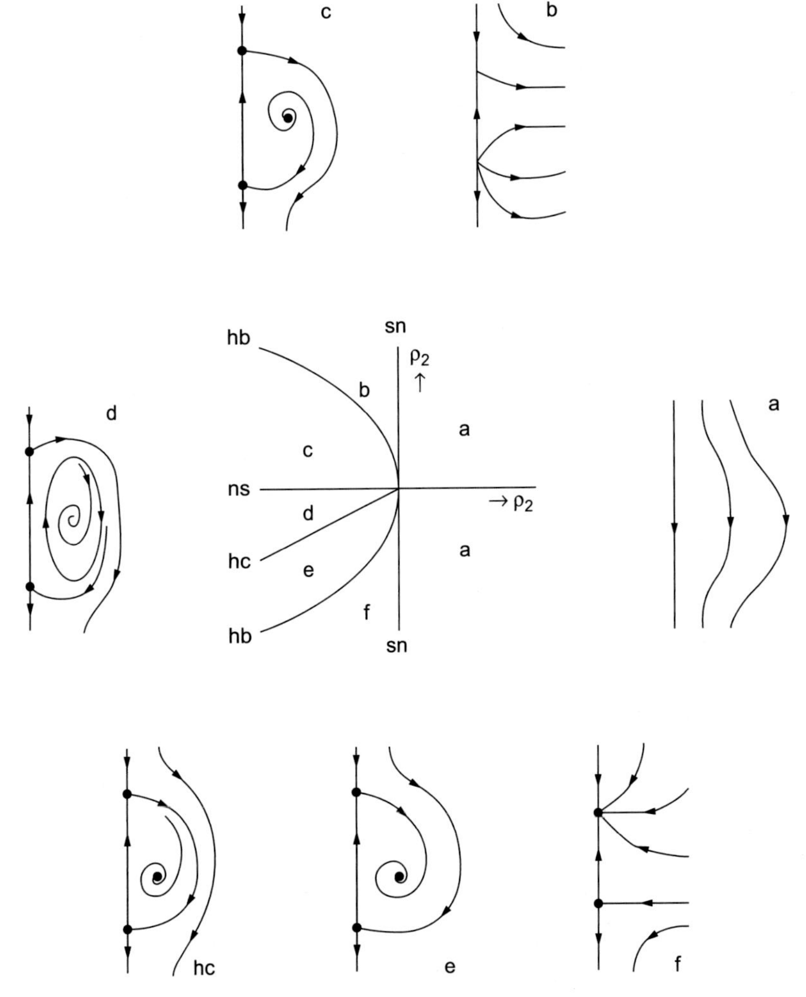

Figure 7.13 Different behaviour (phase-plane pictures) near the fold-Hopf point in a two-parameter (ρ_1, ρ_2) plane (figure from Crommelin et al., 2004).

onto the leading two EOFs; at least sixty-eight fixed points were found (Crommelin, 2003). This does not motivate us to totally clarify the bifurcation behaviour of these models, and one, moreover, wonders how the specific truncation affects the number of fixed points.

Indications for the existence of heteroclinic connections in the minimal model were found in Crommelin (2003). In Fig. 7.14a the projection of trajectories starting near MAX1 (near steady-state A) in the PC1–PC3 plane shows that these end up near B, whereas during the transition, the sign of PC3 is negative. In contrast, in Fig. 7.14b, trajectories started near MAX2 (steady-state B) end up near A, but the sign of PC3 is positive. Using an optimisation technique, approximations of heteroclinic orbits are

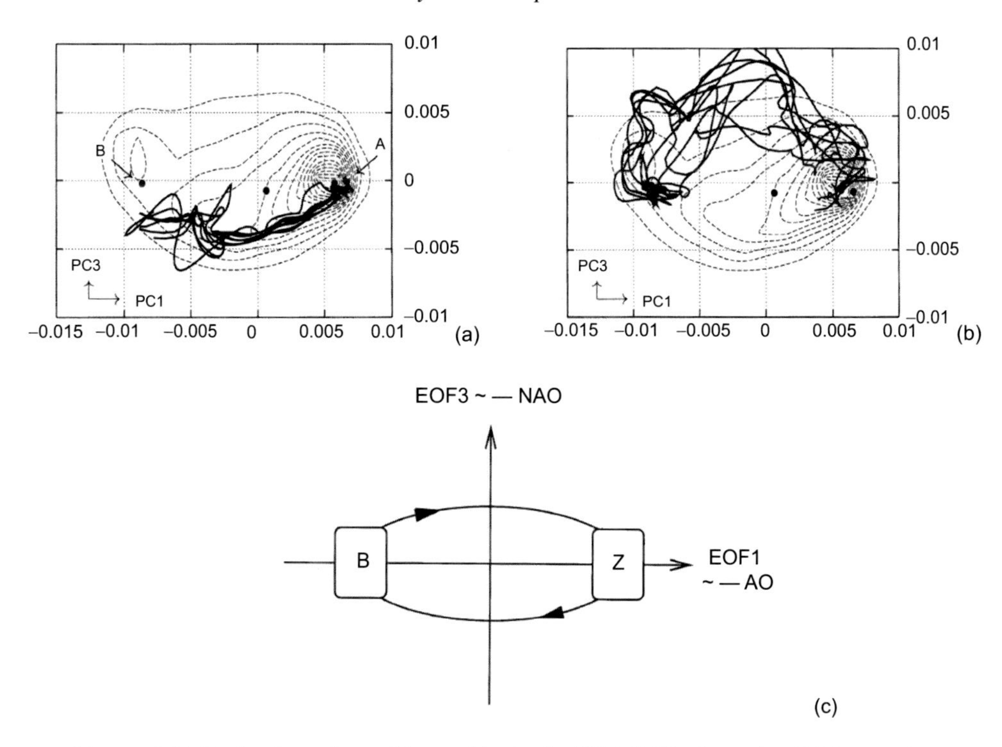

Figure 7.14 Projection of transition orbits in the PC1–PC3 plane. (a) Orbits starting close to steady state A near MAX1. (b) Orbits starting close to steady state B near MAX2. Also shown are the contours (dashed) of the probability density function of the 200-yr data set. (c) Schematic representation of the interrelationships of regime behavior, NAO and AO, as emerging from the barotropic model. Here, B denotes the blocked regime; Z is the zonal regime. Negative PC1 corresponds to a stronger polar vortex (positive AO phase); negative PC3 corresponds to the positive NAO phase (figure from Crommelin, 2003).

computed in Crommelin (2003), providing strong support for the existence of these orbits. These orbits play an important role in the transition between zonal (Z) and blocked regimes (B) of the atmospheric flow in this model.

From these results, a highly simplified, schematic picture of the dynamics of regime behaviour between NAO and AO (Fig. 7.14c) was suggested (Crommelin et al. 2003). A negative value of PC1 corresponds to a stronger polar vortex (positive AO phase), whereas a negative value of PC3 corresponds to a positive NAO phase. Starting in the zonal regime, which shows a polar vortex that is weaker than average, the model evolves via a positive NAO phase to the blocked regime. The blocked regime is characterised by a blocking over Europe and, on a larger scale, a strong polar vortex. Leaving the blocked regime, the block persists for a while but shifts to the west, to the North Atlantic. This transition phase thereby gets the features of a negative NAO phase, and finally, the model atmosphere is back in the zonal regime.

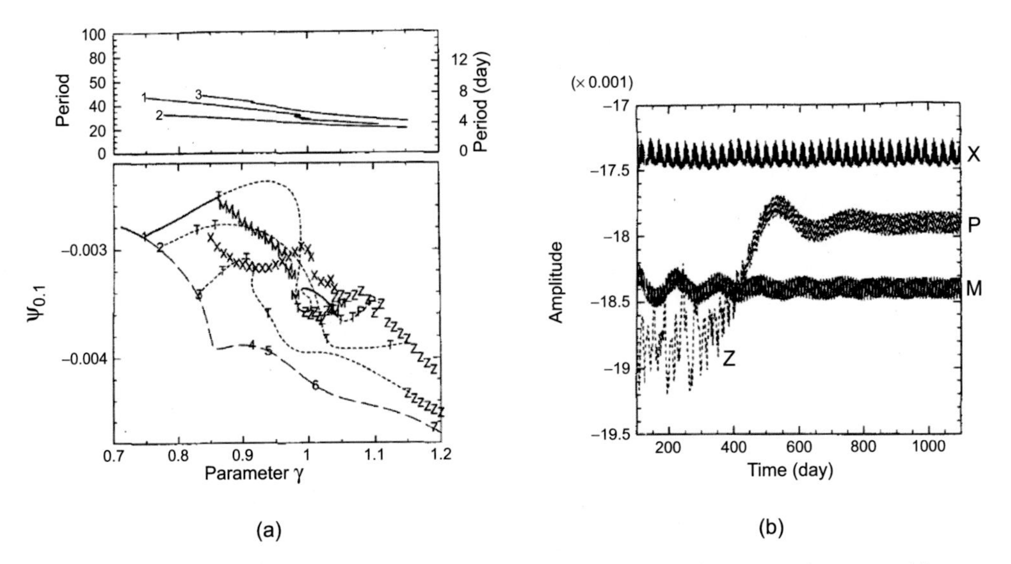

Figure 7.15 (a) Bifurcation diagram of the two-layer linear balance quasi-geostrophic model as in Itoh and Kimoto (1996). The lower branch represents steady states (drawn: stable; dashed: unstable). From the numerals, periodic solutions branch from the steady states (drawn: stable; dotted: unstable). The points labelled T are Neimark-Sacker bifurcations, and the labels X, M, P and Z point to particular time-dependent states. (b) Time variation of $\psi_{0,1}$ for $\gamma = 1$ for several different trajectories, illustrating the states X, P, M and Z.

7.4.2 Baroclinic quasi-geostrophic models

In Mukougawa (1988), a low-order truncation of a two-layer quasi-geostrophic model (having 28 degrees of freedom) in a β channel is analysed. Apart from real steady states, so-called quasi-stationary states, shown to be characterised by a small-phase space velocity, are also investigated. Again, many steady and quasi-stationary states as well as Hopf bifurcations were found.

An extension of the Legras and Ghil (1985) study for a two-level linear balance quasi-geostrophic model was presented in Itoh and Kimoto (1996). A spherical harmonic truncation was applied of the governing equations, leading to a dynamical system of 240 degrees of freedom. As control parameter, the equator to pole temperature gradient parameter γ was chosen, with $\gamma = 1$ corresponding to a realistic value. The bifurcation diagram was computed using the AUTO (Doedel, 1980) code, and details on the time-dependent behaviour were determined with many transient model simulations.

The bifurcation diagram, showing the amplitude of one of the streamfunction mode components $\psi_{0,1}$ versus γ, is plotted in Fig. 7.15a. The steady state, which is stable at a small value of γ, destabilises through a sequence of Hopf bifurcations where periodic solutions appear. The periods of the first three solutions are shown to be in

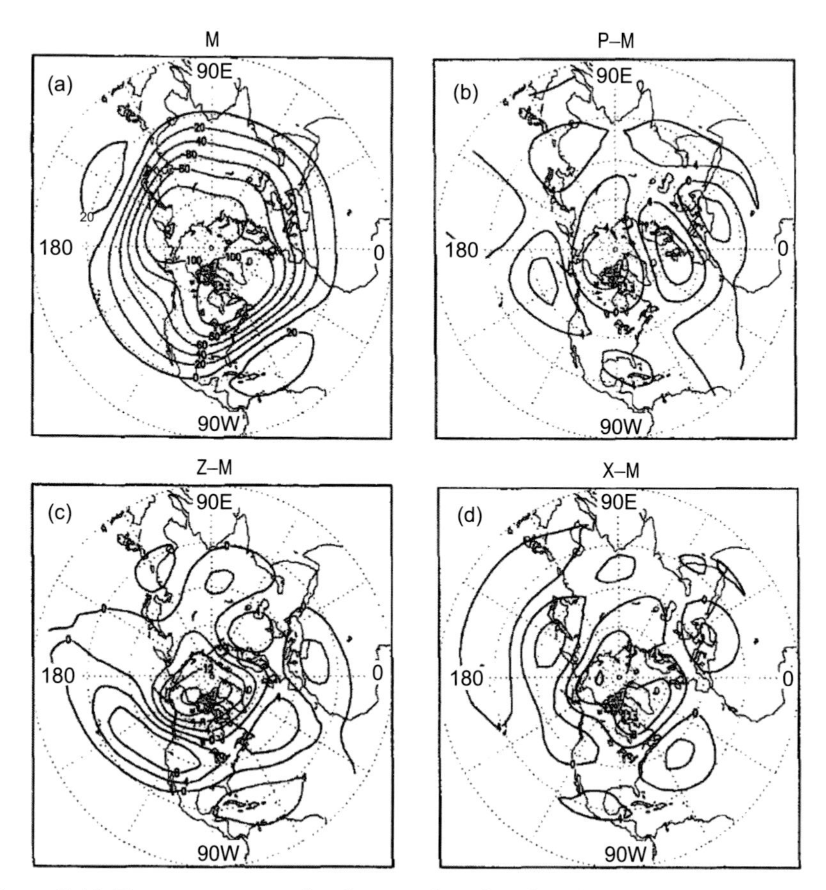

Figure 7.16 Time-mean upper level streamfunction for (a) state M, (b) state P, (c) state Z and (d) state X. Only (a) is the total field (contour interval 20×10^6 m^{2}s^{-1}) and deviations from (a) (contour interval 4×10^6 m^{2}s^{-1}) are plotted in (b–d); figure from Itoh and Kimoto (1996).

the range of 4–10 days (upper panel of Fig. 7.15). The periodic solutions destabilise through Neimark-Sacker (Section 2.4) bifurcations, leading to quasi-periodic and chaotic solutions. Several of the time-dependent equilibrium solutions are labelled in Fig. 7.15a, and trajectories computed are shown in Fig. 7.15b. Clearly X, M and P are equilibrium solutions, whereas the state Z appears as a transient state, which can persist for about 400 days.

The time-mean patterns of upper-level streamfunction for the state M and the differences between M and X, P and Z are shown in Fig. 7.16. The difference patterns P–M, Z–M and X–M are characterised either by wave trains passing near the North Pole (Fig. 7.16b,d) or by a north-south dipolar pattern (Fig. 7.16c). It was shown that when noise is applied to the heat flux in the model, a trajectory may itinerate over the remains of the attractors (e.g., X, Z and P) in what Itoh and Kimoto (1996)

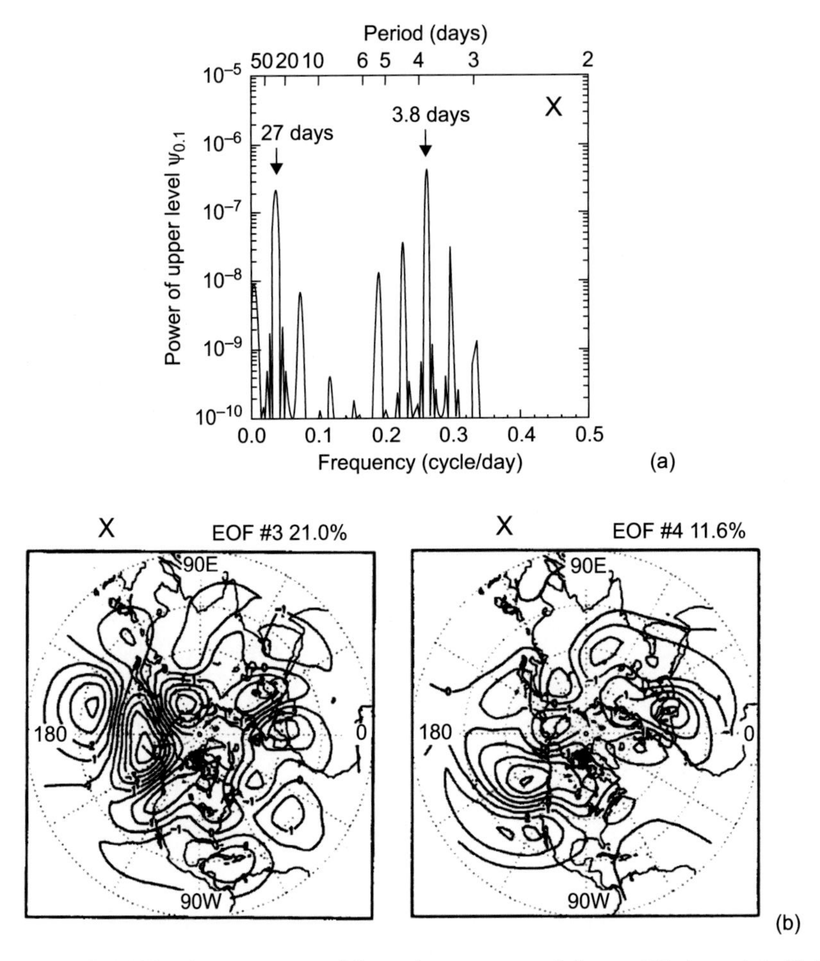

Figure 7.17 (a) Fourier spectrum of the trajectory around the equilibrium state X. (b) Patterns of the EOF3 and EOF4 showing the pattern of the low-frequency oscillation (figure from Itoh and Kimoto, 1996).

call 'chaotic itinerancy'. In Selten and Branstator (2004), using a T21 three-level quasi-geostrophic model, different equilibrium states are also found, which correspond to the so-called Pacific North American (PNA) and the Northern Annular Mode (NAM) pattern. Support is found for the existence of an unstable period that connects the different regimes, and it is suggested to be a remnant of a heteroclinic connection.

As can be seen in Fig. 7.15b, the trajectories around the states Z and X clearly display low-frequency variability. The spectrum of X is shown in Fig. 7.17a. The short-term dominant variability (of 3.8 days) is associated with baroclinic instability. The low-frequency variability is associated with the patterns of EOF3 and EOF4 (shown in Fig. 7.17b).

7.4.3 Baroclinic shallow-water models

The first study to use a two-layer shallow-water model was Keppenne et al. (2000). No detailed bifurcation diagrams were provided, but the focus was on the intraseasonal variability in the atmospheric angular momentum in the model. Spectral analysis of a 100-year long time series led to the identification of oscillations with 35–40 days and 65–70 days. The 70-day variability was not found so far in the quasi-geostrophic models and so was suggested to involve the ageostrophic flow component. The topography was found to be central in all the intraseasonal oscillations found.

In Sterk et al. (2010), a highly truncated version of the two-layer shallow water model in a β-plane channel was constructed. In each layer the velocity field (u, v) is two-dimensional. The thickness h of each layer is variable, which is the only three-dimensional aspect of this model. With the constants H_1 and H_2 denoting the mean thickness of each layer, and the fields η'_1 and η'_2 the deviations from the mean thickness, the thickness fields of the two layers are given by

$$h'_1 = H_1 + \eta'_1 - \eta'_2, \tag{7.24a}$$

$$h'_2 = H_2 + \eta'_2 - h'_b, \tag{7.24b}$$

where the prime indicates dimensional quantities and h_b is the bottom topography profile. The pressure fields are related to the thickness fields by means of the hydrostatic relation

$$p'_1 = \rho_1 g(h'_1 + h'_2 + h'_b), \tag{7.25a}$$

$$p'_2 = \rho_1 g h'_1 + \rho_2 g(h'_2 + h'_b), \tag{7.25b}$$

where the constants ρ_1 and ρ_2 denote the density of each layer.

The governing equations are nondimensionalised using scales L, U, L/U, D and $\rho_0 U^2$ for length, velocity, time, depth and pressure, respectively, and are given by

$$\begin{aligned}
\frac{\partial u_\ell}{\partial t} + u_\ell \frac{\partial u_\ell}{\partial x} + v_\ell \frac{\partial u_\ell}{\partial y} &= -\frac{\partial p_\ell}{\partial x} + (Ro^{-1} + \beta y) v_\ell \\
&\quad - \sigma\mu(u_\ell - u_\ell^*) + Ro^{-1} E_H \Delta u_\ell - \sigma r \delta_{\ell,2} u_\ell, \\
\frac{\partial v_\ell}{\partial t} + u_\ell \frac{\partial v_\ell}{\partial x} + v_\ell \frac{\partial v_\ell}{\partial y} &= -\frac{\partial p_\ell}{\partial y} - (Ro^{-1} + \beta y) u_\ell \\
&\quad - \sigma\mu(v_\ell - v_\ell^*) + Ro^{-1} E_H \Delta v_\ell - \sigma r \delta_{\ell,2} v_\ell, \\
\frac{\partial h_\ell}{\partial t} + u_\ell \frac{\partial h_\ell}{\partial x} + v_\ell \frac{\partial h_\ell}{\partial y} &= -h_\ell \left(\frac{\partial u_\ell}{\partial x} + \frac{\partial v_\ell}{\partial y} \right),
\end{aligned}$$

where u_ℓ and v_ℓ are eastward and northward components of the two-dimensional velocity field, respectively, for $\ell = 1, 2$. In addition, the nondimensional pressure

Table 7.1. *Standard values of the fixed parameters as used in the two-layer shallow water model of Sterk et al. (2010)*

Parameter	Meaning	Value	Unit
A_H	momentum diffusion coefficient	1.0×10^2	$m^2\,s^{-1}$
μ	relaxation coefficient	1.0×10^{-6}	s^{-1}
r	linear friction coefficient	1.0×10^{-6}	s^{-1}
f_0	Coriolis parameter	1.0×10^{-4}	s^{-1}
β_0	planetary vorticity gradient	1.6×10^{-11}	$m^{-1}\,s^{-1}$
ρ_0	reference density	1.0	$kg\,m^{-3}$
ρ_1	density (top layer)	1.01	$kg\,m^{-3}$
ρ_2	density (bottom layer)	1.05	$kg\,m^{-3}$
g	gravitational acceleration	9.8	$m\,s^{-2}$
α_1	zonal velocity forcing strength (top layer)	1.0	
α_2	zonal velocity forcing strength (bottom layer)	0.5	
L_x	channel length	2.9×10^7	m
L_y	channel width	2.5×10^6	m
H_1	mean thickness (top layer)	5.0×10^3	m
H_2	mean thickness (bottom layer)	5.0×10^3	m
L	characteristic length scale	1.0×10^6	m
U	characteristic velocity scale	1.0×10^1	$m\,s^{-1}$
D	characteristic depth scale	1.0×10^3	m

terms are given by

$$p_1 = \frac{\rho_1}{\rho_0} F(h_1 + h_2 + h_b),$$

$$p_2 = \frac{\rho_1}{\rho_0} F h_1 + \frac{\rho_2}{\rho_0} F(h_2 + h_b).$$

Several nondimensional numbers appear in the governing equations: the β parameter, the Rossby number Ro, the horizontal Ekman number E_H and the inverse Froude number F. These parameters have the following expressions in terms of the dimensional parameters:

$$\beta = \frac{\beta_0 L^2}{U}, \quad Ro = \frac{U}{f_0 L}, \quad E_H = \frac{A_H}{f_0 L^2}, \quad F = \frac{gD}{U^2}.$$

Furthermore, σ is the advective time scale L/U. Standard values of the dimensional parameters as used in Sterk et al. (2010) are listed in Table. 7.1.

The dynamical equations are considered on the zonal β-plane channel

$$0 \leq x \leq L_x/L, \quad 0 \leq y \leq L_y/L.$$

All fields are supposed to be periodic in the x direction, and at $y = 0, L_y/L$, the conditions

$$\frac{\partial u_\ell}{\partial y} = \frac{\partial h_\ell}{\partial y} = v_\ell = 0$$

are imposed. The model is forced by relaxation to a westerly wind given by the profile

$$u_1^*(x, y) = \alpha_1 U_0 U^{-1}(1 - \cos(2\pi y L/L_y)), \quad v_1^*(x, y) = 0,$$
$$u_2^*(x, y) = \alpha_2 U_0 U^{-1}(1 - \cos(2\pi y L/L_y)), \quad v_2^*(x, y) = 0,$$

where the dimensional parameter U_0 controls the strength of the forcing, and the nondimensional parameters α_1 and α_2 control the vertical shear of the forcing. For the bottom topography, a profile with zonal wave number 3,

$$h_b(x, y) = h_0 D^{-1}(1 + \cos(6\pi x L/L_x)),$$

is chosen, where the dimensional parameter h_0 controls the amplitude of the topography. The bottom topography is contained entirely in the bottom layer, which implies the restriction $h_0 \leq H_2/2$.

Observational evidence (Benzi et al., 1986) suggests that the fundamental physical processes involved in low-frequency behaviour manifest themselves at zonal wave numbers less than 5. Therefore, wave numbers $m = 0, 3$ in the zonal direction and the wave numbers $n = 0, 1, 2$ in the meridional direction were chosen. Let

$$R = \{(0, 0), (0, 1), (0, 2), (3, 0), (3, 1), (3, 2)\}$$

denote the set of retained wave number pairs. In Sterk et al. (2010), basis functions (for an integer $k \geq 0$ and a real number $\alpha > 0$)

$$c_k(\xi; \alpha) = \begin{cases} \dfrac{1}{\sqrt{\alpha}} & k = 0 \\ \sqrt{\dfrac{2}{\alpha}} \cos\left(\dfrac{k\pi\xi}{\alpha}\right) & k > 0, \end{cases} \tag{7.26}$$

$$s_k(\xi; \alpha) = \sqrt{\frac{2}{\alpha}} \sin\left(\frac{k\pi\xi}{\alpha}\right),$$

are used, where $\xi \in [0, \alpha]$, and the numerical factors serve as normalisation constants.

All nondimensional fields are next expanded as

$$u_\ell(x, y, t) = \sum_{(m,n)\in R} \left[\widehat{u}^c_{\ell,m,n}(t)c_{2m}(x; a) + \widehat{u}^s_{\ell,m,n}(t)s_{2m}(x; a)\right] c_n(y; b),$$

$$v_\ell(x, y, t) = \sum_{(m,n)\in R} \left[\widehat{v}^c_{\ell,m,n}(t)c_{2m}(x; a) + \widehat{v}^s_{\ell,m,n}(t)s_{2m}(x; a)\right] s_n(y; b),$$

$$h_\ell(x, y, t) = \sum_{(m,n)\in R} \left[\widehat{h}^c_{\ell,m,n}(t)c_{2m}(x; a) + \widehat{h}^s_{\ell,m,n}(t)s_{2m}(x; a)\right] c_n(y; b),$$

where $a = L_x/L$ and $b = L_y/L$. In this way the truncated expansions satisfy the boundary conditions, and a 46-dimensional system of ordinary differential equations results. As control parameters, the maximal zonal velocity of the jet U_0 and the maximum height of the topography h_0 are chosen.

For $U_0 \leq 12.2\,\mathrm{ms}^{-1}$ there is a stable equilibrium corresponding to a steady westerly wind. This steady flow becomes unstable through Hopf bifurcations as the forcing U_0 increases. This gives rise to two types of stable waves: for lower orography (about 800 m), the period is about 10 days and there is eastward propagation in the bottom layer; for more pronounced orography, the period is longer (30–60 days), and the waves are nonpropagating. The waves can be identified as mixed baroclinic/barotropic instabilities, where the baroclinicity is not that associated to midlatitude synoptic systems (indeed, wave number 3 is not the most unstable baroclinic mode). Rather, instabilities here bear resemblance to the orographic baroclinic instability (Cessi and Speranza, 1985).

The waves remain stable in relatively large parameter domains and bifurcate into strange attractors through a number of scenarios in the parameter quadrant $U_0 \leq$ 14.5 m/s and $h_0 \geq 850$ m. The Lyapunov diagram (top panel of Fig 7.18) shows a classification of the dynamical behaviour in the different regions of the (U_0, h_0) plane. Bifurcations of equilibria and periodic orbits (bottom panel) explain the main features of the Lyapunov diagram. The two Hopf curves $H_{1,2}$ give birth to stable periodic orbits. In turn, these periodic orbits bifurcate into strange attractors through three main routes to chaos:

1. Period doubling cascade of periodic orbits (the curves P_1, P_2 and P_3 in Fig. 7.18b);
2. Hopf-Neimark-Sacker bifurcation of periodic orbits (the curve T_2 in Fig. 7.18b), followed by the breakdown of a torus (cf. Section 2.4);
3. Saddle-node bifurcation of periodic orbits taking place on a strange attractor (the curve SP_4 in Fig. 7.18b), the so-called intermittency route.

The low-frequency atmospheric behaviour in this model is characterised in terms of intermittency due to bifurcations of waves. Nonpropagating planetary waves arise from the interaction of zonal flow with orography. Low-frequency behaviour with the appropriate time scales (10–200 days, where the lower-frequency components of

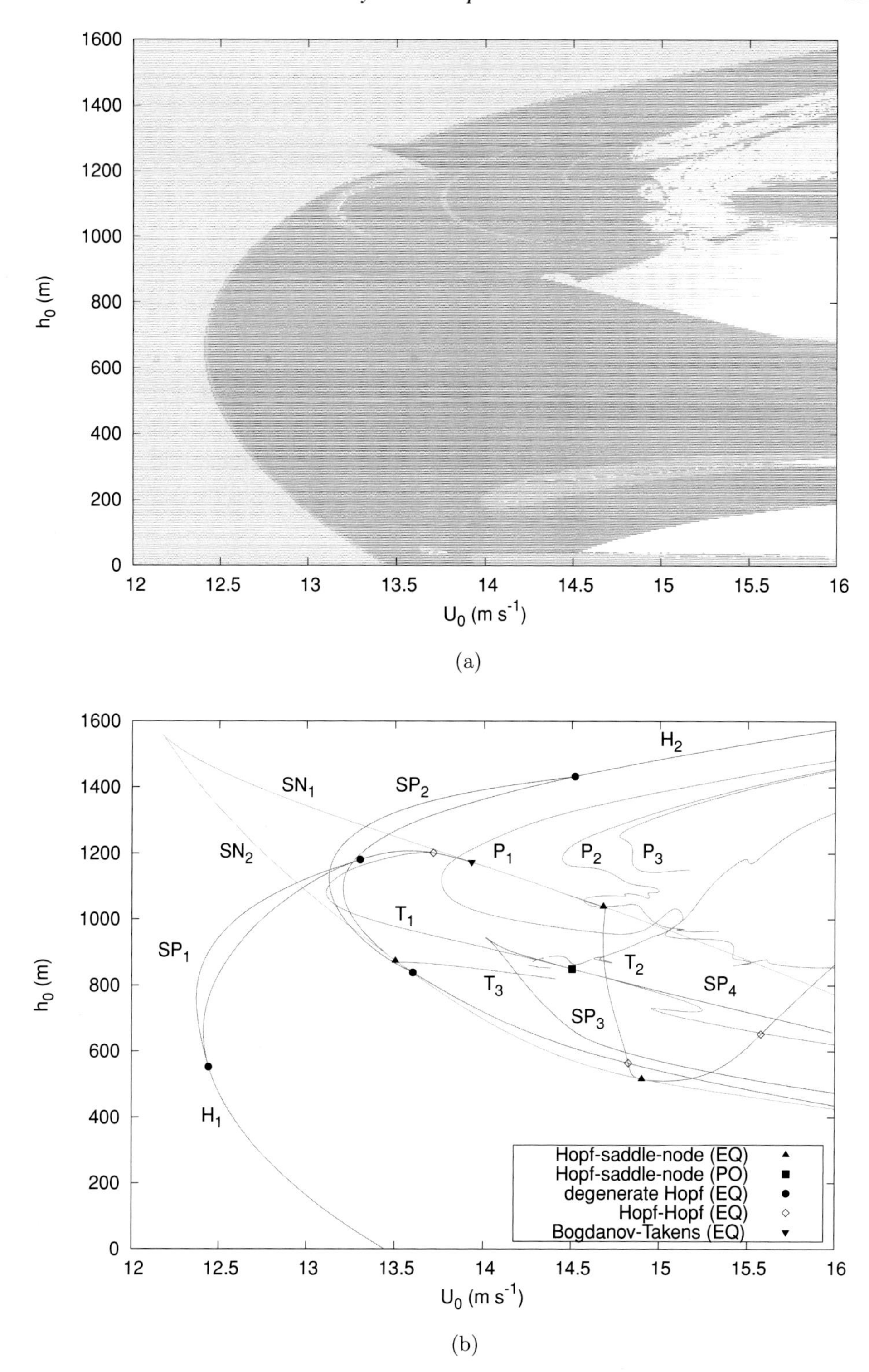

Figure 7.18 (a) Lyapunov diagram for the attractors of the system. (b) Regime diagram of attractors of the low-order model in the (U_0, h_0) parameter plane. The marked locations are codimension-2 bifurcations. The colour coding of both diagrams is provided in Table 7.2. (See Colour Plate.)

Table 7.2. *Colour coding for the Lyapunov diagram and bifurcation diagram in Fig. 7.18*

Colour	Lyapunov exponents	Attractor type
green	$0 > \lambda_1 \geq \lambda_2 \geq \lambda_3$	equilibrium
blue	$\lambda_1 = 0 > \lambda_2 \geq \lambda_3$	periodic orbit
magenta	$\lambda_1 = \lambda_2 = 0 > \lambda_3$	2-torus
cyan	$\lambda_1 > 0 \geq \lambda_2 \geq \lambda_3$	strange attractor
white		escaping orbit
Colour	**Bifurcation type**	**Bifurcating attractor**
green	saddle-node bifurcation	equilibrium
red	Hopf bifurcation	equilibrium
magenta	Hopf-Neimark-Sacker bifurcation	periodic orbit
grey	period doubling bifurcation	periodic orbit
blue	saddle-node bifurcation	periodic orbit

60–200 days can be interpreted as harmonics of the higher-frequency components of 10–60 days) is exhibited for physically relevant values of the parameters (realistic wind speed U_0 and topography height h_0).

7.5 Regime transitions and metastable states

In Berner and Branstator (2007), an analysis is performed of data from seven million days of output of the 500-hPa geopotential height field from an atmospheric GCM (the CCM0). Although they find statistically significant non-Gaussianities in the probability density function (in a reduced space spanned by four EOFs), they find no evidence for multiple local density maxima. They do find local maxima when conditional probability density functions are constructed by making two-dimensional slices through the four-dimensional probability space, but such features are not present in the full space. It is suggested that the non-Gaussianity of the probability density function can be attributed to a mix of a small number of Gaussian components, of which two are dominant (corresponding to a Pacific blocked state and zonal state).

These results can be understood by so-called Hidden Markov Models (HMMs) (Majda, 2006; Franzke et al., 2008). An HMM is designed to provide a description of a stochastic dynamical system for which one has only partial information. Let Y_t be the observed random variable and assume that the statistics of Y_t are dependent on a hidden (non-observable) discrete variable $X_t = X_{t_k}, k = 1, \ldots, N$, which is governed by a so-called Markov chain, that is,

$$P(X_{n+1} = x | X_1 = x_1, \ldots, X_n = x_n) = P(X_{n+1} = x | X_n = x_n). \tag{7.27}$$

The transitions between the different states in the Markov chain are described by a transition probability matrix A.

An HMM hence consists of a set S of N hidden states $S = \{s_1, \ldots, s_N\}$, and the hidden variable X takes values on this set, $X_t \in S$. The coefficients of the transition probability matrix A are given by $a_{ij} = P(X_t = s_j | X_{t-1} = s_i)$. The dependence of the probability distributions of Y_t on X_t is described using the conditional probabilities $B_i = P(Y_t | X_t = s_i)$. Given the initial distribution $P(X_1)$, the joint probability distribution is then given by

$$P(X_1, \ldots, X_T, Y_1, \ldots, Y_T) = P(X_1)P(Y_1|X_1)\prod_{t=2}^{T} P(X_t|X_{t-1})P(Y_t|X_t). \quad (7.28)$$

When only a time series of Y_t is available, parameters in the HMM must be estimated (given a prescribed number of hidden states N), and algorithms to do so are provided in Franzke et al. (2008).

As an example, consider the two-state ($N = 2$) Markov chain with the transition probability matrix (Franzke et al., 2008)

$$A = \begin{bmatrix} 0.99 & 0.01 \\ 0.01 & 0.99 \end{bmatrix}, \quad (7.29)$$

and two univariate distributions $B_1 = N(3.0, 1.0)$ and $B_2 = N(-3.0, 2.0)$, where $N(\mu, \sigma^2)$ is the normal distribution with mean μ and variance σ^2. If the system is in the hidden state X_1, the observed signal Y_t is drawn randomly from B_1; otherwise it is drawn from B_2. When in state X_1, the probability to stay in state X_1 or switch to state X_2 is given by a_{11} and a_{12}, respectively.

A typical realisation of Y_t versus time is presented in Fig. 7.19, where the existence of two distinct regime states can be seen (left panel of Fig. 7.19) and the resulting probability density function is bimodal (right panel of Fig. 7.19). The middle panel of Fig. 7.19 shows the most probable hidden state sequence, the so-called Viterbi path. With overlapping Gaussian distributions $B_1 = N(1.0, 1.7)$ and $B_2 = N(-0.5, 1.2)$, the probability distribution is neither bimodal nor Gaussian.

If the state space S of X can be decomposed into two or more sets with relatively infrequent transitions between these sets, the Markov chain is said to be metastable. The presence of metastability allows for a time scale separation between 'fast' transitions within metastable sets and the 'slow' transitions between metastable sets. The eigenvalues λ of the transition probability matrix A provide information on metastability, as is illustrated using the example that follows.

Consider the Itô SDE given by (Franzke et al., 2008)

$$Y_t = (-4Y_t^3 - c(3Y_t^2 - 1))dt + \sigma dW_t, \quad (7.30)$$

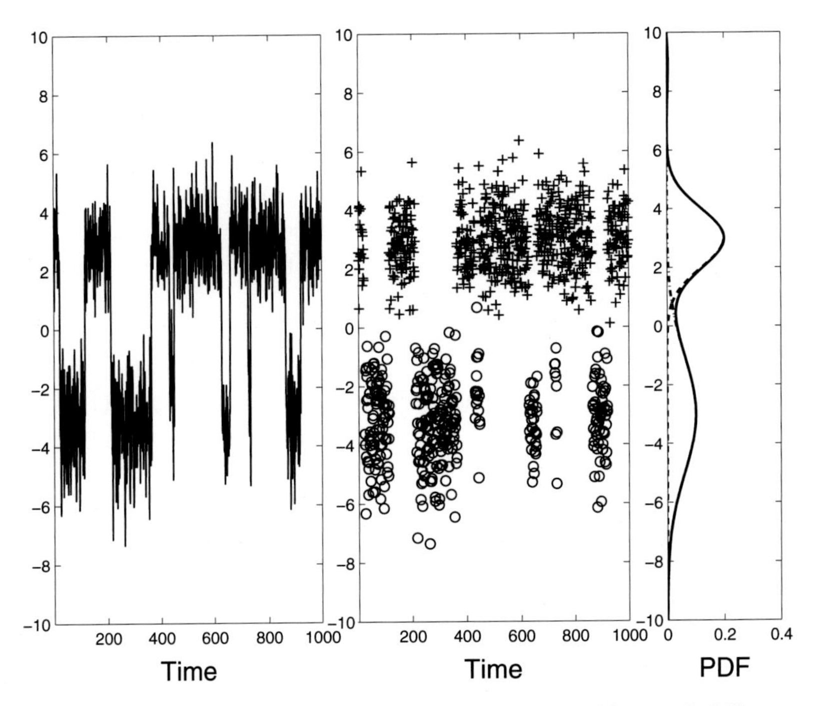

Figure 7.19 Realisation of the two-state HMM with transition probability matrix (7.29) and Gaussian distributions B_1 and B_2 (figure from Franzke et al., 2008).

where the first term is the gradient of the potential $V(y) = y^4 + cy(y-1)(y+1)$, which is shown for $c = 0.22$ in Fig. 7.20a. It is shown in Franzke et al. (2008) that if

$$c_k = \frac{1}{\Delta t} \log \lambda_k, \tag{7.31}$$

where Δt is the time step used to construct A, then metastability is characterised by $0 = c_1 > Re(c_2) \geq Re(c_3) \ldots \geq Re(c_N)$. Such conditions are shown to apply for (7.30) when $c = 2.2$ (Fig 7.20).

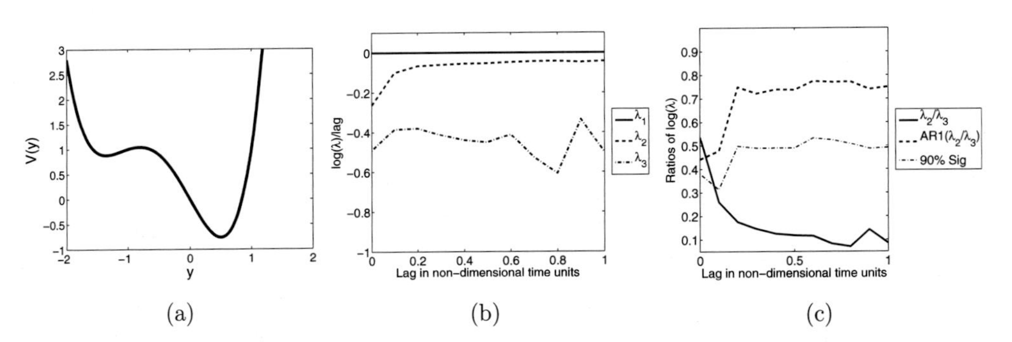

Figure 7.20 (a) Potential $V(y)$ of the system (7.30) for $c = 2.2$. (b) Modified eigenvalues c_k, for $N = 3$, and (c) ratio of eigenvalues (figure from Franzke et al., 2008).

In Franzke et al. (2008), an HMM analysis is also provided of output of a quasi-geostrophic barotropic model of flow over topography. The dimensionless equations for this model (on a $2\pi \times 2\pi$ periodic domain) are given by

$$\frac{\partial q}{\partial t} + \mathbf{u} \cdot \nabla q + U \frac{\partial q}{\partial x} + \beta \frac{\partial \psi}{\partial x} = 0, \tag{7.32a}$$

$$q = \nabla^2 \psi + h, \tag{7.32b}$$

$$\frac{dU}{dt} = \frac{1}{4\pi^2} \int_V h \frac{\partial \psi}{\partial x} d^2x, \tag{7.32c}$$

where q is the potential vorticity, ψ the streamfunction, $\mathbf{u}$ the geostrophic velocity vector, U the mean zonal flow velocity and h the topography. The latter is taken as $h = h_0(\cos x + \sin x)$, and the HMM is constructed from model output of simulations with different values of h_0. For $h_0 = 1.06$ and $\beta = 1$, the HMM analysis with $N = 3$ indicates metastability based on the eigenvalues of the transition probability matrix A.

The first hidden state represents a blocking-like flow, and the other two states represent a zonal circulation with different magnitudes. As can be seen, the probability density function of U is nearly Gaussian distributed, and the three conditional probability density distributions of the hidden Markov states have a substantial overlap (see Fig. 7.21). The skewness in the distribution of U is mainly due to the presence of state 1 (the blocking state).

7.6 Synthesis

In this chapter, we dealt with a classical problem in the theory of the general atmospheric circulation, that is, the characterisation of recurrent flow patterns observed at midlatitudes in the Northern Hemisphere winters. The problem is important in understanding the persistence and predictability of atmospheric motion beyond the time scales of baroclinic synoptic disturbances (2–5 days). Indeed, it is expected that insight into the nature of low-frequency regime dynamics will lead to significant progress in the so-called extended range weather forecasting.

At the same time, the problem is of great relevance in climate science, because it has been suggested that climate change predominantly manifests itself through changes in the atmospheric circulation regimes, that is, ‘changes in the probability density function of the climate attractor’ (Corti et al., 1999). In fact, misrepresentation of the statistics of blocking and planetary waves is widespread in climate models (Lucarini et al., 2007; Palmer et al., 2008): this may have a profound impact on the ability of such models to reproduce both current climate and climate change.

There are different approaches to the problem of low-frequency atmospheric variability. In this chapter, we presented an overview of the dynamical systems approach, where attractors and the behaviour of trajectories is studied within a hierarchy of

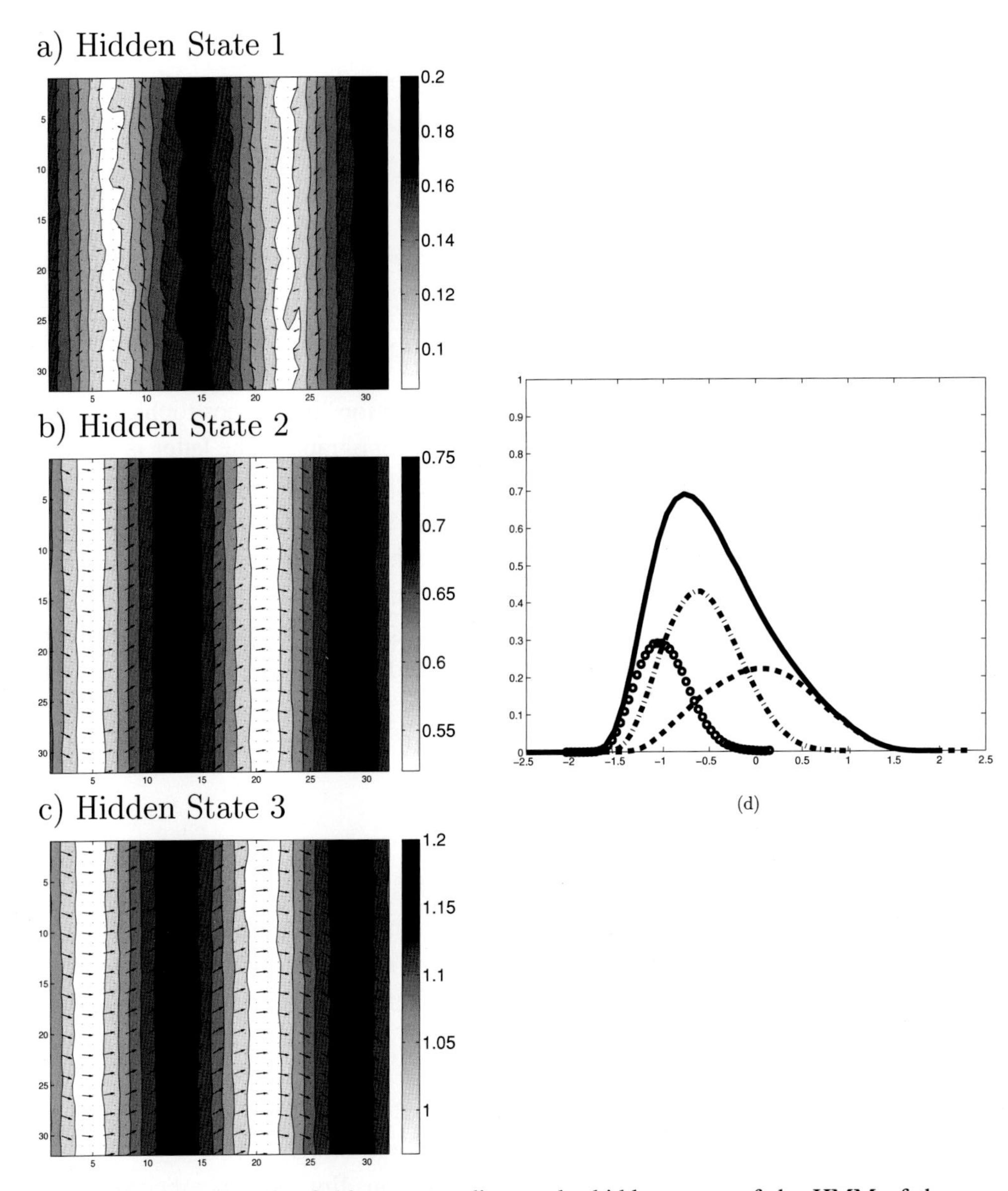

Figure 7.21 Velocity field corresponding to the hidden states of the HMM of the model (7.32), with (a) blocked state 1, (b) zonal flow state 2, and (c) zonal flow state 3. (d) Probability density distribution of U and (weighted) conditional probability density distributions of the hidden states (state 1: dashed; state 2: dashed-dotted and state 3: circles) for a topographic height $h_0 = 1.06$ (figure from Franzke et al., 2008).

models. From the material presented, the atmospheric flow is susceptible to many different type of instabilities (barotropic, baroclinic, topographic/orographic). These instabilities give rise to multiple steady states, multiple limit cycles, tori and more complex attractors, the latter having positive Lyapunov exponents (cf. Fig. 7.18).

A trajectory is expected to display a complicated path through phase space, as is indeed the case in even relatively simple atmospheric models. As we saw in Section 7.1, atmospheric low-frequency variability is characterised by only a few dominant regimes and specific intraseasonal oscillations, with only a few patterns involved. How do these patterns arise from so complex trajectories in phase space?

Orographic resonance theories lend support to the hypothesis that activity of planetary waves possesses a multimodal distribution (Benzi et al., 1986). Highly truncated barotropic models provide a relatively simple view on the origin of the multimodality. From the three-mode model of Charney and DeVore (1979) discussed in Section 7.3.2, it follows that the interaction between zonal flow and wave field causes the occurrence of two equilibria for the amplitude of planetary waves. Legras and Ghil (1983) found intermittent transitions between multiple equilibria representing blocked and zonal flows. Crommelin et al. (2004) explain the transitions in terms of homo- and heteroclinic dynamics near equilibria corresponding to distinct preferred flow patterns (Section 7.3.3). Sterk et al. (2010) add the element of intermittency due to bifurcations of waves to the mechanisms of low-frequency variability (Section 7.4.3).

In the higher-dimensional dynamical systems (quasi-geostrophic and shallow-water) models, a multitude of (mostly unstable) steady states (real fixed points) exists. Trajectories appear to be affected by these unstable fixed points, so clearly seen in the results of Itoh and Kimoto (1996). This is explained in Crommelin et al. (2004) and Sterk et al. (2010) because the multiple steady states are responsible for heteroclinic connections, leading to complicated temporal behavior with only a few patterns involved.

The transitions between different regions in phase space of slow dynamics (or regimes) can be induced by chaotic itinerancy (Itoh and Kimoto, 1996) or can occur along unstable periodic orbits (Selten and Branstator, 2004). A more kinematic view of the transitions is provided by a Markov chain model where a matrix of probabilities for transition from one regime to another is determined based on the expected residence time of the trajectory in each regime. Using this approach, Mo and Ghil (1988), for example, found that transitions between regimes in observations tend to avoid, rather than favour, passages through the climatological mean. Many trajectories in the results previously discussed display specific low-frequency (40–70 days) oscillatory behaviour associated with (stable or unstable) periodic orbits, which can coexist with the regime switches (Itoh and Kimoto, 1996). These prototypes of intraseasonal oscillations essentially depend on the existence of the orography (Keppenne et al., 2000). Patterns of these oscillations also show a close correspondence with those in observations (Kondrashov et al., 2004).

Hence, from the dynamical systems viewpoint, the atmospheric low-frequency variability characterising the Northern Hemisphere midlatitude circulation results from dynamical processes specific to the interaction of zonal flow and planetary waves with orography. The phase space is occupied with clusters of slow dynamics (the regimes) where a trajectory displays low-frequency variability within each cluster with a related pattern. Transitions between the different clusters can occur through noise or deterministic dynamics; hidden Markov chain models as presented in Section 7.5 are a nice tool to model these transitions (Majda et al., 2006, Franzke et al., 2008).

Despite this remarkable research effort, the scientific debate is still very much open on whether a single equilibrium (Ambaum, 2008; Nitsche et al., 1994; Stephenson et al., 2004; Berner and Branstator, 2007) or multiple equilibria (Hansen and Sutera, 1995; Mo and Ghil, 1988; Ruti et al., 2006) characterise the large-scale midlatitude atmospheric circulation. Quite a gap still exists to interpret observations and results of high-resolution atmospheric models with the results of two-layer quasi-geostrophic and shallow-water models. The idealised dynamical models do not capture the time-mean state, all relevant patterns and, consequently, the transitions among them, very well. A nice bridge may be the laboratory work on barotropic flow over topography (Weeks et al., 1997; Tian et al., 2001) where clearly regime-type behaviour is found in a realistic flow (instead of a truncated model).

However, observational time series are still too short to extract transition probabilities between regimes with statistical significance. It is not clear whether nonlinear interactions of waves of different spatial scales play an essential role in the onset or the maintenance of low-frequency atmospheric variability. Alternative mechanisms have also been proposed; the low-frequency large-scale pattern of the North Atlantic Oscillation is found, in Benedict et al. (2004), to result from breaking of synoptic scale waves, where the anticyclonic (cyclonic) wave breaking evolves into the positive (negative) NAO phase. A description of this and other alternative mechanisms (Athanasiadis and Ambaum, 2010; Jin, 2010) is clearly outside the scope of this chapter.

8
El Niño Variability

A swinging theory
DGDGA♯D, California Dreamin', Chris Proctor

About once every four years, the sea-surface temperature in the Eastern Equatorial Pacific is a few degrees higher than normal (Philander, 1990). Near the South American coast, this warming of the ocean water is usually at its maximum around Christmas. Long ago, Peruvian fishermen called it El Niño, the Spanish phrase for the Christ Child.

8.1 Phenomena

During the past several decades, El Niño has been observed in unprecedented detail thanks to the implementation of the TAO/TRITON array and the launch of satellite-borne instruments (McPhaden et al., 1998). The relevant quantities to characterise the state in the equatorial ocean and atmosphere are sea level pressure, sea-surface temperature (SST), sea level height, surface wind and ocean subsurface temperature.

The annual mean state of the equatorial Pacific sea-surface temperature is characterised by the zonal contrast between the western Pacific "warm pool" and the "cold tongue" in the eastern Pacific. The mean temperature in the eastern Pacific is approximately 23°C, with seasonal excursions of about 3°C. What makes El Niño unique

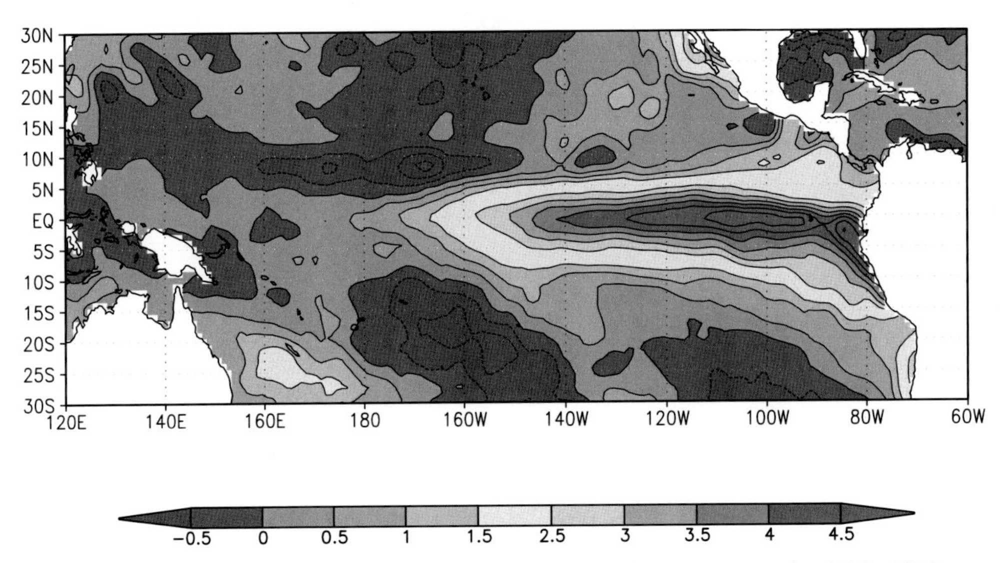

Figure 8.1 Sea-surface temperature anomaly field (with respect to the 1982–2010 mean) of December 1997 at the height of the 1997/1998 El Niño. Data from NOAA, see http://www.emc.ncep.noaa.gov/research/cmb/sst_analysis/. (See Colour Plate.)

among other interesting phenomena of natural climate variability is that it has both a well-defined spatial pattern and a relatively well-defined time scale. The pattern of the sea-surface temperature anomaly for December 1997 is plotted in Fig. 8.1 and shows a large area where the SST is larger than average.

A common index of this sea-surface temperature anomaly pattern is the NINO3 index, defined as the sea-surface temperature anomaly averaged over the region 5°S–5°N, 150°W–90°W. In the time series (blue curve in Fig. 8.2a), the high NINO3 periods are known as El Niños and the low NINO3 periods as La Niñas. There is no clear-cut distinction between El Niños, La Niñas, and normal periods; rather, the system exhibits continuous fluctuations of varying strengths and durations with an average period of about 4 years (blue curve in Fig. 8.2b).

Changes in the tropical atmospheric circulation are strongly connected to changes in sea-surface temperature. The red curve in Fig. 8.2a is the normalised pressure difference between Tahiti (Eastern Pacific region) and Darwin; this index is referred to as the Southern Oscillation Index (SOI). It measures the variations in the tropical surface winds, dominated by the trade winds. When the SOI is negative (positive), the pressure in Tahiti is relatively low (high) with respect to that in Darwin, and hence the trade winds are weakened (strengthened). The anticorrelation of the NINO3 index and the SOI is obvious from Fig. 8.2a, and the spectrum of the SOI in Fig. 8.2b also shows a similar broad peak as the NINO3 index (centered at about 4 years). As El Niño and the Southern Oscillation are one phenomenon, it is referred to as the ENSO phenomenon.

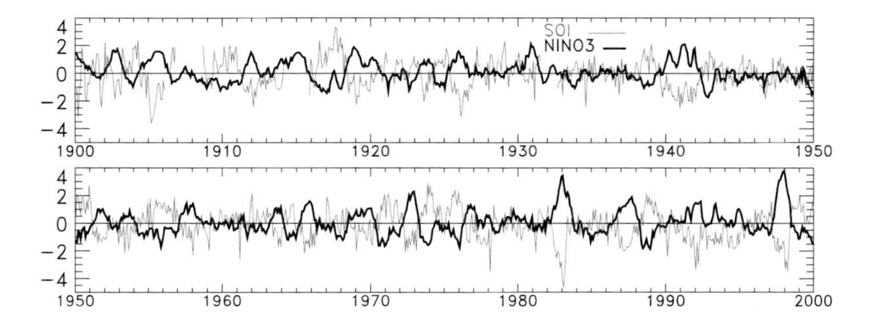

(a)

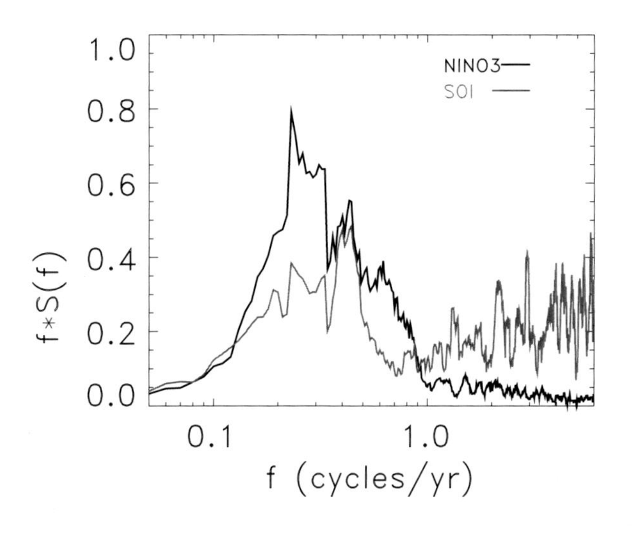

(b)

Figure 8.2 (a) Time series of the NINO3 index (dark) and SOI (light) over the years 1900–2000. (b) Spectrum of the NINO3 index (dark) and SOI (light) in (a), where on the vertical axis the product of frequency f and the spectral power $S(f)$ is plotted (figure based on Dijkstra and Burgers, 2002).

There are relations among the seasonal cycle, the spatial sea-surface temperature pattern of the annual-mean state and the El Niño variability. Large sea-surface temperature anomalies often occur within the cold tongue region. In addition, El Niño is to some extent phase-locked to the seasonal cycle, as most El Niños and La Niñas peak around December (Fig. 8.3). The root mean square of the NINO3 index is almost twice as large in December than in April.

When one considers the spectrum of the NINO3 index (Fig. 8.2b), energy is also found at lower frequencies, in particular in the decadal-to-interdecadal range (Jiang et al., 1995; Zhang et al., 1997; Fedorov and Philander, 2000). The strength of El Niño before the mid-1970s appears to be smaller than that after this period; this transition is sometimes referred to as the 'Pacific climate shift' (Trenberth, 1997). According to

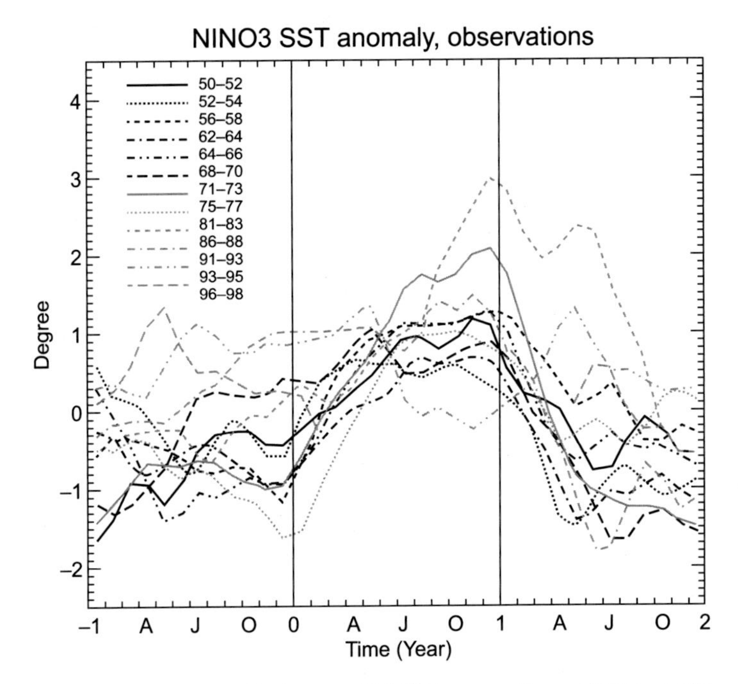

Figure 8.3 NINO3 index for 13 observed ENSO events from 1950 to 1998 with the mean seasonal cycle of this period removed (from Neelin et al. [2000]). The curves are aligned based on the year of the peak warm phase (at year 1).

NCEP data, the standard deviation of the SOI (NINO3) for 1951–1975 is 1.64 (0.81), to be compared for 1976–2000, where it is 1.84 (1.00). The spatial pattern of these (multi)decadal changes is fairly similar to that of the interannual variability, but the sea-surface temperature anomalies at the eastern side of the basin extend from the equator to midlatitudes (Zhang et al., 1997).

8.2 A minimal model

One of the first models that was able to reasonably simulate ENSO was that of Zebiak and Cane (1987). In its original version, an annual mean state or seasonal cycle of both ocean and atmosphere was obtained from observations, and within the model, the evolution of anomalies with respect to this reference state was computed. We refer next to this intermediate complexity model (ICM) as the ZC model.

8.2.1 Formulation

The ZC model captures the evolution of large-scale motions in the tropical ocean and atmosphere in a domain of infinite extent in the meridional direction. The ocean is bounded by meridional walls at the west ($x = 0$) and east ($x = L$) coast. The ocean

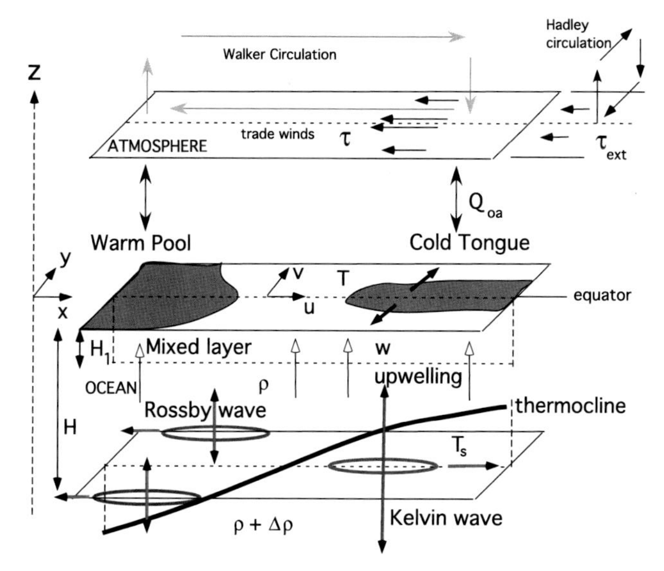

Figure 8.4 Overview of the oceanic and atmospheric processes of the equatorial coupled ocean-atmosphere system, which are represented in the ZC model.

component of the model consists of a well-mixed layer of mean depth H_1 embedded in a shallow-water layer of mean depth $H = H_1 + H_2$ having a constant density ρ (Fig. 8.4). The lower boundary of the shallow-water layer corresponds to the region of a sharp vertical gradient in temperature and is referred to as the thermocline. Only long wave motions above the thermocline are considered, and the deep ocean (having a constant density $\rho + \Delta\rho$) is assumed to be at rest.

We start with the ocean model component. In a first step, the total horizontal velocity field is split into a mean $\mathbf{u}$ over the mixed layer and shallow-water layer and a difference velocity, the surface velocity $\mathbf{u}_s$. The mean velocity $\mathbf{u} = (u, v)$ satisfies a reduced gravity model on the equatorial β plane (with planetary vorticity gradient β_0). With the surface wind stress field given by (τ^x, τ^y), these equations are given by

$$\frac{\partial u}{\partial t} + a_m u - \beta_0 y v + g' \frac{\partial h}{\partial x} = \frac{\tau^x}{\rho H}, \tag{8.1a}$$

$$\beta_0 y u + g' \frac{\partial h}{\partial y} = \frac{\tau^y}{\rho H}, \tag{8.1b}$$

$$\frac{\partial h}{\partial t} + a_m h + c_o^2 \left(\frac{\partial u}{\partial x} + \frac{\partial v}{\partial y} \right) = 0, \tag{8.1c}$$

where h is the total upper layer thickness. In addition, a linear damping coefficient a_m and the reduced gravity $g' = g\Delta\rho/\rho$ are introduced. The quantity $c_0 = \sqrt{g'H}$ is the equatorial Kelvin wave speed. In the long-wave approximation, the boundary conditions are

$$\int_{-\infty}^{\infty} u(0, y, t)dy = 0 \ , \ u(L, y, t) = 0. \tag{8.2}$$

The equations for the surface layer velocities $\mathbf{u}_s = (u_s, v_s)$ are the Ekman-layer balances

$$a_s u_s - \beta_0 y v_s = \frac{H_2}{H} \frac{\tau^x}{\rho H_1}, \tag{8.3a}$$

$$a_s v_s + \beta_0 y u_s = \frac{H_2}{H} \frac{\tau^y}{\rho H_1}, \tag{8.3b}$$

where a_s is a linear damping coefficient. With $u_1 = u_s + u$ and $v_1 = v_s + v$ being the horizontal velocity components in the mixed layer, the vertical velocity component is determined from continuity as

$$w_1 = H_1 \left(\frac{\partial u_1}{\partial x} + \frac{\partial v_1}{\partial y} \right). \tag{8.4}$$

The evolution of the mixed layer temperature T is governed by the equation

$$\frac{\partial T}{\partial t} + a_T(T - T_0) + \frac{w_1}{H_u}\mathcal{H}(w_1)(T - T_s(h)) + u_1 \frac{\partial T}{\partial x} + v_1 \frac{\partial T}{\partial y} = 0, \tag{8.5}$$

where $\mathcal{H}$ is a continuous approximation of the Heaviside function. The second term in (8.5) is usually referred to as the Newtonian cooling term, with inverse damping time scale a_T, representing all processes as horizontal mixing, sensible and latent heat surface fluxes, and long wave and shortwave radiation. T_0 is the temperature of radiative equilibrium, which is realised in the absence of large-scale horizontal motion in the upper ocean and atmosphere. The third term in (8.5) models the heat flux due to upwelling through the total velocity w_1 and the approximate vertical temperature gradient $(T - T_s(h))/H_u$. The subsurface temperature (T_s) depends on the thermocline deviations and models the effect that heat is transported upwards (if $w_1 > 0$) when the cold water is further from the surface. An explicit expression, which is often used, is

$$T_s(h) = T_{s0} + (T_0 - T_{s0}) \tanh \left[\frac{h + h_0}{h_1} \right], \tag{8.6}$$

where h_0 is an offset value, T_{s0} is the subsurface temperature for $h = -h_0$ and h_1 controls the steepness of the transition as h passes through $-h_0$. The last two terms in (8.5) represent horizontal advection of heat.

Sea-surface temperature anomalies cause wind anomalies, and one of the simplest models to represent the wind response can be determined by the Gill (1980) model. In this description, the atmospheric zonal and meridional surface velocities (U, V) and geopotential Θ satisfy

$$\frac{\partial U}{\partial t} + a_M U - \beta_0 y V - \frac{\partial \Theta}{\partial x} = 0, \tag{8.7a}$$

$$\frac{\partial V}{\partial t} + a_M V + \beta_0 y U - \frac{\partial \Theta}{\partial y} = 0, \tag{8.7b}$$

$$\frac{\partial \Theta}{\partial t} + a_M \Theta - c_a^2 \left(\frac{\partial U}{\partial x} + \frac{\partial V}{\partial y} \right) = \alpha_T (T - T_r), \tag{8.7c}$$

where a_M is a damping coefficient and c_a^2 is the atmospheric equatorial Kelvin wave speed. The right-hand side of (8.7c) is proportional to the ocean-atmosphere heat flux Q_{oa}, and T_r is a reference temperature.

Atmospheric surface velocity anomalies give wind stress anomalies to the ocean surface according to

$$\frac{\tau^x}{\rho H} = \gamma U \; ; \; \frac{\tau^y}{\rho H} = \gamma V, \tag{8.8}$$

which completes the coupling between the atmosphere and ocean. In Zebiak and Cane (1987), many more details of the model set-up are provided.

8.2.2 Ocean processes

The effects of the wind stress on the upper ocean are threefold. As described previously, the dominantly easterly wind stress causes water to pile up near the western part of the basin. This induces a higher pressure in the upper layer western Pacific than that in the eastern Pacific and, consequently, a shallowing of the thermocline towards the east. Second, the winds cause divergences and convergences of mass in the upper ocean, due to the frictional (Ekman) boundary layer in which the momentum input is transferred. North of the equator, the trade winds cause an Ekman transport to the right of the wind away from the equator. Similarly, south of the equator the Ekman mass transport is away from the equator. With a wind stress amplitude of 0.1 Pa, a typical value of the vertical velocity is a few meters per day. Finally, the wind stress is responsible for the presence of the upper ocean currents.

When the amplitude and/or the direction of the wind stress changes, the upper ocean adjusts through wave dynamics. The most important waves involved in this adjustment process are as follows:

(i) Equatorial Kelvin waves. For such a wave, the meridional structure of the thermocline is maximal at the equator. Its amplitude decays exponentially in meridional direction with a decay scale λ_0 of about 300 km. The zonal velocity has the same spatial structure as the thermocline, and the meridional velocity is zero. The group/phase velocity of these nondispersive waves is $c_o \approx 2\ \mathrm{m\,s^{-1}}$. It takes about $\tau_K = 3$ months to cross the Pacific basin of width $L = 15{,}000$ km.

(ii) Long equatorial Rossby waves. For such a wave, the meridional thermocline structure has an off-equatorial maximum. For each of these waves, the meridional velocity is much smaller than the zonal velocity. The group/phase velocity of these nondispersive waves c_j, with a meridional spatial structure coupled to the index j, is $c_j = -c_o/(2j+1)$. The $j = 1$ long Rossby wave travels westward with a velocity that is one-third that of the Kelvin wave and hence takes about $\tau_R = 9$ months to cross the Pacific basin.

When a Kelvin wave meets the east coast, the dominant contribution of the reflected signal is long Rossby waves. Similarly, when a Rossby wave meets the west coast, it is reflected dominantly as a Kelvin wave.

The small-amplitude response of the flow in an equatorial ocean basin to a certain wind field can be described as a directly forced response and a sum of free waves to satisfy the boundary conditions (Moore, 1968; Cane and Sarachik, 1977). In the response to a time-periodic wind stress, the ocean reacts not only to the instantaneous wind pattern but also to previous winds through propagation of waves. Along the equator, the ocean adjustment is mainly accomplished by relatively fast eastward Kelvin waves.

Off-equatorial adjustment is accomplished by slower Rossby waves that travel westward. The off-equatorial thermocline pattern consists partly of free Rossby waves that are adjusting to the wind and partly of a forced response that is in quasi-steady balance with the instantaneous wind stress. It is the departure of this quasi-steady balance that is crucial to further evolution of the thermocline and provides the ocean with a memory. A measure of this memory was considered in Neelin et al. (1998) to be the difference between the actual response and that which would be, at every time, in steady balance with the instantaneous wind stress. For the wind patterns over the Tropical Pacific, this 'ocean memory' is largest near the western boundary.

8.2.3 Coupled feedbacks

If the underlying sea surface is warm, the air above it is heated and rises. When one assumes that convection mostly occurs over the warmest water and that the adiabatic heating is dominated by heat that is released during precipitation, latent heat anomalies are proportional to sea-surface temperature anomalies, represented by the proportionality factor α_T in (8.7). In this case, a low-level zonal wind response is found with westerly (easterly) winds to the west (east) of positive (negative) sea-surface temperature anomalies (Fig. 8.5). As the zonal wind stress anomalies are proportional to the zonal wind anomalies, with proportionality factor γ (8.8), the nature of the coupling between ocean and atmosphere can now be explained.

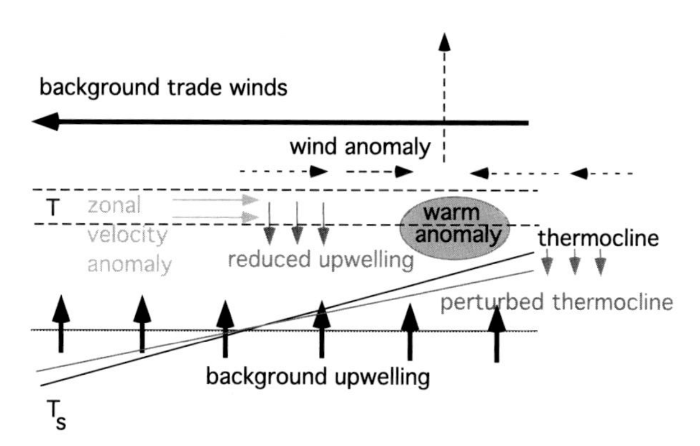

Figure 8.5 The thermocline feedback, the upwelling feedback and the zonal advection feedback.

A sea-surface temperature anomaly will give, through local heating, a lower-level wind anomaly. The resulting wind stress anomaly on the ocean-atmosphere surface will (i) change the thermocline slope through horizontal pressure differences in the upper ocean, (ii) change the strength of the upwelling through the Ekman divergences in the upper layer and (iii) affect the upper ocean currents (u, v) in the mixed layer. The changes in velocity field and thermocline field will affect the sea-surface temperature, according to (8.5).

The strength of the coupling between the ocean and atmosphere is determined by the combined effects of the amplitude of the zonal wind anomaly that is generated by sea-surface temperature anomalies and how much of the momentum of this wind is transferred as stress to the upper ocean layer. The strength of the coupling is measured by the parameter μ, which is a (dimensionless) product of α_T and γ, that is,

$$\mu = \frac{\gamma \alpha_T \Delta T L^2}{c_o^2 c_a^2}, \tag{8.9}$$

where the quantity ΔT is a typical zonal temperature difference over the basin and L is the zonal length of the basin.

The main positive feedbacks identified in this coupled ocean-atmosphere system are referred to as the thermocline, upwelling and zonal advection feedback (Neelin, 1991). They are best illustrated by looking at the growth of very small perturbations (quantities with a tilde) on a background state (quantities with a bar). The linearised temperature equation (8.5), describing the evolution of the temperature perturbations $\tilde{T}$, can be written as

$$\frac{\partial \tilde{T}}{\partial t} + a_T \tilde{T} + \frac{\bar{w}_1}{H_u} \mathcal{H}(\bar{w}_1)(\tilde{T} - \tilde{T}_s) + \frac{\tilde{w}_1}{H_u} \mathcal{H}(\bar{w}_1)(\bar{T} - \bar{T}_s)$$
$$+ \bar{u}_1 \frac{\partial \tilde{T}}{\partial x} + \bar{v}_1 \frac{\partial \tilde{T}}{\partial y} + \tilde{u}_1 \frac{\partial \bar{T}}{\partial x} + \tilde{v}_1 \frac{\partial \bar{T}}{\partial y} = 0, \tag{8.10}$$

The thermocline feedback is best explained by looking at a sloping thermocline in a constant upwelling ($\bar{w}_1 > 0$) ocean, as sketched in Fig. 8.5. The sloping thermocline and the upwelling are caused by the background easterly winds. Now assume that a positive sea-surface temperature perturbation $\tilde{T}$ is present at some location, for example, in the eastern part of the basin. This leads to a perturbation in the low-level zonal wind that is westerly with a maximum located west of the maximum of the sea-surface temperature anomaly. Because the background winds are weakened locally, the slope of the thermocline decreases and it becomes more flat (red arrows in Fig. 8.5). In this case, the colder water will be closer to the surface in the west, but it will be farther down in the east. Hence the subsurface temperature effectively increases at the level of upwelling, giving a positive heat flux perturbation at the bottom of the mixed layer. According to (8.10),

$$\partial \tilde{T}/\partial t \approx -\bar{w}_1(\tilde{T} - \tilde{T}_s)/H_u,$$

and when $\tilde{T}_s - \tilde{T} > 0$, the original disturbance is amplified, as $\bar{w}_1$ is positive. As $\tilde{T}_s \sim \tilde{h}$, a deeper thermocline in the east can amplify a positive temperature anomaly. The thermocline feedback is present in a transient state but also in an balanced (adjusted) state.

To understand the upwelling feedback, consider again a positive sea-surface temperature anomaly in the eastern part of the basin and associated changes in the wind. However, now changes in the upwelling $\tilde{w}_1$, mainly through the Ekman layer dynamics, occur. Weaker easterly winds imply less upwelling, and hence less colder water enters the mixed layer (blue arrows in Fig. 8.5). If $\tilde{w} < 0$ and the background vertical temperature gradient is stably stratified ($\bar{T} > \bar{T}_s$), then the sea-surface temperature perturbation is amplified. The latter can be seen again from (8.10), that is,

$$\partial \tilde{T}/\partial t \approx -\tilde{w}_1(\bar{T} - \bar{T}_s)/H_u.$$

The zonal advection feedback arises through zonal advection of heat. Imagine a region with a strong sea-surface temperature gradient, say $\partial \bar{T}/\partial x < 0$. Such a region occurs for example at the east side of the warm pool. Suppose a positive anomaly in sea-surface temperature occurs, leading to westerly wind anomalies. Consequently, the zonal surface ocean current ($\tilde{u} > 0$) is intensified (green arrows in Fig. 8.5), leading to amplification of the positive temperature perturbation, according to (8.10), that is,

$$\partial \tilde{T}/\partial t \approx -\tilde{u}_1\, \partial \bar{T}/\partial x.$$

Part of the mixed layer zonal velocity will be due to wave dynamics and part due to Ekman dynamics.

8.3 The ENSO mode

In many early model studies on ENSO variability, the annual-mean state was simply prescribed without any consideration of the processes causing this state. For example, a flat thermocline was chosen, with a spatially constant zonal temperature gradient and no-flow in both ocean and atmosphere (Hirst, 1986; Neelin, 1991). Alternatively, an annual-mean state or a seasonal cycle derived from observations was prescribed (Zebiak and Cane, 1987). Later, however, it was recognised that the annual-mean state also results from the same coupled processes involved in ENSO (Neelin and Dijkstra, 1995).

8.3.1 The annual-mean state

To understand the spatial pattern of the warm pool/cold tongue, it was recognised that part of the (annual mean) wind stress in the Pacific basin is in fact related to its zonal temperature gradient (Neelin and Dijkstra, 1995). Only a small part of the wind stress, say $\vec{\tau}_{\text{ext}}$, is determined externally. The external zonal wind stress τ^x_{ext} can be taken constant, imagined as being caused by the zonally symmetric atmospheric circulation.

At zero coupling strength ($\mu = 0$), the ocean circulation and, consequently, the sea-surface temperature is determined by the external zonal wind stress τ^x_{ext}. A small easterly wind stress causes a small amount of upwelling and a small slope in the thermocline. Within the ZC model variant used in Van der Vaart et al. (2000), the equatorial sea-surface temperature increases monotonically from about 26°C (Fig. 8.6a) in the east to about 29°C in the west, in response to $\tau^x_{\text{ext}} = 0.01$ Pa. The thermocline is approximately linear at the equator, its depth is increasing westwards, and it has slight off-equatorial maxima.

Additional easterly wind stress occurs due to coupling because of the zonal temperature gradient. In Neelin and Dijkstra (1995), the zonal wind stress was represented as

$$\tau^x = \tau^x_{\text{ext}} + \mu \mathcal{A}(T - T_0), \tag{8.11}$$

where $\mathcal{A}$ is a short representation of the atmosphere model (8.7) and T_0 is again the radiative equilibrium temperature. Increasing μ leads to larger upwelling and a larger thermocline slope, strengthening the cold tongue in the eastern part of the basin. The temperature of the cold tongue T_C is shown in Fig. 8.6a as a function of μ. The zonal scale of the cold tongue is set by a delicate balance of the different coupled feedbacks. The thermocline field at $\mu = 0.5$ (Fig. 8.6b) displays the off-equatorial maxima and a deeper (shallower) equatorial thermocline in the west (east). This shows that coupled processes are involved in the annual mean spatial patterns of the Pacific climate system (Dijkstra and Neelin, 1995).

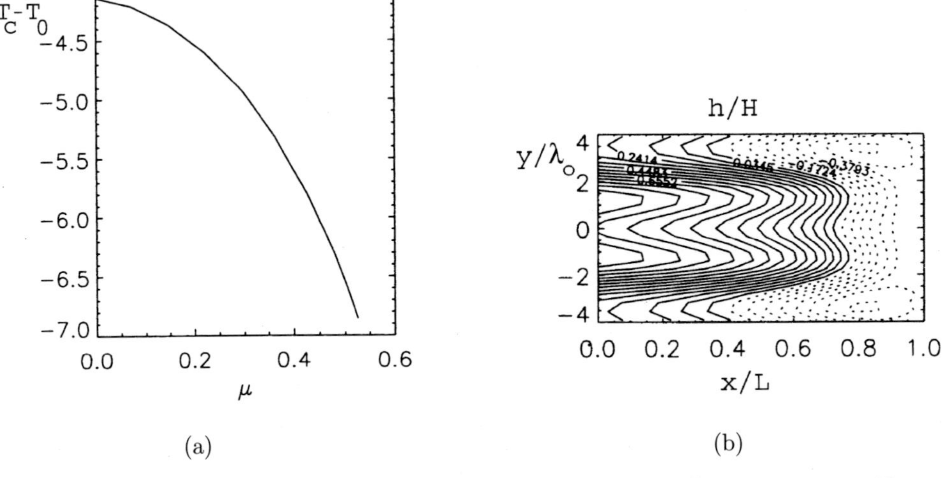

Figure 8.6 (a) Eastern Pacific ($x/L = 0.8$) equatorial sea-surface temperature T_C deviation from $T_0 = 30°$C as a function of the coupling strength μ. (b) Pattern of the thermocline depth anomaly at $\mu = 0.5$. The quantities $\lambda_0 = \sqrt{c_0/\beta_0}$ ($\sim$ 300 km) and $H = 100$ m are typical meridional and vertical scales of the thermocline (figure from Van der Vaart et al., 2000).

8.3.2 Spatial pattern of the ENSO mode

Taking the annual-mean state as a background state, we are now interested in its sensitivity to small perturbations. Necessary conditions for instability can be obtained by determining the linear stability boundary through normal mode analysis. In such an analysis, an arbitrary perturbation is decomposed in modes (e.g., Fourier modes), and the growth or decay of each of these modes is investigated. As we have seen in Chapter 2, when the background state is stationary, the time dependence of each mode is of the form $e^{\sigma t}$, where $\sigma = \sigma_r + i\sigma_i$ is the complex growth rate. With μ as the control parameter, the linear stability boundary μ_c then is the first value of μ where $\sigma_r = 0$ for one particular normal mode. A mode with $\sigma_i \neq 0$ is oscillatory (with a period $2\pi/\sigma_i$), whereas a mode with $\sigma_i = 0$ is called stationary. A Hopf bifurcation occurs when the growth rate of an oscillatory mode changes sign (Section 2.2). If the growth factor of a mode is positive, then sustained oscillations will occur. If the growth factor of one of these modes is negative, it will be damped.

In early studies (Hirst, 1986), the stability of a prescribed spatially constant mean state in a periodic ocean basin was considered. In this case, both uncoupled ocean and atmospheric Rossby and Kelvin waves may destabilise due to the coupled processes. The resulting unstable modes are travelling waves, which have an interannual oscillation period for wavelengths that are in the order of the basin size (Philander et al., 1984). Also a slow westward propagating mode, a so-called SST mode, may become

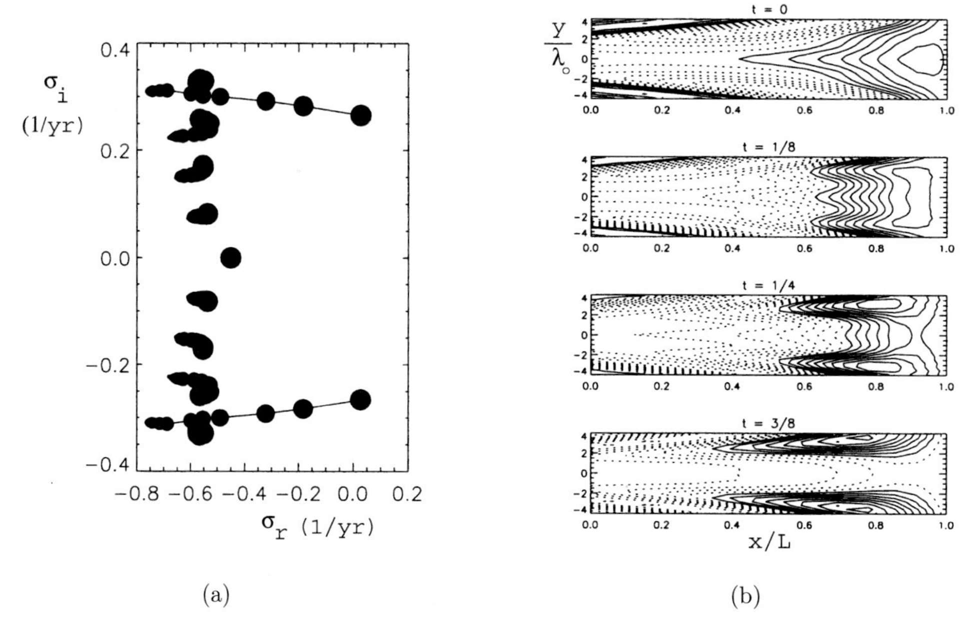

Figure 8.7 (a) Plot of the eigenvalues for the six leading eigenmodes in the (σ_r, σ_i) plane. Values of the coupling strength μ are represented by dot size (smallest dot is the uncoupled case $[\mu = 0]$ for each mode, largest dot is the fully coupled case at the stability boundary $[\mu_c = 0.5]$, the location of the Hopf bifurcation). (b) Planforms of the thermocline depth anomaly at several phases of the 3.7-year oscillation; time $t = 1/2$ refers to half a period. Drawn (dotted) lines represent positive (negative) anomalies (figure from Van der Vaart et al., 2000).

unstable. This mode is not related to wave dynamics in either the atmosphere or the ocean but to adjustment processes of SST (Neelin, 1991).

The linear stability of prescribed zonally varying annual-mean states within a bounded basin was investigated in Jin and Neelin (1993). In the uncoupled case, two distinct sets of modes appear. One set is primarily related to the time scales of sea-surface temperature change (the previously mentioned SST modes), and the other set is related to time scales of ocean adjustment (ocean-dynamics modes). Depending on the other parameters in the model, the mean state can become unstable to stationary instabilities as well as oscillatory ones. In the parameter regime considered 'most realistic', referred to as the 'standing oscillatory' regime, merging of an oscillatory ocean dynamics mode with a stationary SST mode occurs. This leads to slightly growing (so-called mixed SST/ocean-dynamics) modes, which inherit their spatial structure from the stationary SST mode, but for which their interannual oscillation period is set by ocean subsurface dynamics (Neelin et al., 1994).

The linear stability of the fully coupled annual-mean states in Fig. 8.6 was studied in Van der Vaart et al. (2000). In Fig. 8.7a, the path of six modes – which become

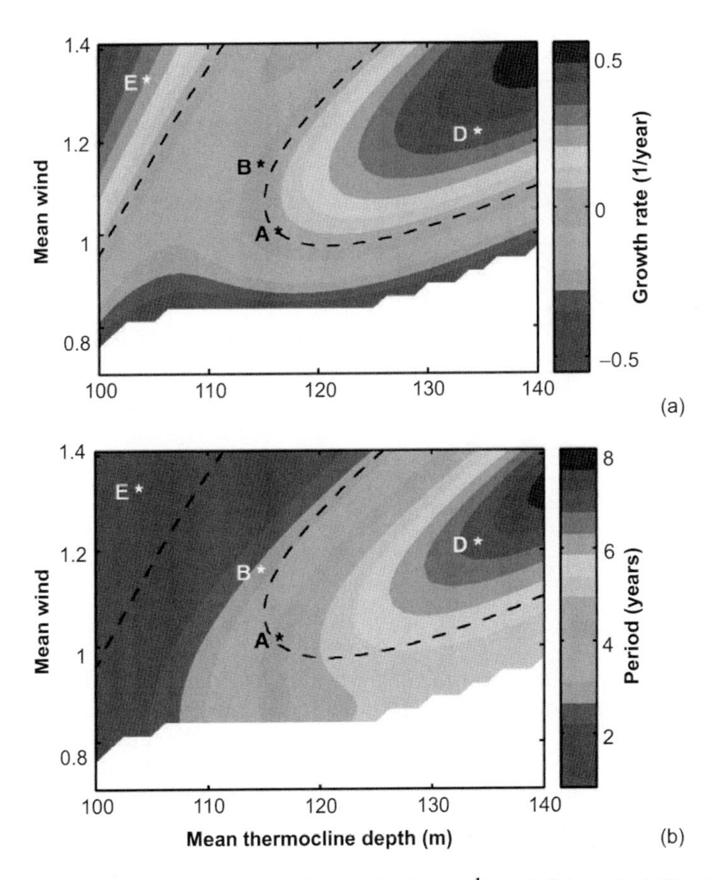

Figure 8.8 Dependence of (a) growth rate (σ_r in yr^{-1}) and (b) period ($2\pi/\sigma_i$ in yr) of the ENSO mode on the mean zonal wind intensity (in units of 0.5 cm^{2}s^{-2}) and the mean thermocline depth (in m) (figure from Fedorov and Philander, 2000). (See Colour Plate.)

leading eigenmodes at high coupling – is plotted as a function of the coupling strength μ. A larger dot size indicates a larger value of μ, and both oscillation frequency σ_i and growth rate σ_r of the modes are given in year^{-1}.

The growth rate of one oscillatory mode becomes positive as μ is increased beyond $\mu_c \approx 0.5$, which is the location of the first Hopf bifurcation. The equatorial sea-surface temperature pattern of this mode displays a nearly standing oscillation for which the spatial scale is confined to the cold tongue of the mean state. The wind response is much broader zonally and is in phase with the sea-surface temperature anomaly. In the spatial structure of the thermocline field (plotted in Fig. 8.7b at several phases of the oscillation), the eastward propagation of equatorial anomalies, their reflection at the eastern boundary and subsequent off-equatorial westward propagation can be distinguished.

Using prescribed annual-mean states, Fedorov and Philander (2000) provide an overview of the dependence of the growth factor (Fig. 8.8a) and period (Fig. 8.8b) of

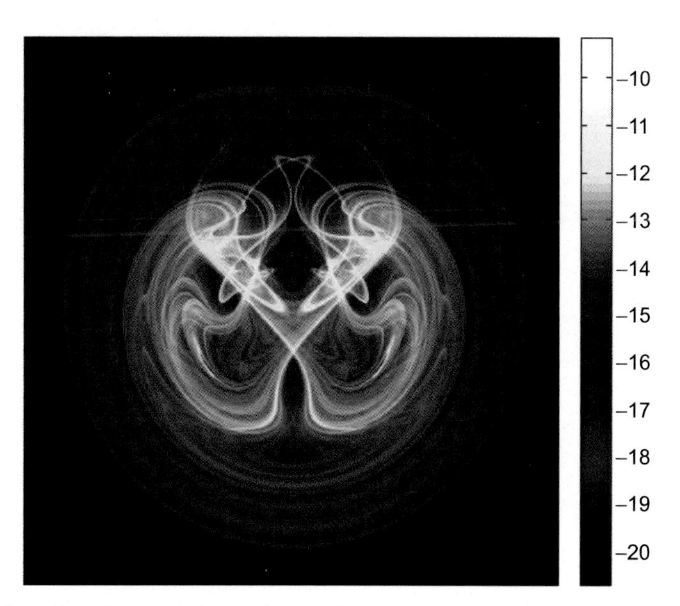

Plate 4.4 Density of trajectories of the random Lorenz system (4.44), projected on the (Y_t, Z_t) plane. The colour scale is logarithmic, with yellow indicating a high density of sample paths; see text for further description (figure from Chekroun et al., 2011).

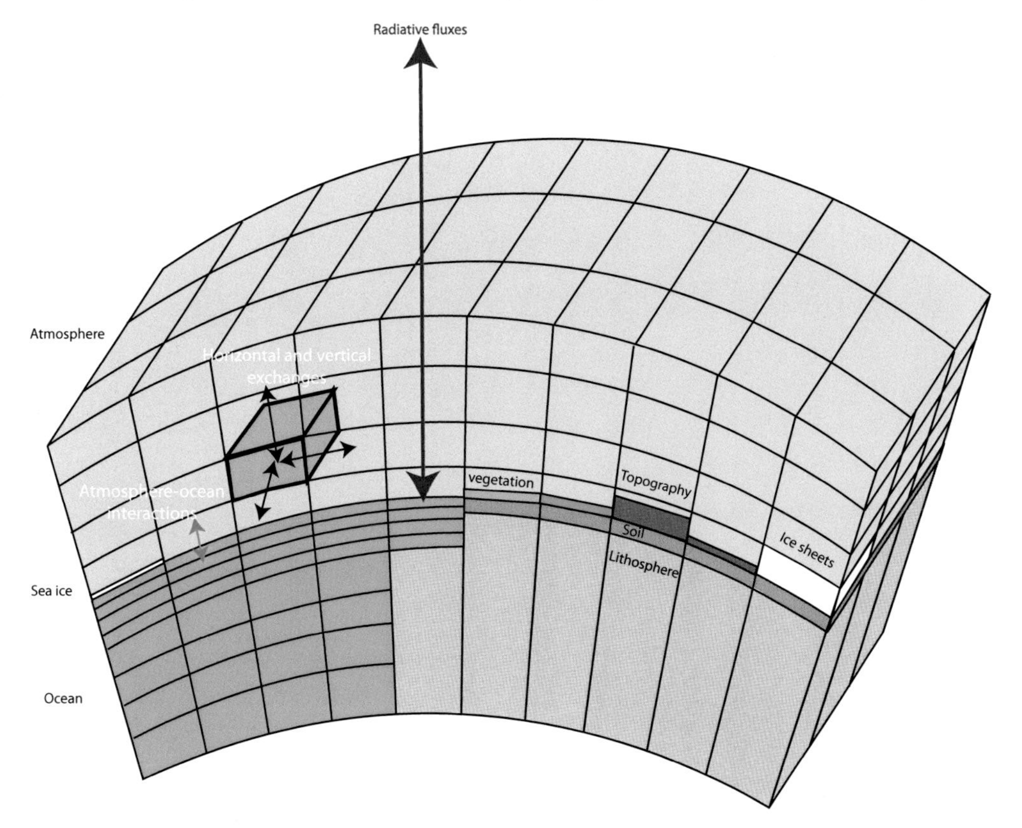

Plate 6.4 A typical structure of a GCM; the number of grid boxes in each of the components determines the spatial resolution of the model (figure from http://stratus.astr.ucl.ac.be/textbook/ and courtesy of Hugues Goosse).

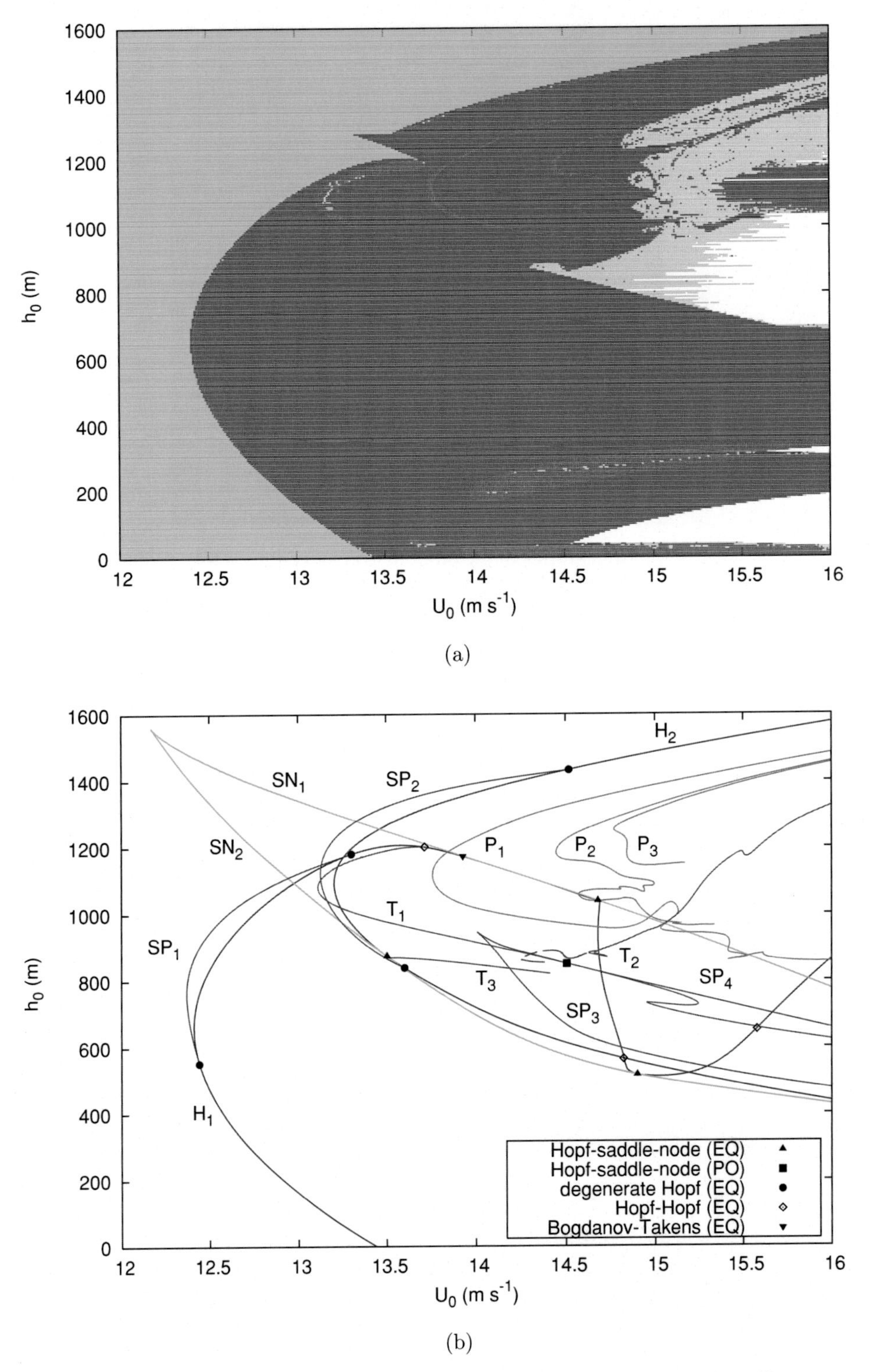

Plate 7.18 (a) Lyapunov diagram for the attractors of the system. (b) Regime diagram of attractors of the low-order model in the (U_0, h_0) parameter plane. The marked locations are codimension-2 bifurcations. The colour coding of both diagrams is provided in Table 7.2.

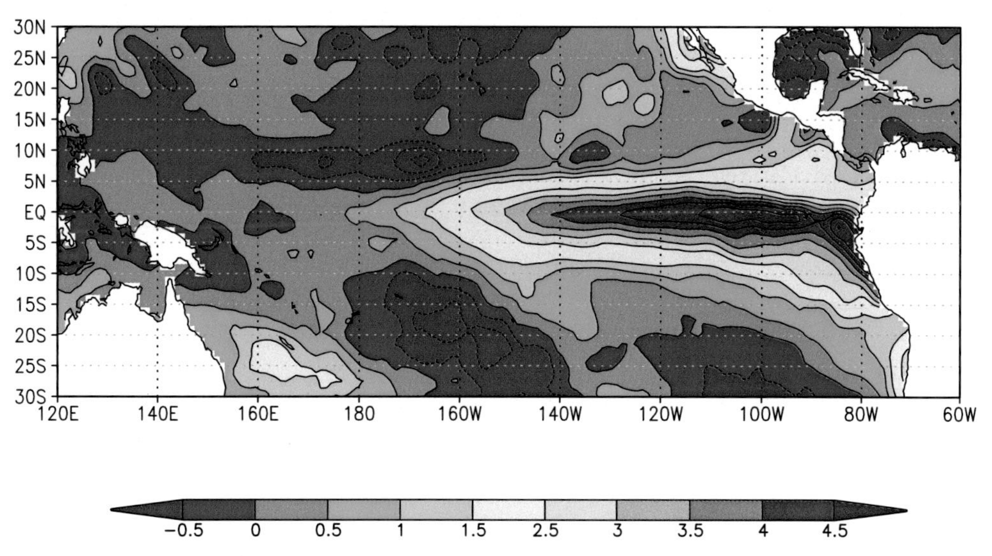

Plate 8.1 Sea-surface temperature anomaly field (with respect to the 1982–2010 mean) of December 1997 at the height of the 1997/1998 El Niño. Data from NOAA, see http://www.emc.ncep.noaa.gov/research/cmb/sst_analysis/.

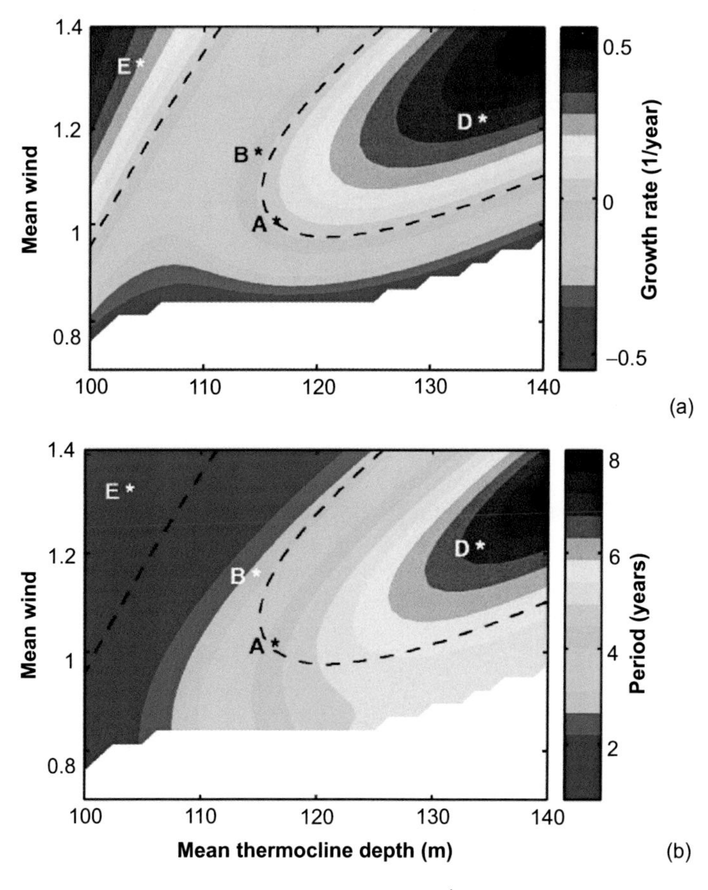

Plate 8.8 Dependence of (a) growth rate (σ_r in yr^{-1}) and (b) period ($2\pi/\sigma_i$ in yr) of the ENSO mode on the mean zonal wind stress (in units of 0.5 $\mathrm{cm}^2\mathrm{s}^{-2}$) and the mean thermocline depth (in m) (figure from Fedorov and Philander, 2000).

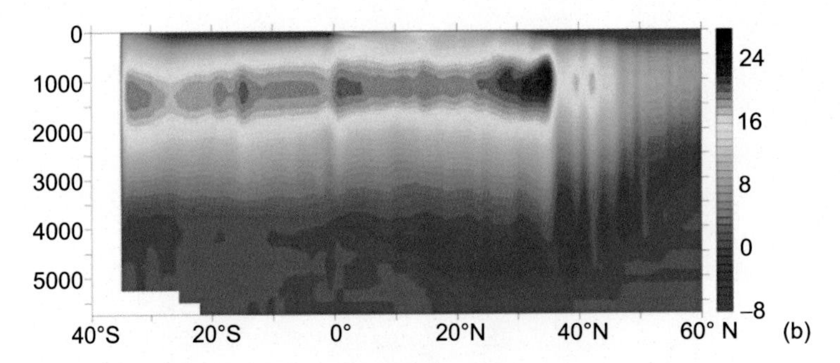

Plate 10.3 (b) Contour plot of the 10-year mean MOC (contours in Sv) in the Atlantic from the control simulation (C-MIXED) with the POP 0.1° high-resolution model. The depth on the vertical axis is in meters (figure from Weijer et al., 2012).

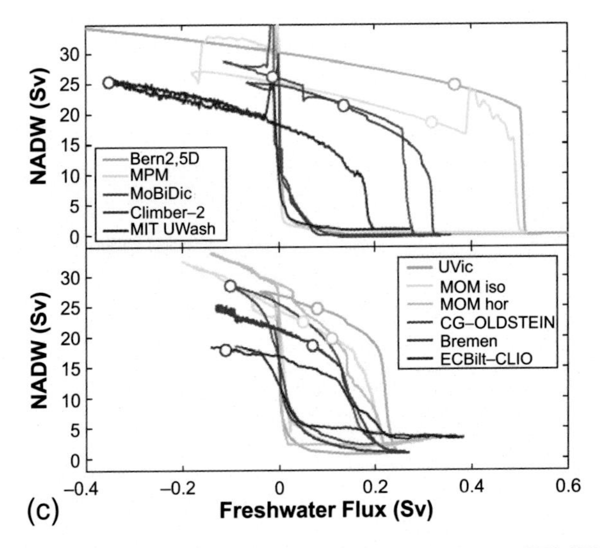

Plate 10.5 (c) Typical hysteresis behaviour as found in a set of EMICs (Rahmstorf et al., 2005) when the freshwater flux is increased slowly in time from zero up to large values and back. On the vertical axis is a measure of the strength of the MOC. The curves are lined up with the recovery, and the equilibrium solutions obtained with zero freshwater perturbation are shown as open circles.

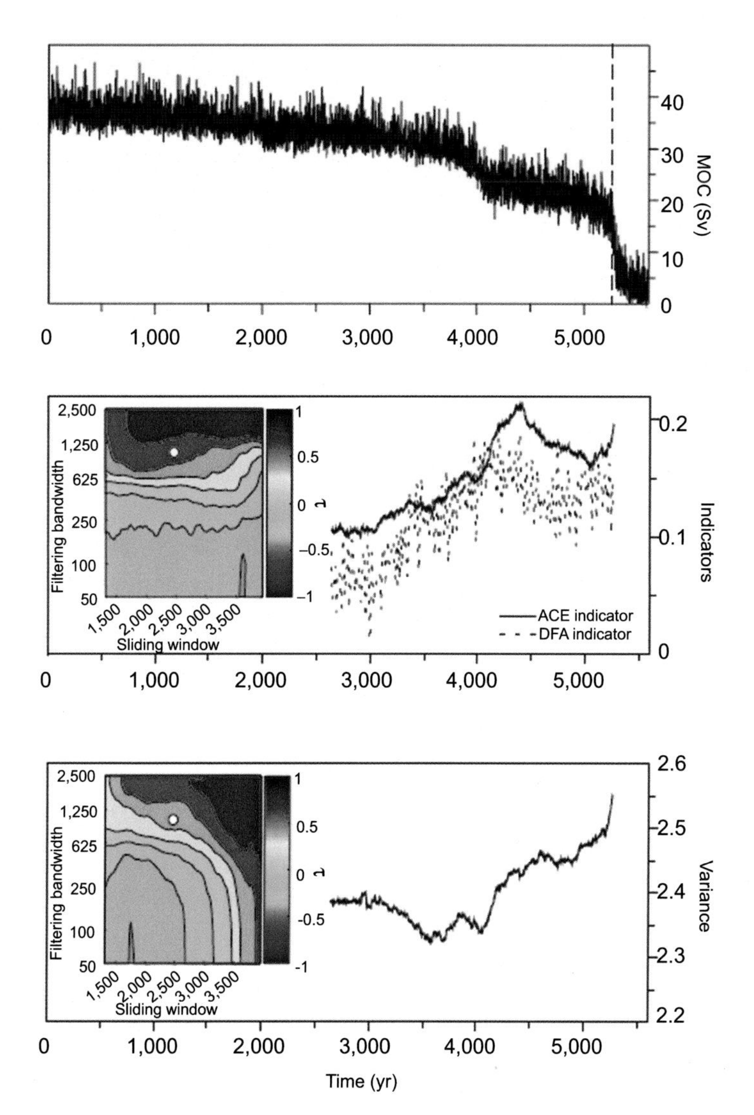

Plate 10.7 Upper panel: time series of the MOC strength as determined from an EMIC where freshwater is slowly released in the northern North Atlantic. Middle panel: Values of the ACF and DFA indicators. Bottom panel: variance of the time series. The values of τ in the inset are results from a statistical test to determine a significant increase in the value of the particular quantity (figure from Lenton, 2011).

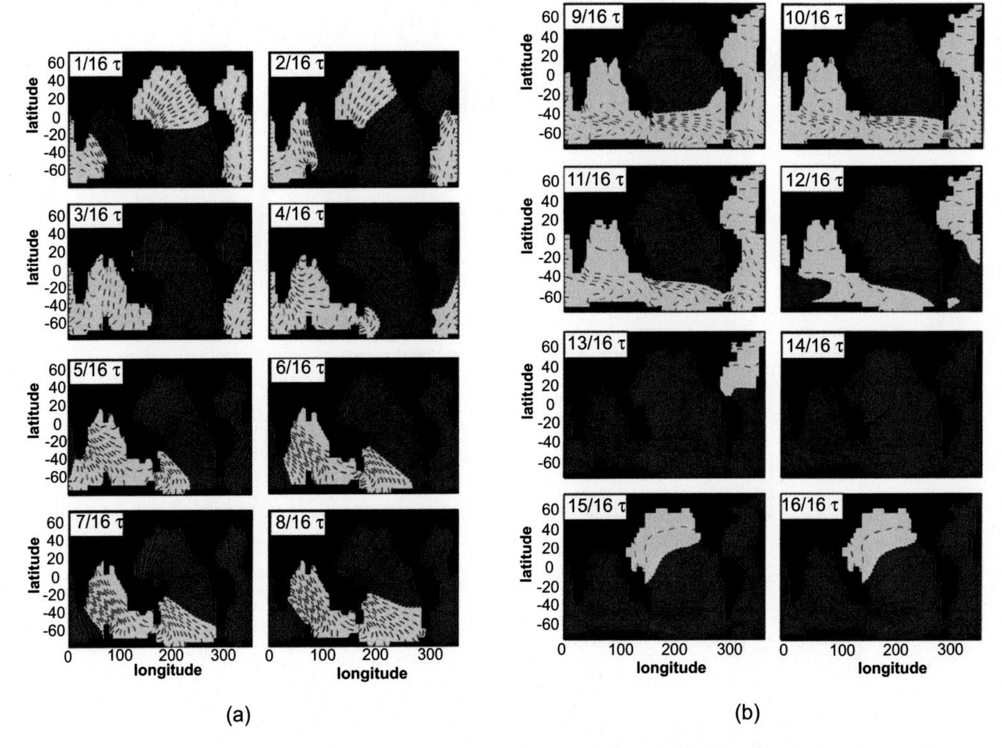

Plate 10.11 (a) Propagation of temperature anomalies at 3,000-m depth for the oscillatory mode 3 in Weijer and Dijkstra (2003) having a period τ of about 2,500 years. Positive (negative) anomalies are denoted by solid (dashed) contours and red (blue) colours. Contour levels are the same for all panels. (b) As Fig. 10.11a, but now at 500 m and showing the second half of the oscillation.

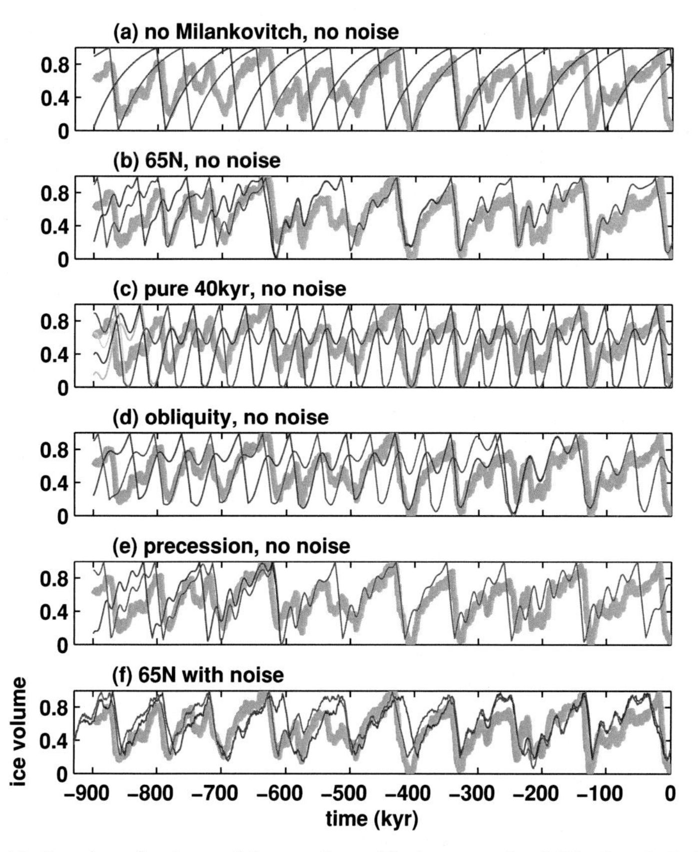

Plate 11.16 Synchronisation of internal oscillations to the Milankovitch forcing. The grey curve is a proxy $\delta^{18}O$ compilation; the thin colour curves are ice volume time series from different model runs using different initial conditions. (a) A model run with no M-forcing. (b) Model forced by M-forcing (65°N summer insolation). (c) Model run forced by a periodic 40-kyr forcing (i.e., not the obliquity time series but a sine wave with a 40-kyr period). (d) Model forced by obliquity variations only. (e) Model forced by precession only. (f) Same as (b) but in the presence of noise (figure from Tziperman et al., 2006).

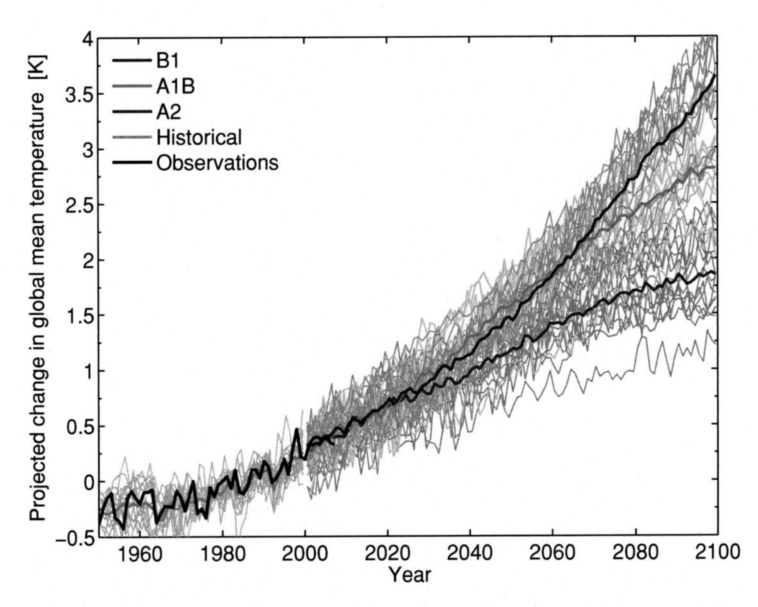

Plate 12.4 Global mean, annual mean, surface air temperature projections from fifteen different global climate models under three different emission scenarios from 2000 to 2100 (thin lines): SRES A2 (red), A1B (green), and B1 (blue), designated as high-, medium-, and low-emission paths, respectively. The same models forced with historical forcings are shown as the thin grey lines, and the observed global mean temperatures from 1950 to 2007 are shown as the thick black line. The multimodel mean for each emission scenario is shown with thick coloured lines (figure from Hawkins and Sutton, 2009).

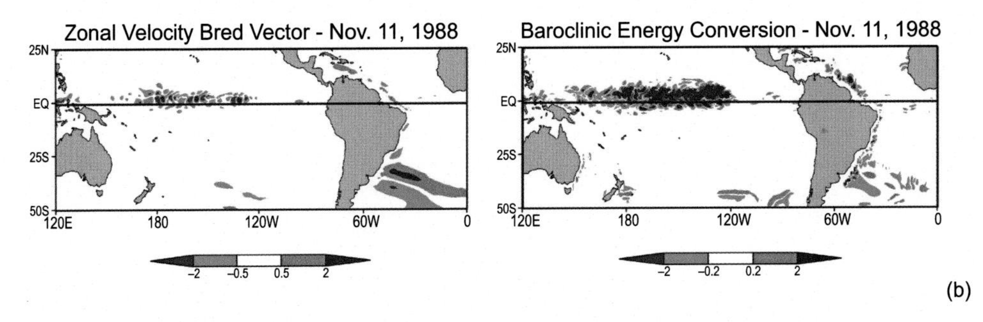

Plate 12.12 (b) Bred vector (left panel) of the zonal velocity starting at 11 November 1988 from the GFDL MOM2 model using a ten-day breeding interval and corresponding baroclinic energy conversion (figure from Hoffman et al., 2009).

the ENSO mode on the background conditions. The dashed curve in Fig. 8.8 represents zero growth (neutral conditions, $\sigma_r = 0$) of the ENSO mode, which is just a path of Hopf bifurcations in the two-parameter plane. Present-day estimates for the Pacific background state correspond to the area near the points A and B, with a period of 2–6 years and near neutral conditions.

8.3.3 Mechanism of the ENSO mode

Conceptual models are available to explain the physics of the oscillation period and the growth of the ENSO mode. Most of these share the ideas that one of the positive feedbacks (as discussed in Section 8.2.3) acts to amplify sea-surface temperature anomalies and that adjustment processes in the ocean eventually cause a delayed negative feedback. These common elements are grouped together in the so-called delayed oscillator mechanism. The differences between the more detailed views are subtle (Jin, 1997b) and are related to the role of the boundary wave reflections, the importance of the adjustment processes of sea-surface temperature, the dominant feedback that is responsible for amplification of anomalies and the view of the dynamical adjustment processes in the ocean. There are four type of ENSO oscillators, the (i) 'classical' delayed oscillator, (ii) the recharge oscillator, (iii) the western Pacific oscillator and (iv) the advective/reflective oscillator. Here, we describe only (i) and (ii); the other two oscillators are described elsewhere (Wang, 2001; Wang and Picaut, 2004).

The classical delayed oscillator

In this view, the eastern boundary reflection is unimportant, the thermocline feedback is dominant and only individual Kelvin and Rossby waves control ocean adjustment. A minimal model (Suarez and Schopf, 1988; Battisti and Hirst, 1989; Munnich et al., 1991) representing this behavior is a differential delay equation of the form

$$\frac{dT(t)}{dt} = a_1 h_e(x_c, t - \tau_1) - a_2 h_o(x_c, t - \tau_2) - a_3 T^3(t). \tag{8.12}$$

In this equation, the $a_i, i = 1, 2, 3$ are constants, and T is the eastern Pacific temperature anomaly, which is influenced by midbasin (at $x = x_c$) equatorial thermocline anomalies h_e and off-equatorial anomalies h_o. Furthermore, $\tau_1 = \frac{\tau_K}{2}$ and $\tau_2 = \frac{\tau_R}{2} + \tau_K$, where τ_K and τ_R are the basin crossing times of the Kelvin wave and the gravest ($j = 1$) Rossby wave (cf. Section 8.2.2).

When the equatorial Kelvin wave, which deepens the thermocline, arrives in the eastern Pacific, local amplification of temperature perturbations occurs through the thermocline feedback, represented by the first term on the right-hand side of (8.12). The wind response excites (off-equatorial) Rossby waves, which travel westwards, reflect at the western boundary and return as a Kelvin wave, which rises the thermocline and provides a delayed negative feedback, represented by the second term in

(8.12). The delay τ_2 is the time taken for the Rossby wave to travel from the center of wind response to the western boundary plus the time it takes the reflected Kelvin wave to cross the basin. The nonlinear term in (8.12) is needed for equilibration of the temperature anomaly to finite amplitude.

The recharge oscillator view

A reduced model of ENSO dynamics was derived from the ZC model in Jin (1997b) and was extended by Timmermann et al. (2003b). We follow here the presentation in Neelin et al. (2000). On the ENSO time scale, the western (h_W) and eastern (h_E) equatorial Pacific thermocline anomalies are approximately described by the equations

$$h_E = h_W + \hat{\tau}, \tag{8.13a}$$

$$\frac{dh_W}{dt} = -rh_W - \alpha\hat{\tau}, \tag{8.13b}$$

where $\hat{\tau}$ is proportional to the zonally integrated wind stress in an equatorial band. Equation (8.13a) represents the equatorial Sverdrup balance constraining the east-west thermocline difference. The equation (8.13b) represents the western thermocline changes during basin-wide adjustment. The first term in (8.13b) represents collective damping processes, and the second term is the effect of the zonally integrated wind stress and represents Sverdrup transport across the basin.

The variation of eastern Pacific equatorial SST is described by

$$\frac{dT_E}{dt} = -\epsilon_T T_E - \frac{\bar{w} + w}{H}(T_E - T_{\text{sub}}) - \frac{w}{H}(\bar{T}_E - \bar{T}_{\text{sub}}). \tag{8.14}$$

The first term on the right-hand side represents damping and the second and third terms the vertical transport of heat, where the barred quantities represent the background state. Again, for the subsurface temperature, a function of the form

$$F(h) = T_{r0} + \frac{\Delta T}{2}\left(1 - \tanh\left[\frac{h + h_0}{H_*}\right]\right) \tag{8.15}$$

is chosen, and

$$\bar{T}_{\text{sub}} = F(\bar{h}_E), \tag{8.16a}$$

$$T_{\text{sub}} = F(h_E + \bar{h}_E) - F(\bar{h}_E). \tag{8.16b}$$

The model is closed by specifying the upwelling as a function of the zonal wind stress τ_E and the zonally integrated wind stress $\hat{\tau}$ as

$$w = -\delta_s \mu a_0 T_E, \tag{8.17a}$$

$$\hat{\tau} = \mu b_0 T_E. \tag{8.17b}$$

Table 8.1. *Values of the parameters in the reduced ENSO model defined by the equations (8.13) and (8.14) developed in Jin (1997a) and used as in Neelin et al. (2000)*

Parameter	Value	Parameter	Value
a_0	0.15 ms^{-1}°C^{-1}	b_0	15 m °C^{-1}
T_{r0}	18 °C	ΔT	12 °C
h_0	25 m	H_*	50 m
ϵ_T	1/150 days^{-1}	H	50 m
r	1/8 months^{-1}	$\alpha = r/2$	1/16 months^{-1}

The parameters of the model used are presented in Table 8.1, and δ_s and μ are two dimensionless control parameters. As a background state, $\bar{T}_E = 25$°C, $\bar{w} = -a_0(\bar{T}_E - 30)$ ms^{-1} and $\bar{h}_E = -0.4b_0(30 - \bar{T}_E)$ m is taken.

When the model is linearised around this mean state, the (initial) evolution of the disturbances $\tilde{T}_E$ and $\tilde{h}_W$ can be written (Jin, 1997a), for $\delta_s = 0$, as

$$\frac{d\tilde{h}_W}{dt} = -r\tilde{h}_W - \mu\alpha b_0 \tilde{T}_E, \tag{8.18a}$$

$$\frac{d\tilde{T}_E}{dt} = (\gamma\mu b_0 - \bar{\epsilon}_T)\tilde{T}_E + \gamma\tilde{h}_W, \tag{8.18b}$$

where $\gamma = \frac{\bar{w}}{H}\frac{dT_{\mathrm{sub}}}{dh_E}$ and $\bar{\epsilon}_T = \epsilon_T + \bar{w}/H$. The terms $\bar{w}$ in the right hand side of (8.18b) describe the effects of the thermocline feedback and damping. The eigenvalues of the linear system in (8.18) are plotted as a function of μ in Fig. 8.9. When the relative coupling coefficient μ is weak, there are two decaying modes: the SST mode and the ocean-adjustment mode. These two modes eventually merge into an oscillatory mode as the coupling coefficient increases. When the coupling is further increased, the oscillator breaks down to give again two stationary modes. One is a purely growing mode because the strong coupling through Bjerknes feedback results in a growth rate being too fast to be linearly balanced by the slow ocean adjustment process. The other is a real SST mode whose growth rate decreases rapidly and becomes negative for large μ. For a wide range of moderate coupling, the model does support an oscillatory mode, with a period mostly in the range of 3.5 years. This oscillatory mode grows as $\mu > \mu_c = 2/3$ and decays for $\mu < \mu_c$. Hence, also in this simple model, a Hopf bifurcation is the main dynamical object causing ENSO-type variability.

The mechanism of the ENSO mode growth and propagation in this model is the recharge-oscillator view (Jin, 1997a). As follows from the preceding results, the ocean adjustment is viewed as being caused by a collective of Kelvin and Rossby waves, and adjustment of sea-surface temperature through surface layer processes is also important. Consider again a positive sea-surface temperature anomaly in the eastern part of

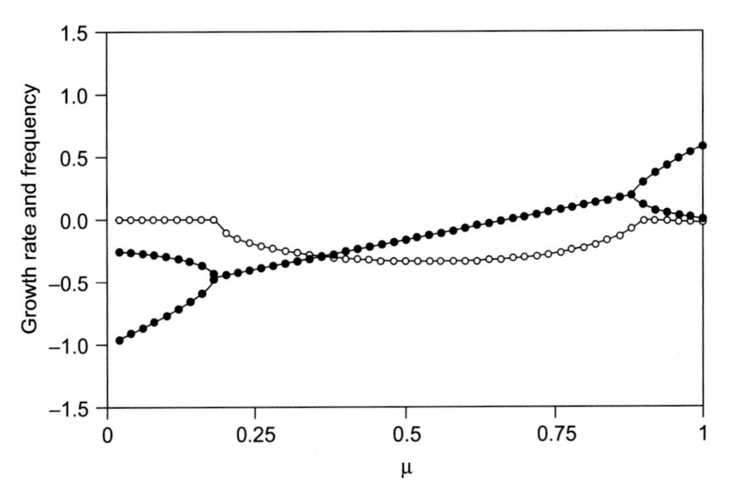

Figure 8.9 Dependence of the eigenvalues of the linearised model (8.18) on the relative coupling coefficient μ. The curves with dots are for the growth rates (in 1/year), and the curves with circles are for the frequencies (in 1/year) when the real modes merge as an oscillatory mode (Jin, 1997a).

the basin that induces a westerly wind response. Through ocean adjustment, the slope in the thermocline is changed, giving a deeper eastern thermocline. Hence through the thermocline feedback, the sea-surface temperature anomaly is amplified, which brings the oscillation to the extreme warm phase (Fig. 8.10a). Because of ocean adjustment, a nonzero divergence of the zonally integrated mass transport occurs, and part of the equatorial heat content is moved to off-equatorial regions. This exchange causes the

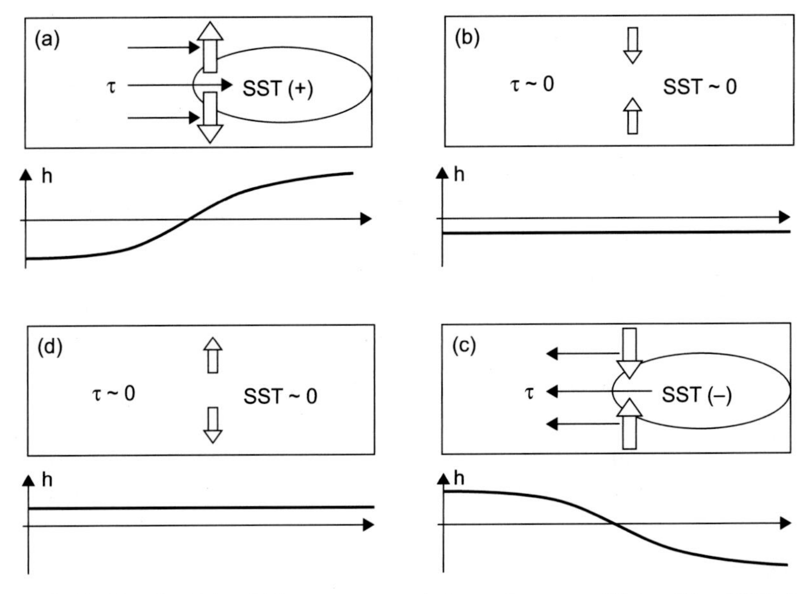

Figure 8.10 Sketch of the different stages of the recharge oscillator (Jin, 1997a).

equatorial thermocline to flatten and reduces the eastern temperature anomaly (again, the thermocline feedback plays a role), and, consequently, the wind stress anomaly vanishes (Fig. 8.10b).

Eventually a nonzero negative thermocline anomaly is generated, which allows cold water to get into the surface layer by the background upwelling. This causes a negative sea-surface temperature anomaly leading through amplification to the cold phase of the cycle (Fig. 8.10c). Through adjustment, the equatorial heat content is recharged (again the zonally integrated mass transport is nonzero) and leads to a transition phase with a positive zonally integrated equatorial thermocline anomaly (Fig. 8.10d). This recharge oscillator view can be easily adapted to include the zonal advection feedback (Jin and An, 1999), as the zonal advection feedback and the thermocline feedback induce the same tendencies of sea-surface temperature anomalies.

8.4 Excitation of the ENSO mode

From the perspective of El Niño, processes that evolve independently on smaller time and space scales can be considered as noise. In the atmosphere, for example, the 30- to 60-day or Madden-Julian oscillation (Madden and Julian, 1994) gives rise to westerly wind bursts. These are events of anomalous westerlies lasting typically a week, which are strongest somewhat west of the region of the El Niño wind response (Vecchi and Harrison, 2000). A comprehensive study of small scale oceanic variability in the equatorial Pacific is presented in Kessler et al. (1996). Kelvin waves are excited very effectively by the Madden-Julian oscillation (MJO) and they travel eastward with a similar speed. Also, Rossby waves are excited by individual westerly wind bursts (Van Oldenborgh, 2000). Inertia-gravity waves, which have periods up to about a week (Philander, 1990), play a role as well.

8.4.1 Red noise wind stress forcing

In Roulston and Neelin (2000), a systematic study is performed on the effect of stochastic forcing on the ENSO variability in the ZC model. Only a stochastic component in the wind stress was considered over the NINO4 (160°E to 150°W and 5°S to 5°N) region. To estimate the spectrum of the wind noise, a linear model relating SST to wind stress was constructed empirically from observations. That part of the wind-stress variance that could be explained by this model was subtracted to leave the residual wind stress. Figure 8.11a shows the amplitude spectrum of the residual zonal wind-stress averaged over the NINO4 region. From this residual wind stress, several noise products were constructed, of which the amplitude spectrum is also seen in Fig. 8.11b–f, with a red noise product (b), a white noise product (c), low- (d) and high- (e) pass-filtered versions of (c) and a product (f) where the amplitude of the 30- to 60-day oscillation was enhanced by a factor five.

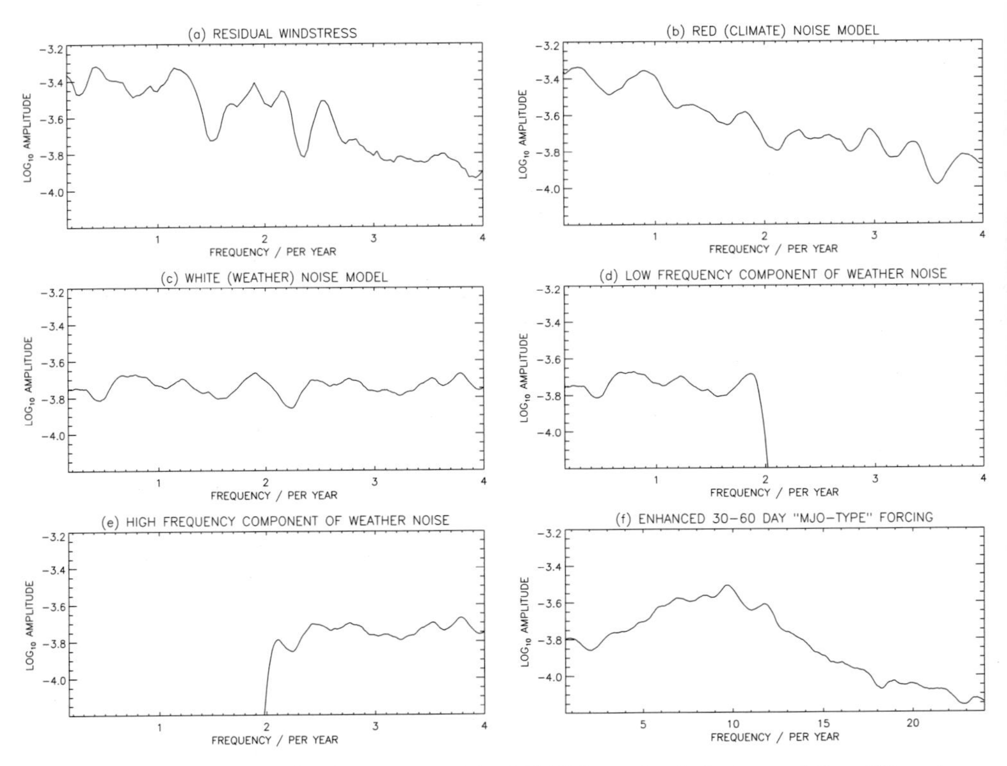

Figure 8.11 The amplitude spectra of the NINO4 zonal wind stress of the different noise products as used in Roulston and Neelin [2000].

The critical coupling strength μ_c in Roulston and Neelin (2000) is approximately known, and the model is forced with the different noise products below (subcritical) and above (supercritical) the Hopf bifurcation. The response of the ZC model to the stochastic wind noise based on the red noise wind stress (Fig. 8.11b) is shown in Fig. 8.12a–b and Fig. 8.12c–d for the subcritical and supercritical cases, respectively. In the subcritical case, the stochastic noise is necessary for interannual variability. This variability, however, has a spectrum that differs markedly from the noise that is driving it. There is a broad peak around 4 years, typical of ENSO. In the supercritical case, the model exhibits regular interannual oscillations with a period of 4.5 years in the absence of noise. The addition of the red noise causes the ENSO peak to broaden.

Fig. 8.13 shows the amplitude response of the model in its subcritical and supercritical regimes for the noise products in Fig. 8.11c, d and e. The white noise forcing excites interannual variability in the subcritical regime (Fig. 8.13a) and broadens the spectral peak of the preexisting interannual variability in the supercritical regime (Fig. 8.13b). In neither case is the interannual peak as broad or high as when the model is driven by red noise. The response of the model to the low-frequency part of the white noise is similar to its response to the full white noise in both regimes (Fig 8.13c–d). In

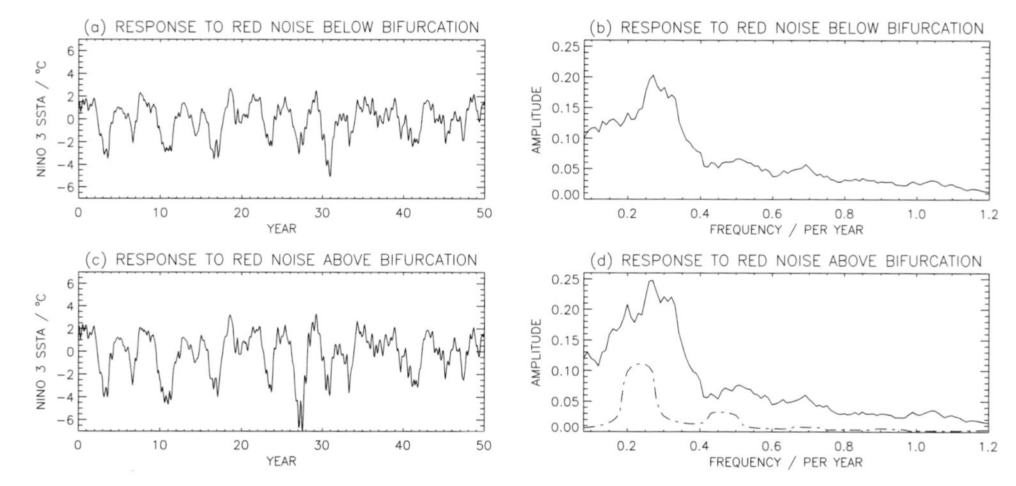

Figure 8.12 The response of the ZC model to the red noise forcing (Fig. 8.11b) conditions. (a) NINO3, subcritical conditions. (b) Spectrum of NINO3, subcritical conditions. (c) NINO3, supercritical conditions. (d) Spectrum of NINO3, supercritical conditions (figure from Roulston and Neelin, 2000). The dashed curve in (d) is the spectrum for the noise-free case.

the subcritical regime, the high-frequency noise excites a small interannual response, but the amplitude of this response is less than 20% of its amplitude when the model is forced by the full white noise (Fig. 8.13e). In the supercritical regime, there is little broadening of the interannual peak, but its amplitude is significantly reduced (Fig. 8.13f). The small amount of power at interannual frequencies in the subcritical case is evidence of nonlinear processes. These processes do not seem to be very efficient and only generate a small amplitude response at low frequencies. The response to the MJO type stochastic forcing (not shown here) is similar to the high-frequency forcing case. The amplitude of the primary interannual peak is reduced, and again there is little transfer of energy to other frequencies.

In Roulston and Neelin (2000), it is concluded that the primary nonlinearities affecting ENSO – the upwelling and thermocline feedbacks – are not effective at transporting energy from intraseasonal to interannual frequencies. In general, the results suggest that low-frequency forcing has the largest impact on the interannual response of the model; therefore, climate noise has an important effect on ENSO variability. In particular, red climate noise communicated by the atmosphere from outside the Pacific domain can have a substantial influence on ENSO.

8.4.2 Stochastic optimals

The linearised operator of the ZC model around a typical Pacific mean state is non-normal due to the horizontal and vertical advective heat transport processes represented. Hence, as we have seen in Section 2.2, the potential for non-normal growth exists in the subcritical case (below the Hopf bifurcation). The general problem is

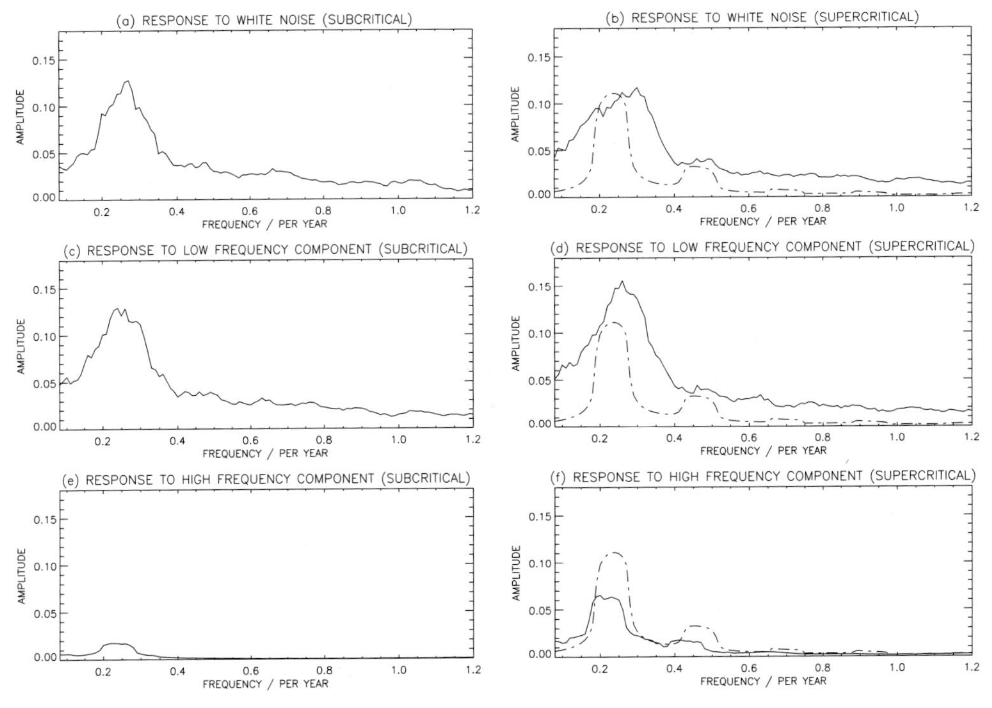

Figure 8.13 The response of ZC model NINO3 index to the white noise products (Fig. 8.11c,d,e) in the subcritical and supercritical cases. The amplitude spectrum of the noise-free supercritical case is shown (dashed curve) for comparison (figure from Roulston and Neelin, 2000).

considered in Moore and Kleeman (1999) and applied to a variant of the ZC model in Kleeman (1993).

When the model is linearised around a seasonal cycle under subcritical conditions, the problem to study the evolution of stochastic perturbations is

$$d\boldsymbol{\psi}_t = A(t)\boldsymbol{\psi}\, dt + F(t)\, d\mathbf{W}_t, \tag{8.19}$$

where $\boldsymbol{\psi}$ is the state vector and $F(t)$ is the time-independent matrix representing the noise forcing. In absence of the noise term, and for constant A, the propagator of the system is indicated by $R = e^{At}$.

Using the Euler-Maruyama scheme (3.96), we find in discrete form that $\boldsymbol{\psi}$ at $t_n = n\Delta t$ is determined from

$$\boldsymbol{\psi}_n - \boldsymbol{\psi}_{n-1} = A_{n-1}\boldsymbol{\psi}_{n-1}\Delta t + F_{n-1}\left(\mathbf{W}(t_n) - \mathbf{W}(t_{n-1})\right), \tag{8.20}$$

where $A_n = A(n\Delta t)$ and $F_n = F(n\Delta t)$. In discrete form, the propagator in the absence of noise can be written as

$$\boldsymbol{\psi}_n = R_{k,n}\boldsymbol{\psi}_k;\ R_{k,n} = \prod_{m=k+1}^{n} (1 + \Delta t A_{n-m+1}). \tag{8.21}$$

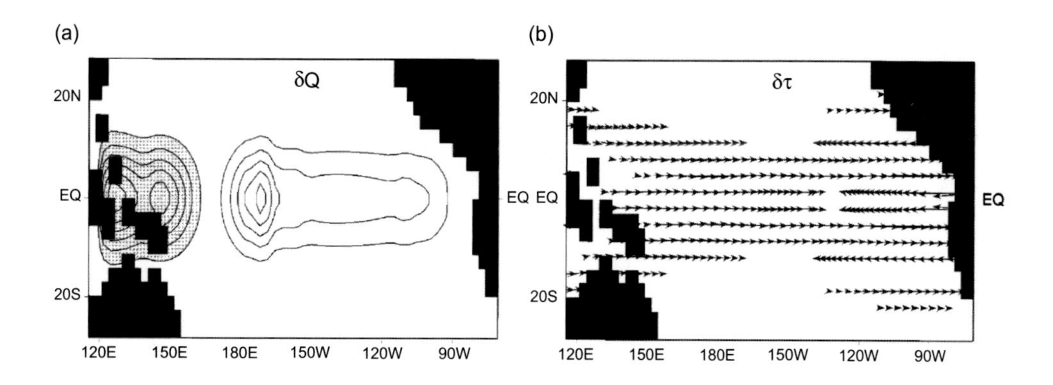

Figure 8.14 The first stochastic optimal of a variant of the ZC model assuming a decorrelation time of 3.5 days for the stochastic noise: (a) the surface heat flux and (b) the surface wind stress. In (a) shaded and unshaded regions are of opposite sign; the contour interval and arrow scaling is arbitrary because stochastic optimals are eigenfunctions (figure from Moore and Kleeman, 1999).

The variance of NINO3 is then given by (Moore and Kleeman, 1999)

$$Var[\psi_n]_X = \Delta t \text{ Trace}\left[\sum_{k=0}^{n-1}\sum_{l=0}^{n-1} D_{l,k} R_{n,l}^T X R_{k,n} C\right] = \text{Trace}\,[BC], \qquad (8.22)$$

where C is the constant spatial covariance matrix of the noise pattern F, $D_{k,l}$ is the decorrelation time of the noise and X is chosen such that the variance in NINO3 is determined from the state vector. The spatial patterns of the so-called stochastic optimals (eigenvectors of B) are the patterns of the forcing required to maximise the variability in NINO3.

For a growth interval of 6 months, the first stochastic optimal (eigenfunction of the matrix B for $t = 6$ months) in the surface heat flux and the surface wind stress is shown in Fig. 8.14. The dominant feature of the stochastic optimal heat flux (Fig. 8.14a) is the dipole centered on the western tropical Pacific. The wind stress component of the stochastic optimal (Fig. 8.14b) takes the form of bands of predominantly zonal winds. So if the spatial structure of the noise forcing resembles that of Fig. 8.14, then, according to (8.22), the stochastic forcing will optimally increase the variability in NINO3.

This signature of a disturbance in Fig. 8.14 is often associated with the MJO and suggests that this large-scale pattern of internal atmospheric variability may be favourably configured to excite the ENSO mode. It also demonstrates that only noise with large-scale spatial coherency will be effective. These results are from one coupled model, but only the general qualitative conclusions hold for many models (for results with different models, see Moore and Kleeman, 2001). The detailed nature of the stochastic optimal can show some variation from model to model and also depends on the background state used. When a model in the subcritical regime (with a decaying oscillation in the absence of noise) is forced with the noise according to the first stochastic optimals, an irregular ENSO oscillation is found with spectral properties as in observations (Kleeman, 2008).

8.5 ENSO's phase locking to the seasonal cycle

Fully coupled mean states with seasonal time dependence having the correct seasonal equatorial temperature distribution are difficult to obtain within ICMs such as the ZC model. There are strong indications that coupled processes, in particular those in the surface layer, are involved in the seasonal cycle (Philander et al., 1996). Conceptual studies have identified the equatorial north-south asymmetry as being a crucial factor in the seasonal cycle (Xie and Philander, 1994), with both evaporation and vertical mixing being able to amplify equatorial asymmetries. The stability of a seasonally varying background state has to be determined using so-called Floquet analysis and has only been applied to prescribed seasonal background states (Jin et al., 1994, 1996). It was shown that the spatial pattern of the ENSO mode does not change much with respect to that determined from a steady background state.

A linear mechanism for ENSO's phase locking to the seasonal cycle within the delayed oscillator view was proposed in Galanti and Tziperman (2000). The coupling strength $\mu(t)$ varies on a seasonal time scale, and it was found in the ZC model to be maximal around June and minimal around December. Consider now the variation of the eastern equatorial Pacific thermocline depth h_e in the 'classical' delayed oscillator picture, connected to (8.12) as described by

$$\frac{dh_e(t)}{dt} = b_1 F_w(\mu(t-\tau_1)h_e(x_c, t-\tau_1)) - b_2 F_w(\mu(t-\tau_2)h_e(x_c, t-\tau_2)) - b_3 h_e(t). \tag{8.23}$$

In this equation, just as in (8.12), $\tau_1 = \frac{\tau_K}{2}$ and $\tau_2 = \frac{\tau_R}{2} + \tau_K$, where τ_K and τ_R are the basin crossing times of the Kelvin wave and the gravest ($j = 1$) Rossby wave (cf. Section 8.2.2). In addition, F_w is in general a nonlinear function, but $F_w(x) = x$ would represent the effect of linear wave propagation with a varying coupling strength. The damping of thermocline anomalies is assumed to be linear (the third term in [8.23]) and the $b_i, i = 1, 2, 3$ are constants.

At peak time of El Niño, the eastern thermocline anomalies are maximal, and hence $dh_e/dt \approx 0$. For small dissipation, the warming by the Kelvin wave (the first term in [8.23]) has to balance the cooling due to the Rossby wave (the second term in [8.23]).

Suppose that the peak of El Niño would occur during June–July, that is, during maximum μ (Fig. 8.15a). In this case, the warm Kelvin wave was excited during April, and because coupling is strong over April–June, it has a relatively large amplitude. The Rossby wave was, however, excited about 6 months earlier, that is, during

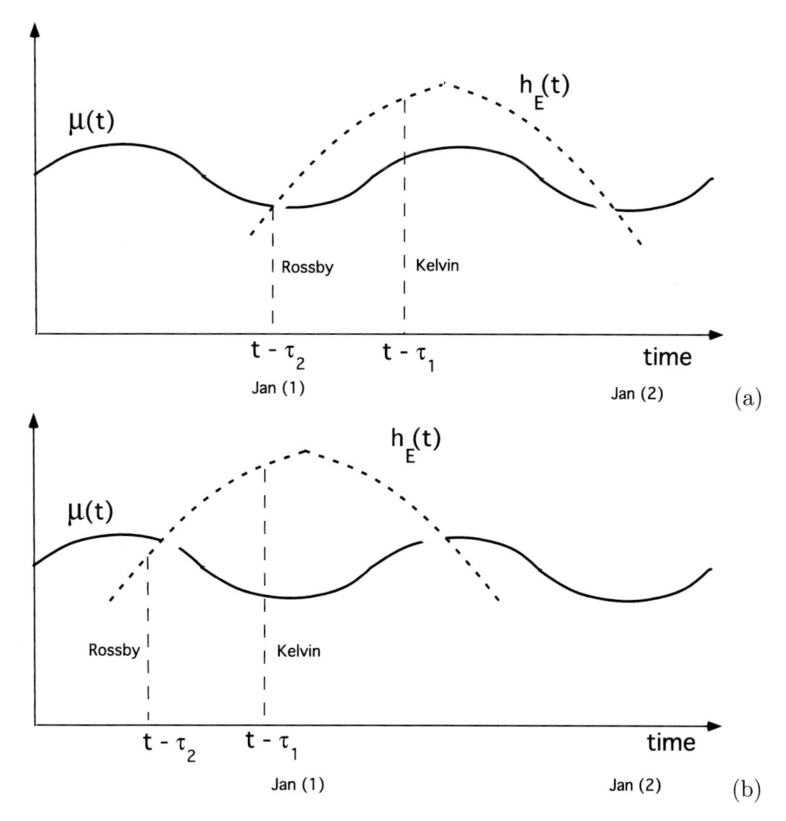

Figure 8.15 Sketch of the phase-locking mechanism as proposed in Galanti and Tziperman (2000). In (a) the maximum El Niño amplitude occurs in June–July, whereas in (b) it occurs in December–January.

November–December when coupling was small. It has, therefore, a small amplitude and cannot balance that of the Kelvin wave, as $\mu(t - \tau_1)h_e(x_c, t - \tau_1)$ in (8.23a) is large, whereas $\mu(t - \tau_2)h_e(x_c, t - \tau_2)$ is small. However, when peak El Niño conditions are in December–January, the Kelvin wave is excited in October during relatively small coupling, and its amplitude is only weakly amplified. The Rossby wave is now excited in June when coupling is large and is amplified to significant amplitude to balance the Kelvin wave thermocline signal (Fig. 8.15b).

Variations in phase-locking behavior were investigated in detail (Neelin et al., 2000) within the recharge oscillator model (8.13)–(8.17). Instead of a constant $\bar{T}_E$ (in °C), the background state was taken seasonally dependent with an annual frequency ω, that is,

$$\bar{T}_E = 25 + 2.5 \sin \omega t. \tag{8.24}$$

Let the internal frequency of the ENSO mode be indicated by ω_0. It was found that the phase locking to the seasonal cycle is dependent on ω_0, and a summary result from Neelin et al. (2000) is shown in Fig. 8.16. The frequency ratio changes in a sequence of so-called frequency-locked discrete steps, where, within each step, the ENSO frequency in the model does not vary with ω_0 (upper panel in Fig. 8.16a). The lower panel in Fig. 8.16a shows a measure of the phase-locking behavior, that is, the month of maximum warming, over one of these steps. Even though the frequency is locked over the interval in ω_0, the phase-locking behaviour changes.

The mechanism for the variations in phase locking is sketched in Fig. 8.16b. The inherent ENSO cycle (thin solid line) that would occur in absence of the seasonal cycle is shown for a case with period slightly longer than 2.5 years. The thick bars indicate the 'favoured season' for maximum warming of the SST anomaly. The length of the ENSO cycle is modified by nonlinear interactions with the seasonal cycle so that warming tends to occur in the favoured season (indicated by thick bars). For example, the first cycle is shortened to about 2 years and the second one is lengthened to 3 years. In this example, this produces an alternation of approximately 2-year (thick solid curve) and 3-year (dashed solid curve) ENSO cycles and hence an average frequency of 5/2 years. When both these cycles are plotted with maximum amplitude at the same time, the onset of the warming in both cycles differs (lower panel of Fig. 8.16b).

8.6 The irregularity of El Niño

From the time series of the NINO3 index such as in Fig. 8.2, it is clear that El Niño events occur irregularly. In the ZC model, only the interaction between large-scale (basin wide) patterns and processes on interannual time scale is captured. In accordance with the stochastic dynamical systems view in Chapter 1, the faster and smaller scale processes such as the MJO and the intraseasonal ocean waves can be only incorporated through 'noise'. There are two different views on the processes that cause the irregularities in El Ninõ occurrence (Kleeman, 2008). In one view, ENSO's irregularity arises because the large-scale dynamics displays complex (maybe chaotic) behaviour because of strong nonlinear interactions. In the other view, the large-scale dynamics does not exhibit this complex behaviour (i.e., it is only weakly nonlinear), and it is the influence of the noise that causes the irregular behaviour.

8.6.1 The weakly nonlinear stochastic view

When additive noise is added to the wind stress, small-scale processes ('noise') inevitably will lead to low-frequency variability (Kirtman and Schopf, 1998; Latif, 1998; Burgers, 1999), but the amplitude is generally small. However, as discussed in Section 8.4.2, if the Tropical Pacific climate system is stable but close to neutral conditions, then non-normal growth of perturbations can lead to irregular behaviour (Penland and Sardeshmukh, 1995; Moore and Kleeman, 1999). In this case, certain

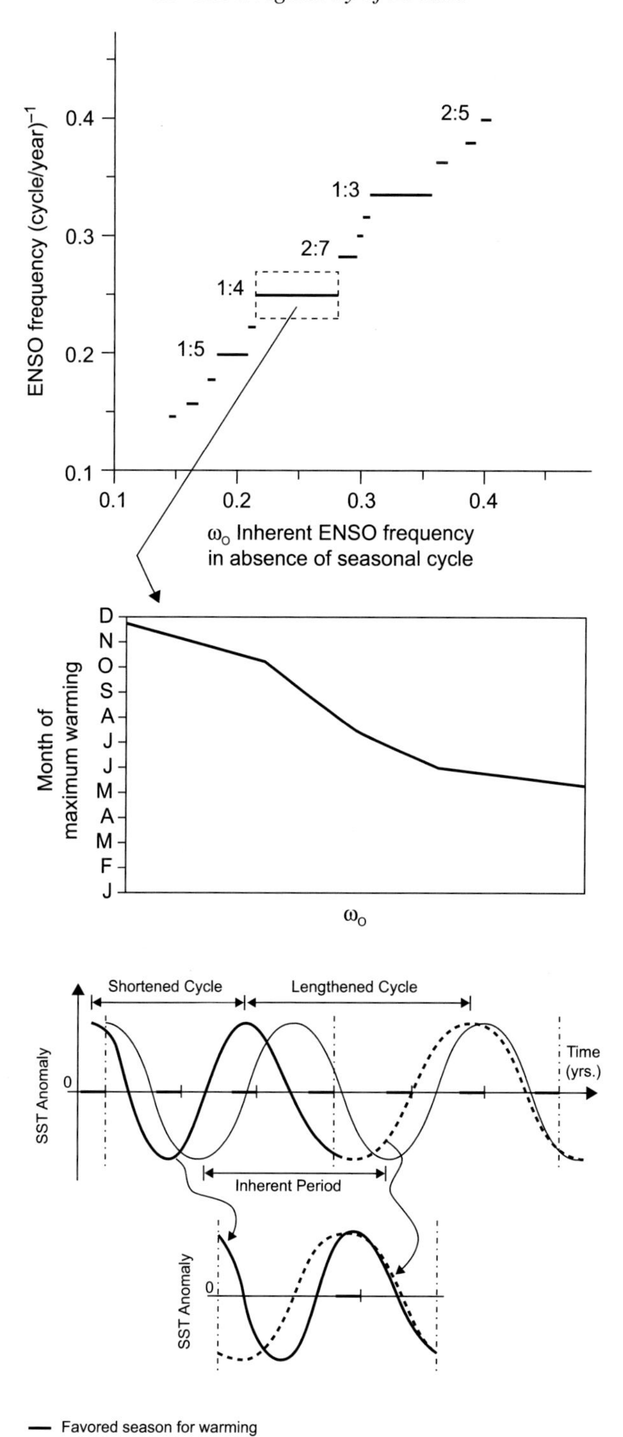

Figure 8.16 (a) Schematic of the difference between phase-locking and frequency-locking behaviour. The frequency ratio of the number of ENSO cycles per annual cycle (or the dominant ENSO frequency in cycles per year) is shown as a function of the inherent ENSO mode frequency, ω_0. (b) The physics of the variations in phase locking (figure from Neelin et al., 2000).

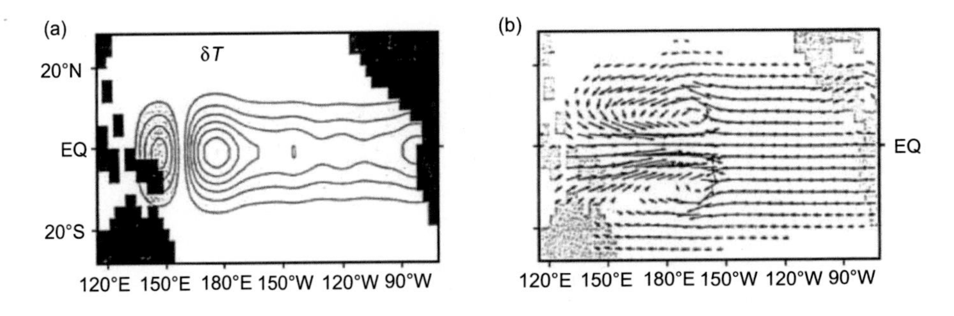

Figure 8.17 The short time-scale response of the Kleeman (1993) model to the forcing shown in Fig. 8.14. (a) The SST anomaly (shading indicates negative values) and (b) the wind stress anomaly (figure from Moore and Kleeman, 1999).

patterns are amplified more rapidly under random initial conditions than the ENSO mode. As shown in Fig. 8.17, patterns of the stochastic optimals initially grow into SST and wind stress disturbances that resemble westerly wind burst (Moore and Kleeman, 1999). If a ZC type model is forced by noise that is temporally white but with spatial correlations according to the stochastic optimals, then irregular behavior in NINO3 is found (Fig. 8.18a). The filled circles show December of each year, indicating that the observed seasonal synchronisation of large warm events, as discussed in Section 8.5, is also achieved in this model. Here, this tends to be greatest in the northern spring and in the lead-up to warm events. At that time, convective anomalies in the atmosphere are able to develop with the greatest zonal extent because the mean SST is high at the mentioned phases of both the seasonal and ENSO cycles. This synchronisation scenario has recently been confirmed in the observational study of Hendon et al. (2007).

The spectrum of the irregular oscillation in Fig. 8.18b is qualitatively the same as the observed spectrum (Fig. 8.2b). The irregular behaviour noted is particularly robust, as one can vary the amplitude of the forcing by some orders of magnitude without a qualitative effect. It has been observed by many other investigators using a range of different models (Blanke et al., 1997; Eckert and Latif, 1997). In addition, long integrations of this stochastic model show that there is a decadal variation of its spectrum qualitatively similar to the observations (Zhang et al., 1997).

8.6.2 The strongly nonlinear deterministic view

Nonlinear resonances with the seasonal cycle

Deterministic chaotic behavior can result from the interaction of the seasonal cycle and the ENSO mode (Tziperman et al., 1994a; Jin et al., 1994). When the coupling strength μ is increased above criticality in ICMs, such as the ZC model, which are

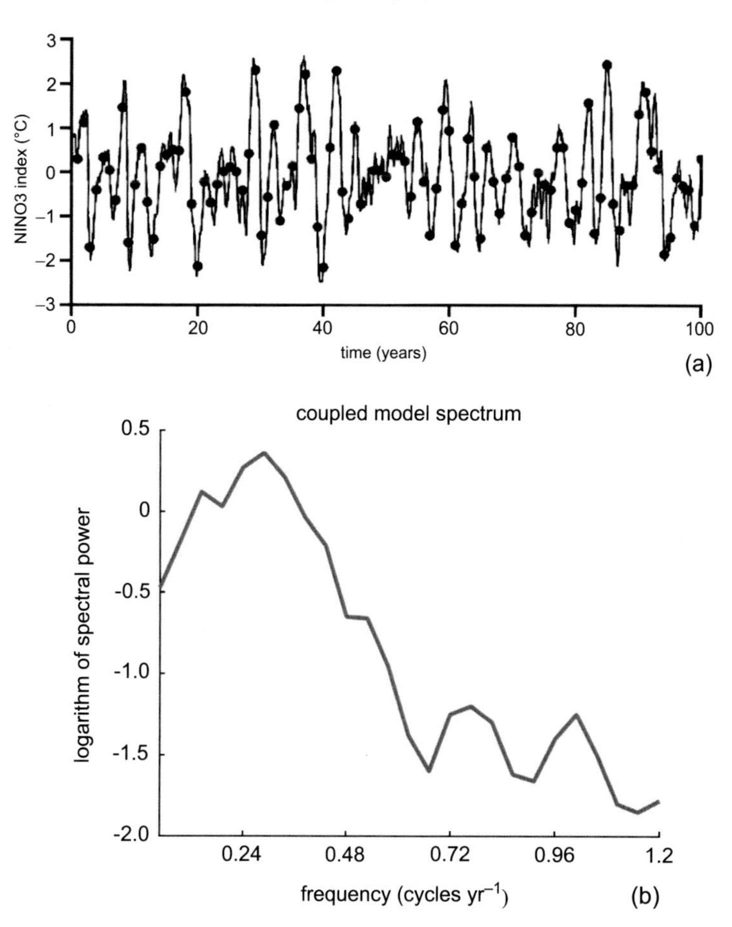

Figure 8.18 (a) A plot of the NINO3 index from a stochastically forced ZC type model (Moore and Kleeman, 1999). The filled circles indicate December. Note the locking of warm event peaks to this time of year. (b) Spectrum corresponding to the time series in (a).

seasonally forced, the NINO3 index displays regimes of frequency locking to the annual cycle alternated by chaotic regimes (Jin et al., 1996). In the frequency-locked regimes, the ENSO period is a rational multiple of the seasonal cycle (cf. Fig. 8.16). Although chaotic solutions are nearly everywhere in parameter space, the locked regime is also very broad. For many solutions found, the phase locking to the annual cycle is near to that observed, with January being the preferred month of the peak of the warm event.

Using a particular variant of the coupled wave oscillator model (Cane et al., 1990; Munnich et al., 1991), the interaction of the ENSO mode and the seasonal cycle has been studied (Tziperman et al., 1994a) by looking at solutions of

$$\frac{dh}{dt} = c_1\ F_w(h(t-\tau_1)) - c_2\ F_w(h(t-\tau_2)) + c_3\ \cos\ \omega t, \tag{8.25}$$

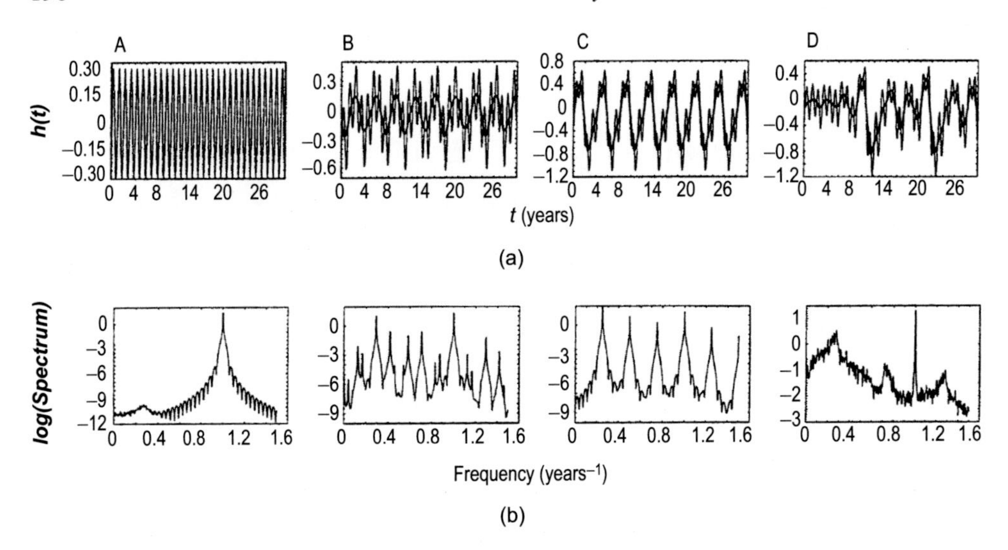

Figure 8.19 (a) Time series and (b) power spectra of the dimensionless thermocline thickness using the wave oscillator model for four values of (A) $\kappa = 0.9$, (B) $\kappa = 1.2$, (C) $\kappa = 1.5$, (D) $\kappa = 2.0$ (figure from Tziperman et al., 1994a).

where F_w is a nonlinear function and τ_1 and τ_2 are as in (8.23). The last term is the seasonal forcing, with frequency ω corresponding to a 1-year period. An example of the shape of the function F_w is a tangent hyperbolic function that saturates both at large negative and positive thermocline anomalies and has a sharp transition near $h = 0$. For this function, κ is the slope at $h = 0$ and measures the strength of the thermocline feedback; this parameter is used as a control parameter.

Results of the function $h(t)$, scaled by the equilibrium thermocline thickness H, of 1,024 years of integration and its spectrum are plotted in Fig. 8.19 for four values of κ. For small κ, there is a simple periodic orbit with annual period induced by the forcing. For $\kappa = 1.2$, the ENSO frequency appears that is incommensurate with the annual frequency, and hence a quasi-periodic signal is obtained (cf. Chapter 2). For larger κ, locking to the frequency of the annual cycle occurs. For even larger nonlinearity, the spectrum becomes broad banded, which is an indication of chaotic behavior in the system. Similar results were obtained in Jin et al. (1996) using a slightly simplified version of the ZC model.

The interaction between seasonal cycle and ENSO mode has been analysed in detail within the Zebiak-Cane model (Tziperman et al., 1995). The coupling strength is varied by changing the drag coefficient in the bulk wind stress formula. When the amplitude of the annual cycle is zero, the annual mean state is stable at standard value of the coupling. With perpetual July conditions as mean state, the system behaves very irregularly with a nearly continuous frequency spectrum; decreasing the coupling gives a nice periodic ENSO signal. It shows that even without the seasonal cycle, chaotic motion can appear in the ZC model. This has been attributed to mode interaction between the ENSO mode and another so-called 'mobile' mode (Mantua and Battisti, 1995). When the amplitude of the seasonal cycle is increased, frequency locking occurs in the same way as seen previously, with chaotic regimes in between. Tziperman et al. (1997) analyse the factors that influence the interaction between seasonal cycle and ENSO in this model and point to the seasonality of the atmosphere as the primary effect. A similar sequence of transitions was also found in a more elaborate intermediate model by Chang et al. (1994), which simulates both the seasonal cycle and ENSO. They stress the relative importance of the amplitude of the seasonal cycle and show that the ENSO frequency gets entrained (disappears from the signal) if the amplitude of the seasonal cycle becomes too large.

Homoclinic orbits

Another possible mechanism is that of complex nonlinear interactions between the basin scale instabilities. Due to these nonlinear interactions, El Niño bursting phenomena can occur (Timmermann et al., 2003a), where ENSO grows to a certain amplitude, then a quick reset takes place, and small ENSO variations grow again.

The model used is a slight extension of the recharge oscillator (8.13)–(8.17) in Section 8.4.2 by considering now the temperatures in a western box T_1 and an eastern box T_2. The model is then formulated by the equations

$$h_2 = h_1 + bL\,\tau, \tag{8.26a}$$

$$\frac{dh_1}{dt} = -r\left(h_1 + \frac{bL}{2}\tau\right), \tag{8.26b}$$

where h_1 and h_2 are the thermocline anomalies in the western and eastern box, respectively. The variation of western T_1 and eastern T_2 Pacific equatorial SST is described by

$$\frac{dT_1}{dt} = -\alpha(T_1 - T_r) - \frac{u(T_2 - T_1)}{L/2}, \tag{8.27a}$$

$$\frac{dT_2}{dt} = -\alpha(T_2 - T_r) - \frac{w(T_2 - T_{sub}(h_2))}{H_m}, \tag{8.27b}$$

where T_r is the radiation equilibrium temperature. The first term on the right-hand side of (8.27a) represents damping and the second term zonal advection. The first term on the right-hand side of (8.27b) represents damping and the second term vertical advection of heat through upwelling. Again for the subsurface temperature, a function of the form

$$T_{sub}(h) = T_r - \frac{T_r - T_{r0}}{2}\left(1 - \tanh\left[\frac{h + H - z_0}{h_*}\right]\right) \tag{8.28}$$

Table 8.2. *Values of the parameters in the reduced ENSO model used in Timmermann et al. (2003a)*

Parameter	Value	Parameter	Value
μ	0.0026 (°C day)$^{-1}$	L	15 10^6 m
T_{r0}	16 °C	T_r	29.5 °C
z_0	75 m	h_*	62 m
α	$1/180$ days^{-1}	H_m	50 m
$\mu bL/\beta$	22 m °C^{-1}	H	100 m
r	$1/400$ days^{-1}	ζ	1.3

is chosen. The model is closed by specifying the zonal velocity, the upwelling velocity and the wind stress τ as

$$\tau = \frac{\mu}{\beta}(T_2 - T_1), \tag{8.29a}$$

$$\frac{u}{L/2} = \epsilon\beta\tau, \tag{8.29b}$$

$$\frac{w}{H_m} = -\zeta\beta\tau. \tag{8.29c}$$

The parameters of the model used are presented in Table 8.2 and ϵ (measuring the strength of the zonal advection feedback) is used as control parameter.

The bifurcation diagram where T_2 is plotted versus ϵ is shown in Fig. 8.20. For a small value of ϵ, a stable equilibrium exists which corresponds to an eastern cold

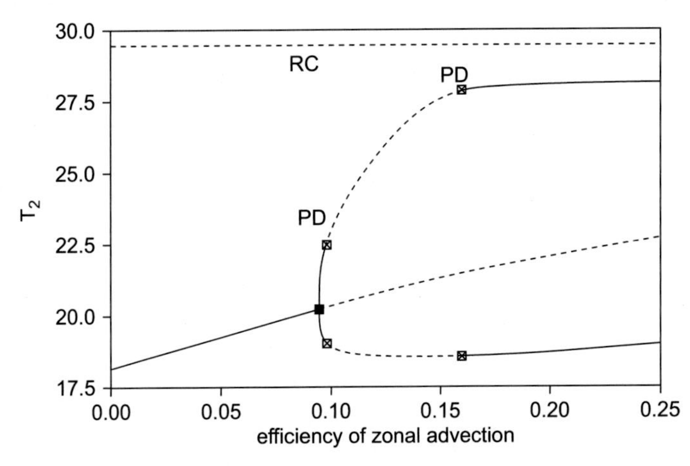

Figure 8.20 Bifurcation diagram of T_2 (°C) versus ϵ displaying stationary and oscillatory solutions as well as their stability. A dashed curve symbolises an unstable periodic orbit. Square filled symbols represent Hopf bifurcation points; PD denotes a period doubling bifurcation. RC denotes the radiative-convective equilibrium saddle point. Stable solutions are marked with solid lines, whereas dashed curves indicate unstable solutions (figure from Timmermann et al., 2003a).

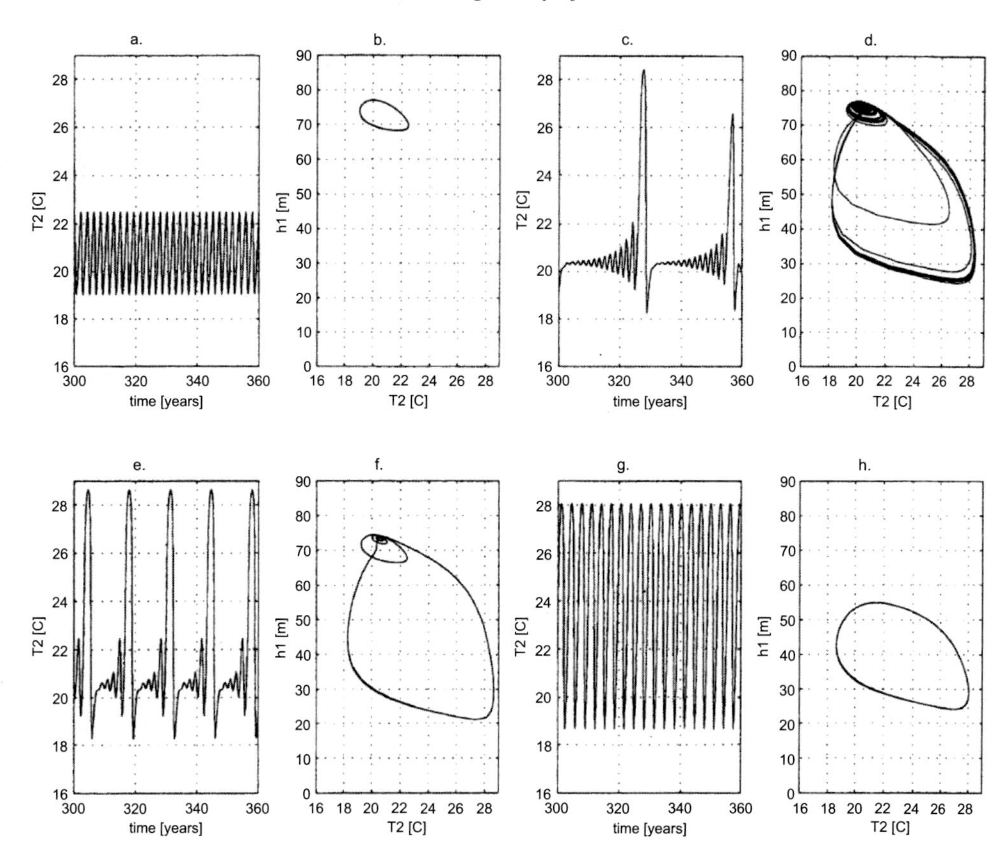

Figure 8.21 (a), (c), (e), (g) Simulated time series of the eastern equatorial temperature T_2 (°C) and (b), (d), (f), (h) phase space plots [eastern equatorial temperature T_2 (°C) and western thermocline depth h_1 (m)] corresponding to different values of the zonal advection efficiency ϵ of (a–b) 0.098; (c–d) 0.1; (e–f) 0.11; (g–h) 0.18 (figure from Timmermann et al., 2003a).

tongue and a western warm pool. From a linear stability analysis, it is found that the least damped eigenmode of the Jacobian matrix is oscillatory and hence corresponds to the damped oscillatory ENSO mode. The steady state becomes unstable via a supercritical Hopf bifurcation at $\epsilon = 0.09$, and a stable periodic orbit appears. There is a relatively large range over which this ENSO limit cycle is stable (two period-doubling bifurcations occur along the way), but it is eventually destroyed (at much larger ϵ, not shown in Fig. 8.20) by a subcritical Hopf bifurcation.

The time-dependent behavior of trajectories at several values of ϵ is shown in Fig. 8.21. For $\epsilon = 0.098$, the system exhibits a regular low-amplitude oscillation with a period of 25 months (Fig. 8.21a,b). The mechanism of oscillation in this model can be interpreted in terms of the recharge oscillator view (Jin, 1997a). An increase of ϵ to values of about 0.1 leads to qualitatively new dynamical behavior. One observes strong amplitude modulations of the ENSO variability that are associated with a bursting of extreme El Niño events occurring on decadal and interdecadal time scales (Fig. 8.21c,d). This bursting is associated with a large positive skewness of the distribution of eastern equatorial Pacific SST anomalies. The mechanism of the low-frequency variability is the interaction of the limit cycle with the unstable fixed (point), which is labelled RC in Fig. 8.20, leading to a Shilnikov type homoclinic bifurcation, as discussed in Chapter 2. The phase space plot reveals that the large temperatures of this amplitude vacillation are associated with a flat thermocline within the tropical west Pacific. Within this regime of amplitude modulated behavior, both periodic and chaotic windows exist. Large values of ϵ, say around 0.18, generate again a large-amplitude regular ENSO oscillation with a period of 3–4 years.

8.7 Synthesis

Clearly, the theory on the dynamics of the ENSO phenomenon is in much better shape than that of the North Atlantic Oscillation, as discussed in Chapter 7. From a dynamical systems perspective, the reason is that the dominant variability is related to only one fixed point (the background climatology) and a few normal modes (variants of the ENSO mode, depending on the dominant feedback), which may interact with the external seasonal forcing. Although this is clear from the ZC-type models, also in GCMs only the behaviour of a few dominant statistical modes (EOFs, MSSA modes) appear to span the range of behaviour of the ENSO phenomena (Guilyardi, 2006).

From the ZC-type models, a Hopf bifurcation related to the destabilisation of the background climatology is central for generating the 3- to 4-year variability in the equatorial Pacific. If the background state is a prescribed seasonal cycle, then this variability arises through the destabilisation of the seasonal periodic orbit (Jin et al., 1996). When, in the ZC models, the background state is also affected by coupled ocean-atmosphere feedbacks (Van der Vaart et al., 2000), then the Hopf bifurcation is the first bifurcation, as no multiple equilibria exist (Neelin and Dijkstra, 1995).

The physics of the destabilisation of the background state can be understood from the Bjerknes' feedbacks, which explain why SST perturbations on the background state can grow in time. Depending on the properties of the background state, the SST pattern of the ENSO mode at Hopf bifurcation may slightly vary, as well the dominant Bjerknes' feedback causing its growth. In the modern literature, two variants appear dominant, the cold-tongue ENSO and the warm-pool ENSO (Hendon et al., 2009), having a different ratio of warming in the central and eastern Pacific. Once the perturbation has grown to sufficient amplitude, ocean adjustment processes (both waves and SST) lead to a delayed response and eventually to an amplitude change of the SST perturbation, leading to the opposite phase of the phenomenon. The different delayed-oscillator variants have been identified in GCMs (Mechoso et al., 2003; Guilyardi et al., 2009) and in observations (Jansen et al., 2009).

If the background conditions are such that perturbations would not be amplified in the absence of noise, then the pattern of the ENSO mode can always be excited by any small-scale noise. This can occur either through non-normal growth mechanisms (which operate in subcritical conditions) or through the direct rectification effects of noise. The stochastic Hopf bifurcation (as explained in Section 4.3.3) provides a general framework to describe this situation from a dynamical systems point of view. This picture unifies the viewpoints of people who prefer the theory that ENSO is noise driven (Penland and Sardeshmukh, 1995; Penland et al., 2000) and of those who prefer that ENSO is a self-sustained oscillation (Tziperman et al., 1994a, 1998). It just depends on the background conditions whether the ENSO mode is damped or will grow and the noise either excites the pattern or influences the pattern of this mode.

Weather noise is certainly the default explanation for a substantial portion of the irregularity of ENSO (Burgers, 1999). However, also nonlinear mechanisms can be responsible for irregular behaviour of ENSO. Several variants have been presented in this chapter, for example, nonlinear resonances with the seasonal cycle and homoclinic orbits. Although both mechanisms do occur in GCMs (Guilyardi et al., 2009), it is difficult to test these mechanisms in observations. Nonlinear mechanisms are certainly at work to cause the phase locking to the seasonal cycle, with variations associated with the inherent period of the ENSO mode (Neelin et al., 2000).

In summary, the dynamical systems view has been very fruitful to put much of the ENSO work into context, in particular concerning (i) the role of coupled processes in the Pacific climatology and (ii) the issue on stochastically driven versus self-sustained ENSO variability. Combined with much modelling and observational work, there is a reasonable solid theoretical framework for the phenomenon, which can be used in issues on ENSO predictability and the behaviour of ENSO under different background states.

9

Multidecadal Variability

Propagation in the subsurface.
DADGCE♭, Western, Alex de Grassi

After the excursion to the equatorial Pacific in the previous chapter, we return to the North Atlantic area to discuss climate variability on multidecadal time scales.

9.1 Introduction

9.1.1 Basic phenomena

The first analyses of multidecadal variability in the North Atlantic (Schlesinger and Ramankutty, 1994; Kushnir, 1994) were based on observed sea-surface temperature (SST) and indicated the existence of variability on a time scale of fifty to seventy years. This variability was named the Atlantic Multidecadal Oscillation (AMO) by Kerr (2000). An AMO index, defined by Enfield et al. (2001) as the ten-year running mean of detrended Atlantic SST anomalies north of the equator, is plotted in Fig. 9.1. Warm periods were in the 1940s and from 1995 to the present, whereas during the 1970s the North Atlantic was relatively cold.

Low-frequency variability in the North Atlantic SST has been determined from proxy data stretching back at least 300 years (Delworth and Mann, 2000), and within

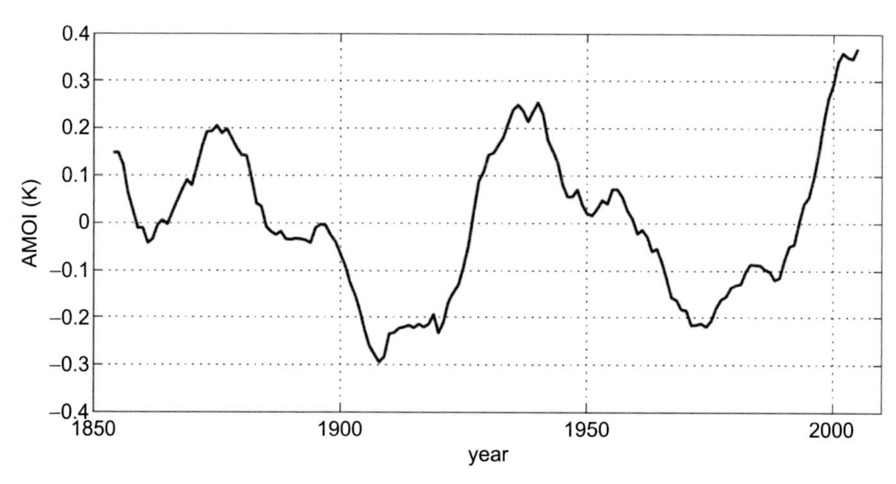

Figure 9.1 The AMO index (the ten-year running mean of detrended Atlantic sea-surface temperature anomalies north of the equator) of the instrumental record (using data from the HadSST2 data set).

this data there is a statistically significant peak above a red-noise background at about fifty years. From recent Greenland ice-core analysis, where five overlapping records between the years 1303 and 1961 are available with annual resolution, significant multidecadal peaks in the spectrum were found (Chylek et al., 2011).

A first impression of the pattern of the AMO was obtained from an analysis of SSTs in the North Atlantic from the instrumental record (Kushnir, 1994). Figure 9.2 shows the difference between the average SST during the relatively warm years 1950–1964 and the relatively cool years 1970–1984. There is a negative SST anomaly near the

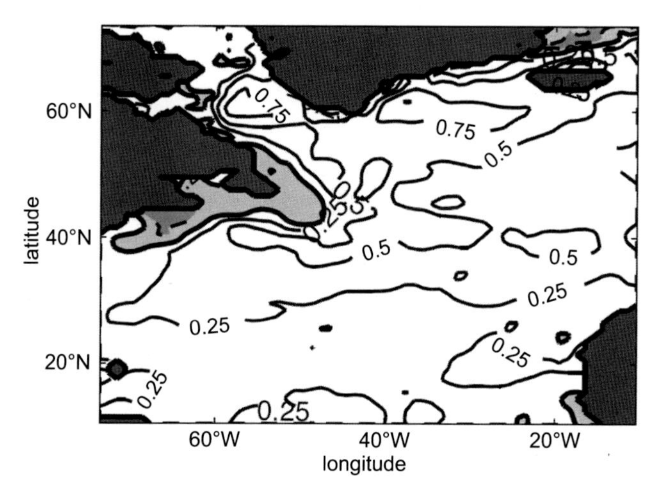

Figure 9.2 An impression of the AMO pattern as shown by the difference in observed average North Atlantic SST between the periods 1950–1964 (warm period) and 1970–1984 (cold period). Units are in °C and negative values are shaded.

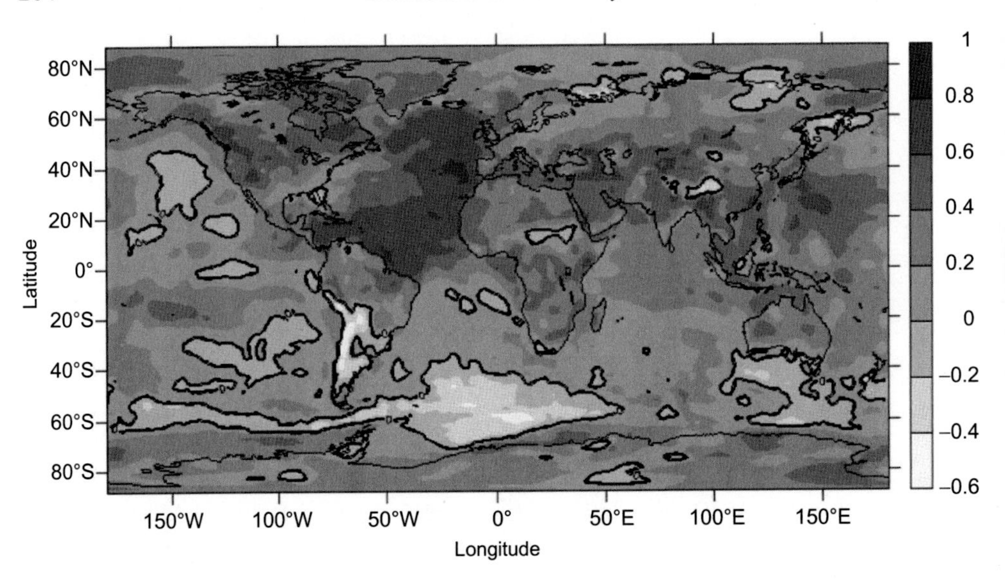

Figure 9.3 Correlation between the AMO index and atmospheric surface temperature (from the NCEP/NCAR reanalysis); the zero contour is bold. Image from NOAA/ESRL Physical Sciences Division, Boulder Colorado (http://www.esrl.noaa.gov/psd/).

coast of Newfoundland and a positive SST anomaly over the rest of the North Atlantic basin. Because of the relatively short observational time series of the instrumental record ($\sim$150 years of SST and SLP), it is difficult to extract a dominant pattern of multidecadal variability with much confidence. Kaplan et al. (1997) and Delworth and Greatbatch (2000) present a reconstruction of a signal with an approximately fifty-year period, which shows a near-standing pattern in SST and SLP. The SST pattern is basin-wide, with the largest anomalies appearing south of Greenland.

A persistent, large-scale temperature anomaly over an ocean basin the size of the North Atlantic represents a significant amount of heat. It comes as no surprise, therefore, that the AMO has an effect on the climate of the surrounding land masses. Fig. 9.3 shows the correlation between the AMO index and surface temperatures over the entire globe. Consistent with this, Sutton and Hodson (2005) found that sea level pressure, precipitation and temperature over Europe and North America, particularly during June–August, are affected by the AMO.

The AMO has also been linked to rainfall and thence to river flows in the United States (Enfield et al., 2001). There is a negative correlation between the AMO index and rainfall over the Mississippi basin and a positive correlation between the AMO index and rainfall in Florida. This means that there is on average less (more) rain over the Mississippi catchment area and more (less) rain over Florida when the AMO index is high (low). Positive correlations have also been found between the AMO index and rainfall in the Sahel and between the AMO index and the strength of the Indian summer monsoon (Zhang and Delworth, 2006; Knight et al., 2006; Feng and Hu, 2008).

A basic theory of the AMO should explain (i) the processes leading to the shape and amplitude of the spatial pattern of the SST anomalies and (ii) the physics of the dominant time scale of variability. In this chapter, the main ingredients of such a theory, based on a stochastic dynamical systems approach, are described.

9.1.2 Towards an understanding of the AMO

A mechanistic understanding of the phenomena of multidecadal variability is important for several reasons. First is the impact of AMO variability on climate over the United States and Western Europe, as briefly described previously (Enfield et al., 2001; Sutton and Hodson, 2005). Second, multidecadal variations may contribute to changes in global mean surface temperature and hence may alternately mask and enhance temperature and precipitation changes due to increasing levels of greenhouse gases (Zhang et al., 2007). Third, if there are preferred patterns of multidecadal variability, then these may play a significant role in climate predictability on these time scales (Griffies and Bryan, 1997; Keenlyside et al., 2008). Finally, understanding this variability is an important component of a more general theory of climate variability and climate change.

In understanding the physics of the AMO, there are two different, and so far quite disjointed, approaches. In what one can call a 'top-down' approach, Global Climate Models (GCMs) are used to simulate the multidecadal variability and then deduce the processes from (a mainly statistical) analysis of the results. From different model simulations (Delworth et al., 1993; Delworth and Greatbatch, 2000; Eden and Jung, 2001; Cheng et al., 2004; Dong and Sutton, 2005; Jungclaus et al., 2005; Zhu and Jungclaus, 2008), several different mechanisms of the AMO have been suggested, which we discuss later in this chapter (Section 9.5). A problem with this type of analysis is that statistically many fields are varying in concert, and the chain of processes causing the multidecadal behavior, in particular the pattern and time scale of the variability, may be difficult to extract.

The other approach to studying the AMO is more 'bottom-up' and recognises that there must be a so-called minimal model that captures the heart of the physics of the AMO. In this context, the term minimal model refers to the simplest possible model in which AMO-like variability can be simulated. Additional physics included in models extending such a minimal model then only quantitatively affects patterns and time scale, without changing the underlying mechanism of variability. When moving up the hierarchy of models towards GCMs, one attempts to identify specific features of the phenomenon, as determined in the minimal model, as signatures of the mechanism. A problem with this approach is that it may be very hard to attribute patterns of variability in observations and GCMs to the processes derived from the minimal model (cf. Chapter 6).

In the context of this book, the top-down approach uses the analysis of time series coming from the GCMs using techniques as discussed in Chapter 5, whereas the bottom-up approach is based on an analysis of the qualitative properties of model solutions in parameter space using techniques as discussed in the Chapters 2–4. So we start with the description of the results of the 'bottom-up' approach in the next section and eventually aim to bridge them with those of the 'top-down' approach at the end of the chapter.

9.2 Stability of the thermally driven ocean flows

Based on results from GCM studies (Delworth et al., 1993), a minimal model of the AMO was formulated by Greatbatch and Zhang (1995) and Chen and Ghil (1996). It consists of a flow in an idealized three-dimensional northern hemispheric sector basin forced only by a prescribed heat flux. It turns out that these flows are susceptible to large-scale multidecadal time-scale oscillatory instabilities (Huck et al., 1999; Huck and Vallis, 2001; Huck et al., 2001; Te Raa and Dijkstra, 2002; Kravtsov and Ghil, 2004). The mechanism of these oscillations and the associated spatial SST patterns are presented in this section.

9.2.1 The minimal model

We consider ocean flows in a model domain on the sphere bounded by the longitudes $\phi_w = 286°\text{E}$ (74°W) and $\phi_e = 350°\text{E}$ (10°W) and by the latitudes $\theta_s = 10°\text{N}$ and $\theta_n = 74°\text{N}$; the ocean basin has a constant depth H. The flows in this domain are forced by a restoring heat flux Q_{rest} (in Wm^{-2}) given by

$$Q_{\text{rest}} = -\lambda_T(T^* - T_S), \tag{9.1}$$

where λ_T (in $\text{Wm}^{-2}\text{K}^{-1}$) is a constant surface heat exchange coefficient. The heat flux Q_{rest} is proportional to the temperature difference between the ocean temperature T^* taken at the surface and a prescribed atmospheric temperature T_S, chosen as

$$T_S(\phi, \theta) = T_0 + \frac{\Delta T}{2} \cos\left(\pi \frac{\theta - \theta_s}{\theta_n - \theta_s}\right), \tag{9.2}$$

where $T_0 = 15°\text{C}$ is a reference temperature and ΔT is the temperature difference between the southern and northern latitude of the domain. The thermal forcing is distributed as a body forcing over the first (upper) layer of the ocean having a depth H_m.

Temperature differences in the ocean cause density differences according to

$$\rho = \rho_0(1 - \alpha_T(T^* - T_0)), \tag{9.3}$$

where α_T is the volumetric expansion coefficient and ρ_0 is a reference density. Inertia is neglected in the momentum equations because of the small Rossby number; we use the Boussinesq and hydrostatic approximations and represent horizontal and vertical mixing of momentum and heat by constant eddy coefficients. With r_0 and Ω being the radius and angular velocity of the Earth, the governing equations for the zonal, meridional and vertical velocity u, v and w, the dynamic pressure p (the hydrostatic part has been subtracted) and the temperature $T = T^* - T_0$ become

$$-2\Omega\, v \sin\theta + \frac{1}{\rho_0 r_0 \cos\theta}\frac{\partial p}{\partial \phi} - A_H L_u(u, v) - A_V \frac{\partial^2 u}{\partial z^2} = 0, \tag{9.4a}$$

$$2\Omega\, u \sin\theta + \frac{1}{\rho_0\, r_0}\frac{\partial p}{\partial \theta} - A_H\, L_v(u, v) - A_V \frac{\partial^2 v}{\partial z^2} = 0, \tag{9.4b}$$

$$\frac{\partial p}{\partial z} - \rho_0 g \alpha_T T = 0, \tag{9.4c}$$

$$\frac{1}{r_0 \cos\theta}\left(\frac{\partial u}{\partial \phi} + \frac{\partial (v\cos\theta)}{\partial \theta}\right) + \frac{\partial w}{\partial z} = 0, \tag{9.4d}$$

$$\frac{DT}{dt} - K_H \nabla_H^2 T - K_V \frac{\partial^2 T}{\partial z^2} - \frac{(T_S - T^*)}{\tau_T}\mathcal{H}(\frac{z}{H_m} + 1) = 0, \tag{9.4e}$$

where $\mathcal{H}$ is a continuous approximation of the Heaviside function, g is the gravitational acceleration and $\tau_T = \rho_0 C_p H_m/\lambda_T$ is the surface adjustment time scale of heat (C_p is the constant heat capacity). In these equations, A_H and A_V are the horizontal and vertical momentum (eddy) viscosity and K_H and K_V the horizontal and vertical (eddy) diffusivity of heat, respectively. In addition, the operators in the equations above are defined as

$$\begin{aligned}
\frac{D}{dt} &= \frac{\partial}{\partial t} + \frac{u}{r_0\cos\theta}\frac{\partial}{\partial\phi} + \frac{v}{r_0}\frac{\partial}{\partial\theta} + w\frac{\partial}{\partial z},\\
\nabla_H^2 &= \frac{1}{r_0^2\cos\theta}\left[\frac{\partial}{\partial\phi}\left(\frac{1}{\cos\theta}\frac{\partial}{\partial\phi}\right) + \frac{\partial}{\partial\theta}\left(\cos\theta\frac{\partial}{\partial\theta}\right)\right],\\
L_u(u, v) &= \nabla_H^2 u + \frac{u\cos 2\theta}{r_0^2\cos^2\theta} - \frac{2\sin\theta}{r_0^2\cos^2\theta}\frac{\partial v}{\partial\phi},\\
L_v(u, v) &= \nabla_H^2 v + \frac{v\cos 2\theta}{r_0^2\cos^2\theta} + \frac{2\sin\theta}{r_0^2\cos^2\theta}\frac{\partial u}{\partial\phi}.
\end{aligned} \tag{9.5}$$

Slip conditions and zero heat flux are assumed at the bottom boundary, whereas at all lateral boundaries, no-slip and zero heat flux conditions are applied. As the forcing is represented as a body force over the first layer, slip and zero heat flux conditions

Table 9.1. *Standard values of parameters used in the minimal model*

2Ω	=	1.4×10^{-4}	[s^{-1}]	r_0	=	6.4×10^6	[m]
H	=	4.0×10^3	[m]	τ_T	=	3.0×10^1	[days]
α_T	=	1.0×10^{-4}	[$°C^{-1}$]	g	=	9.8	[ms^{-2}]
A_H	=	1.6×10^5	[m^2s^{-1}]	T_0	=	15.0	[°C]
ρ_0	=	1.0×10^3	[kgm^{-3}]	A_V	=	1.0×10^{-3}	[m^2s^{-1}]
K_H	=	1.0×10^3	[m^2s^{-1}]	K_V	=	1.0×10^{-4}	[m^2s^{-1}]
C_p	=	4.2×10^3	[$J(kg°C)^{-1}$]	ΔT	=	20.0	[°C]

apply at the ocean surface. Hence the boundary conditions are

$$z = -H, 0: \quad \frac{\partial u}{\partial z} = \frac{\partial v}{\partial z} = w = \frac{\partial T}{\partial z} = 0, \tag{9.6a}$$

$$\phi = \phi_w, \phi_e: \quad u = v = w = \frac{\partial T}{\partial \phi} = 0, \tag{9.6b}$$

$$\theta = \theta_s, \theta_n: \quad u = v = w = \frac{\partial T}{\partial \theta} = 0. \tag{9.6c}$$

The parameters for the standard case are the same as in typical large-scale low-resolution ocean components of (4° horizontally) GCMs, and their values are listed in Table 9.1.

9.2.2 The AMO mode

Following the strategy of determining the qualitative behaviour of solutions as parameters are changed in the preceding minimal model, techniques described in Chapter 2 are applied. The governing equations (9.4) and boundary conditions (9.6) are discretised on an Arakawa B-grid using central spatial differences. In the results in this section, a horizontal resolution of 4° is used. An equidistant grid with sixteen levels is applied in the vertical so that the first layer thickness $H_m = 250$ m. The discretized system of equations can be written in the form (2.20), and a $16 \times 16 \times 16$ grid with five unknowns per point (u, v, w, p and T) leads to a dynamical system of dimension, that is, the number of degrees of freedom, $d = 5 \times 16^3 = 20{,}480$.

The steady equations are solved using a pseudo-arclength continuation method (Keller, 1977). As primary control parameter, we choose the equator-to-pole temperature difference ΔT. For every value of ΔT, a steady solution of the minimal model is calculated under the restoring flux Q_{rest} in (9.1). For each steady flow pattern, the maximum of the meridional overturning streamfunction (ψ_M) is calculated, and it is plotted versus ΔT in Fig. 9.4a. The meridional overturning streamfunction, representing the zonally averaged circulation, for $\Delta T = 20$°C, is plotted in Fig. 9.4b. The

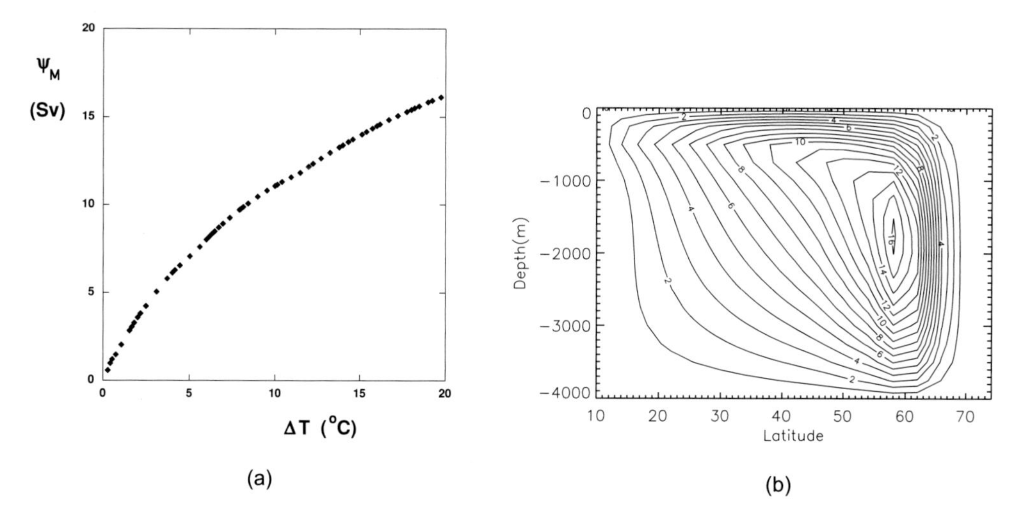

Figure 9.4 (a) Plot of the maximum meridional overturning (ψ_M) of the steady solution (in Sv) versus the equator-to-pole temperature difference ΔT (°C) under restoring conditions. (b) Plot of the meridional overturning streamfunction (contour values in Sv) for $\Delta T = 20$ (°C).

maximum of ψ occurs at about 60°N, and the amplitude is about 16 Sv (1 Sv = 10^6 m^3/s).

Next, the ocean-atmosphere heat flux Q_{pres} (where the subscript refers to 'prescribed') is diagnosed for each of the steady solutions, and the linear stability of these solutions is computed under the heat flux Q_{pres}. The discretised linear stability problem of these steady states is formulated as a generalised eigenvalue problem,

$$J\mathbf{x} = \sigma B\mathbf{x}. \tag{9.7}$$

Here $\sigma = \sigma_r + i\sigma_i$, where σ_r is the growth rate, σ_i is the angular frequency, J is the Jacobian matrix and B is the mass matrix of the equations (9.4). We solve for the 'most dangerous' normal modes, that is, those with σ closest to the imaginary axis. The growth rate and period $2\pi/\sigma_i$ of the mode with the largest growth rate, the AMO mode, are plotted versus ΔT in Fig. 9.5.

For $\Delta T = 20$°C, this mode has a positive growth factor, and hence the background state, of which the meridional overturning streamfunction was shown in Fig. 9.4b, is unstable to the AMO mode. The period of the AMO mode is about sixty-seven years at $\Delta T = 20$°C, and it decreases with increasing ΔT (Fig. 9.5). From Fig. 9.5, we also see that the growth factor of the AMO mode decreases strongly with decreasing ΔT and becomes negative near $\Delta T_c \approx 4$°C, where a Hopf bifurcation occurs. For $\Delta T < \Delta T_c$, the steady states are therefore linearly stable under the prescribed flux Q_{pres}.

It was shown in Dijkstra (2006) that for small ΔT, the angular frequency of the AMO mode becomes zero, and the complex conjugate pair of eigenvalues splits up

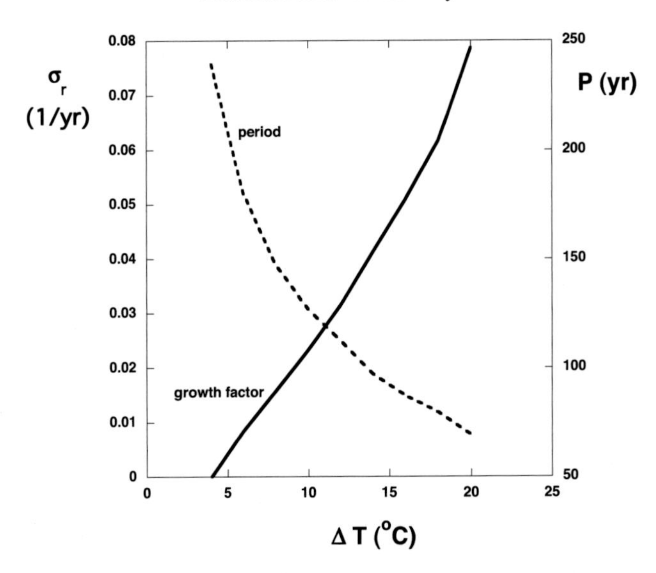

Figure 9.5 Growth factor σ_r (in yr^{-1}, drawn) and period $P = 2\pi/\sigma_i$ (in yr, dashed) versus ΔT (in °C) of the AMO mode in the minimal model under prescribed flux conditions.

into two real eigenvalues. The paths of the two different modes can be followed to the $\Delta T = 0$°C limit, where the eigensolutions connect to those of the diffusion operator of the temperature equation, called SST modes in Dijkstra (2006). These SST modes can be ordered according to their zonal (n), meridional (m) and vertical wave number (l), and it was found that the AMO mode connects to the $(0, 0, 1)$ SST mode and the $(1, 0, 0)$ SST mode near $\Delta T = 0$°C.

For each eigenvalue σ associated with the AMO mode, there is a corresponding eigenvector $\mathbf{x} = \mathbf{x}_r + i\mathbf{x}_i$ according to (9.7). In Fig. 9.6, the sea-surface temperature

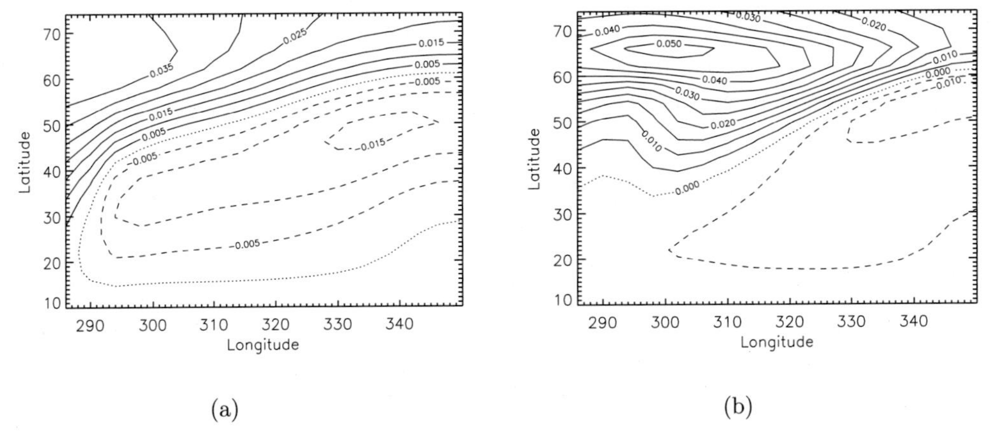

Figure 9.6 SST pattern of the real part the AMO mode for (a) $\Delta T = 4$°C (near Hopf bifurcation) and (b) $\Delta T = 20$°C. Note that amplitudes are arbitrary as these are eigensolutions.

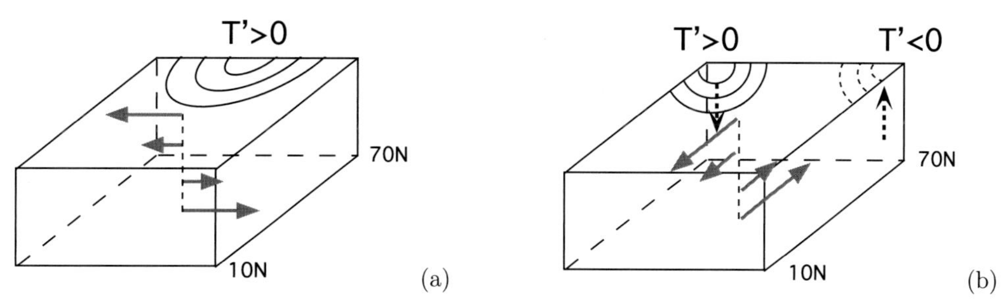

Figure 9.7 Schematic diagram of the oscillation mechanism associated with the multidecadal mode caused by the westward propagation of the temperature anomalies T'. The phase difference between (a) and (b) is about a quarter period. See text and Te Raa and Dijkstra (2002) for a further explanation.

field of the real part of the eigenvector ($\mathbf{x}_r$) of the AMO mode is plotted for $\Delta T = 4°C$ (near the Hopf bifurcation) and for $\Delta T = 20°C$. With increasing ΔT, the pattern becomes more localised in the northwestern part of the basin.

9.2.3 Physical mechanism: the thermal Rossby mode

The physical mechanism of propagation of the AMO mode was presented in Colin de Verdière and Huck (1999) and Te Raa and Dijkstra (2002). This mechanism holds at every ΔT for which an oscillatory AMO mode is present (cf. Fig. 9.5). A sketch of this mechanism is provided in Fig. 9.7. A warm anomaly in the north-central part of the basin causes a positive meridional perturbation temperature gradient, which induces – via the thermal wind balance – a westward zonal surface flow (Fig. 9.7a). The anomalous anticyclonic circulation around the warm anomaly causes southward (northward) advection of cold (warm) water to the east (west) of the anomaly, resulting in westward phase propagation of the warm anomaly. Due to this westward propagation, the zonal perturbation temperature gradient becomes negative, inducing a southward surface meridional flow (Fig. 9.7b). The resulting upwelling (downwelling) perturbations along the northern (southern) boundary cause a negative meridional perturbation temperature gradient, inducing an eastward zonal surface flow, and the second half of the oscillation starts. The crucial elements in this oscillation mechanism are the phase difference between the zonal and meridional surface flow perturbations, and the westward propagation of the temperature anomalies (Te Raa and Dijkstra, 2002). The presence of freshwater forcing and wind stress forcing does not essentially change this mechanism; density anomalies will take over the role of temperature anomalies in the preceding description (Te Raa and Dijkstra, 2003b; Te Raa et al., 2004).

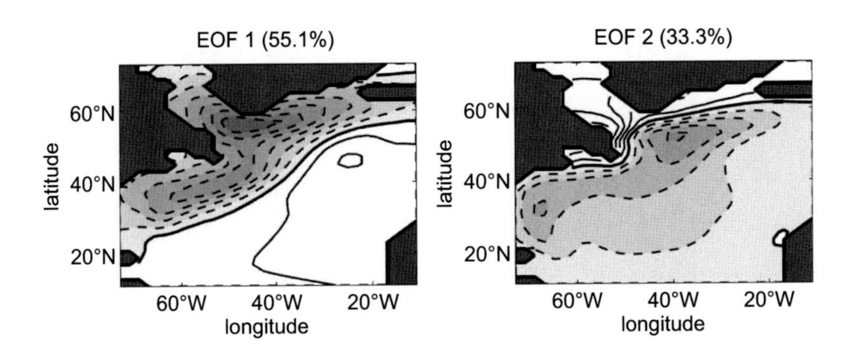

Figure 9.8 First two EOFs of sea-surface temperature for the minimal model configuration with continents, which together explain almost 90% of the variance. Contour interval is 0.5°C, and negative temperature anomaly values are shaded.

9.2.4 AMO mode pattern and continental shape

Oscillatory behaviour in the minimal model will occur for $\Delta T > \Delta T_c$ under prescribed flux conditions, as the steady state is then unstable. A slight extension of minimal model is the inclusion of a more realistic continental shape. In Frankcombe et al. (2010b), the same minimal model setup is used as in Section 9.2.1 but with the horizontal resolution increased to $2^\circ \times 2^\circ$ and using twenty-four nonequidistant levels in the vertical with $H_m = 50$ m. Under prescribed flux boundary conditions, the oscillatory flow has a period of about forty-two years and a spatial pattern of variability, which is shown in the first two EOFs of SST in Fig. 9.8.

The variability is still concentrated in the northwestern part of the basin, but the pattern has been deformed by the continents. To determine the AMO pattern as in Kushnir (1994), the difference in sea-surface temperatures between the warm and cool phases of the AMO in the model is determined. The resulting pattern is shown in Fig. 9.9. The overall effect of including continents in the simple model is to deform

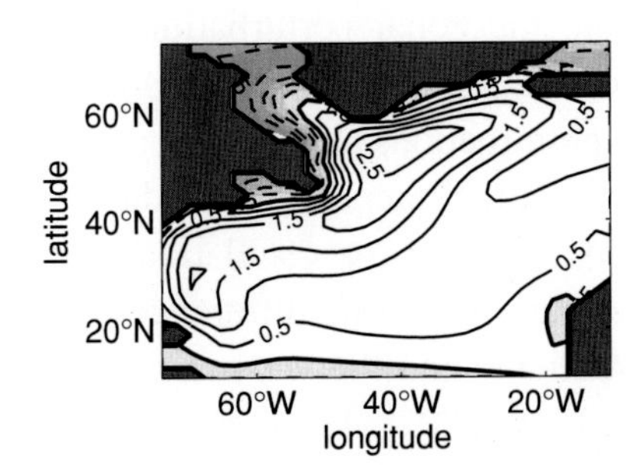

Figure 9.9 The SST pattern of the AMO (determined as the difference between the patterns of SST from years with maximum and minimum AMO index) in the minimal model with continents; contour interval is 0.5°C.

the spatial pattern of the AMO mode into a form that more resembles observations (compare Fig. 9.2).

9.3 Excitation of the AMO mode

Restoring and prescribed flux conditions are the two limits of atmospheric damping of SST anomalies, and to study what happens between these limits, a new general boundary condition for the surface heat flux (Q_D) can be chosen as

$$Q_D = (1 - \gamma)Q_{\text{rest}} + \gamma Q_{\text{pres}}, \tag{9.8}$$

where Q_{rest} is as in (9.1) and Q_{pres} is the diagnosed heat flux of the steady state for $\Delta T = 20°\text{C}$. A value of γ representative of the real ocean can be estimated by examining the damping time scale of SST in the upper layer ocean. Under the forcing (9.8), the damping time scale τ_T is defined as

$$\tau_T = (1 - \gamma)\frac{C_p H_m \rho_0}{\lambda_T}. \tag{9.9}$$

Using variables from the minimal model (with $H_m = 50$ m), $\lambda_T = 20\ \text{Wm}^{-2}\text{K}^{-1}$, and $\tau_T = 30$ days as a representative midlatitude value (Barsugli and Battisti, 1998) gives a value of $\gamma \approx 0.75$.

When γ decreases from $\gamma = 1$ (prescribed flux conditions), the growth factor of the AMO mode (as in Fig. 9.5a) will decrease as the atmospheric damping becomes larger. As for $\gamma = 0$ (restoring conditions), the growth rate of the AMO mode is negative (Te Raa and Dijkstra, 2003b), and the Hopf bifurcation must occur somewhere between $\gamma = 0$ and $\gamma = 1$. Under the heat flux (9.8), equilibrium states were computed using transient integration starting from the $\gamma = 0$ steady solution. For each solution obtained, the standard deviation of the sea-surface temperature over a box B ([46°N – 62°N] × [74°W – 50°W]) is plotted in Fig. 9.10a. For $\gamma \leq 0.85$ there are no oscillations, and near $\gamma_c = 0.85$ the system undergoes the Hopf bifurcation and the multidecadal oscillations appear for $\gamma > \gamma_c$. The amplitude of the oscillations is measured by calculating the standard deviation of the box averaged sea-surface temperature rather than the peak-to-peak amplitude so that comparisons can be made with later simulations where the variability is not regular. The oscillations have periods decreasing from fifty-three years near $\gamma = 0.85$ to forty-five years at $\gamma = 1$, and so the change in period with γ is much smaller than that with ΔT (cf. Fig. 9.5). The first two EOFs of the SST field, which together explain 92.0% of the variance, are shown for $\gamma = 0.9$ in Fig. 9.10b. The variance in temperature at the sea surface is concentrated in the northwest region of the basin, similar to that of the AMO mode (Fig. 9.6b).

This analysis indicates that the atmospheric damping may be too large in reality (as represented by $\gamma \approx 0.75$) for sustained oscillatory behavior to exist. As we know

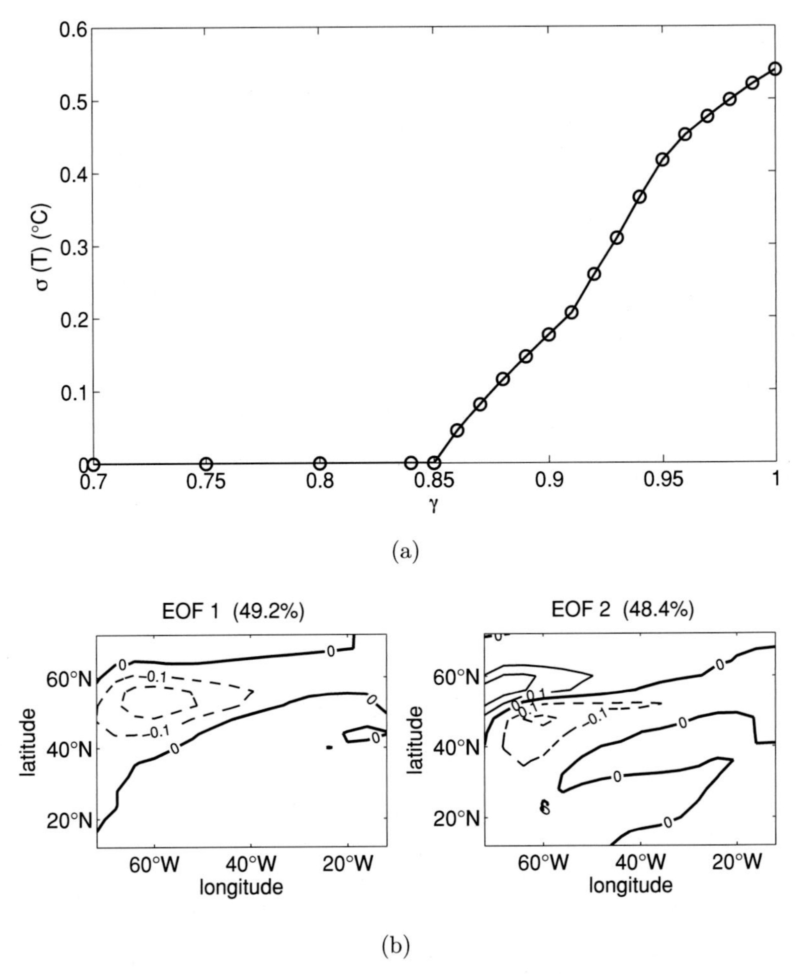

Figure 9.10 Deterministic case, that is, no noise is applied. (a) Standard deviation of the sea-surface temperature over the box [46°N – 62°N] × [74°W – 50°W] as a function of γ. Below $\gamma = 0.85$, the standard deviation is zero, so the mode is damped. (b) First two EOFs of sea-surface temperature for the transient flow at $\gamma = 0.9$.

from Chapters 3 and 4, however, when noise is present in the forcing (in this case in the heat flux), a normal mode pattern can be excited even below the Hopf bifurcation (the stochastic Hopf bifurcation as discussed in Section 4.3.3).

In Frankcombe et al. (2009), the effect of spatial and temporal coherence in the noise forcing under the conditions $\gamma < \gamma_c$ is studied by comparing the responses of the minimal model under the following two heat fluxes:

$$Q_W = Q_D + \lambda Z_{ij}(t), \tag{9.10a}$$

$$Q_{S,W_m} = Q_D + \lambda Z_m(t) \sin\left(\frac{\pi(i - i_e)}{i_w - i_e}\right) \sin\left(\frac{\pi(j - j_s)}{j_n - j_s}\right), \tag{9.10b}$$

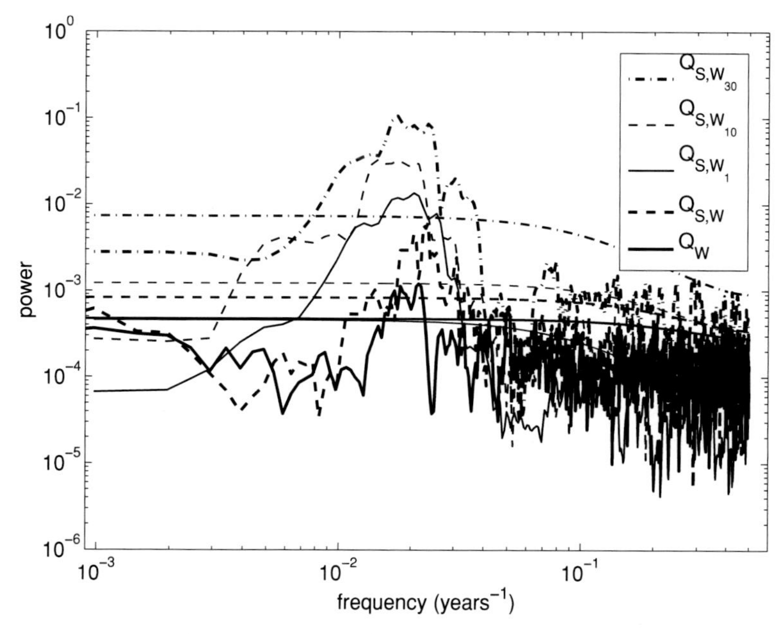

Figure 9.11 Spectra of temperature in the box B for Q_W, $Q_{S,W} = Q_{S,W_0}$ and three others using white noise with time scales of 1 (Q_{S,W_1}), 10 ($Q_{S,W_{10}}$) and 30 ($Q_{S,W_{30}}$) days for $\gamma = 0.8$. The 99% significance levels for each case are also plotted.

where Q_D is given by (9.8). In Q_W, Z_{ij} is a normally distributed random variable that takes on a different value at each grid point (i, j) in space at each time step t. The noise in Q_W is thus uncorrelated in both space and time. In Q_{S,W_m}, $i_e \leq i \leq i_w$ and $j_s \leq j \leq j_n$ are the grid variables in the x and y directions. Furthermore, $Z_m(t)$ is a normally distributed random variable, where m indicates the number of days where this variable is persistent. The spatial pattern in (9.10b) is chosen as a rough approximation to variations in atmospheric heat fluxes seen over the North Atlantic (Cayan, 1992), such as those associated with the North Atlantic Oscillation. In both Q_W and each Q_{S,W_m}, the amplitude (λ) of the noise was taken to be 10% of the difference between the minimum and maximum over the basin of the prescribed heat flux Q_{pres}, which is approximately 20 W/m^2.

In Fig. 9.11, the spectra of the modelled temperature variability for the case of $\gamma = 0.8$ for five different types of noise is shown. Although the noise added to the system has no preferred frequency, the spectrum shows a large peak at multidecadal frequencies. This is in contrast to the $\gamma = 0.8$ case in the absence of noise, where neither the temperature nor the overturning strength vary at all. It can also be clearly seen that both the spatial and temporal correlation of the noise increases the height and breadth of the multidecadal peak. The multidecadal peak increases as the time scale of persistence of the forcing increases. When spatial coherence is removed (so that the noise added to each grid point is independent), the temporal coherence still

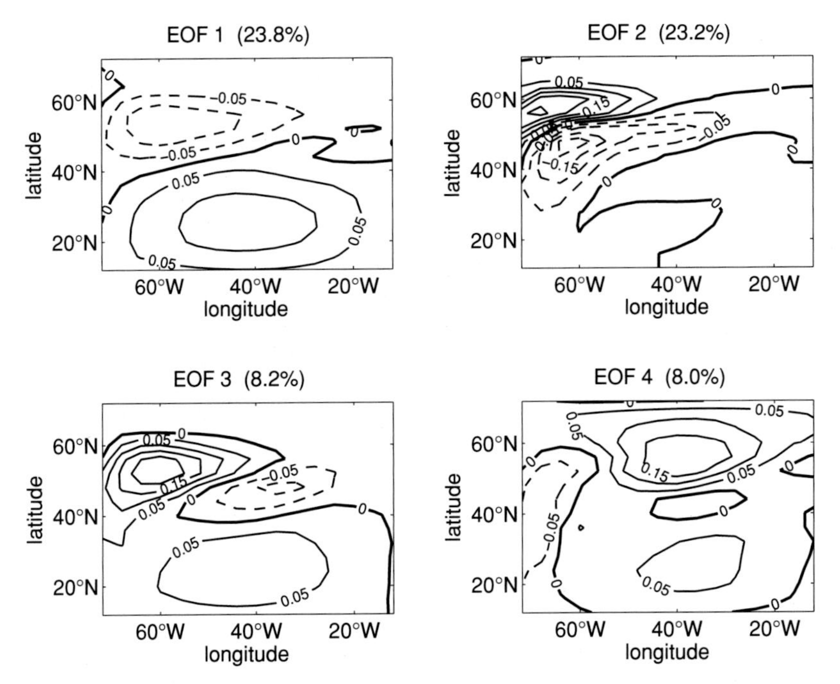

Figure 9.12 First four EOFs of sea-surface temperature for the $Q_{S,W_{30}}$ case, which together explain more than 50% of the variance. The data were low-pass filtered to allow periods between 30 and 100 years.

causes large variations in temperature, but the power at multidecadal frequencies is greatly reduced.

When γ is decreased below the critical value γ_c, noise dominates the patterns seen in an M-SSA analysis (cf. Section 5.3). However, if the data are low-pass filtered to allow periods from 30 to 100 years, then the patterns of multidecadal variability can be seen. Figure 9.12b shows the first four EOFs of SST for the $Q_{S,W_{30}}$ case, with the EOFs accounting for 23.8%, 23.2%, 8.2% and 8.0% of the variance, respectively. Because the eigenvalues of these EOFs are so closely paired, the errors might be large according to (5.37). In this case, however, the model integrations were long enough (2,000 years) so that the approximations for the typical error found using (5.37) are small. Signals from the sinusoidal spatial pattern of the noise forcing are evident, particularly in EOF 4 as well as in the southern part of the basin in EOFs 1 and 3. The EOFs, however, still display the pattern of the AMO mode, which has been excited by the noise forcing (compare Fig. 9.10).

Figure 9.13 shows the effect of the different noise forcings on the standard deviation of the sea-surface temperature in the box B. For values of $\gamma > \gamma_c$, the noise has only a small effect. In contrast, for near and below γ_c, the noise causes the surface temperature to increase substantially. With the flux $Q_{S,W_{30}}$, the largest amplitude of the variability is achieved, showing that spatial and temporal coherence in the noise are important

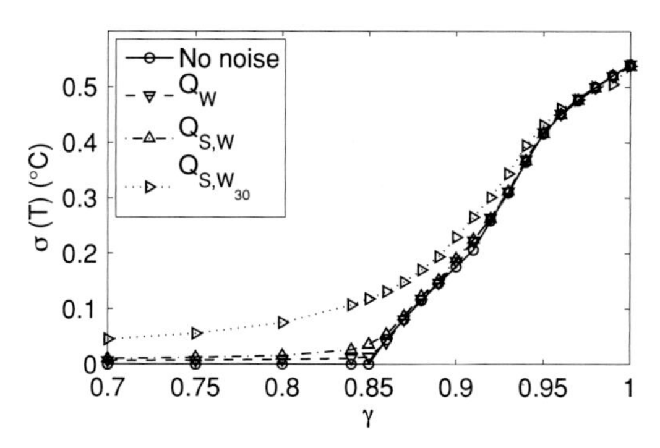

Figure 9.13 Standard deviation of sea-surface temperature (in °C, averaged over the box [46°N – 62°N] × [74°W – 50°W]) as a function of γ, for the no noise, Q_W, the $Q_{S,W}$ and the $Q_{S,W_{30}}$ cases.

to set the amplitude of the multidecadal variability. In addition, the amplitude of the variability versus γ as shown in Fig. 9.13 is qualitatively very similar to that of the elementary stochastic Hopf bifurcation in Section 4.3.3, indicating that the AMO mode is excited by atmospheric noise in the minimal model.

9.4 Dynamical mechanisms of excitation

In the previous section, it was shown that a multidecadal mode that destabilises the background state for values of $\gamma > \gamma_c$ can also be excited under conditions where $\gamma < \gamma_c$ by noisy forcing. In this section, three different mechanisms of excitation are discussed; the first one (rectification) occurs in the minimal model, and two others (non-normal growth and spatial resonance) have been found to be relevant in other models displaying decadal to multidecedal variability.

9.4.1 Rectification

The mechanism of excitation of the AMO mode in the minimal model is examined in detail in Frankcombe et al. (2009). Simulations were performed here with a heat flux similar to Q_{S,W_m} as in (9.10b), but where the $Z_m(t)$ were taken from the NAO index (as discussed in Chapter 7).

Noise can rectify the background state, and the background state has an effect on the stability of the multidecadal mode. Figure 9.14(a) shows the meridional overturning stream function of the background state with steady forcing (i.e., no noise) for $\gamma = 0.8$. In one simulation, the model is then forced with the surface heat flux pattern associated with a permanent NAO+ state until a steady state is reached. Figure 9.14(b) shows the difference between the meridional overturning stream function of this new time-mean state and the steady state under standard no-noise forcing. In another simulation, this

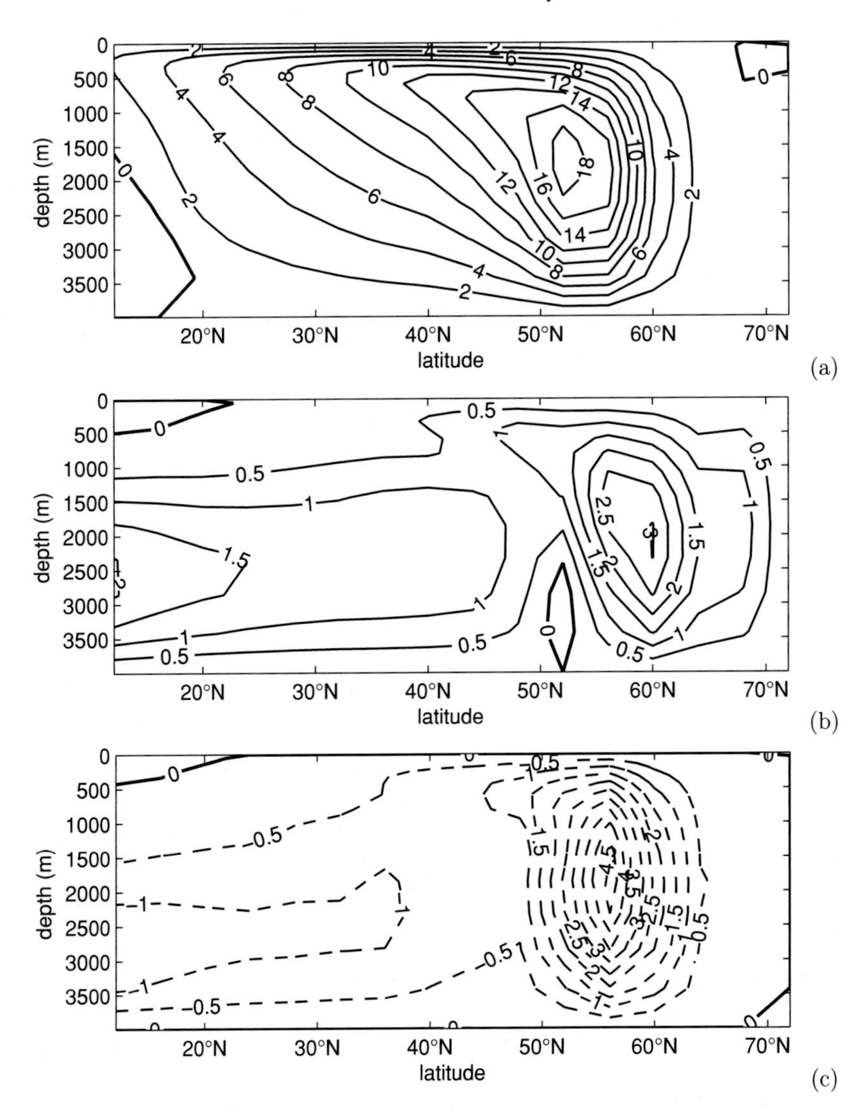

Figure 9.14 (a) Meridional overturning streamfunction of the steady state under steady (no noise) forcing with $\gamma = 0.8$. Contour interval is 2 Sv. (b) Difference between the meridional overturning stream function of the steady state reached under permanent NAO+ forcing and the steady state shown in (a). Contour interval is 0.5 Sv. (c) Same as (b) but for NAO − forcing.

is repeated for a permanent NAO− forcing, and the difference between this time-mean state and the standard no noise case is shown in Fig. 9.14(c).

The pattern of permanent NAO+ forcing has positive heat flux over the southern half of the basin and negative heat flux over the northern half and, thus, acts to increase the north-south surface temperature gradient. This leads to an increase in the meridional overturning streamfunction as well as a slight northward shift of the sinking region, as seen in Fig. 9.14b. The opposite occurs with permanent NAO− forcing, with a decrease in the north-south temperature gradient, leading to a decrease in the meridional overturning stream function (Fig. 9.14c). The growth rate of the multidecadal mode increases with increasing strength of the meridional overturning streamfunction, which in turn increases with increasing ΔT, where ΔT is the north-south temperature difference (Fig. 9.5). This means that the permanent NAO+ and NAO− forcing has alternately decreased and increased the growth factor of the multidecadal mode; that is, under permanent NAO+ forcing, the mean meridional overturning circulation is more unstable to the multidecadal mode than the mean meridional overturning circulation under permanent NAO− forcing. In summary, the rectification of the background state due to the noise is responsible for the excitation of the AMO mode.

9.4.2 Non-normal growth

Another mechanism to excite the multidecadal variability is through non-normal growth (Section 2.2.2.) In a planetary geostrophic three-dimensional ocean model with salinity and temperature as prognostic variables, the non-normal growth of perturbations was studied in Sevellec et al. (2009). They first determined steady states of this model under restoring conditions. Next, the Jacobian matrix J of the model at this steady state was determined under prescribed flux conditions (FBC). All eigenmodes of J have a negative real part, and the least damped one is an oscillatory mode with a period of 34.2 years and a decay time scale of 207 years. The SST pattern of the real and imaginary parts of this eigenmode are plotted in Fig. 9.15a–b.

The patterns of these eigenmodes are similar to those in the minimal model as described in Section 9.2 (cf. Fig. 9.6). As we have seen in Section 2.2, the growth of these normal modes is maximal for an initial condition consisting of the least damped eigenvector of J^T. Although the SST perturbation field has a much larger amplitude in the eigenmode (Fig. 9.15a–b), the amplitude of the sea-surface salinity (SSS) field dominates in the adjoint eigenmode. This indicates that SSS perturbations can optimally excite the AMO mode in this model. The SSS patterns of this adjoint eigenmode are plotted in Fig. 9.15c–d.

Sevellec et al. (2009) computed the optimal initial surface salinity (and salinity flux) perturbations, where optimality refers to a maximum amplitude of excitation of the flow perturbation on a time τ. They used the so-called Bra-Ket notation, where a column vector $\mathbf{x}$ is written as $|x\rangle$, a complex conjugate row vector $\mathbf{x}^\dagger$ as $\langle x|$ and the inner product between two vectors $\mathbf{x}$ and $\mathbf{y}$ is written as $\langle x, y\rangle$. We adopt this notation here for convenience. With the state vector $|u\rangle(t)$, the perturbation equation can be written in Itô form as

$$d\,|u\rangle_t = J\,|u\rangle_t\,dt + |g\rangle\,dW_t, \tag{9.11}$$

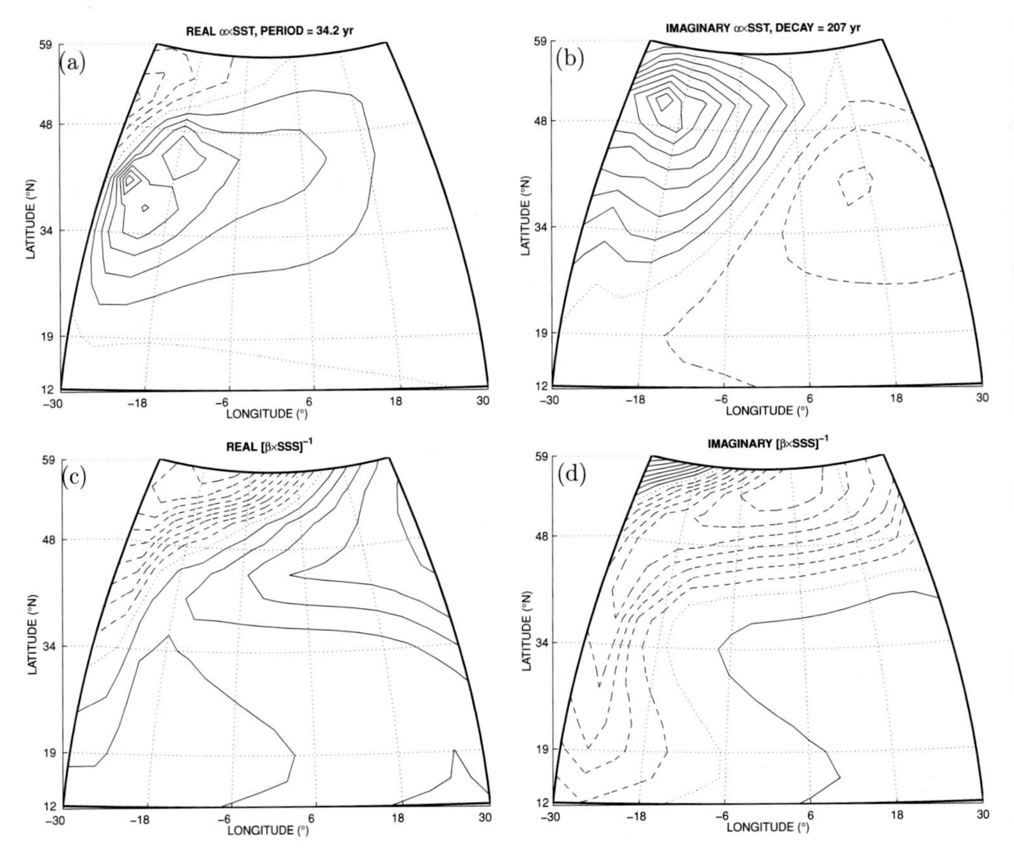

Figure 9.15 Real (a) and imaginary (b) parts of SST of the least damped eigenmode of the Jacobian matrix J in a planetary geostrophic model. Real (c) and imaginary (d) parts of SSS of the least damped eigenmode of the transpose of the Jacobian matrix J^T. Amplitudes are arbitrary (figure from Sevellec et al. [2009]).

where $|g\rangle$ represents the time-independent pattern of the allowed perturbations. The solution to (9.11) with initial condition $|u\rangle = 0$ is given by

$$|u\rangle_\tau = \int_0^\tau R(\tau - s)\,|g\rangle\,dW_s \tag{9.12}$$

where $R(t) = e^{Jt}$ is the propagator of the linearised model.

The situation is complicated by the fact that only surface salinity perturbations are considered, so $|g\rangle = \mathcal{P}\,|g'\rangle$, where $\mathcal{P}$ is a projection of the state vector. In addition, perturbations should be such as to conserve the total salt content, which can be written as a constraint $\langle g|\,\mathcal{S}\,|g\rangle = 1$. In Sevellec et al. (2009), $\mathcal{S}$ is chosen as as

$$\langle u|\,\mathcal{S}\,|u\rangle = \frac{\sum_{i,j}\left(\alpha_T^2(T'_{i,j})^2 + \alpha_S^2(S'_{i,j})^2\right)}{\sum_{i,j} V_{i,j}}, \tag{9.13}$$

where the primes (in temperature T and salinity S) indicate perturbations from the steady state and the sum is over all grid cells (i, j) with volumes $V_{i,j}$. Finally, only the optimal response of the salinity perturbations on the amplitude of the meridional overturning circulation is considered. The latter condition is written as $\langle F, u\rangle$ with the appropriate projection $\langle F|$.

The Lagrangian $\mathcal{L}$ for the optimization problem then becomes

$$\mathcal{L} = Var[\langle F, u\rangle] - \gamma(\langle g|\,\mathcal{S}\,|g\rangle - 1), \tag{9.14}$$

where γ is a Lagrange multiplier, and $Var[\langle F, u\rangle]$ is given by

$$Var[\langle F, u\rangle] = \langle g| \int_0^\tau R^\dagger(\tau - s)\,|F\rangle\,\langle F|\,R(\tau - s)\,|g\rangle\; ds. \tag{9.15}$$

For a Lagrangian of the form $\mathcal{L}(\mathbf{x}, \gamma) = G(\mathbf{x}) - \gamma C(\mathbf{x})$, the first variation, $\delta\mathcal{L}$, is

$$\delta\mathcal{L} = \lim_{\epsilon\to 0} \frac{d}{d\epsilon}(G(\mathbf{x}_* + \epsilon\mathbf{h}) - \gamma C(\mathbf{x}_* + \epsilon\mathbf{h})), \tag{9.16}$$

where $\mathbf{x}_*$ is the desired optimum. This leads to the (component wise) equations

$$\frac{\partial G}{\partial \mathbf{x}}(\mathbf{x}_*) - \gamma\frac{\partial C}{\partial \mathbf{x}}(\mathbf{x}_*) = 0, \tag{9.17a}$$

$$C(\mathbf{x}_*) = 0. \tag{9.17b}$$

When the state vector $\mathbf{x}$ is d-dimensional, these are $d + 1$ equations for $d + 1$ unknowns $x_{*1}, \ldots, x_{*d}$ and γ.

Using the fact that for any quadratic form $G(\mathbf{x}) = \mathbf{x}^\dagger J\mathbf{x}$, the derivative is given by

$$\frac{\partial G}{\partial \mathbf{x}} = \mathbf{x}^\dagger(J + J^\dagger), \tag{9.18}$$

the equations (9.17) lead to an eigenvalue problem for $|g'\rangle$ (the surface salinity perturbations), which can be written (for $\tau \to \infty$) as

$$\mathcal{N}^{-1}\mathcal{H}\,|g'\rangle = \gamma\,|g'\rangle\,. \tag{9.19}$$

Here $\mathcal{N} = \mathcal{P}^\dagger\mathcal{S}\mathcal{P}$ and

$$\mathcal{H} = \lim_{\tau\to\infty} \int_0^\tau \mathcal{P}^\dagger R^\dagger(\tau - s)\,|F\rangle\,\langle F|\,R(\tau - s)\mathcal{P}\; ds. \tag{9.20}$$

Once the stochastic optimal $|g'\rangle$ is known, the surface salinity (or salinity flux) perturbation that optimises the flow response can be determined.

Under FBC, the pattern of the optimal surface salinity perturbation is plotted in Fig. 9.16a. The optimal SSS perturbation has a similar spatial pattern, as the least stable eigenmode of the adjoint normal mode problem (Fig. 9.15c–d) and the time development of the meridional overturning in Fig. 9.16b shows that the perturbation decays with the time scale of the least stable normal mode. When the noisy freshwater

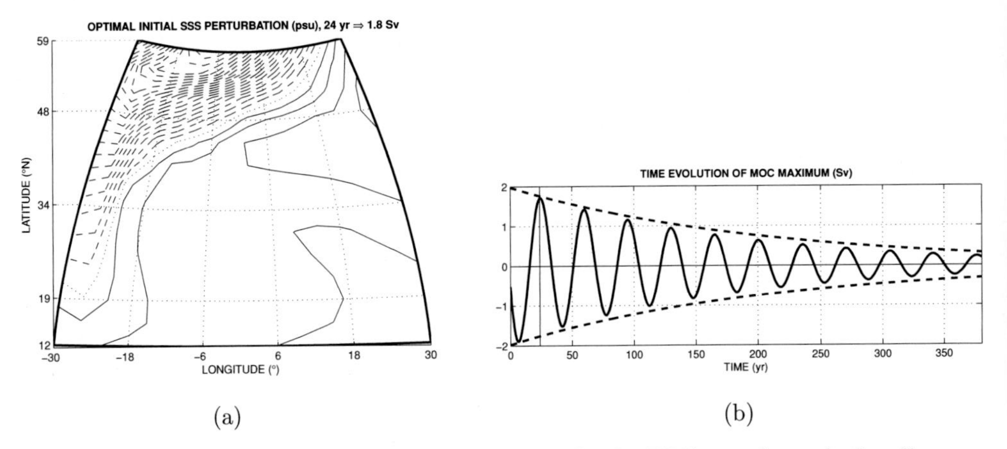

Figure 9.16 (a) Optimal initial SSS perturbation for the FBC experiment in Sevellec et al. (2009). (b) Time evolution of the maximum of the flow perturbation as a response to the SSS perturbation in (a); figure from Sevellec et al. (2009).

flux pattern of Fig. 9.17a is added to the forcing of the model, a response is found as in Fig. 9.17b. The spectrum of the flow shows a red noise background with, in addition, a peak at the period of the normal mode.

In summary, from the study of Sevellec et al. (2009), it follows that specific salinity anomalies can induce growth of the meridional overturning flow under conditions in which the steady state is linearly stable. This growth is due to the non-normal

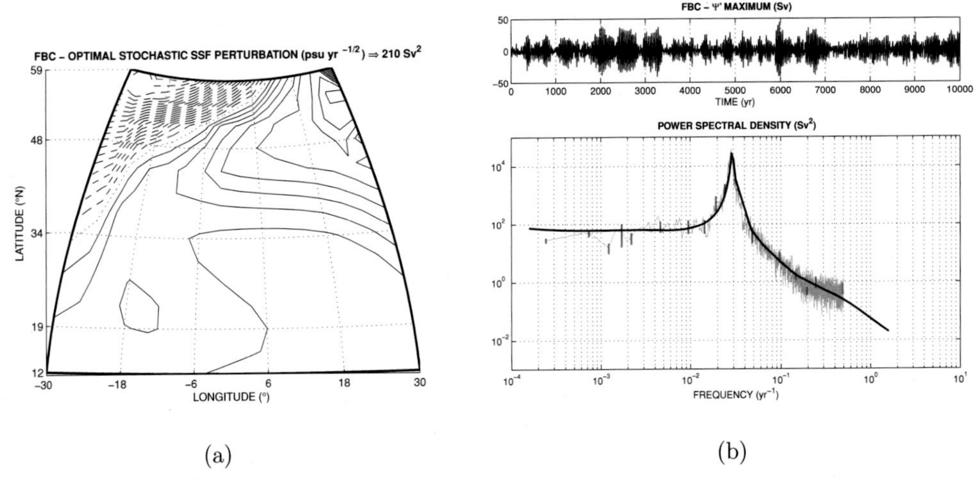

Figure 9.17 (a) Optimal stochastic perturbations for the FBC case. Perturbations induce a variability of the meridional overturning circulation intensity with a standard deviation of 14.5 Sv. (b) Spectrum of the response of the maximum flow to the optimal stochastic freshwater flux perturbations for the FBC case (figure from Sevellec et al. [2009]).

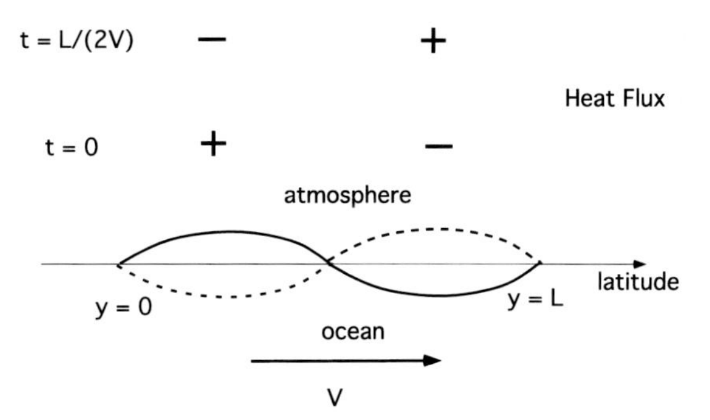

Figure 9.18 Sketch of the situation of a slab ocean coupled to an atmosphere layer as considered in Saravanan and McWilliams (1998) as an illustration of the spatial resonance mechanism. The drawn curve represents the ocean temperature at $t = 0$; the dashed curve shows the ocean temperature at $t = L/(2V)$.

properties of the Jacobian matrix. After a while, however, the time development is controlled by the normal mode pattern.

9.4.3 Spatial resonance

As an extension of the Hasselmann (1976) stochastic climate variability discussed in Section 6.3.2, Saravanan and McWilliams (1998) have considered the response of a slab ocean to atmospheric stochastic forcing in the presence of advection. They considered a domain (Fig. 9.18) with the meridional length L representing the area of interaction between the midlatitude atmosphere and the zonally averaged ocean circulation. The characteristic meridional velocity in the ocean is indicated by V. Assuming that the heat flux F between atmosphere and ocean is linearly related to the difference in temperatures, that is, $F = \kappa(T_a - T_o)$, the heat balances in the upper ocean and lower atmosphere give

$$\frac{\partial T_a}{\partial t} = -\alpha T_a - \frac{F}{C_a} + \epsilon_a, \tag{9.21a}$$

$$\frac{\partial T_o}{\partial t} = -V\frac{\partial T_o}{\partial y} + \frac{F}{C_o}. \tag{9.21b}$$

Here, C_a and C_o are the heat capacities of atmosphere and ocean, respectively, α represents the atmospheric damping and ϵ_a represents the noise in the atmosphere.

With $\mu = \kappa/C_a$ and $\lambda = \kappa/C_o$ and assuming that the atmosphere is in equilibrium on time scales longer than a few months, T_a can be eliminated from the steady equation (9.21a), and an equation for the ocean temperature T_o is obtained as

$$\frac{\partial T_o}{\partial t} + \lambda_e T_o + V\frac{\partial T_o}{\partial y} = \frac{\lambda_e}{\alpha}\epsilon_a, \tag{9.22}$$

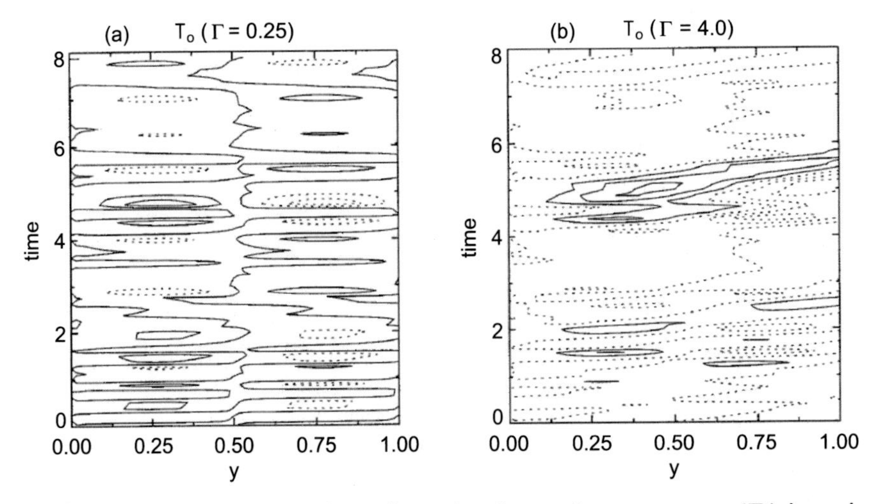

Figure 9.19 Time evolution of nondimensional oceanic temperature (T_o) in a single realisation of the stochastic model (9.23): (a) $\Gamma = 0.25$, contour interval 0.25; (b) $\Gamma = 4.0$ contour interval 0.04. The abscissa corresponds to the nondimensional spatial coordinate [unit length O(5000 km)], and the ordinate corresponds to the nondimensional time [unit time is O(10 yr)]. Dashed contours denote negative values (figure from Saravanan and McWilliams [1998]).

where $\lambda_e = \lambda/(1 + \mu/\alpha)$. When time is scaled with the advective time scale L/V ($t = \hat{t}L/V$), length with L ($y = \hat{y}L$), and temperature with ΔT ($T_o = \hat{T}_o \Delta T$), the dimensionless form of (9.22) becomes

$$\frac{\partial \hat{T}_o}{\partial \hat{t}} + \frac{2\pi}{\Gamma}\hat{T}_o + \frac{\partial \hat{T}_o}{\partial \hat{y}} = \frac{\lambda_e}{\alpha \Delta T}\epsilon_a. \tag{9.23}$$

The response $\hat{T}_o$ depends on the parameter $\Gamma = 2\pi V/(\lambda_e L)$. When $\Gamma \ll 1$, advection is not important, and a red noise response results, just as in the Hasselmann (1976) case discussed in Section 6.3. When $\Gamma \gg 1$, however, the response is dominated by advection, and a peak in the spectrum of the response at the period $\mathcal{P} = L/V$ occurs. Examples of these two cases of the temperature response are plotted in Fig. 9.19.

The physics of this so-called spatial resonance mechanism is sketched in Fig. 9.18. At $t = 0$, the atmospheric heat flux has a simple cosine dependence $\sin 2\pi t/\mathcal{P}$ in meridional direction, and the SST pattern, which is directly forced, is plotted as the drawn curve. At a later time $t = L/(2V)$, the atmospheric forcing has changed sign. Exactly at this time, the SST pattern has also moved a distance $L/2$, and hence the heat flux amplifies the initial SST pattern.

When the minimal model (9.4) is run at a small value of γ in (9.8) such that the background state is very stable, each of the noise forcings (9.10) is able to cause only very small variability (Frankcombe et al., 2009). Hence the presence of the AMO mode and the occurrence of the Hopf bifurcation definitely play a central role in the amplitude of the multidecadal variability. The spatial resonance mechanism hence does not play a role in the multidecadal variability as found in the minimal model.

9.5 Context of the AMO: two time scales?

The minimal model provides a detailed physical mechanism of the possible origin of multidecadal variability in the North Atlantic Ocean. The question remains, however, whether this mechanism is indeed playing an important role in the observed variability. We address this question in Section 9.6 but first provide, in this section, a modest review of observational work and GCM results.

Analysis of multiple data sets of the Atlantic climate system has shown that many more quantities than SST show variations on a multidecadal time scale. Water from the North Atlantic enters the Arctic Ocean through the Barents Sea and Fram Strait. The return flow of water from the Arctic occurs mainly via the East Greenland Current. This exchange forms an oceanic connection between the climates of the Arctic and the North Atlantic. Century-long records of sea-ice extent in the Arctic display multidecadal variability (Venegas and Mysak, 2000), which has been referred to as the low-frequency oscillation (LFO; Polyakov and Johnson, 2000). This variability is strongest in the Kara Sea and decays towards the Canada Basin (Polyakov et al., 2003). There are also multidecadal variations in sea-ice transport through Fram Strait associated with the sea-ice extent variability (Vinje et al., 2002). In addition, much attention has been paid to the intriguing salinity anomalies in the North Atlantic, most notably the Great Salinity Anomaly (GSA) in the early 1970s (Belkin et al., 1998).

The other connection between the North Atlantic and Arctic climate occurs through the atmosphere. As we have seen in Chapter 7, the dominant atmospheric winter variability is the pattern of the North Atlantic Oscillation (NAO), with its Arctic extension, the Northern Annular Mode (NAM; Thompson and Wallace, 2001). Although it cannot be demonstrated that the NAO has any significant preferential frequency, the Atlantic westerlies were relatively weak in the period between 1940 and 1970 and relatively strong from 1980 to 2000. NAO variations impose a relatively well-known tri-polar heat-flux anomaly on the North Atlantic Ocean on seasonal to interannual time scales (Eden and Jung, 2001; Alvarez-Garcia et al., 2008), whereas the low-frequency SST response is more of a hemispheric one-sign pattern (Visbeck et al., 2003).

There are only a few directly measured time series from which multidecadal variability can be reliably determined. One of these is the Central England Temperature (CET) record, which dates back to the second half of the seventeenth century. The SSA spectrum (Ghil and Vautard, 1991) of this time series indicates that the dominant time scales of variability are in the twenty- to thirty-year band (consistent with the analysis of this time series presented in Plaut et al., 1995) and at around seventy years. Both these peaks are significant at the 99% level.

Further signatures of the twenty- to thirty-year variability were found from the analysis of subsurface temperature (XBT) data in Frankcombe et al. (2008). Although there are only subsurface data from 1960–2000, it was suggested that the dominant period of variability is twenty to thirty years. In Frankcombe and Dijkstra (2009), it is shown that tide gauge data around the North Atlantic also support the notion of a dominant time scale of twenty to thirty years. More recently, analysis of 1,000-year long time series of Greenland ice core records with an annual resolution have shown that the twenty- to thirty-year variability is the only significant frequency (Chylek et al., 2011).

So why is it that the CET and Greenland ice cores (Frankcombe et al., 2010a) show a dominant twenty- to thirty-year variability, whereas the basin-wide SST variability is suggested to have a dominant fifty- to seventy-year variability (Schlesinger and Ramankutty, 1994; Enfield et al., 2001)? To explain this, Frankcombe et al. (2010a) considered so-called latitudinal AMO indices, where the ten-year running mean of the North Atlantic SST anomaly in 10° latitude bands is determined using SSTs from the HadSST2 data set (Rayner et al., 2006). Before about 1900, the variability at low latitudes appears to be out of phase with the variability at midlatitudes so that the basin-wide averaged variability is very small. This may also be due to the scarcity of data during this period. After 1900, the variability is much more coherent over latitude, with the cooler period around 1950 being much more pronounced at lower latitudes than at higher latitudes. The twenty- to thirty-year component is dominant in each individual latitudinal band, but although it remains visible in the basin-wide AMO index, it appears overwhelmed by the fifty- to seventy-year component.

In summary, from observations it is clear that, in addition to the previously described fifty- to seventy-year variability found in the North Atlantic and Arctic, a dominant time scale of twenty to thirty years may be found in SST over the North Atlantic as well as in temperatures in central England and ice cores in Greenland.

Several models from the Coupled Model Intercomparison Project (CMIP) suite (e.g., Stouffer et al., 2006) display clear multidecadal variability in the North Atlantic. In particular, the analysis (Zhang, 2008) of the 1,000-year control simulation of the GFDL CM2.1 model shows dominant variability in the twenty- to thirty-year time scale (Fig. 2b in Zhang, 2008). The first EOF shows a dipolar pattern in both SSH and subsurface temperature, with strong positive anomalies south of Greenland and negative anomalies in the Gulf Stream separation region.

The AMO indices (North Atlantic between 10°N and 80°N) for SST and for the 300- to 400-m averaged subsurface temperature in this model simulation display variability on both the twenty- to thirty- and fifty- to seventy-year time scales, both at and below the surface (Frankcombe et al., 2010a). When the spectra of the AMO indices (surface and subsurface) are analysed per latitudinal band, one finds (similar as in observations) shorter periods at lower latitudes and in the subsurface and larger periods at midlatitudes and at the surface. For every latitude, the twenty- to thirty-year subsurface variability has significant energy.

In a 1,000-year control simulation of HadCM3, Dong and Sutton (2005) found variability in the Atlantic MOC with a dominant time scale of about twenty-five years. There is also variability at the fifty- to seventy-year band, but this is not significant (not even at the 90% level, their Fig. 2), although there is significant variability at the 100-year time scale. Vellinga and Wu (2004) analysed this 100-year variability in a 1,600-year control simulation of HadCM3. From the anomaly patterns of the model at this time scale, it is clear that there is no westward propagation and that the time scale is too long for their mechanism to be a plausible candidate for the twenty- to thirty-year variability as found in the observations. The HadCM3 model simulation, however, also shows variability at the twenty- to thirty-year time scale (Fig. 4 in Vellinga and Wu [2004]). Jungclaus et al. (2006) analysed a 500-year control integration with the ECHAM5/MPI-OM model and found a pronounced multidecadal fluctuation in the Atlantic MOC and associated heat transport with a period of seventy to eighty years. The AMO index and its spectrum from a different simulation with the same model (Sterl et al., 2008) were determined in Van Oldenborgh et al. (2009); dominant variability was found in the twenty- to forty-year band. Variability on the longer time scale, fifty to eighty years, also exists, but it is not significant at the 95% level.

9.6 Synthesis

The overview in the previous section motivates us to find explanations for the processes controlling Atlantic multidecadal variability on two time scales: the twenty- to thirty-year variability and the fifty- to seventy-year variability.

9.6.1 Physics of the twenty- to thirty-year variability

The most convincing characteristic of the thermal Rossby mode mechanism, that is, the westward propagation of temperature anomalies, is indeed found in subsurface temperature observations (Frankcombe et al., 2008). In their Fig. 3a, which shows a Hovmöller plot of temperature anomalies, basin (10°N – 60°N) and vertically averaged over 300–400 m, westward propagation is easily identified. For example, the mid-Atlantic subsurface warming (at 60°W) that was at a maximum around 1978 started in the eastern part of the basin around 1970 and reached the west coast around 1981. The phase differences between zonal and meridional temperature differences show a maximum correlation around five years, which leads, according to the thermal Rossby mode mechanism, to an estimate of the period of about twenty years, consistent with the variability in the Central England Temperature record. Phase differences between variability on the eastern and western boundaries of the North Atlantic are also found in sea level anomalies (Miller and Douglas, 2007; Frankcombe and Dijkstra, 2009) and can also be explained by westward propagating anomalies with a time scale of twenty to thirty years.

There is also clear westward propagation in the subsurface temperature field in the GFDL CM2.1 control simulation (Frankcombe et al., 2010a). The SST signal is much more stationary at the surface than in the subsurface. This difference in propagation can be explained by the thermal Rossby mode theory, as the background zonal flow is a parameter in the propagation speed. With so many indications of westward propagation and a mechanism based on clear physics (i.e., only based on thermal wind) explaining this propagation, the excited multidecadal internal (thermal Rossby) ocean mode appears to be a very good explanatory model of the twenty- to thirty-year variability.

9.6.2 Possible causes of the fifty- to seventy-year variability

One possibility is that the fifty- to seventy-year variability is caused by the same processes as the twenty- to thirty-year variability. In idealised models, where an internal ocean mode is excited by atmospheric noise, the spectrum shows a broad peak at multidecadal frequencies (see, for example, Fig. 6b in Frankcombe et al., 2009). This means that both the twenty- to thirty-year and fifty- to seventy-year periods could be encompassed by the same spectral peak. This could be caused by a series of weak oscillations followed by a particularly strong anomaly, making the period of the oscillation appear longer than the underlying twenty to thirty years.

A second possibility is that the fifty- to seventy-year variability is forced by the atmosphere, in particular through multidecadal variability of the North Atlantic Oscillation. This mechanism was considered in the study of Eden and Jung (2001), in which an eddy-permitting ocean GCM was forced by heat, freshwater and momentum anomalies of multidecadal NAO variability. Although it was found that direct forcing by the NAO heat flux anomalies cannot explain the SST cooling in the 1970s, for example, it was shown that a lagged response of the Atlantic ocean circulation to this forcing can explain this behaviour. Clearly, the variability in Eden and Jung (2001) is on a much larger time scale than the twenty- to thirty-year variability (their Fig. 3, for example), and so the NAO forcing combined with a lagged flow response cannot explain this shorter time-scale variability. Eden and Jung (2001) also showed that the variability is mainly temperature controlled; that is, salinity is not essential for its existence.

A third possibility for the fifty- to seventy-year variability is modulation of the twenty- to thirty-year variability by a tropical connection. Vellinga and Wu (2004) found that centennial fluctuations in the strength of the meridional overturning in HadCM3 were caused by processes in the tropics, with the period set by the travel time of salinity anomalies from their generation region near the equator to the convection

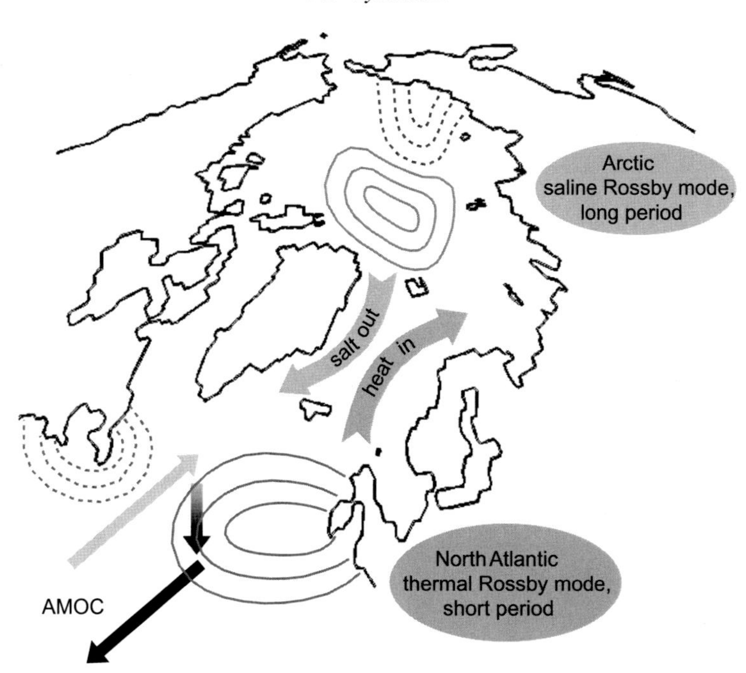

Figure 9.20 Illustration of the interaction of the two internal Rossby modes. The shorter period mode in the North Atlantic appears as a strengthening/weakening of the Atlantic ocean circulation associated with the westward propagation of temperature anomalies near the surface. The longer period mode in the Arctic involves salinity anomalies propagating across the pole (figure from Frankcombe and Dijkstra, 2011).

region in the subpolar North Atlantic. This process, however, is on a time scale that is probably too long to cause variability on the fifty- to seventy-year time scale.

Finally, the origin of the fifty- to seventy-year variability may come from the Arctic, as suggested first by Jungclaus et al. (2005). Hawkins and Sutton (2007) showed three-dimensional patterns of propagation of temperature and salinity anomalies in HadCM3 and found that salinity anomalies in the Arctic may also play an important role in the variability in that model, although in their case the variability was on a centennial rather than on a multidecadal time scale. Frankcombe et al. (2010a) showed that the longer period variability in the GFDL CM2.1 model is most prominent in Arctic Ocean subsurface salinity.

There have been various results on the effect of Arctic salinity anomalies on the Atlantic ocean circulation strength. Zhang and Vallis (2006) and Dima and Lohmann (2007) argue that these Arctic salinity anomalies are linked to temperature and circulation changes in the North Atlantic. However, Haak et al. (2003) found that Arctic salinity anomalies have a minimal impact on the Atlantic meridional overturning circulation. For the GFDL-CM2.1 model (Zhang and Vallis, 2006), it was found that the Atlantic ocean circulation adjusts to salinity changes at high latitudes on the (multidecadal to centennial) overturning time scale.

9.6.3 A schematic framework for the AMO variability

Frankcombe et al. (2010a) presented a mechanistic framework, based on stochastic dynamical systems theory (Fig. 9.20), to explain both time scales of variability, assuming that the fifty- to seventy-year variability originates in the Arctic. The twenty- to thirty-year variability, as observed in North Atlantic temperatures (including westward propagation of temperature anomalies below the surface (Frankcombe et al., 2008) and sea level variations on the European and North American coasts (Frankcombe et al., 2009), is all consistent with the period and pattern of the thermal Rossby mode in the North Atlantic, which is excited by atmospheric noise (Frankcombe et al., 2009). The variability in temperature is transported northward to the Arctic, leading to the shorter period multidecadal variability in the Arctic, as observed in the exchange of salinity at 70°N (Frankcombe and Dijkstra, 2011).

Independently, long time scale (longer than forty years) variability arises due to internal ocean mechanisms associated with saline Rossby modes in the Arctic (analogously to the thermal Rossby mode in the North Atlantic). Frankcombe et al. (2010a) showed in an idealised Arctic model that an internal mode on approximately the right period does exist. The mode is damped in the idealised Arctic models but could be excited by variability in Atlantic inflow as well as atmospheric, sea ice or river runoff variability. This longer period variability is found in the southward salt exchange at 70°N and consequently propagates into the North Atlantic (Frankcombe and Dijkstra, 2011) to give rise to the fifty- to seventy-year variability.

10

Dansgaard-Oeschger Events

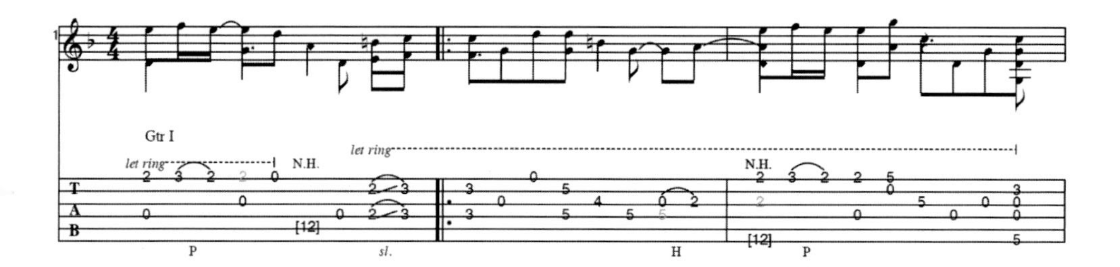

Flip-flops under icy conditions.
CADGAD, Ragamuffin, Michael Hedges

One of the main breakthroughs in climate research over the past decades has been the reconstruction of past temperatures from ice cores. A surprising result was the discovery of millennial time-scale variability during the last glacial period. In this chapter, stochastic dynamical systems theory is used to understand the characteristics of this millennial climate variability.

10.1 Phenomena

At the moment, analyses from ice cores drilled on Greenland (Andersen et al., 2004) and on Antarctica (EPICA, 2006) are available. From the Greenland ice cores, the local temperature could be determined down to about 100-kyr BP. From the European Project for Ice Coring in Antarctica (EPICA) core data, temperatures and greenhouse gas concentrations over the last 800,000 years could be reconstructed.

Water in ice cores contains two isotopes of oxygen, ^{18}O and ^{16}O. The normalised isotope ratio δ^{18}O is calculated as a deviation from a reference sample as

$$\delta^{18}\mathrm{O} = \frac{\left(\frac{^{18}\mathrm{O}}{^{16}\mathrm{O}}\right)_{\mathrm{sample}} - \left(\frac{^{18}\mathrm{O}}{^{16}\mathrm{O}}\right)_{\mathrm{reference}}}{\left(\frac{^{18}\mathrm{O}}{^{16}\mathrm{O}}\right)_{\mathrm{reference}}}, \tag{10.1}$$

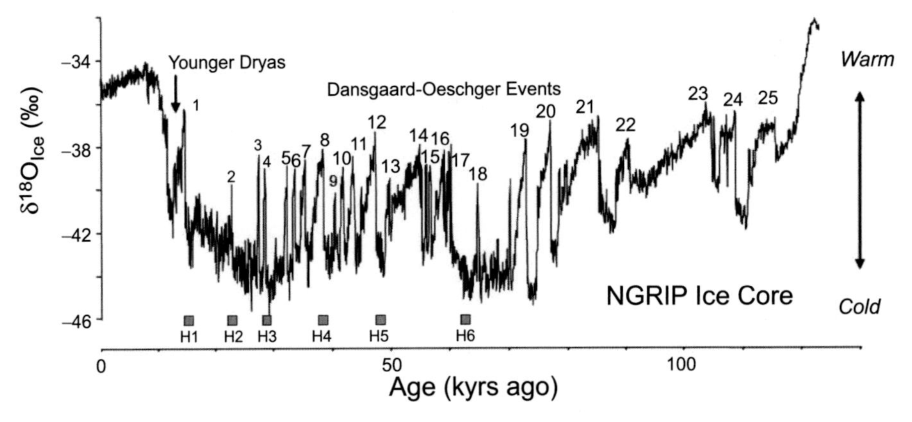

Figure 10.1 Plot of the $\delta^{18}O$ data from the NGRIP ice core record (Andersen et al., 2004). The numbers refer to the interstadials (warm periods), the occurrence of the Heinrich events is shown by the labels H_1 to H_6 and the Younger Dryas is indicated by an arrow (figure from Clement and Peterson, 2008).

where the reference sample is 'standard mean' ocean water. The isotope ^{16}O is lighter than ^{18}O so that water containing ^{16}O is preferentially evaporated and a temperature-dependent fractionation occurs. Changes in $\delta^{18}O$ reflect the combined effect of changes in global ice volume and temperature at the time of deposition of the sampled material. During very cold conditions, global ice volume is relatively large and hence sea level is low, which enriches water in the ocean with ^{18}O. Also because of the colder temperatures, more ^{18}O remains in the ocean and less ^{18}O becomes locked in the ice. Hence the ratio $\delta^{18}O$ in ice cores will decrease (become more negative) under colder conditions.

An example of a $\delta^{18}O$ record from the North Greenland Ice Core Project (NGRIP) ice core data is provided in Fig 10.1. The record spans the last 123,000 years and reveals changes between relatively cold periods (so-called stadials) and warm periods (so-called interstadials). Relatively rapid climatic shifts occurred during the period 70–10 kyr BP, with peak-to-peak temperature changes of about 10°C. These changes are generally referred to as Dansgaard-Oeschger (hereafter D-O) events. Fig. 10.1 shows that most D-O events are characterised by a rapid warming transition and a slow cooling phase.

There has been an extensive discussion on the dominant time scale of the D-O events (Wunsch, 2000). After careful analysis of the GISP2 (Stuiver and Grootes, 2000) record, Schultz (2002) concluded that between 46 and 13 kyr BP, the onset of D-O events was paced by a fundamental period of ~1,470 years. Before 50 kyr BP, the presence of such a dominant period is unclear due to dating uncertainties in the ice core record.

Bond et al. (1997) have shown that individual D-O events are apparently clustered to form a longer sawtooth-shaped cycle with a time scale of about 10,000 years; this is sometimes referred to as the Bond cycle. In the ice core records, discrete layers of ice-rafted debris have also been found, which were deposited during the coldest phases of several D-O events. These so-called Heinrich events (H_1–H_6 in Fig 10.1) are interpreted as large bursts of freshwater into the North Atlantic due to massive iceberg releases through the Hudson Strait. The last rapid shift in Greenland temperatures occurred as a period of significant cooling between 12,500 and 11,500 years ago. The resulting stadial is referred to as the Younger Dryas, during which the apparent warming trend from the Last Glacial to the Holocene was delayed for approximately 1,000 years (Fig 10.1). The Younger Dryas period ends with a rapid shift to warmer temperatures into the beginning of the Holocene.

There are many indications from proxy data that there have been large-scale reorganisations of both the atmosphere and ocean associated with D-O events, and a recent review is provided in Clement and Peterson (2008). In the subpolar North Atlantic, D-O events were matched with corresponding sea-surface temperature changes of at least 5°C. Of special interest is that the temperature anomalies on Antartica are about 180° out of phase with those on Greenland.

The behaviour of the climate system on millennial time scales during the last glacial period is highly interesting for an understanding of feedbacks among the ocean, atmosphere and cryosphere. A theory of D-O events will have to explain at least the processes controlling:

(i) the dominant ∼1,500-year time scale of the variability;
(ii) the asymmetric character of the transition with a rapid warming and a relatively slow cooling phase; and
(iii) the phase difference between the temperature in the northern and southern polar regions.

Finally, the theory should provide an explanation for why there are no D-O events observed in the Holocene and a consistent causal connection among D-O events, the Bond cycle and the Heinrich events. As discussed in Clement and Peterson (2008), several different views have been proposed to explain the millennial climate variability during the last glacial period. Leading theories involve changes in the global ocean circulation and stochastic resonance. In this chapter, we start with the basic ingredients of the present-day ocean circulation, then discuss transitions in ocean circulation patterns and, finally, look at the effect of noise on these transitions.

10.2 The meridional overturning circulation

On the large scale, the ocean circulation is driven by momentum fluxes (wind stress) and affected by fluxes of heat and freshwater at the ocean-atmosphere interface. The

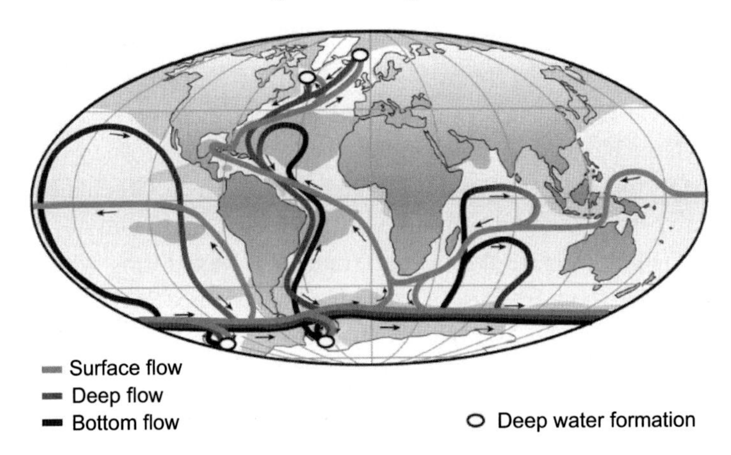

Figure 10.2 Strongly simplified sketch of the global ocean circulation. In the North Atlantic, warm and saline waters flow northward into the deep water formation areas (indicated by small circles). Bottom water is also formed in the Southern Ocean (figure modified Kuhlbrodt et al., 2007).

latter fluxes change the surface density of the ocean water, and through mixing and advection, density differences are propagated horizontally and vertically.

An illustration of the global ocean circulation is provided in Fig. 10.2. In the North Atlantic, the Gulf Stream transports relatively warm and saline waters northwards. The heat is quickly taken up by the atmosphere, making the water denser. When there is strong cooling in winter, the water column becomes unstably stratified in certain areas (e.g., the Greenland Sea and the Labrador Sea), resulting in strong convection. The net result of this is the formation of a water mass called North Atlantic Deep Water (NADW), which overflows the various ridges that are present in the topography and enters the Atlantic basin.

The NADW flows southwards at mid-depth in the Atlantic, enters the Southern Ocean and from there reaches the other ocean basins. Through upwelling in the Atlantic, Southern, Pacific and Indian Oceans, water is slowly brought back to the surface, and the mass balance is closed by transport back to the sinking areas in the North Atlantic. In the Southern Ocean, bottom water is formed (the Antarctic Bottom Water, AABW), which has a higher density than NADW and therefore appears in the abyssal Atlantic. In the North Pacific, no deep water is produced. The deep water formation at high latitudes, the upwelling at lower latitudes and in the Southern Ocean and the horizontal currents together form the global ocean circulation.

A crucial component of the global ocean circulation is the meridional overturning circulation (MOC), which is the zonally integrated volume transport. The MOC is strongly coupled to meridional heat transport. There are no observations available to reconstruct the pattern of the MOC, but its strength at 26°N in the Atlantic is now routinely monitored by the RAPID-MOCHA array (Cunningham et al., 2007). The

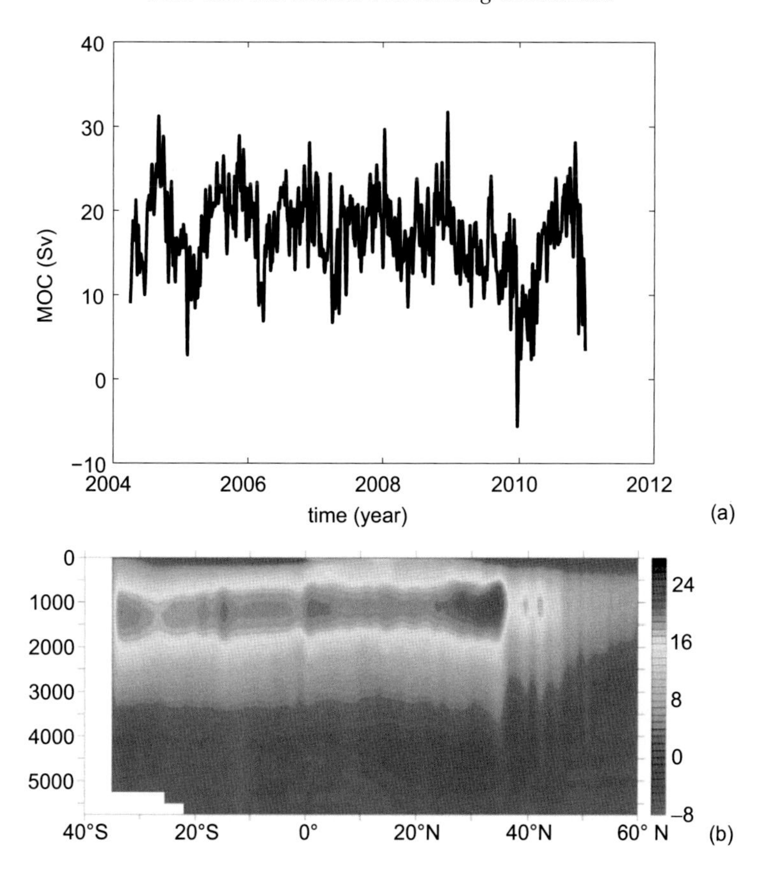

Figure 10.3 (a) Volume transports of the Atlantic MOC at 26°N as measured by the RAPID-MOCHA array (Cunningham et al., 2007). Data are taken from http://www.noc.soton.ac.uk/rapidmoc/. (b) Contour plot of the 10-year mean MOC (contours in Sv) in the Atlantic from the control simulation (C-MIXED) with the POP 0.1° high-resolution model. The depth on the vertical axis is in meters (figure from Weijer et al., 2012). (See Colour Plate.)

currently available time series of the MOC strength is shown in Fig. 10.3a, indicating the large amplitude variability. The pattern of the MOC can be determined from numerical ocean models, and a typical pattern of the Atlantic MOC from a high-resolution ocean model is plotted in Fig. 10.3b. At 26°N the heat transport associated with the Atlantic MOC is estimated to be 1.2 PW (Johns et al., 2011).

In a glacial period, with different insolation and CO_2 levels, of course the forcing of the ocean circulation differed markedly. Proxy data indicate that the NADW cell was weaker during glacial times, in particular during the Last Glacial Maximum, and there was enhanced intrusion of AABW into the Atlantic basin (Clement and Peterson, 2008). These proxy data hence imply that the glacial Atlantic MOC was likely weaker than today.

10.3 Sensitivity of the MOC: hysteresis

The North Atlantic experiences heat input and freshwater loss at low latitudes and heat loss and freshwater input at high latitudes. The surface freshwater flux and heat flux therefore have opposite effects on the meridional density gradient affecting the large-scale ocean circulation. This makes it interesting to investigate what happens when the relative importance of the two surface fluxes varies.

As already mentioned in Chapter 6, Stommel (1961) proposed a box model that can be used to study this problem in its simplest form. Under appropriate scaling (see Example 6.1), the dimensionless equations become

$$\frac{dT}{dt} = \eta_1 - T(1 + \mid T - S \mid), \tag{10.2a}$$

$$\frac{dS}{dt} = \eta_2 - S(\eta_3 + \mid T - S \mid). \tag{10.2b}$$

Here T and S monitor the dimensionless equatorial-to-pole temperature and salinity difference, $\Psi = T - S$ is the dimensionless MOC strength (positive when there is sinking in the northern box) and t indicates dimensionless time. The parameters η_1 and η_2 represent the strength of the thermal and freshwater forcing, respectively, and η_3 is the ratio of the thermal and freshwater surface restoring time scales (Example 6.1).

The bifurcation diagram for this model is shown (with the steady-state value $\bar{\Psi}$ plotted versus η_2) in Fig. 10.4a. For small η_2, a unique steady state with $\bar{\Psi} > 0$ is reached for all initial states. This state is called the thermally dominated state or TH state (left panel in Fig. 10.4b). For large η_2, again a unique steady state exists with $\bar{\Psi} < 0$, which is called the salinity dominated state or SA state (right panel in Fig. 10.4b). For values of η_2 up to the point L_1, only the TH state is linearly stable, whereas for values beyond L_2, only the SA state is linearly stable. On the branch that connects the solutions at L_1 and L_2, the steady states are unstable. Between the points L_1 and L_2, the TH and SA solutions coexist and are both linearly stable.

The multiple equilibria arise due to a positive feedback between the flow and the salt transport, called the salt-advection feedback (Walin, 1985). The surface forcing salinifies and warms the low-latitude region, whereas it freshens and cools the high-latitude region. If the circulation strengthens, then more salt is transported poleward. This enhanced salt transport will increase further the density in high latitudes and, consequently, amplify the original perturbation of the circulation. The strengthening of the circulation also transports more heat northward; this will weaken the flow by lowering the density in high latitudes. This negative feedback is, however, weak, as the atmosphere strongly dampens surface temperature anomalies. On the contrary, ocean water salinity does not affect the freshwater flux at all.

Although the bifurcation diagram as in Fig. 10.4a is obtained from a relatively simple model, it turns out that this diagram is found in bifurcation studies using a

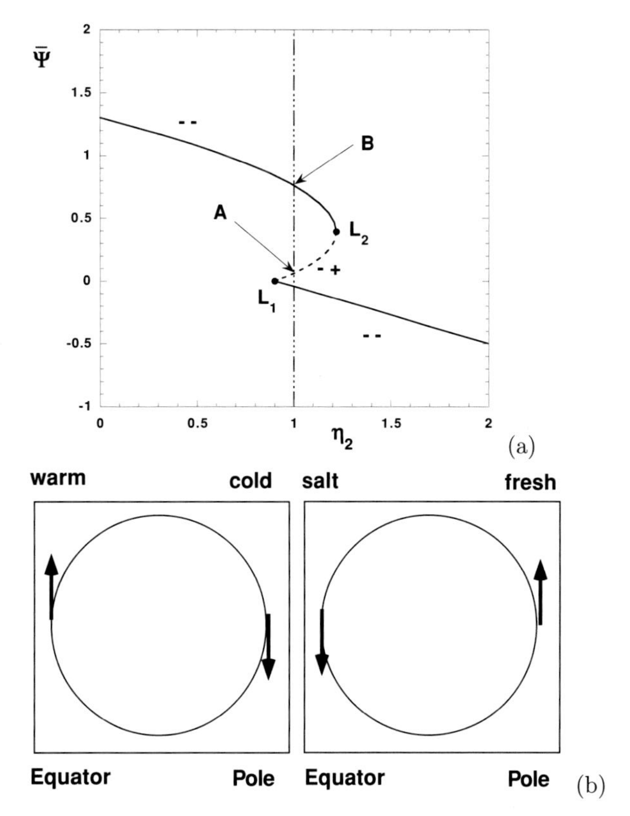

Figure 10.4 (a) Bifurcation diagram of the Stommel two-box model (10.2) for $\eta_1 = 3.0$ and $\eta_3 = 0.3$. (b) Patterns of the thermally driven (TH) MOC state (left panel) and the salinity driven (SA) MOC state (right panel).

hierarchy of two-dimensional and three-dimensional ocean models. For example, the bifurcation diagram found in the global ocean model of Dijkstra and Weijer (2005) is plotted in Fig. 10.5a. Here the strength of the Atlantic MOC (ψ) of each steady state is plotted versus the strength (γ_p in Sv) of a freshwater flux change in the northern North Atlantic. Again, a solid line style along the branch indicates that steady solutions are stable, whereas steady states are unstable on the dashed part of the branch. There are also two saddle-node bifurcations, indicated again by L_1 and L_2, respectively, which separate the stable and unstable parts of the branch (Fig. 10.5b).

In many models such as EMICs and GCMs (cf. Section 6.1), however, the bifurcation diagrams cannot be directly computed. In this case, so-called quasi-equilibrium transient simulations are carried out, in which a parameter such as γ_p is changed very slowly in time (Fig. 10.5c); such simulations are usually referred to as 'hosing experiments' and are examples of fast-slow systems as discussed in Section 2.3.3. Starting at the reference solution for $\gamma_p = 0$, the upper solution branch in Fig. 10.5b is then followed until L_2, where no nearby steady-state solution exists anymore. Hence the

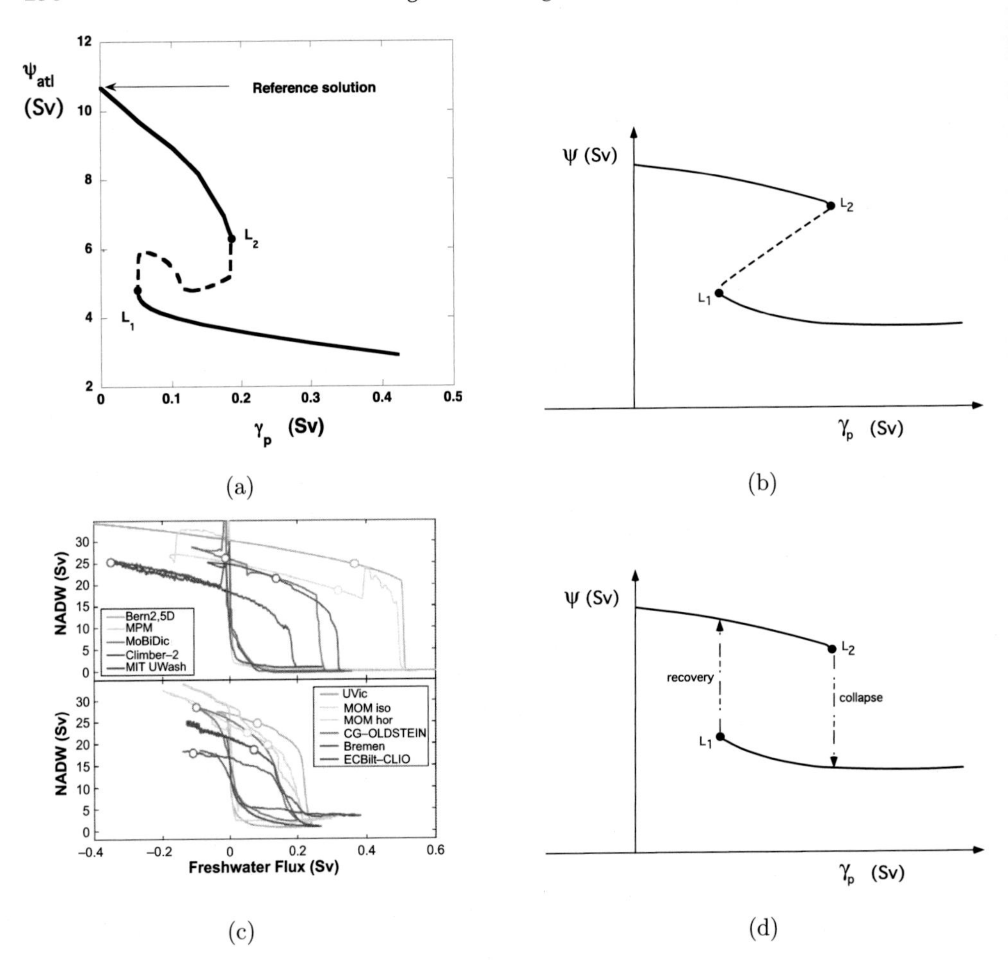

Figure 10.5 (a) Bifurcation diagram as found in the global ocean model in Dijkstra and Weijer (2005) with two saddle-node bifurcations labelled L_1 and L_2. The parameter γ_p represents an additional freshwater flux in the northern North Atlantic. Solid parts of the branch indicate stable steady solutions, and steady states are unstable along the dashed part of the branch. (b) Sketch of this bifurcation diagram to compare with (d). (c) Typical hysteresis behaviour as found in a set of EMICs (Rahmstorf et al., 2005) when the freshwater flux is increased slowly in time from zero up to large values and back. On the vertical axis is a measure of the strength of the MOC. The curves are lined up with the recovery, and the equilibrium solutions obtained with zero freshwater perturbation are shown as open circles. (d) Sketch of the quasi-equilibrium simulations with γ_p. Near the saddle-node bifurcations in (a), the solution jumps from one stable steady state to another. The direction of these transient jumps is indicated with an arrow. (See Colour Plate.)

solution changes rapidly ('collapse' in Fig. 10.5d) to that on the lower branch and follows that branch with increasing γ_p. If from a large value of γ_p the solution is followed with decreasing γ_p, then the lower branch is followed down to γ_p at L_1, where a transition ('recovery' in Fig. 10.5d) occurs to the solution on the upper branch.

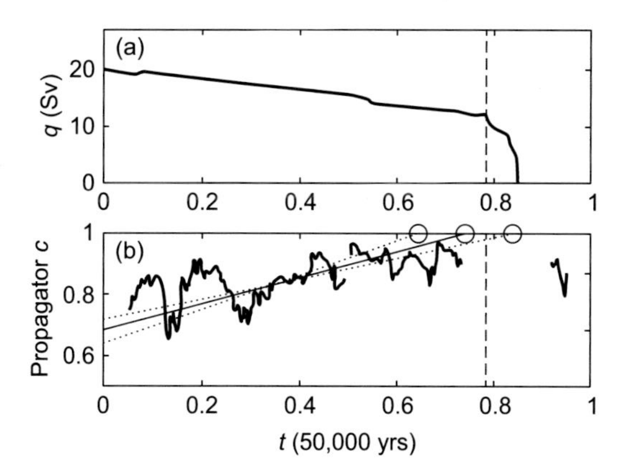

Figure 10.6 (a) Transport of the Atlantic meridional overturning and (b) AR(1) propagator c over time. Here c is obtained from a moving 10% time window. Although (a) shows artificially low variability, no early warning sign for a threshold behaviour is observable. The value of (b) is noisy but is much more informative on the approach to the critical transition (figure from Held and Kleinen [2004]).

Let the value of γ_p at L_i be indicated by γ_p^i; the width of the hysteresis, say Δ_H, then is given by $\Delta_H = \gamma_p^2 - \gamma_p^1$. In typical ocean model studies, where γ_p is varied with about 0.01 Sv/1,000 years, one finds approximations of the value of Δ_H because the jumps are not really 'vertical' as in Fig. 10.5d; values of Δ_H are also strongly model dependent (Rahmstorf et al., 2005), as can be seen from Fig. 10.5c.

In Fig. 10.6, the degenerate fingerprinting technique (cf. Section 5.5.3) is applied to a time series from an EMIC to detect the saddle-node bifurcation L_2. The freshwater input is slowly changing, decreasing the value of the meridional overturning transport q (Fig. 10.6a) over a period of 50,000 years. The value of the AR(1) coefficient c in (5.80) is obtained from a moving 10% window and shows an approach to unity as the transport of q rapidly decreases (Held and Kleinen, 2004).

Both the degenerate fingerprinting or autocorrelation function technique (ACF) and the detrended fluctuation analysis (DFA) were applied to similar simulations by another EMIC, and the results are summarized in Fig. 10.7. For this example, both indicators as well as the variance first increase when the saddle-node is approached, but the DFA and ACF do not show monotonic behaviour up to the critical point. Hence the signal from the indicators may in this case lead to false alarms (Lenton, 2011).

Model intercomparison studies (Gregory, 2005; Stouffer et al., 2006) show no systematic differences in MOC behaviour and climate response between EMICs and GCMs (cf. Chapter 6). The multiple equilibria regime is certainly present in EMICs (De Vries and Weber, 2005; Rahmstorf et al., 2005; Weber et al., 2007) and in somewhat more complex GCMs (Hawkins et al., 2011). However, the simulations to systematically address the existence of multiple equilibrium regimes in state-of-the-art GCMs have not been performed yet.

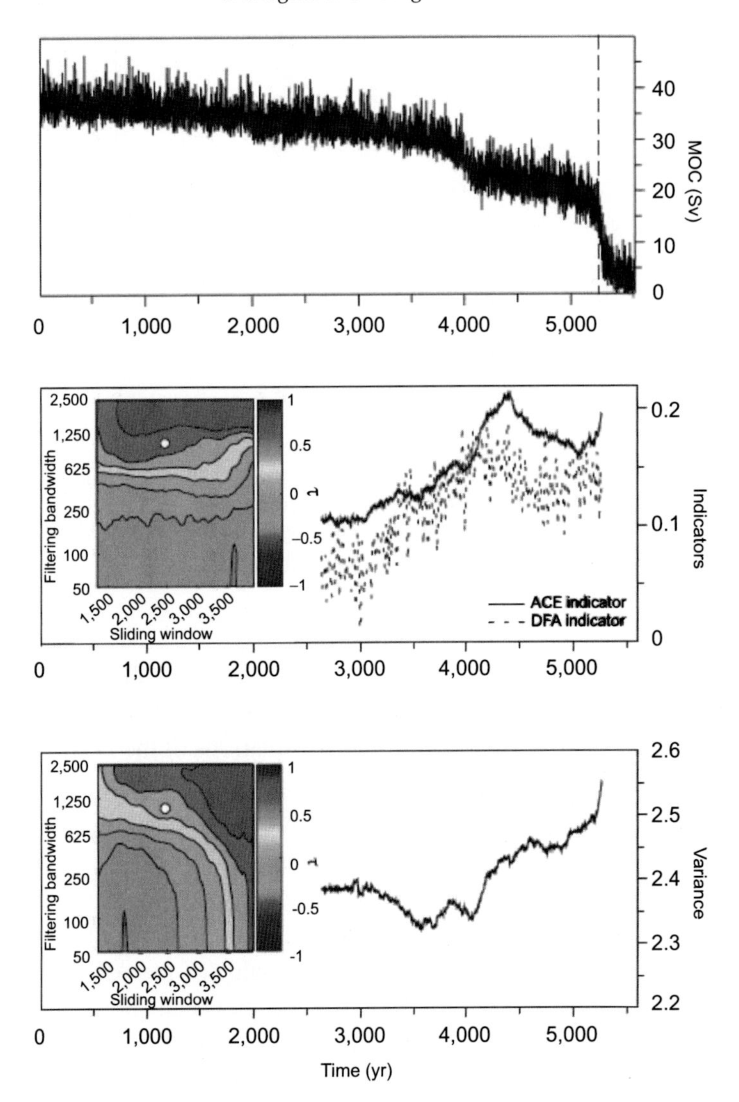

Figure 10.7 Upper panel: time series of the MOC strength as determined from an EMIC where freshwater is slowly released in the northern North Atlantic. Middle panel: Values of the ACF and DFA indicators. Bottom panel: variance of the time series. The values of τ in the inset are results from a statistical test to determine a significant increase in the value of the particular quantity (figure from Lenton, 2011). (See Colour Plate.)

10.4 Variability of the MOC: oscillations

Apart from the possibility of multiple equilibria of the MOC, the MOC is also known to have several modes of internal variability. Two essentially different types of oscillatory modes are discussed in this section: overturning (or loop) oscillations and deep-decoupling oscillations (or flushes).

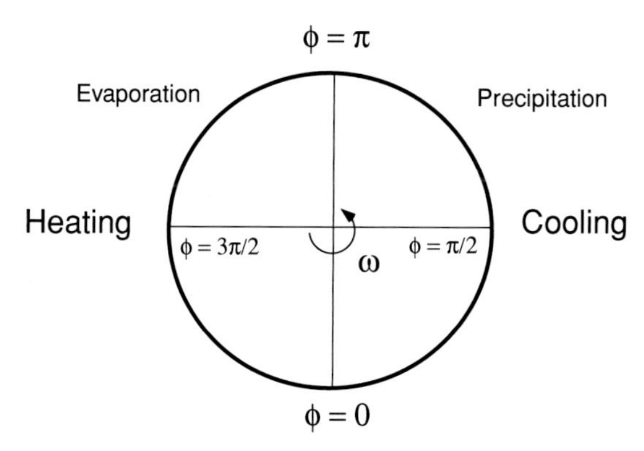

Figure 10.8 Sketch of the Howard-Malkus loop model with the anti-clockwise coordinate ϕ. The domain $\phi \in [0, \pi]$ represents the northern part of the North Atlantic. The freshwater flux is only prescribed over the domain $\phi \in [\pi/2, 3\pi/2]$.

10.4.1 Overturning oscillations

As we have seen in Chapter 9, for thermally driven flows in a single-hemispheric basin, oscillatory normal modes appear in the linear stability analysis of the North Atlantic MOC. One of these has a clear three-dimensional pattern, a multidecadal time scale and was suggested to be associated with the Atlantic Multidecadal Oscillation. However, there is another class that has a centennial time scale and which can be represented in two-dimensional models; it was called an overturning (or loop) oscillation because a central element is the propagation of the anomaly pattern along the overturning loop (Winton and Sarachik, 1993).

The Howard-Malkus loop model

Overturning (or loop) oscillations can already be found in relatively simple box models. The most elementary box model, which includes a loop oscillation, is the four-box model originally used by Huang et al. (1992) and analysed in more detail in Tziperman et al. (1994b). However, by far the best analysis of the mechanism of these oscillators can be done by using the so-called Howard-Malkus loop model (Malkus, 1972; Welander, 1986). This is a one-dimensional model of the overturning circulation representing the ocean in a circular loop, as shown in Fig. 10.8 (Sevellec et al. 2006). The angular coordinate is indicated by ϕ (positive in an anti-clockwise direction) and starts from the northern polar region. Hence the domains $\phi \in [0, \pi]$ and $\phi \in [\pi, 2\pi]$ represent the northern and southern part of the North Atlantic basin, respectively, and $\phi \in [\frac{\pi}{2}, \frac{3\pi}{2}]$ represents the upper ocean.

Mixed boundary conditions are specified along the loop, with cooling and net precipitation in the northern part and heating and net evaporation in the southern part of the basin. The spatial distribution of the restoring temperature $I^T(\phi)$ and the

Table 10.1. *Parameters of the Howard-Malkus loop model as used in Sevellec et al. (2006)*

Parameter	Meaning	Value	Unit
r_T	inverse thermal relaxation time scale	1	year^{-1}
H	mean ocean depth	1,000	m
γ	buoyancy torque parameter	34.4	year^{-1}
K_ϕ	lateral mixing coefficient	2.2 10^{-3}	year^{-1}
F_0	freshwater flux intensity	0.8	m year^{-1}
α_T	thermal expansion coefficient	2.2 10^{-4}	K^{-1}
α_S	haline contraction coefficient	7.7 10^{-4}	–
S_0	reference salinity	35	g kg^{-1}
T_0	amplitude of restoring temperature	10	K

freshwater flux $I^S(\phi)$ are given by

$$I^T(\phi) = -I_{[\frac{\pi}{2},\frac{3\pi}{2}]} \sin\phi, \tag{10.3a}$$

$$I^S(\phi) = -I_{[\frac{\pi}{2},\frac{3\pi}{2}]} \sin 2\phi, \tag{10.3b}$$

where $I_{[a,b]}$ is the indicator function on an interval $[a, b]$.

Advection is represented by an angular velocity ω, and the angular momentum balance over the loop gives (Maas, 1994) that ω is proportional to the buoyancy torque integrated over the loop. With a linear equation of state, the equations of the Howard-Malkus loop model for the temperature T and salinity S of the fluid are given by

$$\frac{\partial T}{\partial t} + \omega\frac{\partial T}{\partial \phi} = r_T(T_0 I^T(\phi) - T) + K_\phi \frac{\partial^2 T}{\partial \phi^2}, \tag{10.4a}$$

$$\frac{\partial S}{\partial t} + \omega\frac{\partial S}{\partial \phi} = -\frac{F_0 S_0}{H} I^S(\phi) + K_\phi \frac{\partial^2 S}{\partial \phi^2}, \tag{10.4b}$$

$$\omega = -\gamma \int_0^{2\pi} (-\alpha_T T + \alpha_S S) \sin\phi \, d\phi. \tag{10.4c}$$

An explanation of the parameters of the model (and their standard value) is provided in Table 10.1. The main control parameter in this model is the freshwater flux strength F_0.

In this model, oscillatory behaviour arises through a Hopf bifurcation as F_0 is increased, and an example of a limit cycle with a period of 170 years (for the parameters as in Table 10.1) is shown in Fig. 10.9. The period of the oscillation is very close to the mean overturning time $(2\pi/\bar{\omega})$, which indicates that the propagation of the anomalies

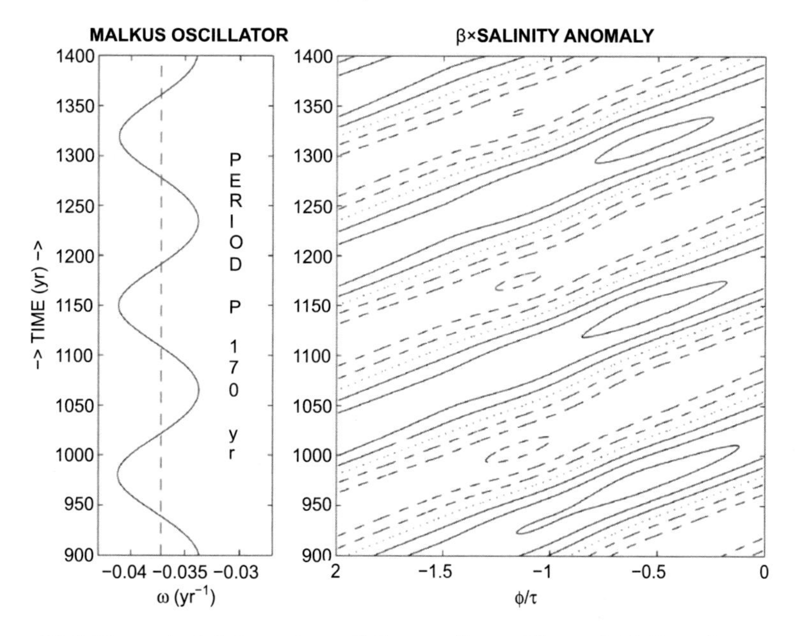

Figure 10.9 Numerical integration of the nonlinear Howard-Malkus loop model (10.4). The left panel shows the overturning ω as a function of time. In the right panel, a Hovmöller diagram of the salinity anomaly (multiplied by α_S) is plotted. The solid, dashed and dotted lines respectively correspond to positive, negative and zero anomalies; the contour interval is 0.75×10^{-4} (figure from Sevellec et al., 2006).

is governed by advection of salinity (buoyancy) anomalies by the mean overturning circulation.

The growth mechanism of the centennial oscillation was presented in detail in Sevellec et al. (2006). Consider in Fig. 10.10a a positive salinity anomaly, which is advected with the thermally driven circulation $\omega < 0$. This anomaly reduces the buoyancy torque and hence reduces the meridional overturning, represented by ω. The residence time of the anomaly in the evaporation zone is increased, which strengthens the anomaly (Fig. 10.10a). When the anomaly leaves the evaporation zone to enter the precipitation zone (Fig. 10.10b), it accelerates the meridional overturning. The residence time in the precipitation zone is hence shortened, which also amplifies the perturbation. The oscillation arises because the salinity anomaly is advected along the loop (Sevellec et al., 2006).

Ocean-only models

The overturning (or loop) oscillations as in the Howard-Malkus model are found in latitude-depth models of the double-hemispheric meridional overturning circulation (Mysak et al., 1993; Quon and Ghil, 1995; Sakai and Peltier, 1995; Dijkstra and Molemaker, 1997) and in three-dimensional models of single- and double-hemispheric ocean flows (Winton and Sarachik, 1993; Te Raa and Dijkstra, 2003a; Sevellec et al.,

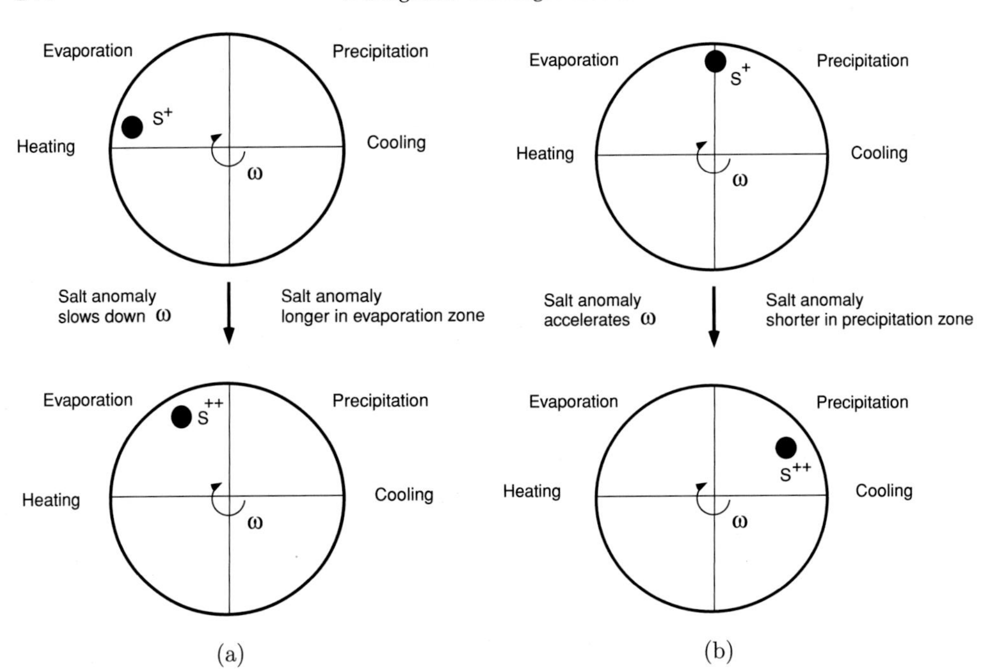

Figure 10.10 Sketch of the time evolution of salinity anomalies in the Howard-Malkus loop model leading to growth of the initial perturbation. (a) A positive salinity anomaly in the evaporation zone. (b) A positive salinity anomaly in the precipitation zone (figure redrawn from Sevellec et al., 2006).

2006). In the two-dimensional models, the oscillations arise through a Hopf bifurcation on the thermal branch, as in the Howard-Malkus model. When these oscillatory modes are damped, they can easily be excited by noise in the freshwater flux (Te Raa and Dijkstra, 2003a; Sevellec et al., 2006).

Loop oscillations are also found in multi-basin models such as those that consists of three latitude-depth ocean models, which are coupled in the south through a circumpolar channel. Sakai and Peltier (1996) find small-amplitude overturning oscillations with a centennial time scale and show that the period increases when freshwater is added to the North Atlantic. This behaviour is robust when the ocean model is coupled to an atmospheric energy balance model (Sakai and Peltier, 1997) and is consistent with their results in a single-basin model (Sakai and Peltier, 1995).

In Weijer and Dijkstra (2003), the explicit identification of multiple internal ocean modes of the global ocean circulation within a low-resolution ocean general circulation model was presented. An equilibrium state of the global ocean model was determined, and next the linear stability problem of this state was solved. It was found that all eigenvalues have negative real parts and hence that the steady solution is linearly stable. Three of the normal modes found are oscillatory and have oscillation periods of 2,000–3,000 years.

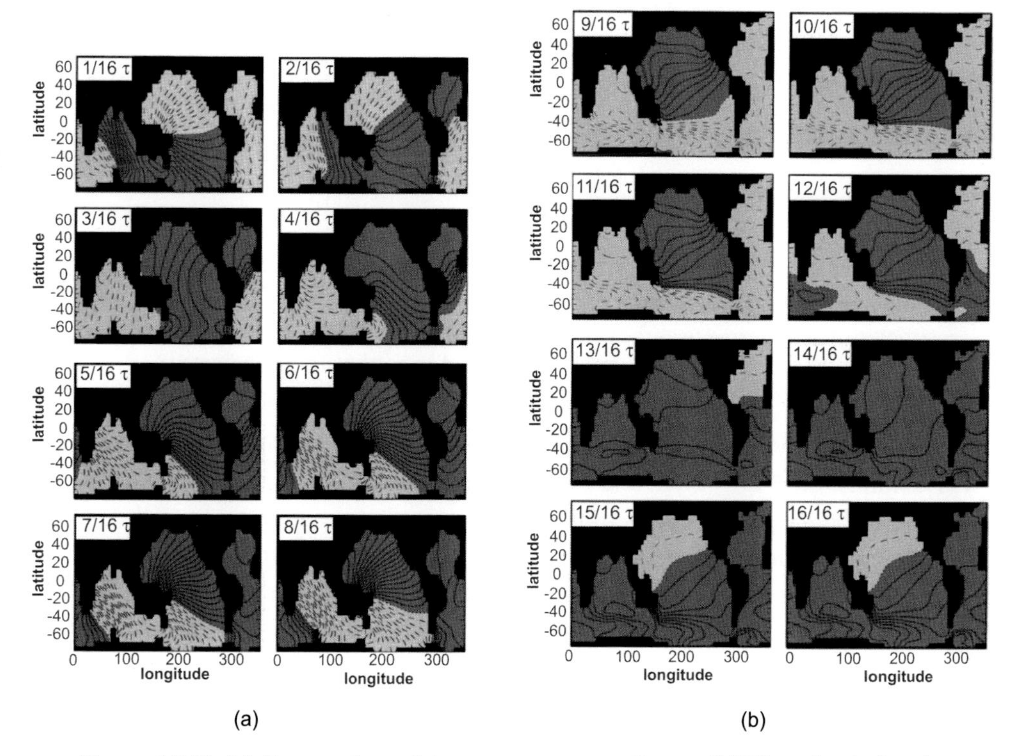

Figure 10.11 (a) Propagation of temperature anomalies at 3,000-m depth for the oscillatory mode 3 in Weijer and Dijkstra (2003) having a period τ of about 2,500 years. Positive (negative) anomalies are denoted by solid (dashed) contours and red (blue) colours. Contour levels are the same for all panels. (b) As Fig. 10.11a, but now at 500 m and showing the second half of the oscillation. (See Colour Plate.)

For the oscillatory modes in Weijer and Dijkstra (2003), the propagation of salinity anomalies displays the same pattern as that of their thermal counterparts. In fact, thermal and saline anomalies are of the same sign and largely, but not completely, density-compensating throughout the oscillations. Hence for the shortest period mode (mode 3 in Weijer and Dijkstra 2003), only the thermal anomalies are shown in Fig. 10.11.

The journey of a thermal anomaly (with oscillation period τ) can best be followed through the zero-phase line that separates negative anomalies (denoted blue) from positive anomalies (red). The anomaly (Fig. 10.11) starts off in the deep North Atlantic ($t = 2/16\ \tau$), and when it reaches the Southern Ocean, it propagates eastward ($t = 6/16\ \tau$) and fills the deep Indian Ocean ($t = 9/16 - 11/16\ \tau$) and the Pacific Ocean ($t = 14/16 - 19/16\ \tau$). However, rather than a single anomaly travelling through the deep ocean, two anomalies are present in the abyss at the same time. So when a positive anomaly sets off in the deep North Atlantic ($t = 2/16\ \tau$, Fig. 10.11), another anomaly is still working its way north in the deep Pacific.

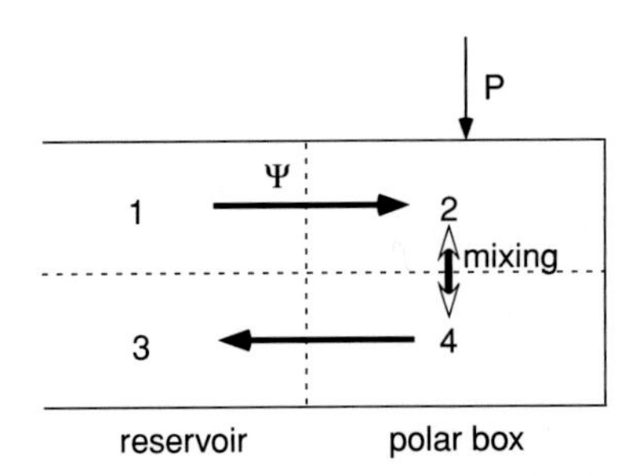

Figure 10.12 Sketch of the box model with a reservoir and a polar box with a vertical structure, here referred to as the CdV model (Colin de Verdière, 2007). The circulation Ψ is positive when the surface volume transport is northward (as indicated).

From the analysis of the propagation of temperature and salinity anomalies, it was demonstrated in Weijer and Dijkstra (2003) that these modes are the global version of the loop oscillations. Important here is that the time scale of the oscillations depend on advective transport processes, whereas the growth rate depends on diffusive processes. The internal modes in Weijer and Dijkstra (2003) are strongly damped because of the large diffusivity in their model, so noise is needed to excite them.

10.4.2 Deep-decoupling oscillations ('flushes')

Oscillatory behavior does not only occur through advective processes, but may also be induced by convective processes. This oscillatory behavior has been found in a hierarchy of ocean and climate models. The basics of the behaviour can be understood from the so-called 'flip-flop' oscillation (Welander, 1982). The origin of these, also referred to as deep-decoupling oscillations, in a single-hemisphere ocean basin is, however, best explained by a box model that combines the Welander (1982) box model with a Stommel box model of the meridional overturning (Colin de Verdière, 2007).

The box model in Fig. 10.12 consists of a reservoir (boxes 1 and 3) in which temperatures and salinities are fixed and an active polar box (boxes 2 and 4). Heat and salt are exchanged with the reservoir by advection and diffusion, and mixing occurs in the polar box when the density in box 2 becomes larger than that in box 4.

It is assumed that the surface restoring of heat is so strong that the temperature in the upper polar box T_2 is constant. Likewise, it is assumed that the salinity in the lower polar box S_4 is fixed. The meridional overturning Ψ (Ψ is taken positive when the circulation is clockwise) depends on the mean density difference between the polar box and the reservoir. With the linear equation of state,

$$\rho = \rho_0(1 - \alpha_T(T - T_0) + \alpha_S(S - S_0)),$$

where α_T and α_S are constant thermal expansion and haline contraction coefficients and T_0 and S_0 reference values, it is given by

$$\Psi = C(\alpha_T(T_1 - T_2) + \alpha_T(T_3 - T_4) - \alpha_S(S_1 - S_2) - \alpha_S(S_3 - S_4)), \qquad (10.5)$$

where C is a hydraulic constant.

The equations for the temperature T_4 and S_2 follow from heat and salt balances and are given by

$$V\frac{dT_4}{dt} = \Psi^+(T_2 - T_4) + \Psi^-(T_4 - T_3) + D(T_3 - T_4), \qquad (10.6a)$$

$$V\frac{dS_2}{dt} = P + \Psi^+(S_1 - S_2) + \Psi^-(S_2 - S_4) + D(S_1 - S_2), \qquad (10.6b)$$

where $\Psi^+ = (\Psi + |\Psi|)/2$ and $\Psi^- = (\Psi - |\Psi|)/2$, P is the virtual salt (freshwater) flux into box 2, V is the volume of the box 2 and of box 4 (taken equal) and D is the lateral diffusion coefficient.

The dimensionless vertical density difference in the polar box $R = (\rho_2 - \rho_4)/\rho_0$ is given by

$$R = -\alpha_T(T_2 - T_4) + \alpha_S(S_2 - S_4). \qquad (10.7)$$

For the stable stratification, $R < 0$ and when $R > 0$ mixing will occur. By introduction of the variables $x = -\alpha_T(T_2 - T_4)$ and $y = -\alpha_S(S_2 - S_4)$, we can write

$$\Psi = C(r_1 - (x + y)), \qquad (10.8a)$$

$$R = x - y, \qquad (10.8b)$$

with $r_1 = \alpha_T(T_{12} + T_{32}) - \alpha_S(S_{14} + S_{34})$, $T_{ij} = T_i - T_j$ and $S_{ij} = S_i - S_j$. The equations for x and y follow directly from (10.6) and then the equations for Ψ and R follow immediately from (10.8) as

$$V\frac{d\Psi}{dt} = F - (\Psi^+ - \Psi^- + D)\Psi + \psi_T\Psi^+ - \psi_H\Psi^-, \qquad (10.9a)$$

$$V\frac{dR}{dt} = G - (\Psi^+ - \Psi^- + D)R + r_2\Psi^+ - r_3\Psi^-. \qquad (10.9b)$$

When $R > 0$, the streamfunction is reset to the convective value. In the model this is represented by adding a term $-\mathcal{H}(R)R/\tau$ on the right-hand side of (10.9b) with a small time scale τ and $\mathcal{H}$ being the Heaviside function. When $R > 0$, the streamfunction is then given by $\Psi = Cr_1$. In the equations, $F = C(\alpha_S P + D(\alpha_T T_{12} - \alpha_S S_{34}))$, $r_2 = \alpha_S S_{14}$, $r_3 = \alpha_T T_{32}$, $G = \alpha_S P + D(r_2 + r_3)$, $\psi_T = C(r_1 + r_2)$ and $\psi_H = C(r_1 - r_3)$. The standard values of the parameters in (10.9) are presented in Table 10.2; the main control parameter is the freshwater forcing strength P.

Table 10.2. *Parameters of the CdV-box model (Colin de Verdière, 2007)*

Parameter	Meaning	Value	Dimension
C	hydraulic constant	10^4	Sv
D	'lateral mixing' coefficient	4.7	Sv
V	volume of polar box	$1.24\ 10^{13}$	km^3
α_T	compressibility of heat	$2\ 10^{-4}$	$^\circ$C^{-1}
α_S	compressibility of salt	$8\ 10^{-4}$	–
S_1	salinity in box 1	35.4	g kg^{-1}
T_1	temperature in box 1	9	$^\circ$C
T_2	temperature in box 2	3	$^\circ$C
T_3	temperature in box 3	8.5	$^\circ$C
S_3	salinity in box 3	35.2	g kg^{-1}
S_4	salinity in box 4	34.9	g kg^{-1}
τ	convective time scale	10^4	s
P	evaporation minus precipitation	0.65	m year^{-1}

In this model, a convective branch of steady solutions (independent of P) is present in addition to the Stommel solution branch connecting the TH and SA solutions (Fig. 10.13a). In a certain interval of P, where the SA solution is unstable, millennial time-scale oscillations occur between this SA state and the unstable convective state (Fig. 10.13b). At the boundaries of this parameter interval, the period of these oscillations approaches infinity. There is first a slow decrease of the meridional overturning, during which the static stability in the polar box decreases because of a warming in the lower box. At some point, the stratification becomes unstable, which induces convection and a rapid resumption of the overturning. This type of deep-decoupling or flushing oscillatory behaviour has also been found in more extensive box models and two-dimensional models (Winton and Sarachik, 1993; Colin de Verdière et al., 2006).

10.5 Effects of noise: stochastic resonance

As can be anticipated from the material presented so far, changes in heat transport associated with transitions or oscillations in the Atlantic MOC may be a good ingredient for explaining rapid temperature changes in the northern North Atlantic associated with the D-O events. Indeed, changes in the MOC play an essential role in the two mechanisms discussed in detail next. Both mechanisms differ in their role attributed to oscillatory internal modes. The (nonautonomous) stochastic resonance theory is based only on the presence of the multiple equilibrium regime, whereas the central ingredient of the so-called coherence resonance (or autonomous stochastic resonance) mechanism is the existence of oscillatory modes with a millennial time scale.

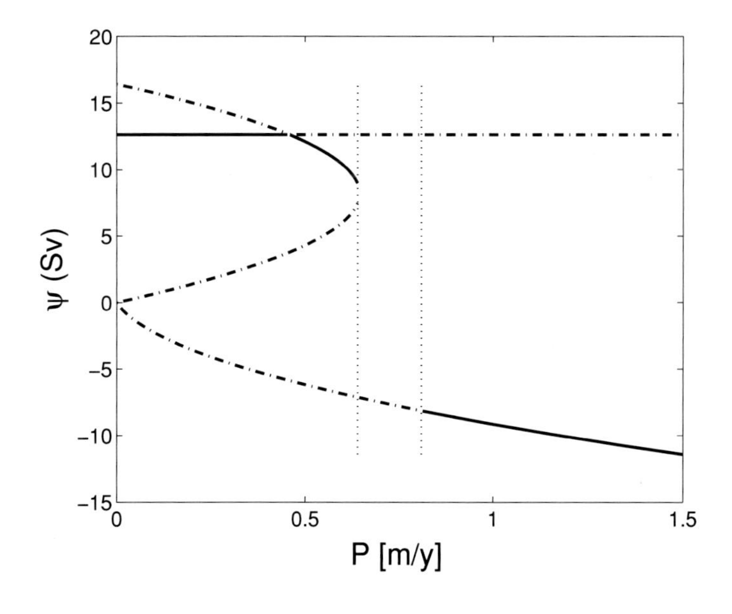

(a)

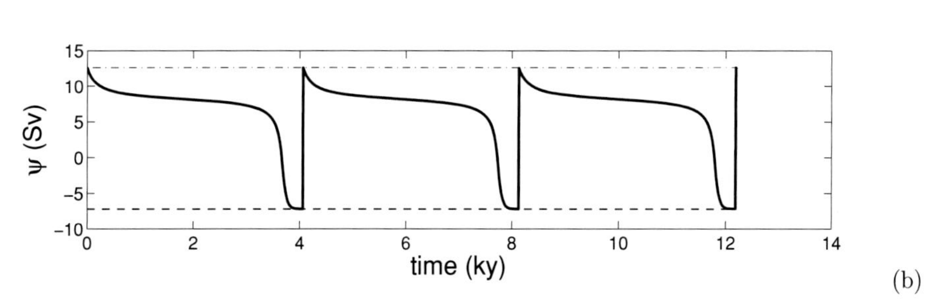

(b)

Figure 10.13 (a) Bifurcation diagram using P as a control parameter for the standard values of the parameters. (b) Temporal evolution of Ψ for $P = 0.65$ m/yr; time is in kyr. The horizontal lines represent the values of the convective and salinity-driven steady states in (a) (figure from Colin de Verdière, 2007).

10.5.1 (Nonautonomous) stochastic resonance

We first describe the basic features of stochastic resonance using results of the stochastic Stommel two-box model. Next, the mechanism is discussed in more detail using the periodically forced stochastic pitchfork bifurcation.

Box models

From the bifurcation diagram in Fig. 10.5a, it is clear that for the range of γ_p values between L_1 and L_2, there are two stable states. The one with a strong MOC, and with the larger northward heat transport, is usually referred to as the 'on-state', whereas that with the weak MOC is called the 'off-state'. Clearly, both on-state and off-state are linearly stable, so a finite amplitude perturbation is certainly needed to induce a transition between both states. These finite amplitude perturbations are provided by background noise, for example, in the freshwater flux.

The effect of additive noise on the transition behaviour between on and off states of the MOC was considered in a box model by Cessi (1994). This model is a variant of the Stommel two-box model presented in Example 6.1. Again, a polar box (with temperature T_p and salinity S_p) and an equatorial box (with temperature T_e and salinity S_e) having the same volume V are connected by advective flow and exchange heat and freshwater with the atmosphere. The heat and salt balances are

$$\frac{dT_e}{dt} = -\frac{1}{t_r}\left(T_e - \left(T_0 + \frac{\theta}{2}\right)\right) - \frac{1}{2}Q(\Delta\rho)(T_e - T_p), \tag{10.10a}$$

$$\frac{dT_p}{dt} = -\frac{1}{t_r}\left(T_p - \left(T_0 - \frac{\theta}{2}\right)\right) - \frac{1}{2}Q(\Delta\rho)(T_p - T_e), \tag{10.10b}$$

$$\frac{dS_e}{dt} = \frac{F_S}{2H}S_0 - \frac{1}{2}Q(\Delta\rho)(S_e - S_p), \tag{10.10c}$$

$$\frac{dS_p}{dt} = -\frac{F_S}{2H}S_0 - \frac{1}{2}Q(\Delta\rho)(S_p - S_e), \tag{10.10d}$$

where F_S is the fresh water flux, H is the ocean depth, S_0 is a reference salinity, T_0 is a reference temperature, t_r is the surface temperature restoring time scale and θ is the equator-to-pole atmospheric temperature difference. A linear equation of state of the form

$$\rho = \rho_0(1 - \alpha_T(T - T_0) + \alpha_S(S - S_0)), \tag{10.11}$$

is assumed, where the subscript '0' refers to the reference values, and α_T and α_S are constant thermal expansion and haline contraction coefficients, respectively. In Cessi (1994), the transport function Q is chosen as

$$Q(\Delta\rho) = \frac{1}{t_d} + \frac{q}{\rho_0^2 V}(\Delta\rho)^2, \tag{10.12}$$

where q is a transport coefficient, t_d a diffusion time scale and $\Delta\rho = \rho_p - \rho_e$.

Subtracting (10.10b) from (10.10a) and (10.10d) from (10.10c) and introducing $\Delta T = T_e - T_p$ and $\Delta S = S_e - S_p$ leads to

$$\frac{d\Delta T}{dt} = -\frac{1}{t_r}(\Delta T - \theta) - Q(\Delta\rho)\Delta T, \tag{10.13a}$$

$$\frac{d\Delta S}{dt} = \frac{F_S}{H}S_0 - Q(\Delta\rho)\Delta S. \tag{10.13b}$$

Table 10.3. *Parameters of the stochastic Stommel box model (slightly changed from those in Cessi [1994])*

Parameter	Meaning	Value	Unit
t_r	temperature relaxation time scale	25	days
H	mean ocean depth	4,500	m
t_d	diffusion time scale	180	years
t_a	advective time scale	29	years
q	transport coefficient	1.92×10^{12}	$m^3 s^{-1}$
V	ocean volume	$300 \times 4.5 \times 8{,}250$	km^3
α_T	thermal expansion coefficient	10^{-4}	K^{-1}
α_S	haline contraction coefficient	7.6×10^{-4}	–
S_0	reference salinity	35	$g\ kg^{-1}$
θ	meridional temperature difference	25	K

With the scales $\Delta T = x\,\theta$, $\Delta S = y\,\alpha_T\theta/\alpha_S$ and time scaled with t_d, the nondimensional system of equations becomes

$$\frac{dx}{dt} = -\alpha(x-1) - x(1+\mu^2(x-y)^2), \tag{10.14a}$$

$$\frac{dy}{dt} = F - y(1+\mu^2(x-y)^2), \tag{10.14b}$$

where $\alpha = t_d/t_r$ is the ratio of the diffusive time scale and the temperature relaxation time scale. Furthermore,

$$\mu^2 = \frac{q t_d (\alpha_T\theta)^2}{V}, \tag{10.15}$$

is the ratio of the diffusion time scale t_d and the advective time scale $t_a = V/(q(\alpha_T\theta)^2)$. Finally, the dimensionless freshwater flux parameter F is given by

$$F = \frac{\alpha_S S_0 t_d}{\alpha_T \theta H} F_S. \tag{10.16}$$

Typical values of the dimensional parameters as motivated in Cessi (1994) are shown in Table 10.3. The volume V is based on the area of transport near the western boundary, and the q is determined from the strength of the southward branch of the MOC.

From these values, we see that α is large, as indeed the diffusion time scale is much larger than the temperature-restoring time scale. For large α, one can approximate $x = 1 + \mathcal{O}(1/\alpha)$, and the equation (10.14b) then becomes

$$\frac{dy}{dt} = F - y(1+\mu^2(1-y)^2). \tag{10.17}$$

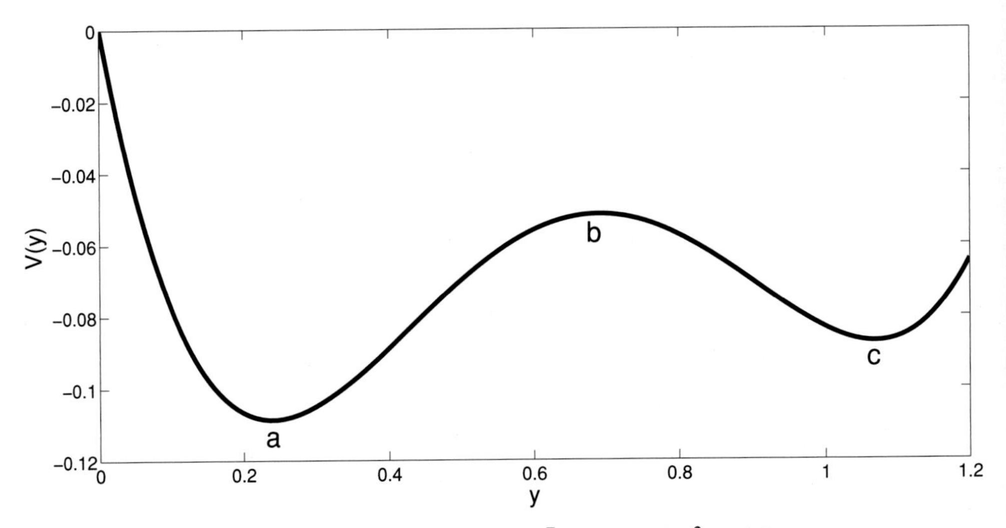

Figure 10.14 Potential $V(y)$ as in (10.19) for $\bar{F} = 1.1$ and $\mu^2 = 6.2$.

When the freshwater flux is time independent, say $F = \bar{F}$, this equation can be written as

$$\frac{dy}{dt} = -V'(y), \tag{10.18}$$

with the potential function

$$V(y) = -\bar{F}y + \frac{y^2}{2} + \mu^2\left(\frac{y^4}{4} - \frac{2y^3}{3} + \frac{y^2}{2}\right), \tag{10.19}$$

with the prime in (10.18) indicating differentiation to y.

For $\bar{F} = 1.1$ and $\mu^2 = 6.2$, this potential is plotted in Fig. 10.14. It is a typical so-called double-well potential with minima at $y_a = 0.24$ and $y_c = 1.07$ and a maximum at $y_b = 0.69$. As extrema of $V(y)$ are the steady states of (10.18), we see immediately that the dynamical system (10.17) has three steady states. The linear stability of these steady states is directly related to the sign of the second derivative of the potential at the extrema, and hence y_a and y_c are linearly stable steady states, and y_b is an unstable steady state.

Consider now that the freshwater flux has a stochastic component, say $F = \bar{F} + \tilde{F}$, with $\tilde{F}$ being additive white noise (i.e., represented by a Wiener process) with amplitude σ; then (10.18) generalises into the Itô equation,

$$dY_t = -V'(Y_t)dt + \sigma dW_t. \tag{10.20}$$

A trajectory of (10.20) for $\bar{F} = 1.1$, $\mu^2 = 6.2$ and $\sigma = 0.2$ is plotted in Fig. 10.15a. As can be seen, the trajectory undergoes transitions between the two stable states (y_c and y_a). When a histogram is made of the y position, the probability density function in Fig. 10.15b results (dashed curve), which indeed shows a bimodal distribution.

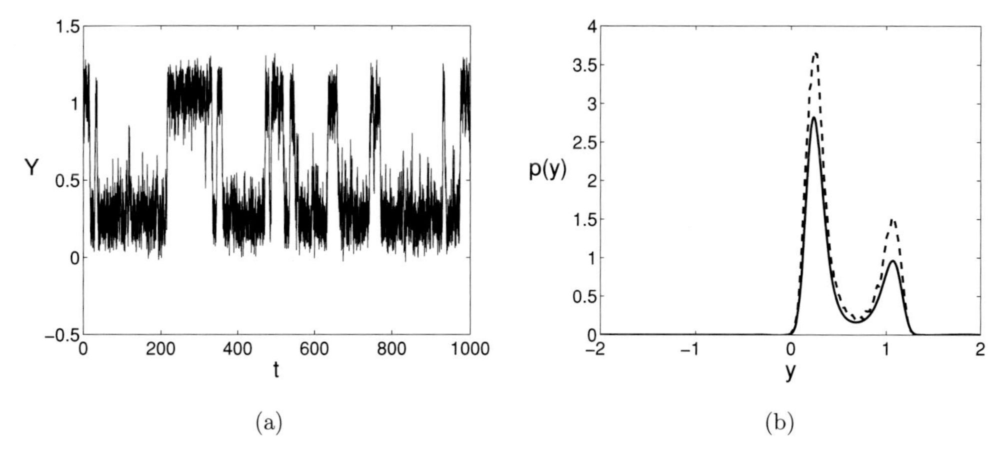

Figure 10.15 (a) Realisation of the system (10.20) using $dt = 0.01$ in a Euler-Maruyama scheme (see Section 3.6) for $\bar{F} = 1.1$, $\mu^2 = 6.2$ and $\sigma = 0.2$. (b) Non-normalised probability density function (dashed) of the simulation in (a) using 100 bins to sample the histogram. The drawn curve is the normalised analytical solution of the steady probability density function (10.22).

The probability density function $p(y, t)$ is determined from the forward Fokker-Planck equation of (10.20), see Section 3.5, which is

$$\frac{\partial p}{\partial t} - \frac{\partial (V'(y)p)}{\partial y} - \frac{\sigma^2}{2}\frac{\partial^2 p}{\partial y^2} = 0. \tag{10.21}$$

The stationary solution $p(y)$ can be easily solved as

$$p(y) = C\, e^{-\frac{2V(y)}{\sigma^2}}, \tag{10.22}$$

where C is a normalisation constant such that $\int p(y)\, dy = 1$. This stationary probability density function is also plotted in Fig. 10.15b as the drawn curve.

In Section 3.5.3, we showed that it was possible to determine how long it takes to leave the neighbourhood of one of the potential wells from a certain initial condition. Because this time is controlled by the time to go over the potential barrier at y_b, the precise specification of the initial condition and the final position is not needed. All points to the left of y_b, for example, can be considered as an ensemble, and one can define a mean escape time $\langle t_{a\to c}\rangle$ from the potential well with a minimum at y_a to the one with the minimum at y_c and vice versa ($\langle t_{c\to a}\rangle$). Expressions of these type of transition times were provided in Section 3.5.4 (Example 3.4) and for the problem (10.20) become (Cessi, 1994)

$$\langle t_{a\to c}\rangle \approx \frac{2}{\sigma^2}\int_{y_a}^{y_c} e^{\frac{2V(y)}{\sigma^2}}\, dy \int_{-\infty}^{y_b} e^{-\frac{2V(z)}{\sigma^2}}\, dz, \tag{10.23a}$$

$$\langle t_{c\to a}\rangle \approx \frac{2}{\sigma^2}\int_{y_a}^{y_c} e^{\frac{2V(y)}{\sigma^2}}\, dy \int_{y_b}^{\infty} e^{-\frac{2V(z)}{\sigma^2}}\, dz. \tag{10.23b}$$

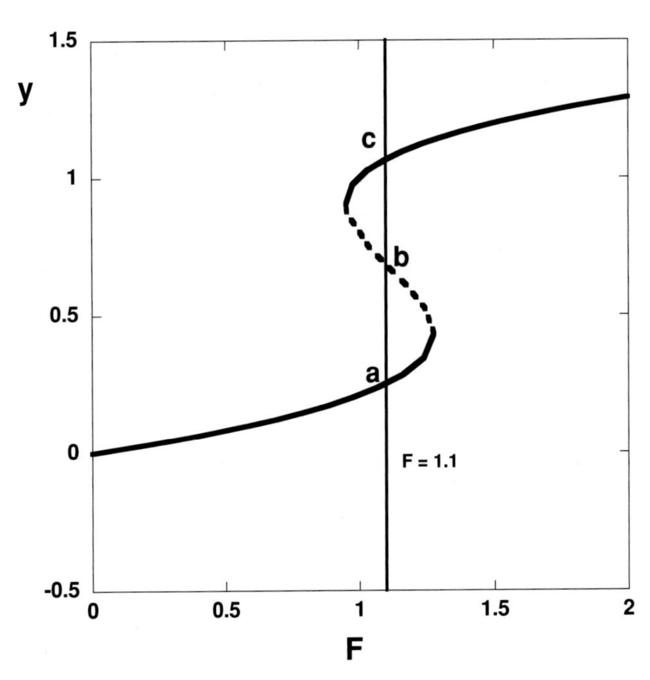

Figure 10.16 Bifurcation diagram of the model (10.14) for $\alpha = 360$ and $\mu^2 = 6.25$, with $\bar{F}$ as control parameter (figure based on Velez-Belchi et al., 2001).

We now turn now to the full system (10.14) and take $\alpha = 360$ and $\mu^2 = 6.25$ (Velez-Belchi et al., 2001). The bifurcation diagram of this model is plotted in Fig. 10.16, where the steady-state value of y is plotted versus $\bar{F}$. This is qualitatively the same diagram as that of the original Stommel model, akin to Fig. 10.4a, with two stable states: the on-state (small salinity difference y) and the off-state (large salinity difference y).

With noise added to the freshwater flux, the transitions between the on- and off-states (cf. Fig. 10.15a) are not very regular and cannot serve as a prototype to explain a specific period such as that appears in the D-O events. The box model (10.14) was therefore extended to include a periodic forcing (Velez-Belchi et al., 2001) by choosing ($\bar{F}$ again indicates the noise component)

$$F(t) = \bar{F} + A \sin\left(2\pi \frac{t}{T}\right) + \tilde{F}, \tag{10.24}$$

with $\bar{F} = 1.1$. The periodic forcing has an amplitude A in the freshwater flux and a dimensionless period T (the dimensional value can be obtained by multiplying with the diffusion time scale t_d).

The time series of the dimensionless transport q from this model for $T \times t_d = 42$ kyr (corresponding to variations associated with the obliquity cycle) and $A = 0.05$ are shown in Fig. 10.17a. Here the stochastic component $\tilde{F}$ is zero mean Gaussian

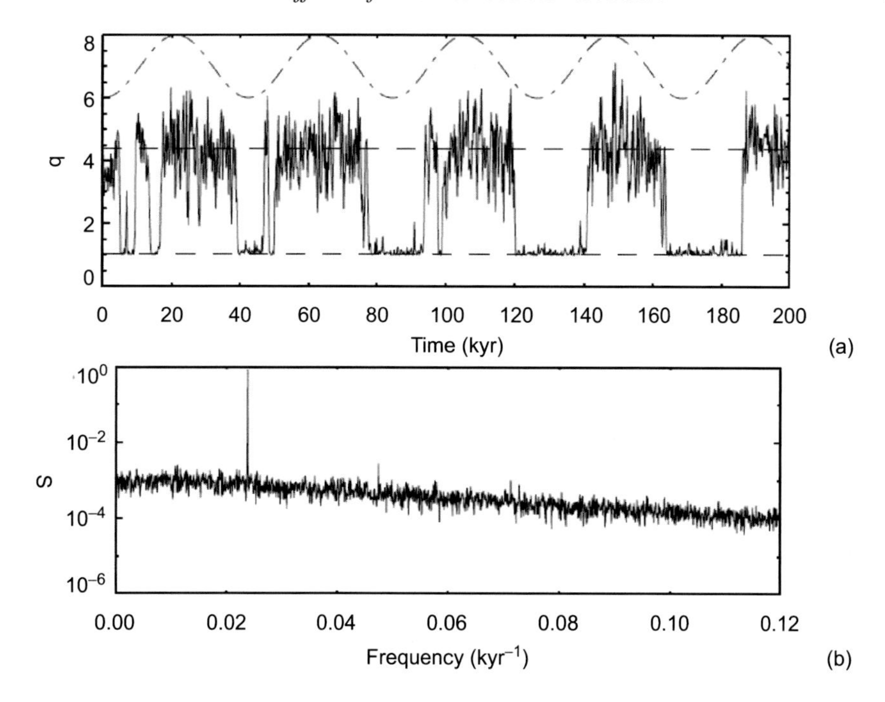

Figure 10.17 (a) Time series (drawn curve) of the transport q between the basins for the periodically and stochastically forced model (10.14), for $T \times t_d = 42$ kyr, $A = 0.05$ and $\epsilon = 0.022$. (b) Power spectrum of the trajectory in (a) (figure from Velez-Belchi et al., 2001).

noise with variance ϵ. The results show that the transitions between the stable states are clearly influenced by the periodic forcing and now become more regular. The spectrum in Fig. 10.17b indicates a red noise background with a sharp spectral peak at the forcing frequency. Apparently the very small periodic signal is amplified by the noise in the nonlinear system having a multiple equilibrium regime; such an amplification is called stochastic resonance.

A characteristic of stochastic resonance is the signal to noise ratio Σ, as measured by

$$\Sigma = 10 \log_{10} \frac{S(\omega_0)}{B}, \tag{10.25}$$

where $\omega_0 = 2\pi/T$ is the forcing frequency, $S(\omega_0)$ the power associated with that frequency and B the background spectrum of the noise. The value of Σ first increases with the noise variance, reaching a maximum value corresponding to the maximum cooperation between the periodic forcing and the noise. For large values of the noise, the transitions will be noise dominated, which is shown as a decay in Σ. In the computations in Velez-Belchi et al. (2001) with the model (10.14), the result for Σ is plotted in Fig. 10.18 for different values of the forcing period showing this characteristic

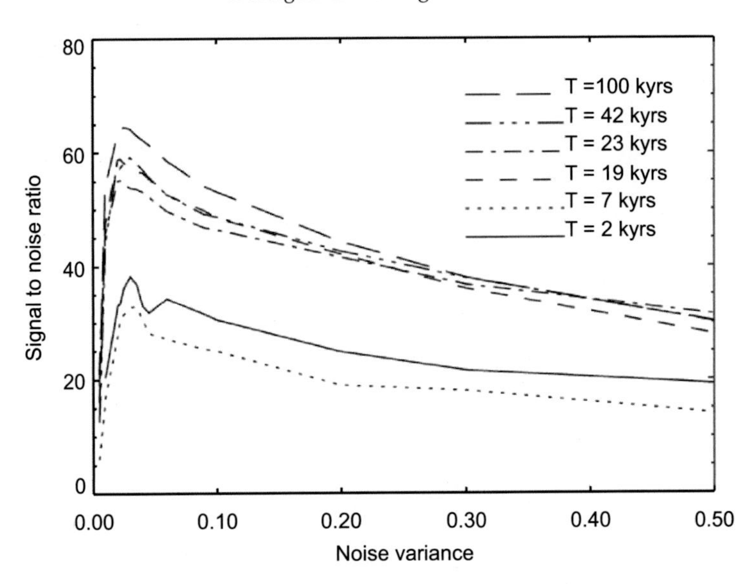

Figure 10.18 Curves of the signal to noise ratio Σ for the periodically and stochastically forced model (10.14) versus the noise variance ϵ as computed in Velez-Belchi et al. (2001).

signature of stochastic resonance. The amplification of the signal happens at small noise amplitude and for a large range of forcing frequencies.

Mechanism of stochastic resonance

To understand the stochastic resonance mechanism in more detail, we consider a sinusoidal temporal deterministic forcing in the problem of the pitchfork bifurcation with additive stochastic noise (cf. Section 4.3). The problem becomes the Itô SDE

$$X_t = X_0 + \int_0^t (X_s - X_s^3 + A\cos\Omega\tau)ds + \int_0^t \sigma\, dW_s, \qquad (10.26)$$

where τ is considered as a parameter. The potential $V(x)$ associated with (10.26) is

$$V(x) = -\frac{1}{2}x^2 + \frac{1}{4}x^4 - Ax\cos\Omega\tau. \qquad (10.27)$$

For $A = 0.1$, this potential is plotted for three values $\tau = 0$, $\tau = \pi/(2\Omega)$ and $\tau = \pi/\Omega$ in Fig. 10.19. Although the fixed point locations do not depend strongly on τ, the actual values of $V(x)$ at these fixed points do.

Central to stochastic resonance is the relative magnitude of the mean transition times with respect to the periodic forcing. The transition times are given by (10.23),

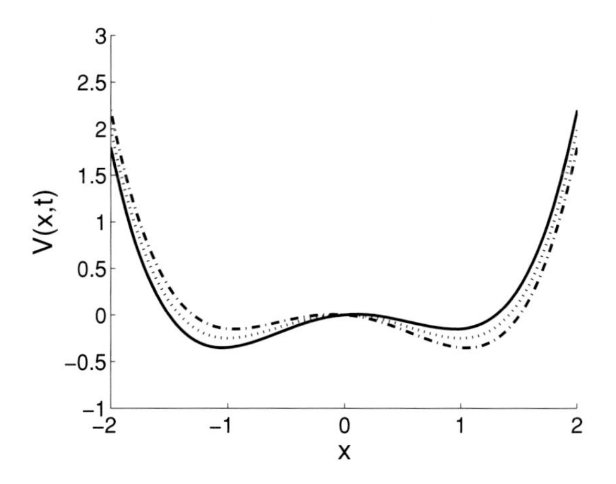

Figure 10.19 The potential $V(x)$ for three values of $\tau = 0$ (dash-dotted), $\tau = \pi/(2\Omega)$ (dotted) and $\tau = \pi/\Omega$ (solid).

with $y_a = -1$, $y_b = 0$ and $y_c = 1$, and become

$$\langle t_{-1\to 1}\rangle \approx \frac{2}{\sigma^2}\int_{-\infty}^{0} e^{-\frac{2V(x)}{\sigma^2}}\,dx \int_{-1}^{1} e^{\frac{2V(x)}{\sigma^2}}\,dx, \tag{10.28a}$$

$$\langle t_{1\to -1}\rangle \approx \frac{2}{\sigma^2}\int_{0}^{\infty} e^{-\frac{2V(x)}{\sigma^2}}\,dx \int_{-1}^{1} e^{\frac{2V(x)}{\sigma^2}}\,dx. \tag{10.28b}$$

Here it is assumed that the equilibration time scale of the probability density function is much faster than the change in the potential. Because σ^2 is a small parameter, the integrals in (10.28) can be approximated using a method due to Laplace.

The integrals are of all of the form

$$\int_a^b e^{Mf(x)}dx, \; M \to \infty. \tag{10.29}$$

When $f(x)$ has a positive maximum $x_0 \in (a, b)$, then the largest contribution of the integral will come from function values near x_0. When $f(x)$ is expanded in a Taylor series near x_0, we find (note that $f'(x_0) = 0$ and $f''(x_0) < 0$)

$$f(x) = f(x_0) - \frac{1}{2}|f''(x_0)|(x - x_0)^2 + \cdots ,$$

and the integral (10.28) can be approximated by

$$\begin{aligned}\int_a^b e^{Mf(x)}dx &\approx e^{Mf(x_0)}\int_a^b e^{-\frac{M}{2}|f''(x_0)|(x-x_0)^2}dx,\\ &\approx e^{Mf(x_0)}\int_{-\infty}^{\infty} e^{-\frac{M}{2}|f''(x_0)|(x-x_0)^2}dx,\end{aligned} \tag{10.30}$$

because the integrand decays very rapidly from x_0. Using

$$\int_{-\infty}^{\infty} e^{-\alpha(x-x_0)^2} dx = \sqrt{\frac{\pi}{\alpha}},$$

it is eventually found (with $\alpha = \frac{M}{2}|f''(x_0)|$) that

$$\int_a^b e^{Mf(x)} dx \approx \sqrt{\frac{2\pi}{M|f''(x_0)|}} e^{Mf(x_0)}. \tag{10.31}$$

Using the Laplace approximation in the integrals in (10.28), we find that (with $f(x) = -V(x)$, $M = 2/\sigma^2$ and $p(x)$ as in (10.22))

$$\int_{-\infty}^{0} p(x)\, dx = C\sqrt{\frac{\pi\sigma^2}{V''(-1)}} e^{-\frac{2V(-1)}{\sigma^2}}, \tag{10.32a}$$

$$\int_{-1}^{1} \frac{1}{p(x)} dx = C^{-1}\sqrt{\frac{\pi\sigma^2}{-V''(0)}} e^{\frac{2V(0)}{\sigma^2}}, \tag{10.32b}$$

and, finally,

$$\langle t_{-1\to 1}\rangle \approx 2\pi\sqrt{\frac{1}{-V''(0)V''(-1)}} e^{\frac{2(V(0)-V(-1))}{\sigma^2}}, \tag{10.33a}$$

$$\langle t_{1\to -1}\rangle \approx 2\pi\sqrt{\frac{1}{-V''(0)V''(1)}} e^{\frac{2(V(0)-V(1))}{\sigma^2}}. \tag{10.33b}$$

The transition times for the potential (10.27), with $V(0) = 0$, $V(-1) = -1/4 + A\cos\Omega\tau$, $V(1) = -1/4 - A\cos\Omega\tau$, $V''(\pm 1) = 2$ and $V''(0) = -1$, are finally given by

$$\langle t_{-1\to 1}\rangle \approx \frac{\pi}{\sqrt{2}} e^{\frac{1}{2\sigma^2}(1-4A\cos\Omega\tau))}, \tag{10.34a}$$

$$\langle t_{1\to -1}\rangle \approx \frac{\pi}{\sqrt{2}} e^{\frac{1}{2\sigma^2}(1+4A\cos\Omega\tau))}. \tag{10.34b}$$

The important element is that the transition times vary with τ as the depth of the potential wells deepens and shallows. Suppose now that, at $\tau = 0$, the state of the system is near $x = 1$. At that time, the transition time $\langle t_{1\to -1}\rangle$ is maximal (as the potential well is deepest), but it decreases with time until it becomes a minimum at $\tau = \pi/\Omega$. Consequently, over the time interval $[0, \pi/\Omega]$ the probability to exit the potential well near $x = 1$ increases. When the transition time $\langle t_{1\to -1}\rangle$ at $\tau = 0$ is on the order of the time scale of the periodic forcing and the transition time at $\tau = \pi/\Omega$ is much smaller than the time scale of the periodic forcing, then the system will surely exit the well near $x = 1$ at $\tau \sim \pi/\Omega$. Once in the other well, the same reasoning can be applied using the transition time $\langle t_{-1\to 1}\rangle$.

Because the variance in the transition time is much smaller than the transition time itself, the transition occurs over a small, well-defined time interval. Consequently, the Fourier spectra of the sample paths have a strong peak at the forcing frequency Ω. Hence the noise amplifies the periodic signal by establishing a coherent transition from one well to the other. Stochastic resonance provides a mechanism by which a small-amplitude periodic signal and noise work together to induce well-defined transitions between different states. Neither noise nor the weak periodic forcing can do this by itself.

10.5.2 Coherence resonance

Again, we discuss the basic features using the results of a box climate model and then describe the mechanism in more detail using a two-dimensional random dynamical system.

A climate box model

Using an ocean box model coupled to a simple (box type) representation of the atmosphere and sea ice, the mechanisms of millennial climate variability were analysed in Timmermann et al. (2003b). This model is discussed in more detail in Section 11.4 as it was originally used (Gildor and Tziperman, 2000) for studying the Pleistocene ice ages.

The response of the model to freshwater input in the most northern ocean box was studied. In the left panel of Fig. 10.20, the hysteresis diagram of the MOC is shown for the deterministic case (no noise in the freshwater flux) for two different rates of quasi-equilibrium change (0.001 Sv/1,000 years and 0.01 Sv/1,000 years). The very slow variation (0.001 Sv/1,000 years) results in behaviour expected from the Stommel-type bifurcation diagram with transitions between the on-state and the off-state of the MOC. Using a faster change (0.01 Sv/1,000 years) results in a transition to a large-amplitude MOC state. Under noise in the freshwater forcing, one also observes oscillatory behaviour near the transition from the on- to the off-state with an amplitude of about 4 Sv in the MOC (right panel of Fig. 10.20).

The character of these oscillations was studied by releasing a meltwater pulse of a specific amplitude for 300 years in the northern ocean box (Fig. 10.21). Under a large-amplitude (0.45 Sv) meltwater input and in the presence of noise, oscillations appear in the MOC. These oscillations have a regular, relaxation type behaviour with a rapid warming of the atmosphere and a slow cooling. Timmermann et al. (2003b) showed that the time scale of the oscillations is strongly dependent on the vertical diffusivity in their model and indicate that the mechanism can be described by a deep-decoupling oscillation.

A very interesting result is found when the response to different amplitudes of the freshwater noise is considered (again for the 0.45 Sv meltwater pulse). When

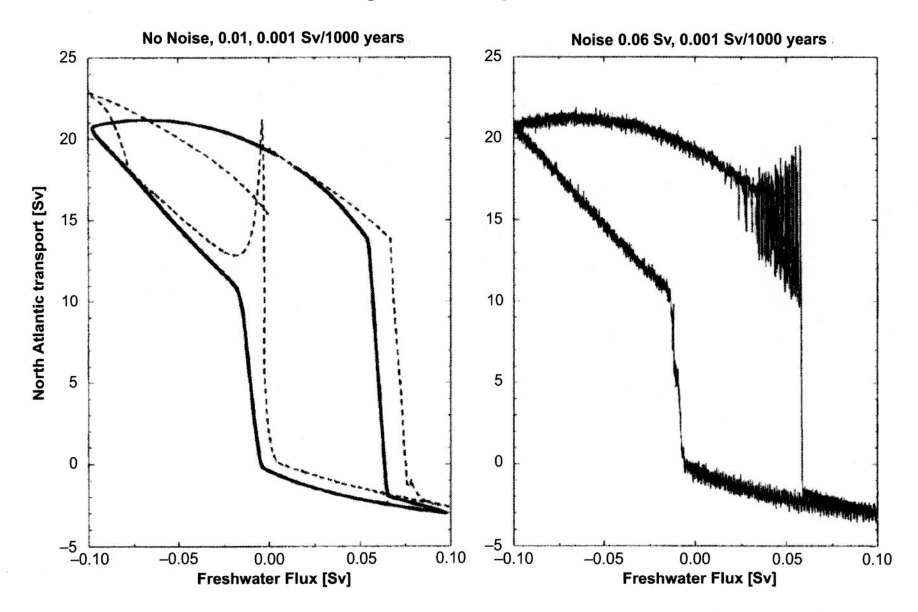

Figure 10.20 Solid lines show North Atlantic MOC transport as a function of slowly modified freshwater flux (left) without stochastic forcing and (right) with a stochastic freshwater component with an amplitude of 0.06 Sv. Transient simulations are performed in which the freshwater flux is changed linearly at a rate of 0.001 Sv per 1,000 years. Dashed line at left shows transient simulations performed in which the freshwater flux is changed linearly at a rate of 0.01 Sv per 1,000 years (figure from Timmermann et al., 2003b).

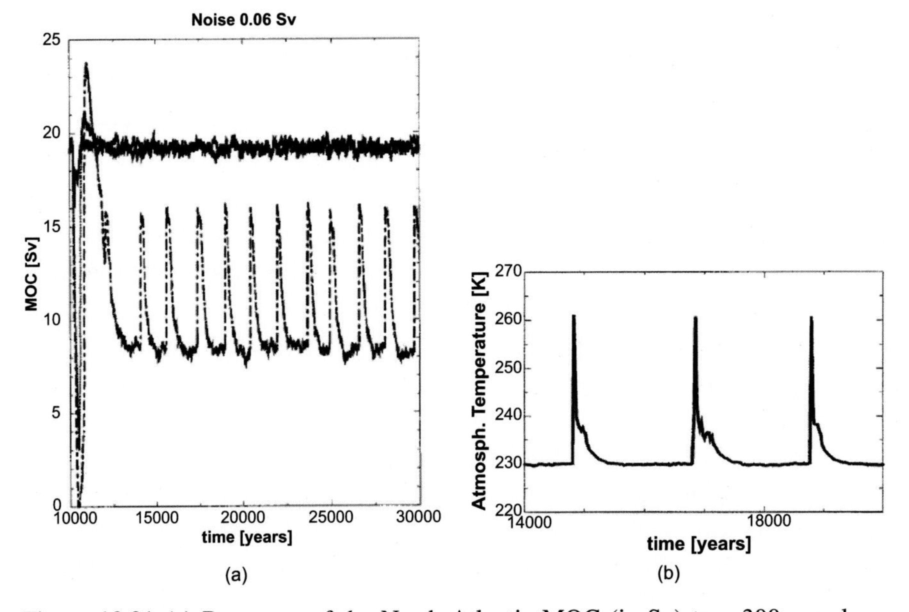

Figure 10.21 (a) Response of the North Atlantic MOC (in Sv) to a 300-year-long meltwater pulse of varying amplitude in the presence of a 0.06-Sv amplitude noise forcing in the freshwater flux. (b) January atmospheric surface temperature in the northernmost box during the oscillation in (a) for a meltwater pulse of 0.45 Sv (figure from Timmermann et al., 2003b).

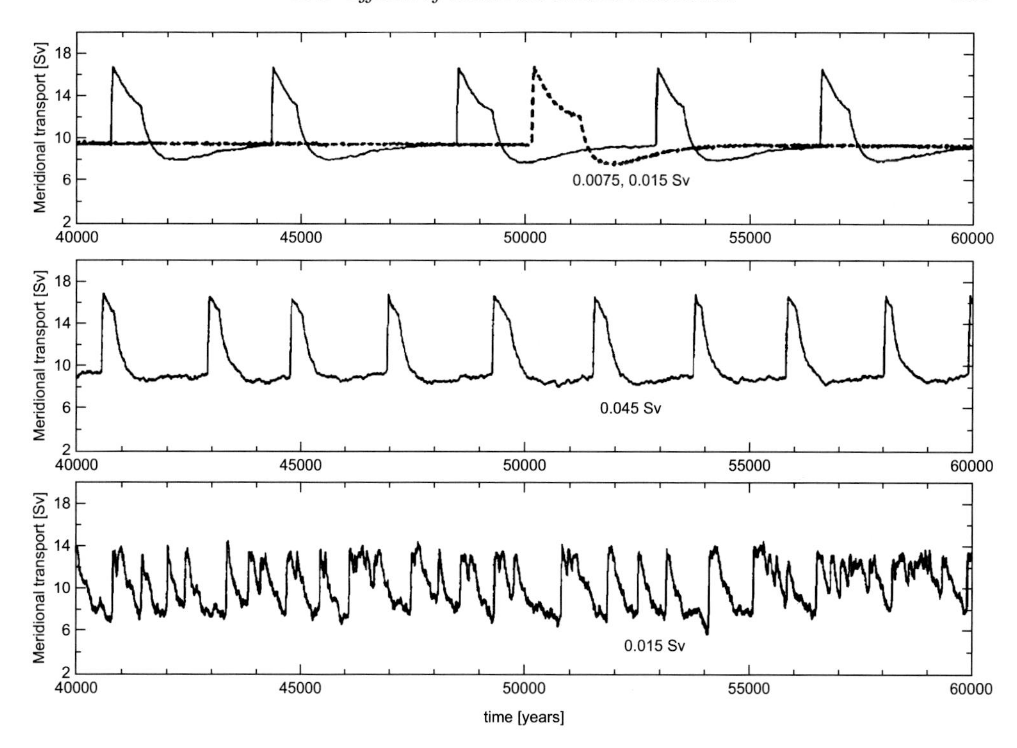

Figure 10.22 Response of the North Atlantic MOC to different stochastic freshwater forcing amplitudes (for a meltwater pulse of 0.45 Sv) (figure from Timmermann et al., 2003b).

noise is absent, there are no millennial oscillations, but they appear under a small noise amplitude (Fig. 10.22, upper panel). When the noise amplitude is very large, the millennial oscillations disappear (Fig. 10.22, lower panel), or at least the period becomes much shorter. From the middle panel in Fig. 10.22, it is seen that there seems to be an optimal noise amplitude to give the most regular millennial oscillations. These characteristics are typical for a phenomenon called coherence resonance, which is discussed in more detail next. Note that no periodic forcing is needed in coherence resonance, in contrast to the classical (nonautonomous) stochastic resonance.

Basic mechanism of coherence resonance

The phenomenon of coherence resonance was first found in the noise-driven fast-slow systems (Section 3.5.5.), in particular in the so-called FitzHugh-Nagumo model (Pikovsky and Kurths, 1997). The equations of this model are the two-dimensional Itô system (for $a > 0$),

$$\epsilon dX_t = \left(X_t - \frac{1}{3}X_t^3 - Y_t\right) dt, \tag{10.35a}$$

$$dY_t = (X_t + a)dt + D\, dW_t. \tag{10.35b}$$

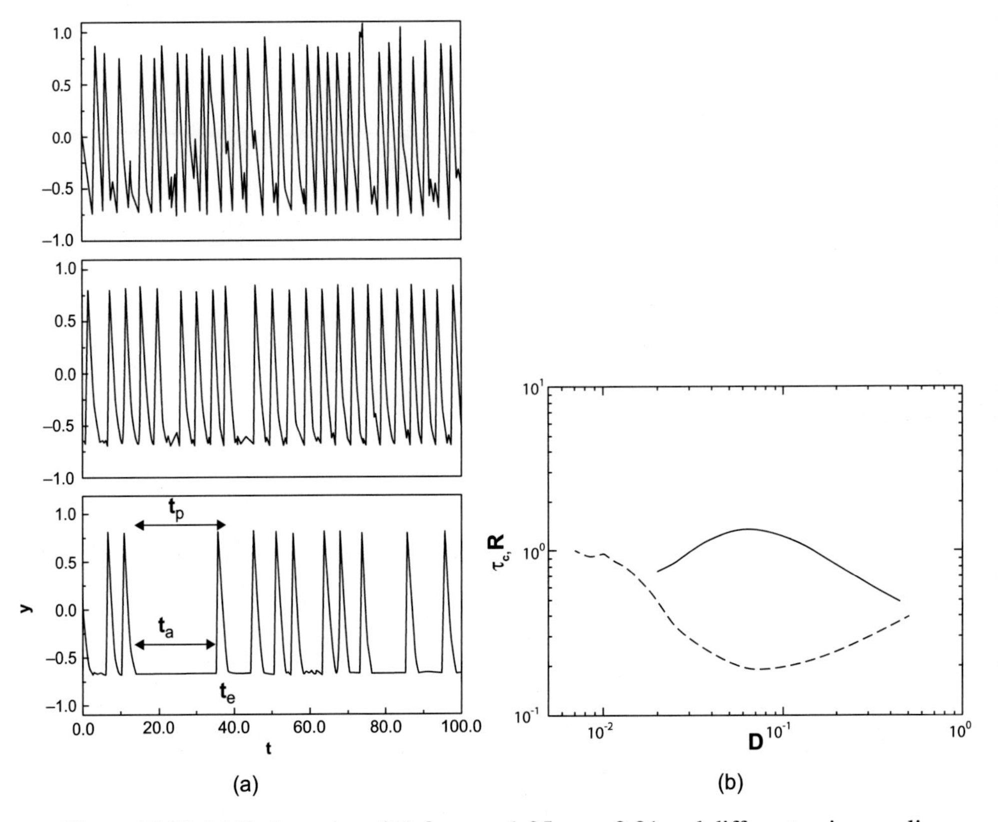

Figure 10.23 (a) Trajectories of Y_t for $a = 1.05$, $\epsilon = 0.01$ and different noise amplitudes D, from top to bottom given by $D = 0.02$, $D = 0.07$ and $D = 0.25$. (b) Characteristic correlation time τ_c (solid) and normalised fluctuations of pulse durations R (dashed) of the signal in (a) (figure from Pikovsky and Kurths, 1997).

The deterministic system has a fixed point $x = -a$, $y = a - a^3/3$, and the linear stability of the fixed point indicates that it is stable for $a > 1$ and unstable for $a < 1$. The eigenvalues of the Jacobian matrix are a complex conjugate pair near $a = 1$, and hence a Hopf bifurcation occurs at $a = 1$.

For a slightly larger than 1, the system is excitable by the noise, as the limit cycle is 'nearby'. For $a = 1.05$ and $\epsilon = 0.01$, the behaviour of Y_t is shown for different noise amplitudes D in Fig. 10.23a. To investigate the degree of coherence in the time series, the characteristic correlation time τ_c and the normalised fluctuations of pulse durations R defined by

$$\tau_c = \int_0^\infty C^2(\tau)d\tau \; ; \; C(\tau) = \frac{\langle \tilde{Y}_t \tilde{Y}_{t+\tau} \rangle}{\langle \tilde{Y}_t^2 \rangle}, \tag{10.36a}$$

$$R = \frac{\sqrt{Var(t_p)}}{\langle t_p \rangle}, \tag{10.36b}$$

are computed, where $\tilde{Y}_t = Y_t - \langle Y_t \rangle$ is the anomaly with respect to mean (indicated by the brackets). There seems to be an optimal noise amplitude for τ_c to be a maximum and R to be minimum (Fig. 10.23b).

The appearance of this coherence resonance is tightly coupled to the excitable nature of this system: the development of the system has two characteristic time scales. An activation time scale t_a is needed to excite the system from the stable fixed point, whereas the excursion time t_e is the time needed to return from the excited state to the fixed point. Both times are indicated in Fig. 10.23a (bottom panel). The pulse duration is the sum of these times, $t_p = t_a + t_e$. For small noise, it takes very long to excite the system, and hence $t_a \gg t_e$ and the pulse time is controlled by the activation time. For large noise amplitude, the activation time is very small, and hence the pulse time is dominated by the excursion time t_e. The coherence resonance appears under a noise level where the threshold of excitation is small (such that $t_a \ll t_e$) but not very large so that fluctuations in the excursion time are small.

10.6 Climate models

The box models of which the results were discussed in the previous sections have relatively few degrees of freedom. These models can represent the multiple equilibrium regime and basic oscillation mechanisms of the Atlantic MOC but, for example, not the effect of different external climate conditions such as the difference in MOC stability under present-day and glacial climate conditions (relevant for the D-O events).

The effect of background climate conditions on MOC stability was explored in Ganopolsky and Rahmstorf (2001) using an EMIC (Climber-2), cf. Chapter 6. By varying the freshwater forcing in two different regions, the hysteresis signature was explored in this model for both the present-day climate and the climate of the Last Glacial Maximum (LGM). The hysteresis is broad for the present-day situation (Fig. 10.24a,c), for which there is a transition between the on- and off-states (Fig. 10.25a,c). However, the hysteresis is nearly absent for the LGM (Fig. 10.24b,d). The off-state does not seem to exist, and the only stable circulation is the weak MOC pattern, in Fig. 10.25d referred to as the ‘cold’ state. However, for slightly negative freshwater forcing, the system can jump to an LGM ‘warm’ state (Fig. 10.25b), but this state does not appear to be an equilibrium state of the model.

For the LGM climate, a periodic forcing in the North Atlantic (over the domain 50–80°N) with an amplitude of 0.03 Sv was applied. When the freshwater flux is slightly negative, a transition from the cold to the warm state is induced (the transition B in Fig. 10.24), which leads to a rapid warming of the North Atlantic climate (Fig. 10.26). Because the warm state is not stable, it decays due to the input of freshwater, which weakens the MOC. This gives the slow cooling phase, which lasts until the freshwater anomaly is negative again, leading to a new transition to the warm state. The MOC changes over this cold-warm transition are enormously large (about 30 Sv), as are the

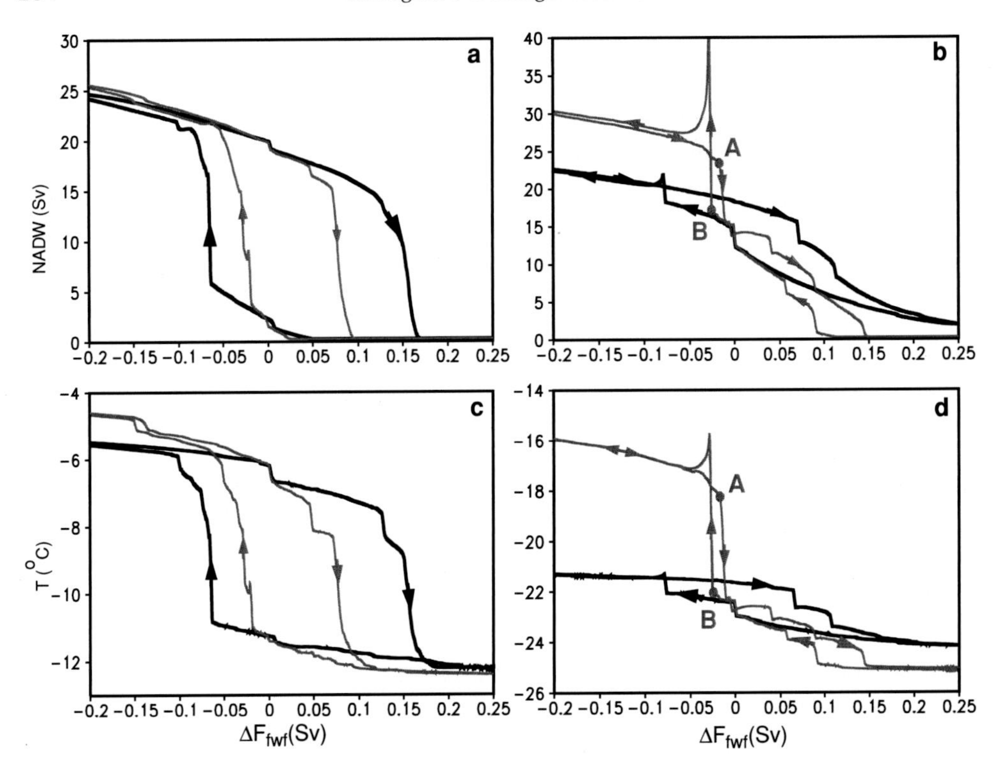

Figure 10.24 Results of quasi-equilibrium computations with the Climber-2 (EMIC) model where a freshwater anomaly is applied in two different regions (dark: 20–50°N, light: 50–70°N). The curves in (a) and (c) are for the present-day climate, whereas curves (b) and (d) are for the LGM (figure from Ganopolsky and Rahmstorf, 2001).

changes in surface heat flux, which provides the strong changes in surface temperature over the North Atlantic sector (Fig. 10.26d). The temperature changes over Antarctica (Fig. 10.26e) are smaller and for the large freshwater amplitude event labelled 3 out of phase with those in the North Atlantic.

The combined effect of noise and a weak periodic forcing (0.01 Sv) was studied within the same model (Ganopolski and Rahmstorf, 2002). When only noise in the freshwater forcing is considered (with a standard deviation of 0.035 Sv), the transitions are irregularly spaced. However, when the weak periodic forcing is included, the combined effect with the noise leads to a synchronisation of the events (Fig. 10.27). Although the periodic forcing here is too weak to cause transitions from the cold state to the warm state, the noise is able to induce these transitions and to synchronise them according to the stochastic resonance mechanism. This type of stochastic resonance is not the same as that in double-well potential discussed in Section 10.5.1, as here there are no two stable equilibrium states.

In Ganopolski and Rahmstorf (2002), the rate of change of the freshwater flux is relatively large. Hence the excitable warm state they find is likely due to a transition

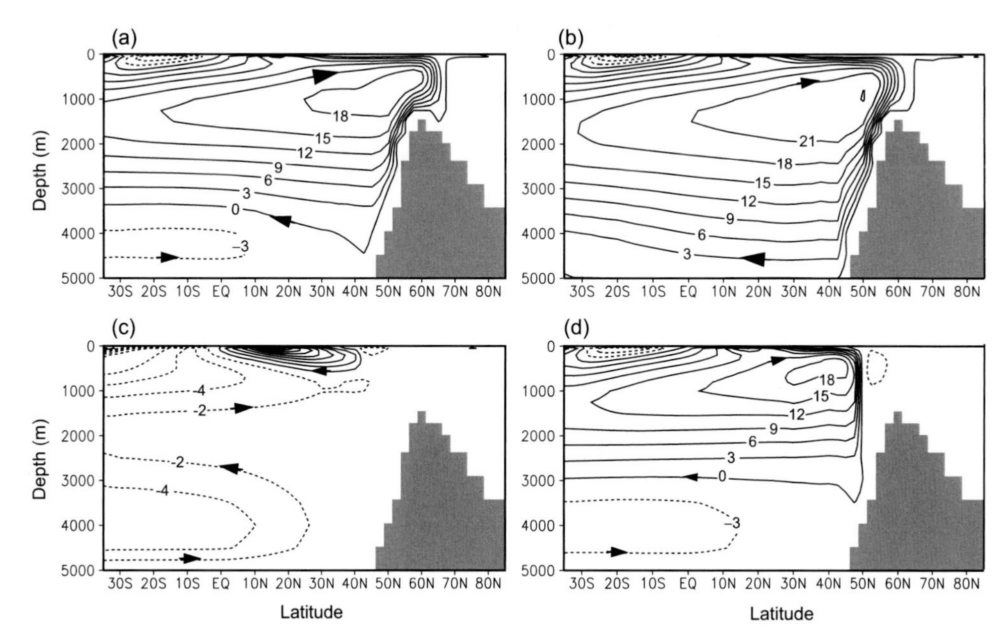

Figure 10.25 Patterns of the equilibrium Atlantic MOC (contours in Sv) in the Climber-2 model. (a) On- and (c) off-state of the Atlantic MOC in the present-day climate. (b) MOC of the excited warm state of the LGM climate and (d) MOC of the cold state of the LGM (figure from Ganopolsky and Rahmstorf, 2001).

to a transient excursion to an orbit associated with the deep-decoupling oscillation. The amplitude of the signal and its spatial pattern seems indeed to correspond to this type of oscillation. In this case, the effect of noise is to synchronise the transitions between the cold and the excited warm state, which is not the traditional stochastic resonance case but dynamics referred to as 'ghost' resonance (Braun et al., 2007). So in this model there likely is oscillatory variability associated with deep-decoupling oscillations. The periodic forcing is needed to cause this 'ghost' transition, and a possible origin of this forcing was suggested to be solar variability (Braun et al., 2005).

10.7 Synthesis

Here we summarise and discuss what dynamical systems theory has taught us about the dynamics of the D-O events, in particular about the time scale and spatial pattern. We restricted the discussion to mechanisms in which the Atlantic MOC was involved; mechanisms for this variability that do not involve changes in the MOC have been suggested as well (for an overview, see Clement and Peterson, 2008), but have major problems in explaining the dominant time scale and spatial pattern of the variability, and hence we have not discussed them here.

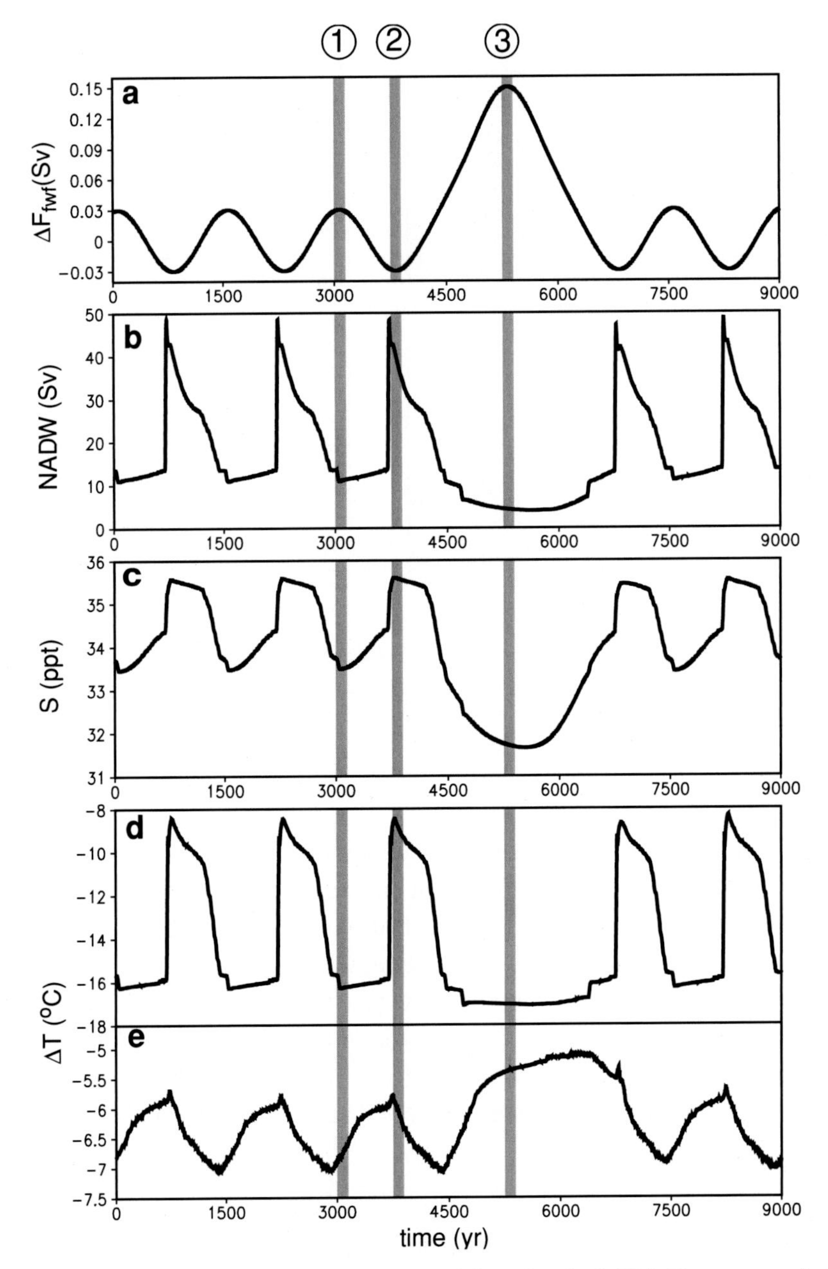

Figure 10.26 Response of the Climber-2 model, under the LGM climate to periodic freshwater forcing. (a) Freshwater forcing amplitude, (b) Atlantic MOC, (c) salinity at 60°N, (d) air temperature over the North Atlantic section and (e) air temperature over Antarctica (from Ganopolsky and Rahmstorf, 2001).

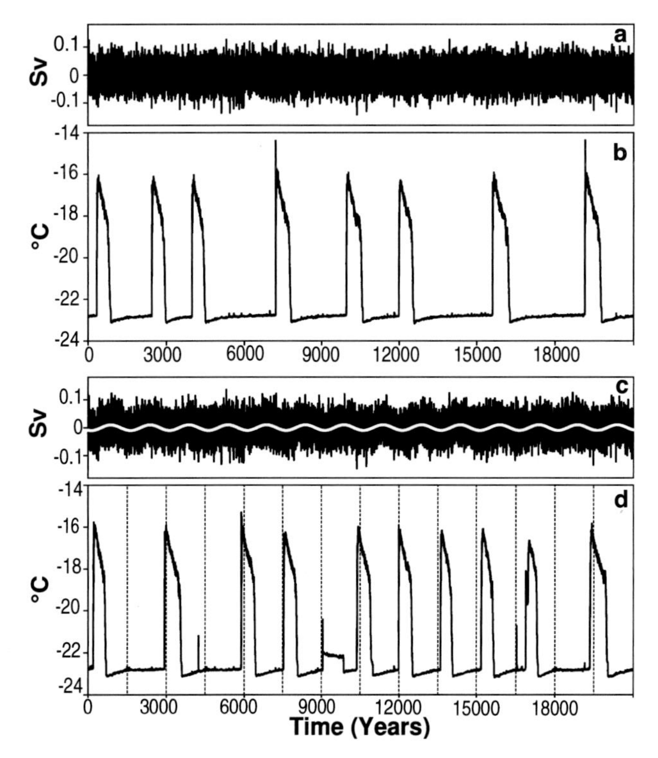

Figure 10.27 Response of the Climber-2 model to periodic freshwater forcing and noise. (a) Noisy freshwater forcing, (b) air temperature over the North Atlantic section under the forcing (a), (c) periodic plus noisy freshwater forcing and (d) air temperature over the North Atlantic section under the forcing (c) (figure from Ganopolski and Rahmstorf, 2002).

With the MOC as a central component in the mechanism, the hypotheses to explain the D-O events gather around two possibilities. The first ascribes these events to switches between equilibria or 'excitable states' of the MOC (Ganopolsky and Rahmstorf, 2001), where internal oscillations are not relevant, but an external periodic forcing is essential. The second involves interactions of noise with internal oscillatory behavior, which can be either loop oscillations or deep-decoupling oscillations.

The regularity of the D-O cycles in the first view results from an amplification of the periodic signal by the noise (Alley et al., 2003; Ganopolski and Rahmstorf, 2002). The waiting times between warm events in the Climber-2 model appear to show the particular features of the stochastic resonance (Fig. 10.28) and are similar to those obtained from the GRIP ice core data (Alley et al., 2003).

The immediate question, however, that arises from these results is related to the origin of the periodic forcing and, in particular, its 1,500-year time scale. The stochastic resonance theory of D-O events embraces the idea that it is related to solar forcing. Although there are no 1,500-year periodic signals known in the solar variability, there

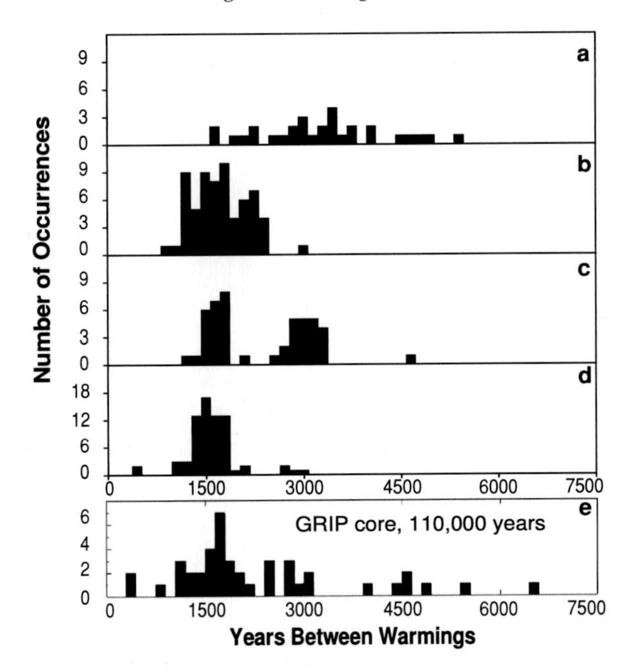

Figure 10.28 Histograms of the waiting time between warm events for 'noise' only situations (a,b) and for simulation with periodic forcing and noise (c,d). The standard deviation of the noise is 0.035 Sv in panels (a) and (c) and 0.05 Sv in the panels (b) and (d). Panel (e) was obtained from the GRIP ice core record (Alley et al., 2003) (figure from Ganopolski and Rahmstorf, 2002).

are the DeVries-Suess and the Gleissberg cycles, with periods of 210 and 87 years, respectively. Forcing the Climber-3 model with periodic perturbations in the freshwater flux with periodic perturbations on these frequencies can indeed provide fairly regular 1,470-year D-O events (Braun et al., 2005), even without the addition of noise (and, consequently, no stochastic resonance mechanism involved). An explanation using a conceptual model was provided in Braun et al. (2007). An open problem is, however, how such a solar forcing leads to a periodic forcing (of the same time scale) in the freshwater flux over the North Atlantic.

The second hypothesis relates the variability on millennial time scales to internal ocean modes, in particular, the overturning oscillations and the deep-decoupling oscillations. In this case, no external periodic forcing is needed, as the time scale of the variability is derived from internal instabilities. In Timmermann et al. (2003b), the coherence resonance arising through the effect of noise in the presence of a nearby limit cycle of a deep-decoupling oscillation gives rise to D-O type behaviour.

Timmermann et al. (2003b) also investigated the effect of a periodic large meltwater pulse such as associated with Heinrich events. In the presence of background noise in the freshwater flux, a Heinrich event leads to a large decrease of the MOC and a sequence of D-O events (Fig. 10.29). It is proposed that there may be a feedback

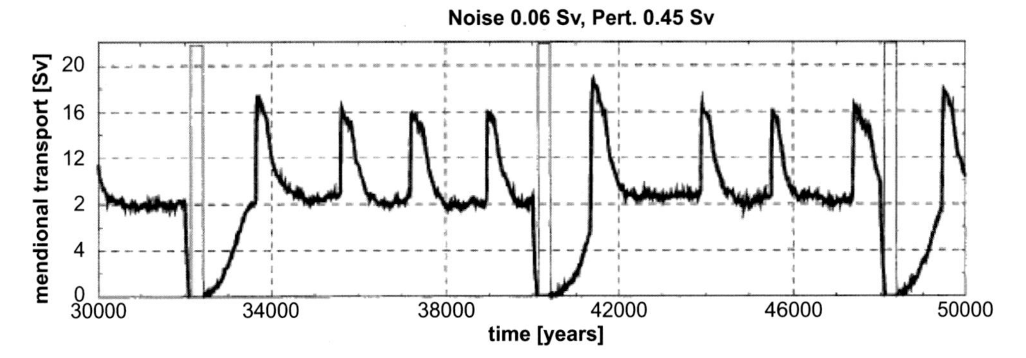

Figure 10.29 Response of the North Atlantic MOC to repeated 300-year-long melt-water pulses of 0.45 Sv in the presence of a 0.06-Sv amplitude noise forcing in the freshwater flux (figure from Timmermann et al., 2003b).

between the presence of the D-O events and the buildup of land-ice, eventually affecting the Heinrich event timing.

The coherence resonance theory, although attractive, also has problems. First, the fact that convection and vertical diffusion play a dominant role in the deep-decoupling oscillations is not very attractive, as these are elements of a climate model that are represented with minimal confidence. An oscillation mechanism based on the overturning oscillations (which are advectively controlled) would have much more preference. Second, the transition from the cold state to the excited warm state is very sensitive to parameters and is likely to be restricted to a small region in parameter space.

Coherence resonance involving the overturning oscillations has not been explored in detail. The difficulty is that the time scale of this type of oscillation depends on the spatial pattern of the MOC, being centennial for Atlantic-size basins and only millennial scale for the global ocean (Weijer and Dijkstra, 2003). In the series of studies by Sakai and Peltier (Sakai and Peltier, 1996, 1997, 1999), millennial oscillatory behaviour was found in a three-basin zonally averaged ocean model coupled to an energy-balance atmosphere model. For both the ocean-only model (Sakai and Peltier, 1996), as well as the coupled model (Sakai and Peltier, 1997), oscillatory regimes are found when the freshwater flux in the northern North Atlantic is changed. When this freshwater input exceeds a certain threshold, the dominant mode of variability appears to change from centennial scale variability into large-amplitude oscillations with millennial timescales. Unfortunately, no patterns of the MOC anomalies in the different basins are presented, and no sensitivity of the time scale to vertical diffusion is investigated. It is therefore difficult to assess which oscillation mechanism is driving the D-O type variability in these models.

11

The Pleistocene Ice Ages

Showing different possible worlds.
DGDGBD, Larry's world, Russ Freeman

The climate variability associated with the Pleistocene Ice Ages is one of the most fascinating puzzles in the earth sciences still awaiting a satisfactory explanation.

11.1 Phenomena

Much of our knowledge of the Pleistocene Ice Ages has been obtained from the analysis of isotope ratios and many other properties of ice cores and marine sediment cores. It is impossible to give even a modest overview of all of this work, so we limit ourselves here to the main results from these analyses.

Very detailed information on past temperatures on Earth has been obtained from marine benthic records. As explained in Section 10.1, the ratio δ^{18}O is a proxy of combined temperature and global ice-volume changes. In Fig. 11.1, a time series is shown of a composite δ^{18}O ocean sediment (benthic) record over the last 2 Myr (Lisiecki and Raymo, 2005). High values indicate a colder climate (and hence the axis is plotted positive downward), as the ocean is enriched with O^{18} during the colder conditions (and ice sheets are enriched with O^{16}). The time series in Fig. 11.1 indicates a cooling trend on which variability in ice cover is superposed. Analysis reveals that this variability is first dominated by a 41-kyr period, and after the so-called Mid Pleistocene Transition (MPT) at about 700 kyr, it is dominated by a 100-kyr period.

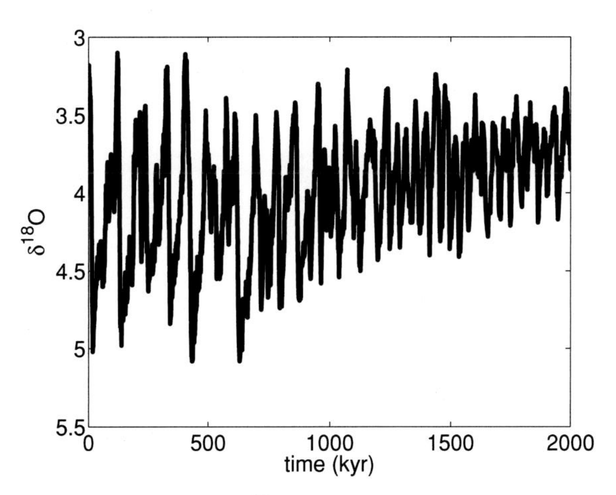

Figure 11.1 (a) The LR04 benthic $\delta^{18}O$ stack over the Pleistocene, constructed by the graphic correlation of fifty-seven globally distributed benthic $\delta^{18}O$ records versus time (in kyr ago) (data from Lisiecki and Raymo, 2005).

The European Project for Ice Coring in Antarctica (EPICA) has provided two deep ice cores in East Antarctica from which climate conditions can be reconstructed back to 800 kyr BP (Jouzel et al., 2007). From the reconstructed temperature anomaly time series (Fig. 11.2) on Antarctica, one observes the asymmetry between the slow glaciation and the rapid deglaciations. From the ice-core data, the marine-core records and other evidence, it is clear that the glacial-interglacial transitions have a global expression, with northern hemispheric temperatures varying approximately in phase with those in the southern hemisphere. One important player in the climate system responsible for the globalisation of these transitions is believed to be the atmospheric CO_2 concentration. A composite CO_2 record is shown in Fig. 11.2, created from a combination of records from the Dome C and Vostok ice cores. The atmosphere CO_2 concentration ($p^a_{CO_2}$) varies from about 180 ppm to 280 ppm during a glacial-interglacial transition and an optimal correlation with the $\delta^{18}O$ time series occurs near lag zero. As CO_2 is well mixed over these large time scales, there is no north-south asymmetry in $p^a_{CO_2}$. This record demonstrates that the atmospheric concentration of CO_2 did not exceed 300 ppmv for at least 800,000 years before the industrial era (Siegenthaler et al., 2005).

Glacial-interglacial transitions have affected all components of the climate system and induced relatively large-amplitude changes of many variables in these components. The results immediately lead to three fascinating main questions:

(i) Why did glacial-interglacial cycles appear in the Pleistocene?
(ii) Which processes in the climate system caused the glacial-interglacial changes in global mean temperature, $p^a_{CO_2}$, and ice sheet extent?
(iii) What caused the transition (the MPT) from the 41-kyr world to the 100-kyr world about 700 kyr ago?

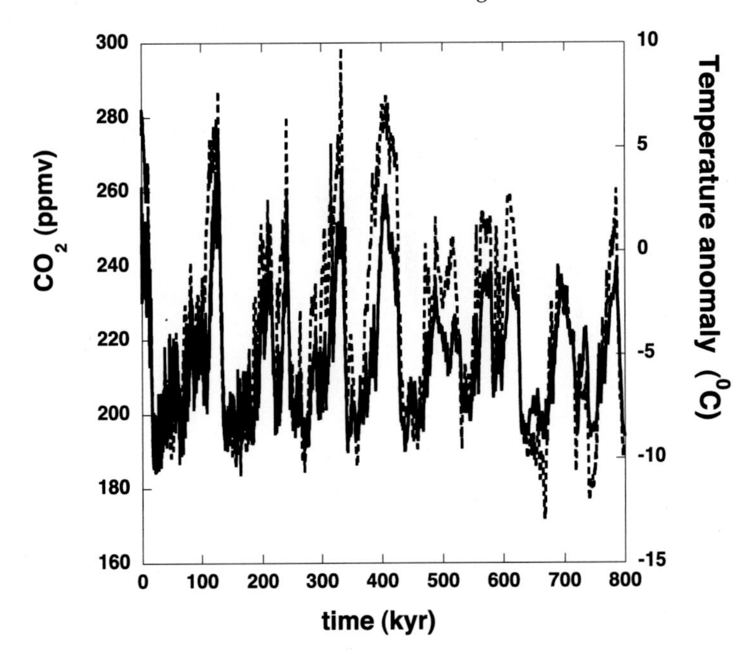

Figure 11.2 Reconstructed temperature and atmospheric CO_2 concentration ($p^a_{CO_2}$) from ice cores on Antarctica. The drawn curve is the temperature anomaly with respect to the mean temperature over the last 100 years. The dashed curve is a composite CO_2 record.

Any theory of the Pleistocene Ice Ages should at least contain a satisfactory answer to each of these main questions. It is clear that time scales are introduced by external forcing (next subsection), but the direct response to this forcing appears insufficient. Hence we explore in subsequent sections the relevant climate feedbacks and then review how these have been incorporated in theories for the Pleistocene Ice Ages. We finally discuss the amplifying role of CO_2 and conclude with a synthesis of the material presented.

11.2 The 'null hypothesis': Milankovitch theory

Approaches to answers on the Pleistocene Ice Ages problem have a very interesting history, which is nicely described in Imbrie and Imbrie (1986). A connection with the orbital characteristics of the Earth-Sun system was already made in the nineteenth century, but in the 1930s, Milankovitch (1930) suggested that glaciations occur when the insolation intensity is weak at high northern latitudes during summer. When insolation at 65°N is small, ice can persist throughout the year, leading to the growth of ice sheets. Favourable conditions for this to happen are when Earth's spin axis is less tilted and the aphelion (point in the orbit, where the Earth is farthest from the Sun) coincides with summer in the Northern Hemisphere.

The variations in insolation are caused by the changes in orbital characteristics of the Earth, and there are three types of motion relevant for the amount of radiation received at a particular point on Earth. First, the spin axis of the Earth undergoes precession, which induces a shift of the seasons along the orbit. About 12 kyr ago, the Earth was closest to the Sun in June and hence the seasonal contrast was larger in the Northern Hemisphere. One full cycle of precession has a period of 27 kyr, but coupled to the movement of the long end of the ellipse around the Sun (in 105 kyr) the net effect is a fluctuation in insolation with a period of 23 kyr. In addition, both the obliquity and the eccentricity of the Earth's orbit undergo periodic variations. The tilt angle changes in 41,000 years between 22° and 24°, leading to variations in seasonal contrast, and the eccentricity varies from 0.0 (perfect circle) to about 0.05, with periodicities of 100 kyr and 450 kyr.

Tilt changes are felt more strongly at high latitudes (if there was no tilt, the poles would receive no radiation), whereas the variations in eccentricity are felt over all latitudes. Precession and obliquity variations do not cause any substantial change in the annual mean insolation, but they give changes in the seasonal contrast. Precession changes have opposite effects in the Northern and Southern hemispheres, but the effects of obliquity changes are similar in both hemispheres. This results in an equatorially asymmetric effect of orbital changes on the insolation.

A time series of the insolation at 60°N (Fig. 11.3a) clearly shows variations of about 100 Wm^{-2} over the last 1 Myr. When a spectrum of this insolation curve (Fig. 11.3b) is compared with the spectrum (Fig. 11.3d) of a $\delta^{18}O$ record from an ocean sediment core over the last 1 Myr (as shown in Fig. 11.3c), the results are reason for excitement. There are clear signatures of the 19- and 23-kyr precession and of the 41-kyr obliquity variations of the Earth's orbit in the Ocean Drilling Program record. However, one raises one's eyebrows: at the 100-kyr time scale, there is hardly any forcing amplitude, whereas the climate signal in the $\delta^{18}O$ record has the largest amplitude. Comparing $\delta^{18}O$ records with the insolation time series (Raymo and Huybers, 2008), it is interesting that the amplitude of the ice-cover variations can be very large, whereas the insolation variation is very small. Furthermore, the variations in insolation provide no clue on the transition from the 41-kyr world to the 100-kyr world, as they have the same temporal characteristics through the transition.

Although it is clear that the orbital insolation variations must play a role, a simple linear forcing-response relation does not apply. The 100-kyr variations in insolation due to eccentricity are very weak (and it is the only forcing with a nonzero annual mean insolation signal), and the 41-kyr and 23-kyr provide only low-frequency variations on the seasonal variations, not on the annual mean insolation. Hence processes internal to the climate system must be crucial in the amplification of the orbitally induced insolation variations. Several feedbacks are available in the climate system to accomplish this, as we see in the next section.

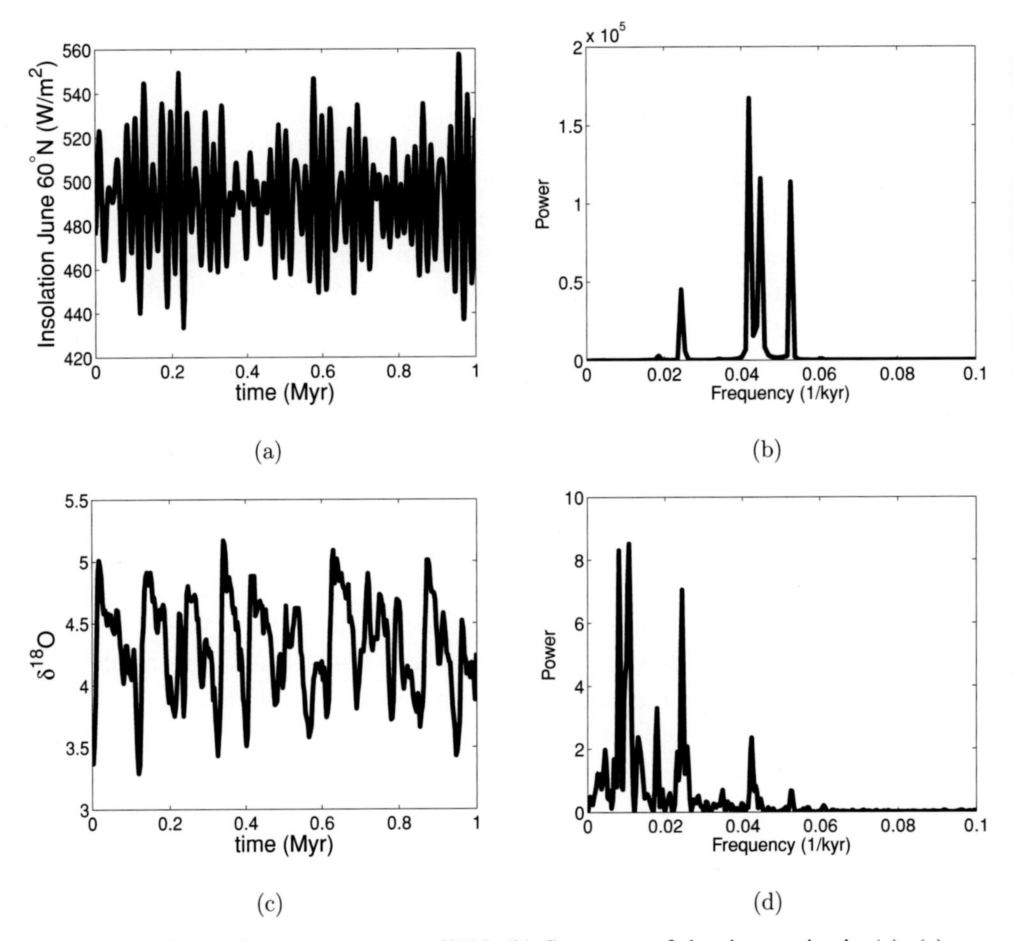

Figure 11.3 (a) June insolation at 60°N. (b) Spectrum of the time series in (a). (c) Time series of δ^{18}O at ODP677 (83°W, 1°N) over the last 1 Myr. (d) Spectrum of the time series in (c). Note that $1/23 = 0.043$, $1/19 = 0.053$, $1/41 = 0.024$ and $1/100 = 0.01$.

11.3 Potentially important feedbacks

In this section, we systematically analyse the interactions among atmosphere, ice, land and ocean for feedbacks that may be relevant to an understanding of the glacial-interglacial cycles. Conceptual models are used, and the main aim is to investigate whether the nonlinear interactions introduce instabilities that may lead to multiple equilibria or oscillatory behaviour in the climate system.

11.3.1 The ice-albedo feedback

We start with the atmosphere only and specifically with the zero-dimensional energy balance model, which was introduced in Example 6.2. With T being the global

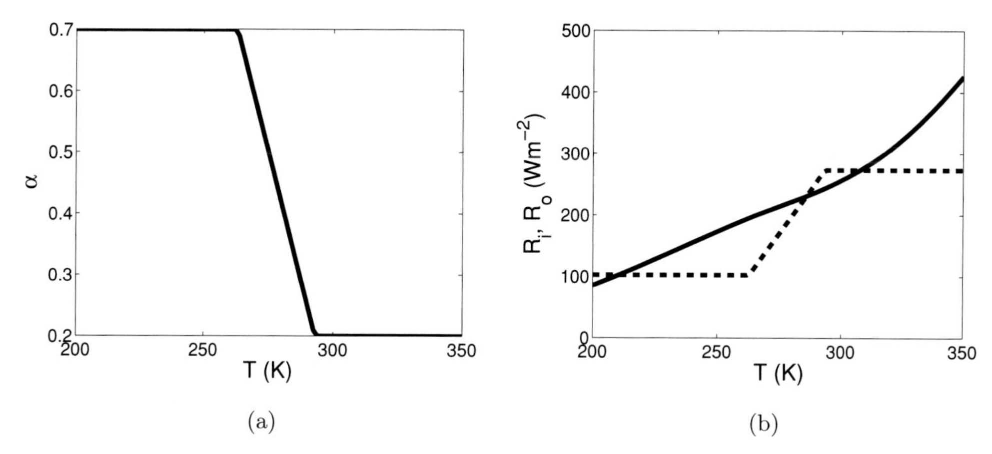

Figure 11.4 (a) Albedo α versus temperature T. (b) Plot of the incoming radiation R_i (dashed) and outgoing radiation R_o (drawn) versus temperature T for the model (11.1).

averaged temperature of the Earth, the governing equation is

$$\rho_a H_a C_p^a \frac{dT}{dt} = Q_0(1 - \alpha(T)) - \sigma\gamma(T)T^4, \tag{11.1}$$

where the incoming radiation $R_i = Q_0(1 - \alpha(T))$ and the outgoing radiation $R_o = \sigma\gamma(T)T^4$. Here, $Q_0 = \Sigma_0/4$, where Σ_0 is the solar constant; furthermore, σ is the Stefan-Boltzmann constant and α is the albedo.

The dependence of the albedo α on T represents the effect of temperature on both cloud and surface properties. When T is below a certain value T_l, the Earth will be ice-covered, and hence there will be a constant surface albedo α_l. Similarly, when $T > T_u$, there is no ice and a constant albedo α_u. In Sellers (1969), $\alpha(T)$ is assumed to vary linearly with T for $T_l < T < T_u$ (Fig. 11.4a). The dependence of $\gamma(T)$ should contain the greenhouse effect, and in Sellers (1969), a function of the form

$$\gamma(T) = 1 - m \tanh\left(\frac{T}{T_*}\right)^6, \tag{11.2}$$

is considered. Here m represents the cloud cover fraction and $T_* = 284$ K a reference temperature. When the temperature increases, $\gamma(T)$ decreases, mimicking the reduced outgoing long-wave radiation due to the presence of greenhouse gases and clouds.

A plot of the incoming and outgoing radiation is provided in Fig. 11.4b for the parameters given in Table 11.1. For these values, one can see that three equilibrium temperatures of (11.1) are possible, as the curves have three intersections. If we take the heat flux Q_0 as a control parameter, then the bifurcation diagram (T versus Q_0), shown in Fig. 11.5 for different values of α_l, shows the two stable equilibrium states (high temperature, hence 'ice-free' and low-temperature, hence 'ice covered'). Clearly, the multiple equilibrium regime depends strongly on the value of α_l. For larger α_l, the multiple equilibrium regime widens, whereas for decreasing α_l, the

Table 11.1. *Parameters of the energy balance model given by (11.1)*

Parameter	Meaning	Value	Dimension
ρ_a	air density	1.25	kg m^{-3}
H_a	atmospheric scale height	8,400	m
C_p^a	atmospheric heat capacity	10^3	J $kg^{-1}K^{-1}$
m	cloud fraction	0.5	–
Σ_0	solar constant	1,367	W m^{-2}
σ	Stefan-Boltzmann constant	5.67×10^{-8}	W $m^{-2}K^{-4}$
T_*	reference temperature	284	K
T_l	ice-covered temperature	263	K
T_u	ice-free temperature	293	K
α_l	ice albedo	0.7	–
α_u	ice-free albedo	0.2	–

saddle-node bifurcations merge into a cusp, and the multiple equilibrium regime disappears.

Although there is no explicit representation of an ice sheet in the model (11.1), its effect on the radiation properties was introduced by the temperature dependence of the albedo. Transitions between different steady states are possible due to finite amplitude perturbations in the multiple equilibrium regime; these transitions are purely

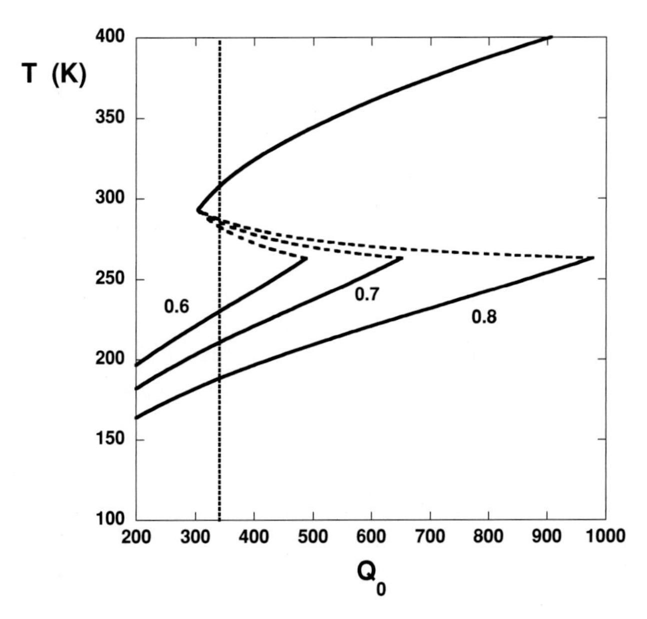

Figure 11.5 Bifurcation diagram of T versus Q_0 for three different values of α_l (provided as labels with the curves) for the model (11.1). Drawn (dashed) curves represent (un)stable steady states.

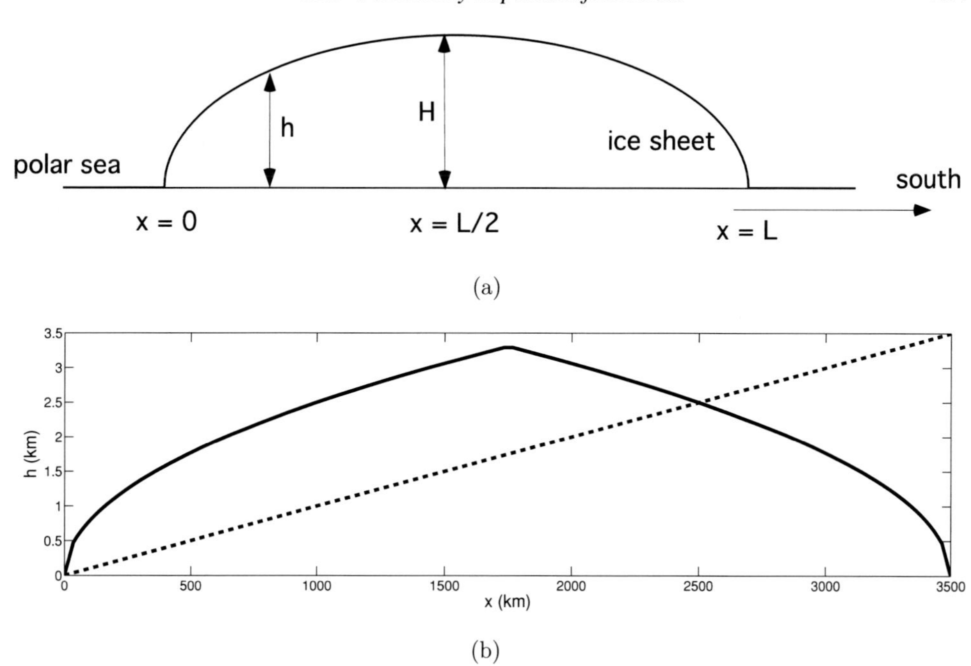

Figure 11.6 (a) Sketch of a continental ice sheet bounded by a polar sea at the northern end. (b) Plot of the solution $h(x)$ (drawn curve) from (11.4) for fixed $L_0 = 3500$ km. The dashed curve is the equilibrium line $h_E(x)$ defined through (11.6), with $x_P = 0$.

stationary; that is, there is no oscillatory behaviour. On the unstable branch, there is exponential growth of small perturbations to either the ice-free or ice-covered state. The physics of this behaviour is called the ice-albedo feedback. When the temperature increases, the albedo decreases (due to the melting of ice), and, consequently, the incoming radiation will increase. This will lead to a further increase in temperature, amplifying the original perturbation.

11.3.2 The height-mass balance feedback

The development of ice sheets is governed by nonlinear processes. Consider an idealised ice sheet as sketched in Fig. 11.6a. We choose the coordinates such that the point $x = 0$ corresponds to the boundary with the polar sea (such as the Arctic Ocean for the Greenland ice sheet) and the ice sheet has length L (Källén et al., 1979; Oerlemans and Van der Veen, 1984).

For a perfectly plastic material, there exists a yield stress τ_0 such that the rate of deformation would be zero (infinite) for horizontal stresses smaller (larger) than τ_0. The ice, consequently, will deform in such a way that the yield stress (at the bottom) is obtained. The horizontal stress balance in the ice is, in such an approach, given by

$$\tau_0 = \rho_i g h \frac{\partial h}{\partial x}, \tag{11.3}$$

Table 11.2. *Parameters of the conceptual ice sheet model given by (11.4), (11.5) and (11.6)*

Parameter	Meaning	Value	Unit
ρ_i	ice density	0.9	kg m^{-3}
L_0	ice cap length	3,500	km
σ	yield stress parameter	6.25	m
β	coefficient in the linear mass balance	10^{-3}	yr^{-1}
χ	coefficient in the linear mass balance	10^{-3}	–

where $h(x,t)$ is the ice thickness. From (11.3) we obtain

$$2h\frac{\partial h}{\partial x} = \frac{\partial h^2}{\partial x} = \frac{2\tau_0}{\rho_i g} \equiv \sigma \rightarrow h^2(x,t) = \sigma x + C(t).$$

On $[0, L/2]$, we use $h(0) = 0$, and it follows that $C = 0$, whereas on $[L/2, L]$, we use $h(L) = 0$, and it follows that $C = -\sigma L$. Using that $L/2 - |x - L/2| = L - x$ when $x > L/2$ and $L/2 - |x - L/2| = x$ when $x < L/2$, we can write the total solution as

$$h(x,t) = \sqrt{\sigma}\left(\frac{L(t)}{2} - |x - \frac{L(t)}{2}|\right)^{\frac{1}{2}}. \tag{11.4}$$

The maximum height of the ice sheet is indicated by $H(t)$ and given by $H(t) = h(L/2,t) = \sqrt{\sigma L(t)/2}$. A solution $h(x)$ is plotted for fixed $L = L_0$ in Fig. 11.6b for parameters as in Table 11.2, and, by construction, it is symmetric with respect to $x = L(t)/2$

Although $h(x,t)$ is a solution to the momentum equations, it must be noted that $L(t)$ is still unknown, as this quantity is determined from the continuity equation

$$\rho_i\frac{\partial h}{\partial t} = P_i - M, \tag{11.5}$$

where the right-hand side is the mass balance for the ice cap (the difference between accumulation P_i and ablation M). The mass balance depends on the distance from the polar ocean, represented by x_P, and the height of the ice sheet and often a linear relation of the form

$$P_i(x,t) - M(x,t) = G(x,t) = \rho_i\beta(h(x,t) - \chi(x - x_P)), \tag{11.6}$$

is used, where β and χ are constants. With increasing distance from the polar sea, the ablation increases. The equilibrium height $h_E(x)$, where $P_i - M = 0$ is hence given by $h_E(x) = \chi(x - x_P)$ and is also plotted as the dashed curve in Fig. 11.6a for $x_P = 0$. In this case, the line intersects the profile of $h(x,t)$ at some location $(x_E, h_E(x,t))$, and the mass balance G is positive for $h > h_E$ and negative for $h < h_E$.

It is now assumed that all the snow accumulating on the northern half of the ice sheet flows into the Arctic Ocean or melts close to it. The evolution of the ice sheet is then governed by the mass balance conditions on the southern half of the ice sheet. To determine the length $L(t)$ of the ice sheet, we substitute $G(x, t)$ into the continuity equation (11.5) and integrate over the southern part of the ice sheet. This gives (with $h = (\sigma(L - x))^{1/2}$)

$$\int_{L/2}^{L} \frac{\partial h}{\partial t}\, dx = H\frac{dL}{dt} = \int_{L/2}^{L} \left[\beta(\sqrt{\sigma(L-x)} - \chi(x - x_P))\right] dx, \qquad (11.7)$$

and hence the equation for L becomes

$$\frac{dL}{dt} = F_1\sqrt{L} + F_2 L + F_3 L\sqrt{L}, \qquad (11.8)$$

with $F_1 = \sqrt{2/\sigma}\,\beta\chi x_P$, $F_2 = 2\beta/3$ and $F_3 = -3\sqrt{2/\sigma}\,\beta\chi/4$. Equilibrium ice sheets (note the mass balance is not in equilibrium at every location but only integrated) are found from $dL/dt = 0$, which leads to

$$F_1\sqrt{L} + F_2 L + F_3 L\sqrt{L} = 0, \qquad (11.9)$$

and hence the solution $L = 0$ (no ice sheet) is always an equilibrium solution. For $x_P = 0$, there is only one additional solution $L = F_2^2/F_3^2$, but for $x_P < 0$, there are two additional solutions given by

$$L_{1,2} = \frac{F_2^2 - 2F_1F_3 \pm F_2\sqrt{F_2^2 - 4F_1F_3}}{2F_3^2}. \qquad (11.10)$$

Note that the equilibria do not depend on β, but of course the linear stability of each solution does. The bifurcation diagram showing the equilibrium length of the ice sheet versus negative x_P therefore displays a multiple equilibrium regime (Fig. 11.7). The first saddle node occurs at $x_P = 0$, $L = 0$, whereas there exists a second saddle-node bifurcation at a negative value of x_P, where $F_2^2 - 4F_1F_3 = 0$.

The stability of the equilibrium states can be easily calculated by a linear stability analysis, and it turns out that one of the branches (with the small ice cap) is unstable. The transition from one equilibrium to another is also brought about by a positive feedback, which is usually referred to as the height-mass balance feedback (or ice-elevation feedback (Ruddiman, 2001)). On the small ice sheet, assume that a perturbation lengthens it, then also its height H (maximum at $L/2$) increases and hence the surface extends more above the equilibrium line (cf. Fig. 11.6). As a consequence, the mass balance becomes more positive, which induces the growth of the ice sheet amplifying the original perturbation.

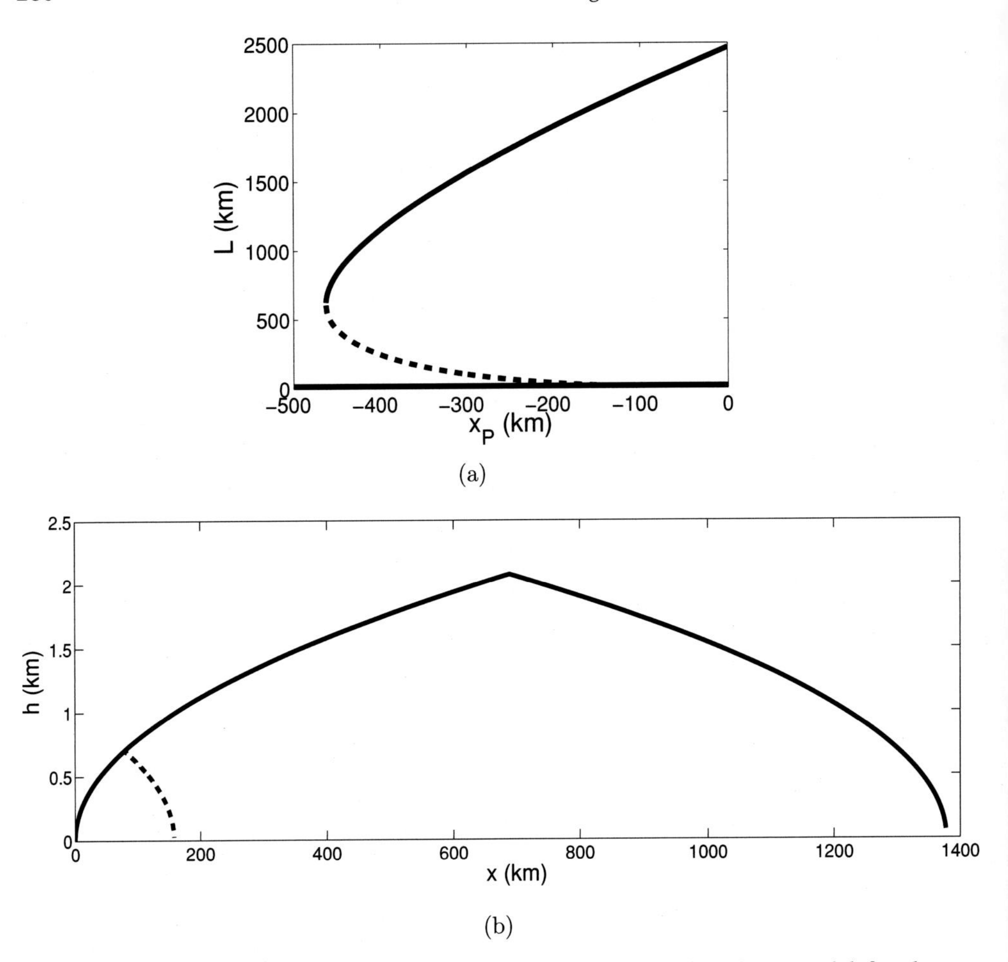

Figure 11.7 (a) Bifurcation diagram (L versus x_P) of the ice sheet model for the parameters as in Table 11.2. (b) Nontrivial ice sheet solutions (drawn: stable and dashed: unstable) for $x_P = -250$ km.

11.3.3 The load-accumulation feedback

A third feedback that has been identified is the so-called load-accumulation feedback, and it involves the coupling of the ice sheet to the solid Earth below (Fig. 11.8). With the deforming bedrock, the mass balance equations (11.5) and (11.6) now become

$$\frac{\partial h}{\partial t} = \beta(h(x,t) + b(x,t) - \chi(x - x_P)), \tag{11.11}$$

where $b < 0$ is the depression of the bedrock. A commonly used model for deformation of the crust is given by

$$\rho_b \frac{\partial b}{\partial t} = -\frac{1}{\tau_b}(\rho_i h + \rho_b(b - b_0(x))), \tag{11.12}$$

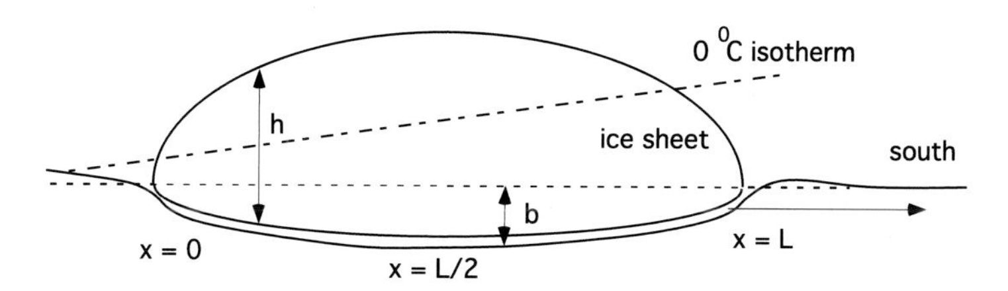

Figure 11.8 Sketch of an ice sheet including bedrock deformation. The total thickness of the ice sheet is h, and the bedrock is below the reference surface ($b < 0$).

where $b_0(x)$ is a bedrock profile in equilibrium with the overlying atmosphere when there is no ice, τ_b is a relaxation time scale and ρ_b is the density of the bedrock.

From the equations (11.12) and (11.11) for b and h, respectively, the evolution of small perturbations $(\tilde{h}, \tilde{b})$ on an equilibrium state $(\bar{h}, \bar{b})$ are

$$\frac{\partial \tilde{h}}{\partial t} = \beta(\tilde{h} + \tilde{b}), \tag{11.13a}$$

$$\frac{\partial \tilde{b}}{\partial t} = -\frac{1}{\tau_b}\left(\frac{\rho_i}{\rho_b}\tilde{h} + \tilde{b}\right). \tag{11.13b}$$

The height-mass balance feedback is captured by the first term in the right-hand side of (11.13a) when $\tilde{b} = 0$. The load-accumulation feedback is represented by the coupling of both equations. When $\tilde{h} > 0$, then it follows that $\tilde{b}$ will decrease (become more negative). Indeed, when an ice sheet grows, it sinks into the bedrock below over a time scale τ_b. This will lead to a decrease in the height of the ice sheet and hence a decrease in accumulation, which leads to a decrease in $\tilde{h}$ and hence oscillatory behaviour. By considering only the coupling terms, we deduce that

$$\frac{\partial^2 \tilde{h}}{\partial t^2} \sim -\frac{\beta}{\tau_b}\frac{\rho_i}{\rho_b}\tilde{h}, \tag{11.14}$$

which shows the explicit oscillatory behaviour arising through the phase difference in the response of the bedrock and the ice sheet.

11.3.4 The precipitation-temperature feedback

The feedbacks in the previous two subsections were due to internal dynamics in the ice sheet (the height-mass balance feedback) or due to the coupling of the ice sheet and the underlying bedrock (the load-accumulation feedback). An additional feedback, the precipitation-temperature feedback, may arise due to the interaction of an ice sheet with the overlying atmosphere (Källén et al., 1979).

To illustrate this feedback, we use the equation for the atmospheric temperature T of the energy balance model (11.1), in which the long-wave radiation component has been linearised around a temperature T_0, that is,

$$c_T \frac{dT}{dt} = Q_0(1 - \alpha(T, L)) - k(T - T_0), \qquad (11.15)$$

where $c_T = \rho_a H_a C_p^a$. The albedo α depends on the extent of the ice sheet and is now given by

$$\alpha(T, L) = \gamma \alpha_{\text{land}} + (1 - \gamma)\alpha_{\text{ocean}}, \qquad (11.16a)$$

$$\alpha_{\text{land}} = \alpha_0 + \alpha_1 L, \qquad (11.16b)$$

and α_{ocean} is again piecewise continuous as in Fig. 11.4a.

The accumulation P_i in the mass balance of the ice sheet now depends on the global mean temperature T (Ghil and Le Treut, 1981). Increasing temperatures imply that evaporation is also larger, and hence snowfall on the ice sheet increases. When we now write the equations for perturbations $(\tilde{h}, \tilde{T})$; we find for the mass balance equation and the energy balance model the equations

$$c_T \frac{d\tilde{T}}{dt} = -Q_0 \tilde{\alpha} - k\tilde{T}, \qquad (11.17a)$$

$$\frac{d\tilde{h}}{dt} = \tilde{P}_i. \qquad (11.17b)$$

where $\tilde{\alpha}$ is the change in albedo and $\tilde{P}_i$ the change in accumulation (both depend implicitly on $\tilde{h}$ and $\tilde{T}$). Here the ablation term M in the mass balance is assumed to be constant.

Suppose there is an increase in temperature $\tilde{T} > 0$, which induces an increase in accumulation $\tilde{P}_i \sim \tilde{T} > 0$. The ice sheet grows ($\tilde{h} > 0$), and hence its length increases ($\tilde{L} \sim \tilde{h} > 0$). The albedo therefore increases as $\tilde{\alpha} \sim \tilde{L} > 0$, and hence the temperature decreases due to ice albedo feedback (represented by the first term on the right-hand side of [11.17a]). The oscillatory behaviour arises due to the time lag of the response of the ice-sheet changes, and the response of the accumulation to the temperature changes. The oscillatory behaviour can again be illustrated by

$$\frac{d^2\tilde{h}}{dt^2} \sim \frac{d\tilde{P}_i}{dt} \sim \frac{d\tilde{T}}{dt} \sim -\tilde{\alpha} \sim -\tilde{L} \sim -\tilde{h}, \qquad (11.18)$$

and the frequency of oscillation is determined by the product of all proportionality coefficients. For the representation of the precipitation-temperature feedback, a coupled model of ice-sheet dynamics and atmospheric temperature is needed.

11.4 Basic theories of interglacial-glacial cycles

In the next subsections, an attempt is made to systematically present the proposed theories of the ice-age variability from a stochastic dynamical systems perspective.

The orbital variations in insolation over the globe are at the heart of all these theories. Traditionally, the June insolation at 65°N (such as shown in Fig. 11.3a) has been used as the most important part of this forcing, because this determines whether snow will be left at the end of the Northern Hemisphere Summer season. Below, we refer to this orbital component of the insolation as the M-forcing.

The theories in the following subsections differ significantly with regards to the importance of the M-forcing in answering the main questions posed at the end of Section 11.1. In Section 11.4.1, we describe the quasi-equilibrium response theory of the climate system to the M-forcing, without any role for multiple equilibria and/or internal oscillations. Here the time scales of the climate response are directly related to those of the M-forcing. The following subsection (Section 11.4.2) deals with theories in which switches between multiple steady states induce time scales in the response, which are not directly related to the M-forcing. Theories in which internal oscillations provide additional time scales in the response, interacting with the M-forcing, are described in Section 11.4.3.

11.4.1 Quasi-equilibrium response

In these theories, the behaviour of the climate system is considered as a transient deviation from a single steady equilibrium due to the M-forcing. One of the simplest models that illustrates this idea is the atmospheric energy balance model as described in Section 11.3.1. With the M-forcing present, we write the equation for the global mean temperature T as

$$c_T \frac{dT}{dt} = Q_0(1 + \mu M(t))(1 - \alpha(T)) - \sigma\gamma(T)T^4, \qquad (11.19)$$

where $c_T = \rho_a H_a C_p^a$, Q_0 again represents the mean insolation and $\mu M(t)$ the M-forcing, with μ being its amplitude and $M(t)$ its time dependence. When $\mu = 0$, the model temperature will equilibrate to the equilibrium temperature $\bar{T}$ within a few years (Fig. 11.9a). For small-amplitude time-dependent forcing, the trajectory of this model will fluctuate around $\bar{T}$ (Fig. 11.9b).

The temperature fluctuations ΔT around this equilibrium are determined by the sensitivity of the value of the equilibrium versus the variations in the forcing $\Delta\mu$ and hence

$$\Delta T \approx \frac{\partial \bar{T}}{\partial \mu} \Delta\mu. \qquad (11.20)$$

From this model, it is found that a 1% change in solar insolation leads to about a 1°C temperature change. The dominant amplitude of eccentricity variations is at a period of about 400 kyr, whereas the next strongest variations occur on about a 100-kyr time

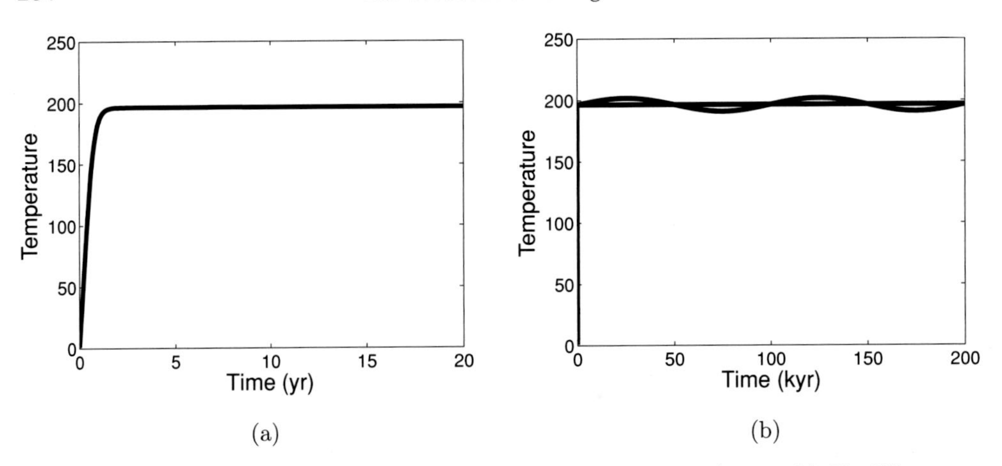

Figure 11.9 Central idea of the quasi-equilibrium response theory. (a) Equilibration of the temperature T (in K) of the energy balance model (11.19) for $\alpha_u = 0.6$ and $Q_0 = 200$ Wm^{-2} (the left endpoint of the bifurcation diagram in Fig. 11.5). (b) Time-dependent temperature solution for $\mu = 0.1$ and $M(t) = \sin 2\pi t/\tau$ with $\tau = 100$ kyr.

scale. The variations in eccentricity do modify the globally and annually averaged amount of insolation, but the amplitudes are very small, in the order of 0.1% of the solar constant. The variations in eccentricity can, therefore, only account for a temperature variation of at most 0.1 K. When considering many other processes (e.g., ice sheets, bedrock) in the direct response to the M-forcing, the sensitivity does not increase enough to explain the climate signal (Ghil, 1994).

Quasi-equilibrium response theories have problems with explaining not only the amplitude of the glacial-interglacial variations but also with the dominant frequency of the cycles. As mentioned in Section 11.1, before about 700 kyr BP the glacial oscillations followed a 41-kyr cycle, which would attribute a dominant role to obliquity variations. During this time, however, the amplitude of the precession insolation variations was much larger than that of the obliquity signal, as can be seen in Fig. 11.3b. The quasi-equilibrium theories try to cope with this problem by referring to other aspects in the orbital forcing than simply the M-forcing. Huybers (2006) argues that the land-ice mass balance depends on whether the temperature is above or below the freezing point, and hence the integrated summer insolation is a more relevant climate forcing on glacial-interglacial time scales than the M-forcing. Indeed, this parameter shows a dominant 41-kyr variability. Raymo et al. (2006) have suggested that the dominant 41-kyr occurred because of the cancellation of the antisymmetric (north-south) climate response to precession (insolation) variations in the $\delta^{18}O$ record. Neither of those adaptations to the basic quasi-equilibrium theory, however, can explain the switch from the 41-kyr cycle to the 100-kyr cycle associated with the MPT.

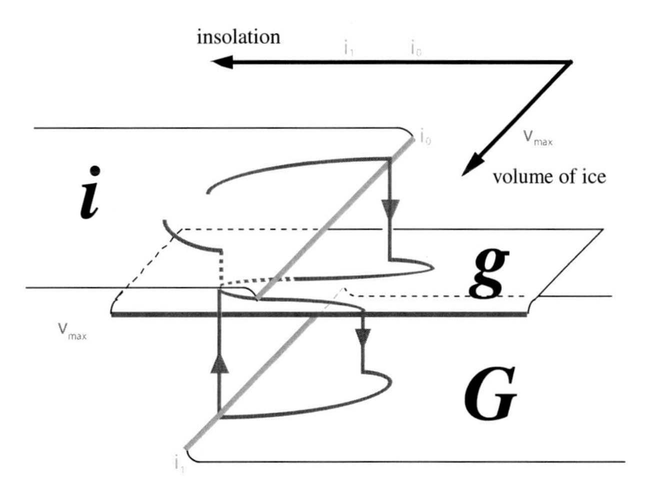

Figure 11.10 Sketch of the transitions between the three different climate states as in the model of Paillard (1998).

11.4.2 Switches between different climate equilibria

As we have seen in Section 11.3.1, a simple energy-balance model can have multiple equilibrium states under steady forcing (Fig. 11.5). When the forcing becomes time-dependent (cf. the M-forcing), we may ask whether switches between these equilibria are possible and, if so, whether they introduce additional response time scales and amplitudes into the climate system.

A remarkably simple model that illustrates the new behaviour introduced by the presence of multiple equilibria is that due to Paillard (1998). In his model, illustrated by Fig. 11.10, there are three equilibrium states in the climate system: an interglacial state **i**, a weak glacial state **g** and a strong glacial state **G**. Transitions from **i** to **g** occur when the summer insolation at 65°N drops below a value i_0. Furthermore, transitions from **g** to **G** occur when the ice volume V increases above some critical level V_c. Finally, a transition from **G** to **i** occurs when the insolation increases above a level i_1. These are the only transitions that are allowed in this model.

The change in ice volume is described as

$$\frac{dV}{dt} = \frac{V_R - V}{\tau_R} - \frac{F}{\tau_F}, \tag{11.21}$$

where R refers to the actual state (**i**, **g** or **G**). In Paillard (1998), the choice $V_{\mathbf{g}} = V_{\mathbf{G}} = V_{\max} = 1$ is made and $V_{\mathbf{i}} = 0$. The function F is a truncation of the M-forcing taking into account the lower sensitivity of ice volume during colder periods. The quantities $\tau_R(R = \mathbf{i}, \mathbf{g} \text{ or } \mathbf{G})$ and τ_F are time constants.

One of the results of this model is shown in Fig. 11.11 where the ice volume V is plotted as the drawn curve in the middle panel. The overall agreement between model and observations is remarkable considering the simplicity of the model. By allowing

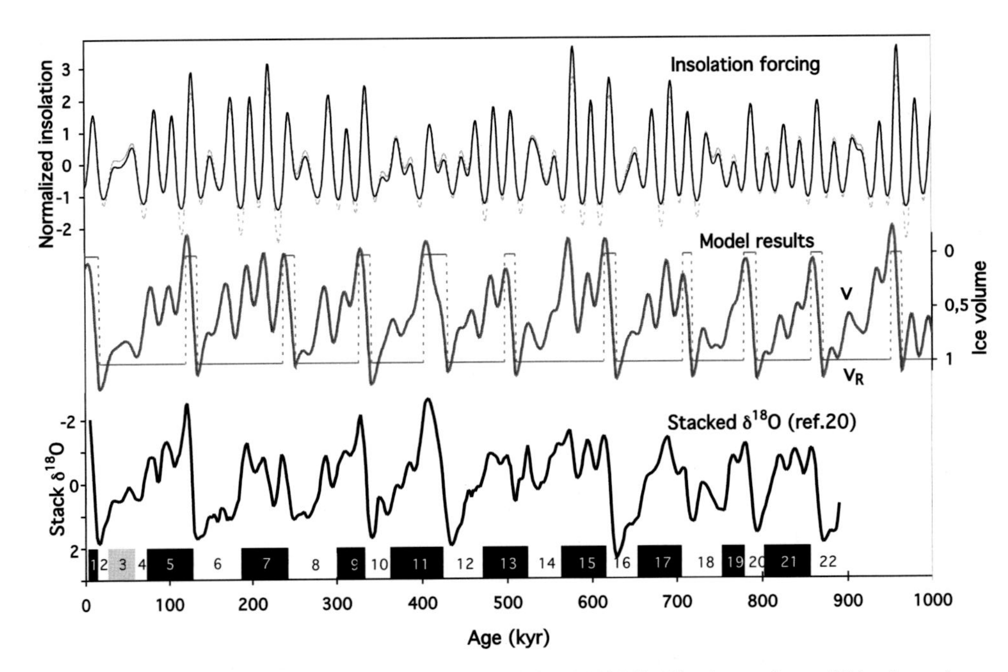

Figure 11.11 Results from the model in Paillard (1998). The ice volume V is plotted as the drawn curve in the middle panel. The value V_R is also plotted (dashed middle curve) to differentiate between **i** regimes ($V_R = 0$) and g or G regimes ($V_R = 1$). The model enters a G regime when the ice volume exceeds the value $V_{\max} = 1$. The forcing F used here (continuous line) is a smoothed truncation of the insolation associated with the M-forcing (dashed curve in the upper panel). A δ^{18}O record is plotted below for comparison. The thresholds are, in variance units, $i_0 = -0.75$, $i_1 = 0$. The time constants are $\tau_i = 10$ kyr, $\tau_G = \tau_g = 50$ kyr and $\tau_F = 25$ kyr.

for a slight linear trend in $V_{\max}$ (from 0.35 to 1.1) and one in the insolation forcing (3 W m^{-2} Myr^{-1}), Paillard (1998) also finds the MPT at around the correct time, and the spectra of his model and typical δ^{18}O data correspond reasonably well to each other.

When multiple equilibria, a weak periodic forcing and noise are present, there is also the possibility of stochastic resonance (as discussed in the previous chapter). In fact, the discovery of stochastic resonance actually occurred (Benzi et al., 1982, 1983; Nicolis, 1982) while trying to explain the 100-kyr dominant glacial cycles using the energy balance atmosphere model (11.19). The central element of this theory is the amplification of the weak periodic eccentricity component of the M-forcing by noise in the presence of multiple equilibria.

11.4.3 Internal oscillations

We have already seen in Sections 11.3.3 and 11.3.4 that coupled atmosphere-cryosphere and cryospheric-lithosphere processes can give rise to internal oscillations.

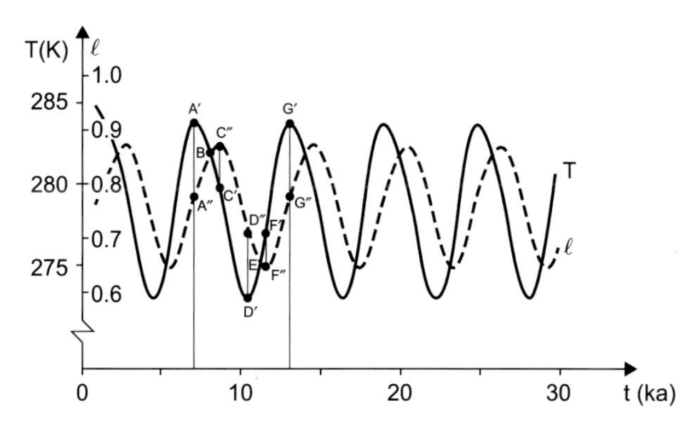

Figure 11.12 Internal oscillation in the temperature T (solid curve) in K and normalised ice cap length $l = L/L_*$ (dashed curve) with $L_* = 1{,}300$ km in an atmosphere-cryosphere model such as discussed in Section 11.3.4 (Ghil and Childress, 1987).

The question is, of course, whether these processes can also give rise to sustained oscillations (through Hopf bifurcations). If so, the time scales and amplitude ranges of these oscillations with 'realistic' values of the parameters may be of interest regarding the glacial cycles.

Interactions among atmosphere, land ice and bedrock

A typical result for a coupled atmosphere-cryosphere model (as discussed in Section 11.3.4) is shown in Fig. 11.12 under steady forcing (Ghil and Childress, 1987). Here the temperature T (in K) and the normalised ice sheet length $l = L/L_*$ are plotted versus time t (in kyr). A sustained oscillation is indeed present under steady forcing, which indicates that the time scale arises from internal nonlinear processes through a Hopf bifurcation. Note, however, that the time scale of the oscillation is about 6–7 kyr and not 100 kyr, which shows that the processes captured in this simple model are not able to generate an internal time scale of 100 kyr.

The mechanism of the oscillation in this system can be described as follows (Fig. 11.12). When T is near its maximum (point A′), snow accumulation increases, the length L. As a result, the albedo increases, and hence the temperature drops. When the temperature drops, snow accumulation decreases, and the ice sheet reaches the maximal extend (point C″) while the temperature is still decreasing. As the ice sheet shrinks, the albedo decreases and the minimum temperature is reached (point D′). The temperature now increases again, leading also to an increased accumulation, and the cycle starts again with negative sign. This mechanism of the oscillation is the combined effect of the ice-albedo feedback and the precipitation-temperature feedback as described in Section 11.3.4.

In Ghil and Le Treut (1981), the behaviour of a coupled atmosphere-cryosphere-lithosphere model is investigated. Compared with the lithosphere model in Section

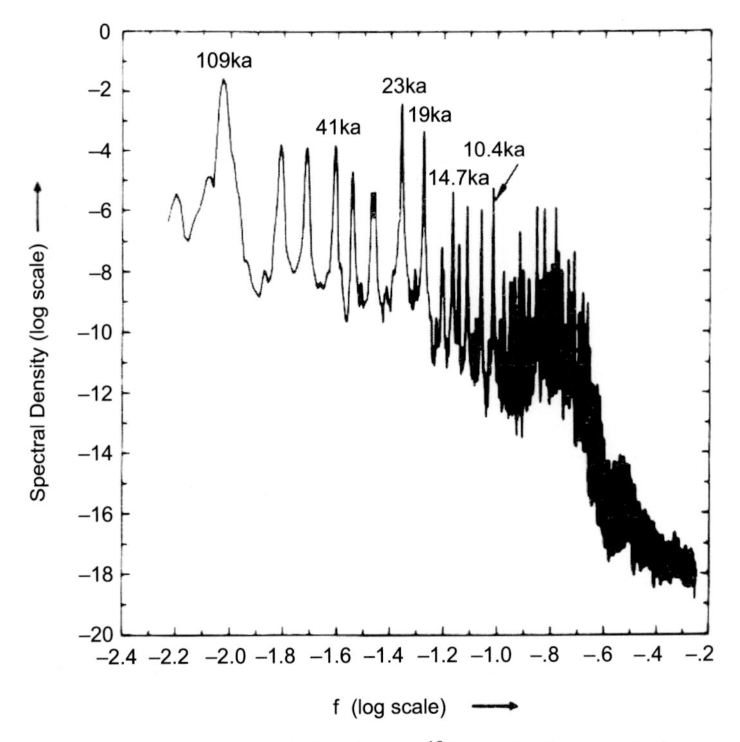

Figure 11.13 Power spectrum of simulated δ^{18}O as in the coupled atmosphere-cryosphere-lithosphere model of Le Treut et al. (1988) under an M-type forcing.

11.3.3, $b_0 = 0$ and $b(x)$ is assumed to have a specific profile (as a function of L) with a time-dependent amplitude ζ. In that case, the dynamical system reduces to three ODEs for T, L and ζ. Once again, oscillatory behaviour is found through Hopf bifurcations, with comparable amplitude in T and L, but the period remains in the range of 6–7 kyr, despite the long time scale τ_b in the lithospheric model (see Section 11.3.3).

In Le Treut et al. (1988), the model in Ghil and Le Treut (1981) is forced by an M-type forcing in the following way. The insolation is varied according to

$$Q(t) = Q_0 \left[1 + \delta_4 \sin \frac{2\pi}{\tau_4} + \delta_5 \sin \left(\frac{2\pi}{\tau_5} + \psi_5 \right) \right], \tag{11.22}$$

where $\tau_4 = 100$ kyr and $\tau_5 = 400$ kyr are the dominant periods of the eccentricity variations. Also the accumulation-to-ablation ratio is forced on time scales $\tau_1 = 19$ kyr, $\tau_2 = 23$ kyr and $\tau_3 = 41$ kyr. When the model is forced under appropriate values of the amplitude parameters δ_4 and δ_5, a spectrum as in Fig. 11.13 is found. The 100-kyr period here arises due to a combination of difference tone excitation and nonlinear resonances of the external frequency of the forcing and the internal frequency of oscillation (cf. Section 8.6.2).

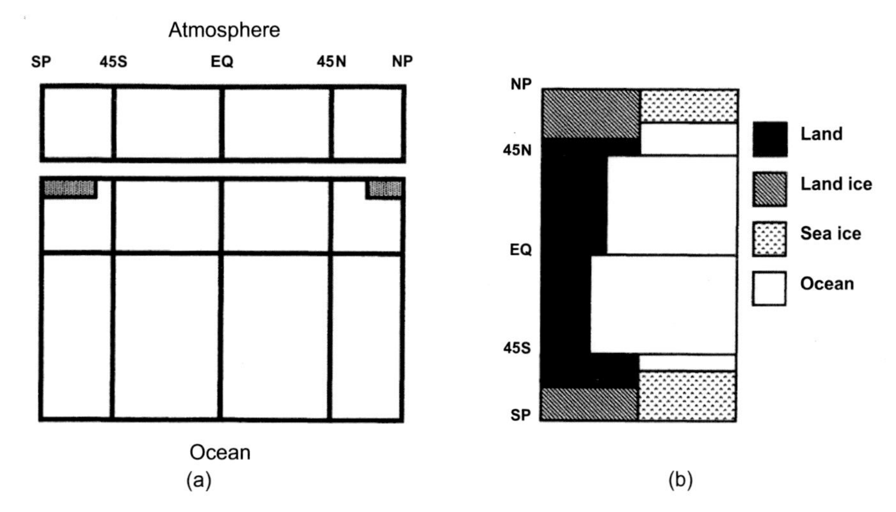

Figure 11.14 The box model used in Gildor and Tziperman (2000) with (a) a meridional cross section and (b) a top view.

In summary, interactions of ice-albedo, load-accumulation and temperature-precipitation feedbacks in an idealised climate model display self-sustained oscillations with a dominant time scale of about 6–7 kyr but not with a 100-kyr period under steady forcing. When forced with orbital variations, a response at 100 kyr may arise through a difference frequency excitation here of the two precession forcing frequencies (Ghil, 1994).

The sea-ice switch

A relatively simple model, where an internal oscillation exists with a time scale of about 100 kyr, is that of Gildor and Tziperman (2000). The model is a box model of the climate where a similar atmospheric-cryospheric model as in Section 11.3.3 is coupled to an ocean model with a sea-ice component (Fig. 11.14).

The ocean is represented by four surface boxes and four deep boxes whose latitudinal extends are South Pole to 45°S, 45°S to equator, equator to 45°N and 45°N to North Pole. The meridional overturning circulation is buoyancy driven; advection and diffusion of temperature and salinity are balanced by surface fluxes from the atmosphere and land ice components. The atmospheric model consists of four boxes representing the same latitude bands as the ocean boxes. The surface below the atmospheric boxes is a combination of ocean, land, land ice and sea ice, each with its specified albedo.

The averaged temperature in the box is calculated on the basis of the energy balance in the box, consisting of incoming solar radiation, using box albedo and with the possibility to include Milankovitch variations, outgoing long-wave radiation, heat flux into the ocean and meridional atmospheric heat transport. The meridional atmospheric moisture transport is proportional to the meridional temperature gradient and the humidity of the box to which the flux is directed. Precipitation is calculated as the convergence of the moisture fluxes.

The land ice model assumes a simple, zonally symmetric ice sheet of perfect plasticity. The ice sheet grows due to precipitation in the box, which is assumed to be distributed evenly. It decreases as a result of ablation, evaporation and calving processes. This sink term is taken to be constant in time, as it is expected that the source term dependence on temperature dominates that of the sink term (Gildor and Tziperman, 2001). The sea ice controls, via its albedo and insulating effects, the atmospheric moisture fluxes and precipitation that enable the land ice sheet growth. This control and the rapid growth and melting of the sea ice allow it to rapidly change the climate state from a growing to a retreating ice sheet phase.

In Gildor and Tziperman (2001), results of the model forced by constant annual mean insolation (no seasonal or M-forcing) are presented to assess the degree to which the internal processes (particularly sea ice) may control glacial cycle variability. A typical result (Fig 11.15) for (near-)standard values of the parameters in Gildor and Tziperman (2001) shows that oscillations with a time scale of about 100 kyr are found.

The proposed mechanism of the variability is as follows (with reference to Fig. 11.15a through Fig. 11.15h, from −160 kyr to −60 kyr): the cycle begins with land ice slowly growing because precipitation in the polar box exceeds the rate of melting (ablation); as land ice increases, albedo also increases and leads to cooling of atmospheric and oceanic temperatures. Eventually a threshold sea-surface temperature is reached, and sea ice starts to form. Within a very short period of time, sea-ice coverage reaches its maximum extent (land ice is at a maximum at this time also). Due to the extensive sea-ice cover in the polar box, the moisture flux from the northern ocean is essentially cut off, and precipitation is significantly reduced. With reduced precipitation, ablation exceeds accumulation, and the land and sea-ice extent begins to decrease (deglaciation has begun). Then, as land ice decreases, the albedo also decreases, which leads to warmer atmospheric and oceanic temperatures; eventually, all the sea ice melts and the pre–ice-age moisture fluxes are restored. With the increased fluxes, another cycle begins as precipitation exceeds ablation and the land ice can grow. Hence, in this model, rapid sea ice growth and decay can act as a switch for the precipitation-temperature feedback affecting the growth and/or decay of ice sheets.

The 100-kyr time scale is due to the growth and decay of ice sheets, which is coupled with relatively rapid sea ice changes. When V is the volume of the ice cap, the growth time scale τ_G of the ice cap from $V_{\min}$ to $V_{\max}$ is determined by

$$\tau_G \approx \frac{V_{\max} - V_{\min}}{a_{\max} - a'}, \tag{11.23}$$

where a' represents the (constant) ablation and a is the accumulation. Once the maximum volume is reached, the sea ice grows and switches the accumulation to a

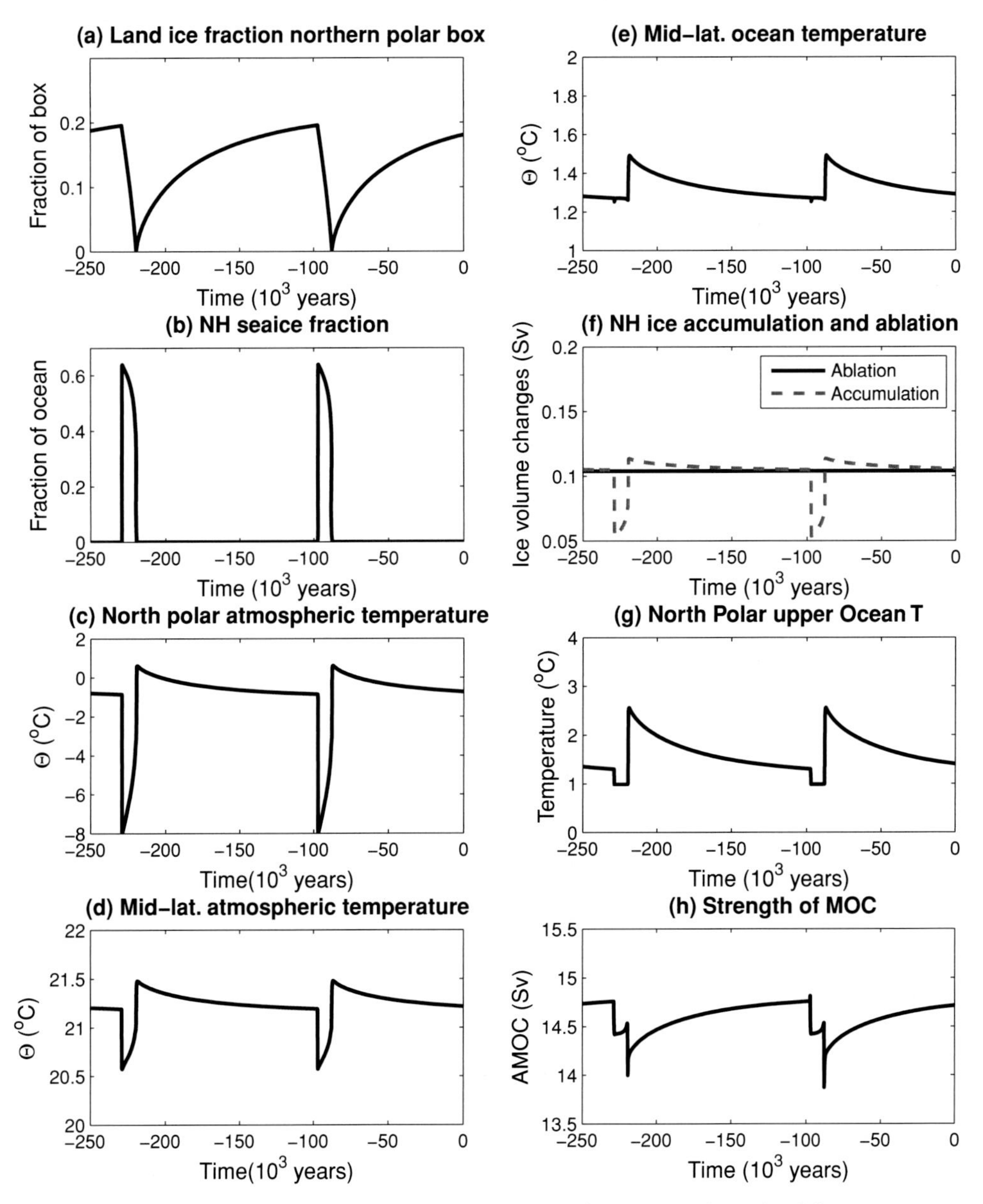

Figure 11.15 Model time series for 250 kyr showing (a) northern land ice extent as a fraction of the polar land box area, (b) northern sea ice extent as a fraction of polar ocean box area, (c) northern atmospheric temperature (°C), (d) midlatitude atmospheric temperature (°C), (e) temperature in the northern midlatitude deep oceanic box (°C), (f) source term (solid line) and sink term (dashed line), for northern land glacier (10^6 m^3 s^{-1})), (g) temperature in the northern upper polar oceanic box (°C) and (h) MOC transport through the northern polar boxes (1 Sv = 10^6 m^3 s^{-1}).

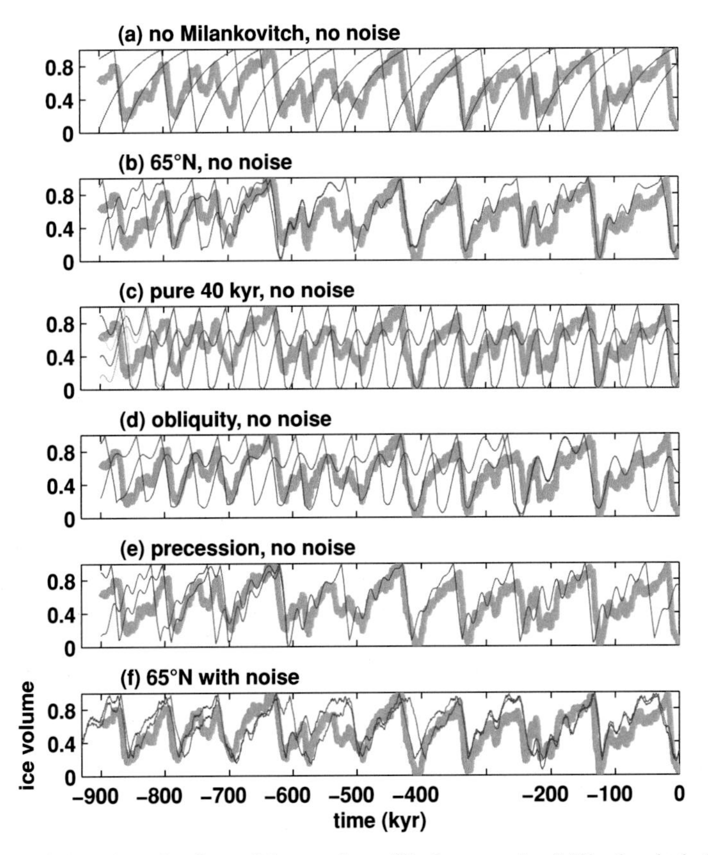

Figure 11.16 Synchronisation of internal oscillations to the Milankovitch forcing. The grey curve is a proxy δ^{18}O compilation; the thin colour curves are ice volume time series from different model runs using different initial conditions. (a) A model run with no M-forcing. (b) Model forced by M-forcing (65°N summer insolation). (c) Model run forced by a periodic 40-kyr forcing (i.e., not the obliquity time series but a sine wave with a 40-kyr period). (d) Model forced by obliquity variations only. (e) Model forced by precession only. (f) Same as (b) but in the presence of noise (figure from Tziperman et al., 2006). (See Colour Plate.)

minimum $a_{\min}$. The decay time scale τ_D of the ice cap back to $V_{\min}$ is therefore

$$\tau_D \approx \frac{V_{\max} - V_{\min}}{a' - a_{\min}}, \tag{11.24}$$

and hence the total time scale is determined by $\tau_G + \tau_D$. With reasonable estimates of these quantities (Gildor and Tziperman, 2001), one arrives at a time scale of about 100 kyr.

When this model is forced with Milankovitch insolation changes, nonlinear resonances (Section 8.6.2) may occur between the internal oscillation and the orbital forcing, leading to time series that qualitatively resemble the observed records (Tziperman et al., 2006). An example is shown in Fig. 11.16, which demonstrates how phase locking to M-forcing affects glacial cycles in this idealised model. Panel (a) shows glacial oscillations in this model in the absence of any 14 forcing. The different time series shown by the thin colour lines correspond to model results using different initial conditions for the ice volume, whereas the thicker grey curve is a δ^{18}O proxy data compilation. Panel (b) shows two ice volume time series from the same model, with the only difference being that M-forcing is now included in the model dynamics. Note that in this case the initially different model solutions converge to a single time series within a few hundred thousand years (see merging of thin coloured lines). The model glacial cycles therefore are becoming locked to the external Milankovitch 'clock'. Of interest, of course, is that this model time series also fits the observed proxy ice volume record (thick grey line) quite well for such a simple model.

These nonlinear resonances are likely to be present in every model in which a strong nonlinear interaction is represented, explaining, for example, the good agreement between very conceptual models, where only a multiple state switch (Paillard, 1998) is represented (Section 11.4.2), and observations. In other words, even if the mechanism of the glacial-interglacial variability is incorrect, there may still be a good fit with the isotopic record. Due to the synchronisation, a comparison between time series of simple models and isotope records is not mechanistically selective.

In conclusion, internal oscillations may be important for the glacial-interglacial transitions. Combined sea-ice and land-ice dynamics introduce a preference for a 100-kyr time scale (Gildor and Tziperman, 2000), whereas the atmosphere-cryosphere-lithosphere interactions appear to have a preference for smaller time scales (Ghil and Childress, 1987). As soon as a model displays nonlinear behavior (which can be a switch or the existence of multiple equilibria), there is a possibility for synchronisation to the M-forcing.

11.5 The role of the carbon cycle

As we have seen in Section 11.1, the atmospheric concentration of carbon dioxide $p^a_{CO_2}$ also varied over glacial-interglacial cycles, being about 180 ppmv during ice ages and 280 ppmv during interglacials (Fig. 11.2). The phase lag between changes in $p^a_{CO_2}$ and global mean temperature T is relatively small, and it appears difficult to determine which one is leading (although there appears to be a consensus that the temperature is slightly leading). The interesting issue regarding the role of $p^a_{CO_2}$ in the dynamics of the glacial cycles is whether processes affecting $p^a_{CO_2}$ are actively involved in the time scale, the phasing with the M-forcing or in setting the amplitude of the glacial cycles. Alternatively, $p^a_{CO_2}$ changes may just be a passive response to global climate changes controlled by physical processes other than those affecting $p^a_{CO_2}$. From a dynamical systems point of view, we can rephrase this as whether the carbon cycle processes are qualitative changing the dynamical behavior or not, for instance, by introducing additional (e.g., saddle-node and/or Hopf) bifurcations.

Now the processes influencing $p^a_{CO_2}$ do not only involve the physics of the climate system, but also its chemistry and biology. Many reviews (Archer et al., 2000) and textbooks (Sarmiento and Gruber, 2006) exist that cover the chemical and biological processes in great detail, and this is not the place to even attempt to provide an overview of these processes. However, in the first subsection that follows, a short summary is given of the most elementary processes that have been considered in models of glacial-interglacial cycles that aim to address the preceding question. Then, in the second subsection, we discuss two relatively simple models, one model indicating a central role for carbon cycle in the 100-kyr cycle and one model in which $p^a_{CO_2}$ variations only affect the amplitude of the glacial-interglacial variability.

11.5.1 Glacial-Interglacial $p^a_{CO_2}$ variations

As described in Chapter 11 in Ruddiman (2001), it is estimated that during the Last Glacial Maximum, the carbon content of vegetation (and soil) of the atmosphere and of the upper ocean was reduced by 25%, 30% and 30%, respectively, with respect to the present day. The total amount from these reservoirs (about 1,000 Gt C) must have been added to the deep ocean, which is an increase of about 2.7% with respect to the present day.

To understand the processes that are responsible for these changes, we first consider how $p^a_{CO_2}$ in the atmosphere is coupled to physical, chemical and biological processes in the ocean. The partial pressure of carbon dioxide in seawater, indicated by $p^o_{CO_2}$, depends on the solubility α_s and the concentration of dissolved CO_2, indicated by $[CO_2]$, such that

$$p^o_{CO_2} = \frac{[CO_2]}{\alpha_s}. \tag{11.25}$$

When CO_2 is dissolved in seawater, there are two dissociation reactions

$$CO_2 + H_2O \rightleftharpoons HCO_3^- + H^+, \tag{11.26a}$$

$$HCO_3^- \rightleftharpoons CO_3^{2-} + H^+. \tag{11.26b}$$

The equilibrium dissociation constants k_1 and k_2 of these reactions are given by

$$k_1 = \frac{[HCO_3^-][H^+]}{[CO_2]}; \; k_2 = \frac{[CO_3^{2-}][H^+]}{[HCO_3^-]}. \tag{11.27}$$

The concentrations of the components CO_2, HCO_3^-, H^+ and CO_3^{2-} can be expressed in k_1 and k_2 and the total carbon content Σ_C and the carbonate alkalinity A_C, with

$$\Sigma_C = [CO_2] + [HCO_3^-] + [CO_3^{2-}], \tag{11.28a}$$

$$A_C = 2[CO_3^{2-}] + [HCO_3^-]. \tag{11.28b}$$

The alkalinity is equal to the stoichiometric sum of the concentrations of the bases in the solution; as CO_3^{2-} has a charge -2, there is a factor 2 in front of the concentration $[CO_3^{2-}]$.

With $[CO_2] \ll [HCO_3^-]$ and $[CO_2] \ll [CO_3^{2-}]$, and using (11.25), we can express $p^o_{CO_2}$ into Σ_C and A_C through

$$p^o_{CO_2} \approx \frac{k_2}{\alpha_s k_1} \frac{(2\Sigma_C - A_C)^2}{A_C - \Sigma_C}. \tag{11.29}$$

The flux of CO_2 across the air-sea interface (positive when into the ocean) depends on the difference between $p^a_{CO_2}$ and surface values of $p^o_{CO_2}$ and can be formulated as

$$F_{CO_2} = K_g \left(p^a_{CO_2} - p^o_{CO_2}\right), \tag{11.30}$$

where K_g is a gas exchange coefficient.

According to (11.29), processes affecting alkalinity A_C and/or total carbon content Σ_C change $p^o_{CO_2}$ and hence $p^a_{CO_2}$. In addition, the solubility α_s depends on temperature and salinity, the dissociation constants k_1 and k_2 depend on temperature and the gas exchange coefficient K_g depends (mainly) on atmospheric surface wind speed. This provides a basis to discuss the three so-called carbon pumps:

(i) *The solubility pump.* When the temperature of the upper ocean decreases, the solubility α_s increases. Hence warm water contains less carbon (decrease in the equilibrium value Σ_C) than cold water. The solubility also decreases with increasing salinity, and hence higher salinity water contains less carbon than lower salinity water.

(ii) *The organic (or soft tissue) pump.* Phytoplankton utilise carbon during photosynthesis to produce soft tissue material, which reduces Σ_C. A fraction of this material is transported to the deep ocean, where it is remineralised, and in this process, carbon is pumped from the surface ocean to the deep ocean. This reduces $p^o_{CO_2}$ in the surface ocean and hence $p^a_{CO_2}$ (in equilibrium).

(iii) *The carbonate pump.* When plankton produces calcium carbonate skeletons, the stoichiometry of the water is changed as Ca^{2+} is incorporated, and hence the alkalinity A_C decreases. According to the denominator of (11.29), the equilibrium value of $p^o_{CO_2}$ is increased and, consequently, the carbonate pump slightly weakens the soft tissue pump.

When a glacial period starts, the temperature decreases, and sea level falls, increasing the salinity of the ocean. By the solubility pump, it is estimated that this leads to about a decrease of 10 ppmv in $p^a_{CO_2}$. In addition, the colder conditions lead to a stronger organic pump due to enhanced ocean mixing (primarily in the southern ocean), which may be responsible for a decrease of up to 30 ppmv (Toggweiler, 1999). In addition, sea-ice extent increases, which strongly decreases the flux F_{CO_2}; this effect may be responsible for a 70-ppmv decrease in $p^a_{CO_2}$. However, the vegetation on land decreases, which gives rise to about a 20-ppmv increase in $p^a_{CO_2}$. In summary, there are plenty of mechanisms by which $p^a_{CO_2}$ can vary due to changes in the physical, chemical and biological processes in the climate system. The important issue is now whether the carbon cycle processes can induce new time scales of variability in the climate system.

11.5.2 Carbon cycle processes and glacial cycles

In a series of studies (Saltzman, 1984; Saltzman and Maasch, 1988) and summarised in Saltzmann (2001), idealised low-dimensional models were proposed in which carbon cycle processes were actively involved in generating an internal oscillation with a time scale of 100 kyr. In this approach, rather than deriving the reduced model directly from governing equations, plausible ad hoc equations were formulated based on considerations of feedback processes.

The equations of the model in Saltzman (1984) are

$$\frac{dX}{dt} = -\alpha_1 Y - \alpha_2 Z - \alpha_3 Y^2, \tag{11.31a}$$

$$\frac{dY}{dt} = -\beta_0 X + \beta_1 Y + \beta_2 Z - \left(X^2 + \beta_3 Y^2\right)Y + F_Y, \tag{11.31b}$$

$$\frac{dZ}{dt} = X - \gamma_2 Z, \tag{11.31c}$$

where X is the total ice mass, Y the deep ocean temperature and Z the $p^a_{CO_2}$. The coefficients and different nonlinearities are based on plausible feedbacks, and parameters are chosen such that a 100-kyr internal oscillation is found. It was shown that with M-forcing, the results can give a reasonable fit with the δ^{18}O records that were available at the time. However, specific mechanisms on what processes determine the time scale cannot be easily extracted from such a model (Crucifix, 2012).

The reason for the oscillatory behaviour was provided in Saltzman and Maasch (1988). During interglacials, $p^a_{CO_2}$ and the Atlantic meridional overturning (MOC) strength are large, but a strong MOC implies a drawdown of CO_2 due to the solubility pump. When CO_2 decreases, ice masses grow, sea level falls and nutrients on land are eroded back to the ocean, increasing productivity, which in turn further decreases CO_2 by the organic carbon pump. The increased ice extent slows down the MOC and a state is reached where both MOC and $p^a_{CO_2}$ are small and the opposite phase of the cycle starts. During this cycle, there is an out-of-phase response between $p^a_{CO_2}$ and the rate of deep water formation causing the oscillation.

The effect of carbon cycle processes on the glacial-interglacial cycles in the model of Gildor and Tziperman (2001) were considered in Gildor et al. (2002). The biochemistry includes total carbon Σ_C, alkalinity A_C and phosphate P (taken to be the limiting nutrient) as prognostic variables that are used to calculate $p^a_{CO_2}$. The prognostic equations for these quantities are similar to the advection-diffusion equations used for the temperature and salinity, except for an added source/sink term due to export, production and remineralisation. A typical result when carbon cycle processes are included is plotted in Fig. 11.17, showing that over a glacial-interglacial cycle, $p^a_{CO_2}$ varies by about 50 ppmv (Fig. 11.17f).

The stratification in the Southern Ocean is composed of cold fresh upper water above salty, warmer water whose source is the North Atlantic Deep Water (NADW). During the stage of Northern Hemisphere ice sheet buildup (Fig. 11.17a), from -100 kyr to -30 kyr, the deep water formed in the North Atlantic becomes gradually colder (Fig. 11.17b) as a result of general Northern Hemisphere cooling due to the increased albedo of the growing land glaciers. Similarly, the mixing of the NADW on its way south with cooler glacial-period surface water in the midlatitudes results in the NADW arriving at the Southern Ocean colder and denser. Because Antarctica is covered by ice even during interglacial periods, the temperature of the adjacent surface water in the southern box stays close to the freezing point. The denser NADW makes the model stratification in the Southern Ocean gradually more stable during glaciation (Fig. 11.17c). According to the stratification-dependent parameterisation of vertical mixing in the model (Gargett, 1984), this more stable stratification reduces the mixing between the deep water and the surface water (Fig. 11.17d). Note that a symmetrically opposite effect happens during deglaciation: the NADW warms, stratification in the Southern Ocean weakens and the vertical mixing strengthens.

Although Southern Ocean sea ice exists throughout the model glacial cycle, its meridional extension varies between glacial and interglacial states (Fig. 11.17e), as well as with the seasonal cycle. During all phases of the glacial cycle, the NADW upwelling in the Southern Ocean is warmer than the surface water and acts to limit the sea ice extent there. During the glaciation phase, the cooling of the NADW cools the water upwelling in the Southern Ocean (which is still warmer than the surface water). This cooling of the upwelling water, as well as the reduced vertical mixing of the surface water with the warm deep water, both contribute to the growth of the Southern Ocean sea ice during glaciation (Fig. 11.17e). The change in sea ice cover during the glacial-interglacial oscillation explains part of the atmospheric variations via the insulating effect of sea-ice cover on gas air-sea exchange. The variations in the vertical mixing and in the sea-ice extent in the Southern Ocean induce together, via the carbon pumps, a difference of about 50 ppm in this atmospheric model between glacial and interglacial periods (Fig. 11.17f).

Although the glacial oscillations in this model exist due to a self-sustained internal variability of the physical climate system (Section 11.4.3), there is a role for $p^a_{CO_2}$. The Southern Ocean mixing and sea ice changes induced by the glacial-interglacial cycle of the physical climate system cause variations in $p^a_{CO_2}$ (Fig. 11.18a) that, in

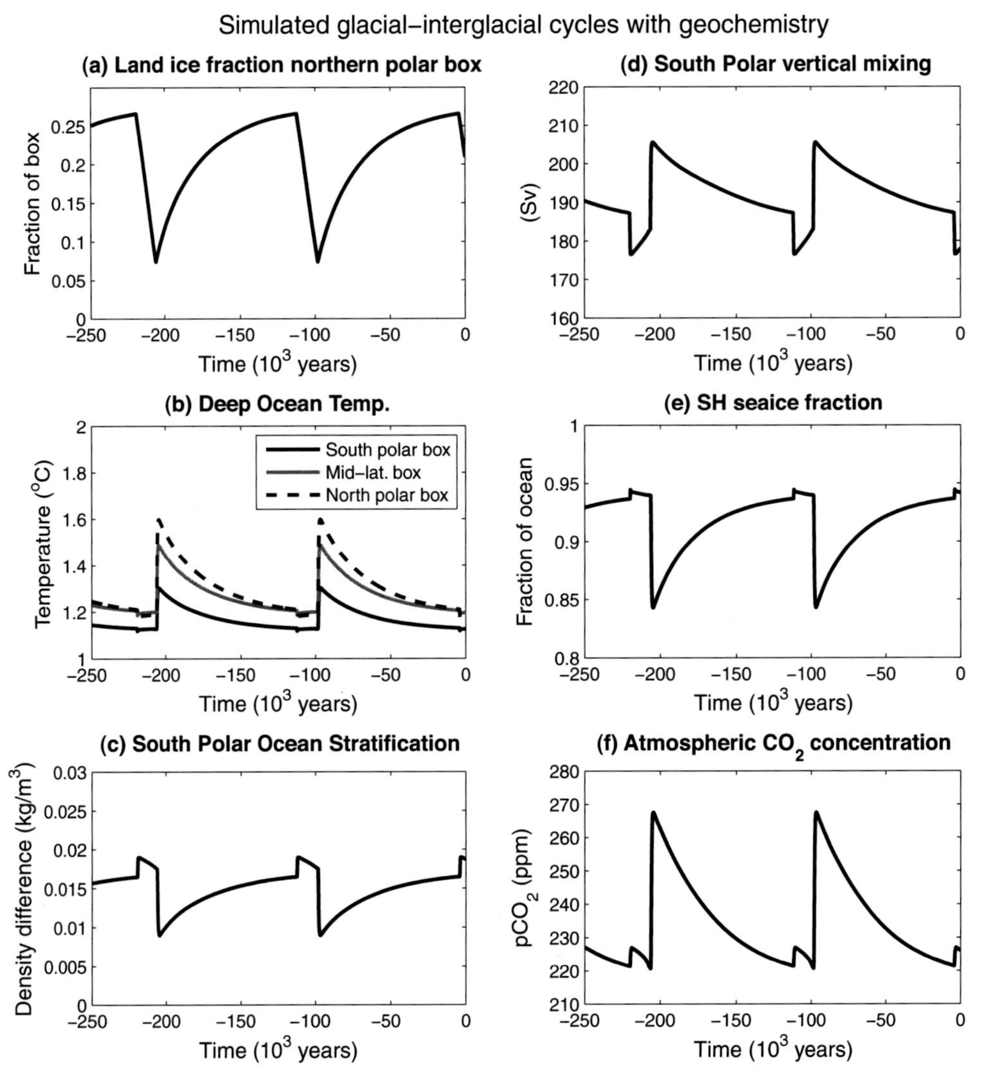

Figure 11.17 Model results for a complete glacial-interglacial cycle for the model used in Gildor and Tziperman (2001). (a) Land ice in the Northern Hemisphere (fraction of northern land box area covered by glaciers); (b) deep ocean temperatures (°C); (c) the stratification in the Southern Ocean (density difference between upper and deep water, in units of 100 kg/m^3; (d) vertical mixing in the Southern Ocean (Sv); (e) sea-ice extent in the Southern Ocean (fraction of the Southern Ocean area); (f) $p^a_{CO_2}$ in ppm.

turn, significantly amplify the glacial-interglacial cycle in the Southern Ocean sea-ice variability (Fig. 11.18b). Through the sea-ice albedo effect, this amplifies the variability of the southern atmospheric temperature relative to a fixed $p^a_{CO_2}$ case (Fig. 11.18c). Midlatitude atmospheric and ocean temperatures (Fig. 11.18d) are also directly affected by the radiative forcing of the variable atmospheric $p^a_{CO_2}$.

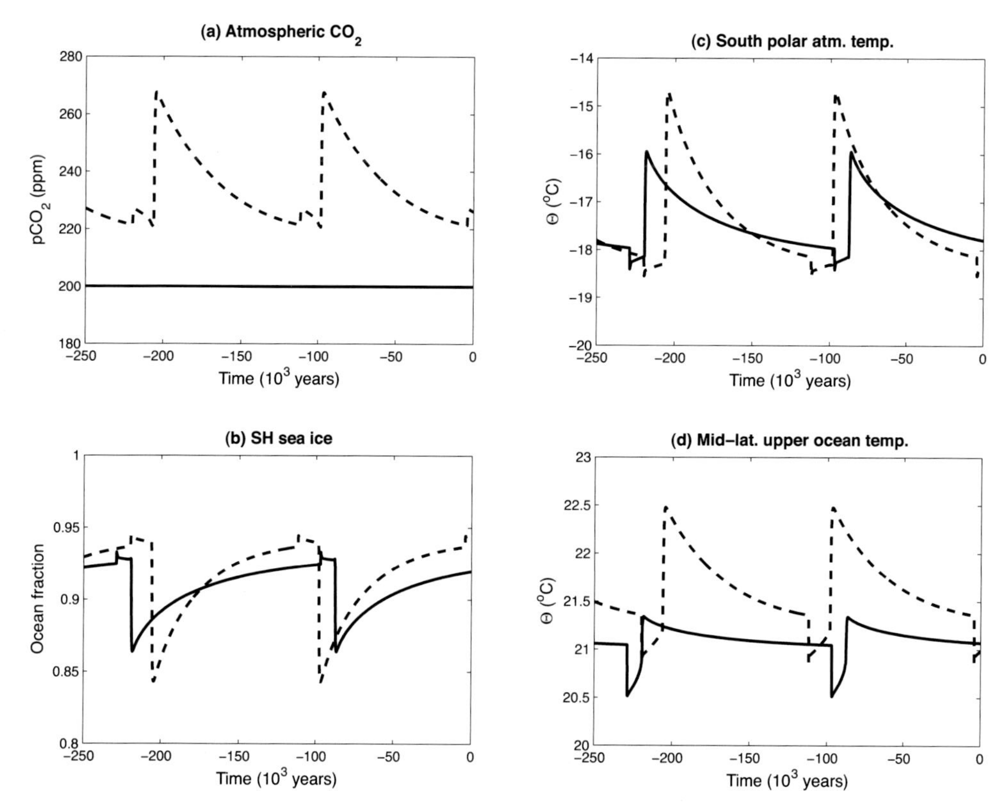

Figure 11.18 The amplification of the glacial-interglacial signal of the physical climate system by the atmospheric variations of $p^a_{CO_2}$. Shown are results from a physics-only model (Gildor and Tziperman, 2001) using a fixed atmospheric concentration (thick line) and results from a fully coupled model with an active ocean biochemistry model (dashed line). (a) $p^a_{CO_2}$ (ppm); (b) Southern Ocean sea-ice extent (fraction of the Southern Ocean area); (c) atmospheric temperature above the Southern Ocean (°C); (d) midlatitude upper ocean temperature.

11.6 Synthesis

The mechanisms of how the Earth's climate system generated the large swings in temperature associated with the glacial-interglacial cycles in the Pleistocene is an intriguing issue that clearly tests our understanding of the climate system. The orbitally induced changes in insolation (Milankovitch forcing) are crucial, but, interestingly, these hardly induced any change in annual mean insolation (over the Pleistocene). So, the large seasonal changes in insolation due to precession and obliquity together with the very small annual mean changes due to eccentricity differences can have generated the large climate changes only through a strong interaction with internal processes in the climate system.

The conceptual models presented in this chapter have indicated that there are several feedback mechanisms that may lead to multiple equilibria (and subsequent switches) and oscillatory behaviour. Although the results where such models are forced with localised M-forcing (e.g., at 65°N) likely overestimate the interaction between radiative forcing and internal processes, they clearly show that a small-amplitude forcing can be substantially amplified in the climate system. They provide the view that before the MPT (Mid-Pleistocene transition, at ~700 kyr BP), the Milankovitch forcing is amplified in the climate system to give dominant 41-kyr variations. After the MPT, a dominant 100-kyr cycle appeared, and from the conceptual models, there appears only one candidate dynamical mechanism behind the MPT: a Hopf bifurcation. In Tziperman et al. (2006), such a Hopf bifurcation is introduced by the sea-ice switch, but other processes may also be responsible for such a bifurcation to occur. Processes in the cryosphere, both regarding land ice and sea ice, are centrally important in the amplification of variability on the relevant time scales and synchronisation to the orbital forcing. Although sea-ice processes are fast by nature, it was shown that, when combined with land ice processes, they can play a role in the long-term response of the climate system to Milankovitch forcing. The weakest link in the sea-ice switch is the assumed importance of the precipitation-temperature feedback.

Despite the decades of work on the problem, from a dynamical systems point of view, the theory of the Pleistocene Ice Ages is still in a comparatively primitive stage (Crucifix, 2012). Current results have led, of course, to many useful concepts, such as multiple equilibria and synchronisation, but have not provided a solid theory of the phenomena. First of all, although there is a multitude of highly parameterised low-order models available (Saltzmann, 2001), each capturing one aspect of the problem, we do not have a good candidate for a minimal model of the glacial cycles that also captures the spatial aspects of the variability. Such a minimal model should at least include a model of atmospheric $p^a_{CO_2}$, as this is considered to be an important amplification mechanism of temperature perturbations. Second, it is not known whether there are normal modes (probably encompassing not only atmosphere and ice but also ocean and carbon cycle) that have an internal time scale of about 100 kyr. Such modes could naturally be involved in the amplification of the weak (annual mean nearly zero) 100-kyr Milankovitch forcing but also could be involved in a Hopf bifurcation, possibly associated with the MPT.

In making a theory of the Pleistocene Ice Ages, a renewed attempt has to be made to develop minimal (spatially extended) models that contain the essential processes and that can still be analysed within the dynamical systems framework. This will then complement the many studies on the problem that have been done and are still ongoing with EMICs (e.g., Ganopolski et al. [2010]).

12

Predictability

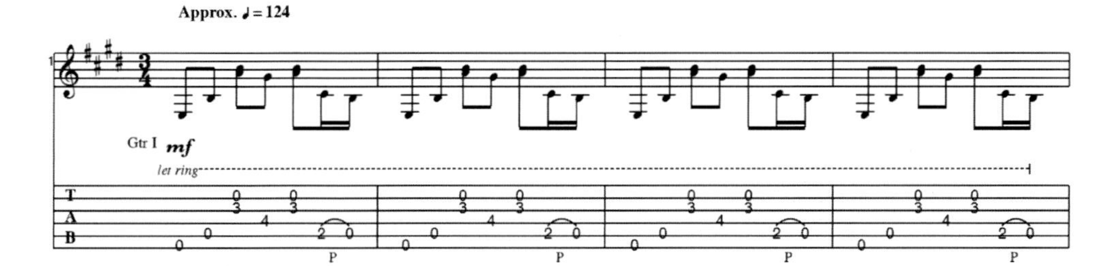

Optimal temporal propagation.
EBEF♯BE, Turning, Alex de Grassi

All of us are familiar with day-to-day weather phenomena. The typical time scale of the processes associated with the weather is a few days. Most of us also know that weather predictions with a lead time longer than about ten days are not very reliable because of the chaotic nature of the atmospheric flow on these time scales. As Chuck Leith, referring to the Lorenz 'butterfly effect', put it: 'It is not all butterflies, even talking about the weather can change the weather...' (see e.g., http://www.archives.ucar.edu/exhibits/washington/science/early_atmos_sci).

As we have seen in Chapter 6, the state vector of the climate system involves not only the atmospheric flow but also the ocean circulation, land and sea ice and, for example, land-surface properties. Climate change refers to the long-term (50–100 years) development of the climate state due to internal variability and the variations in external (e.g., solar) forcing. Over the past several decades, the trajectory of this system is also influenced by the release of CO_2 due to human activities, which can be seen as an external radiative forcing on these time scales.

Regardless of these changes in the radiative forcing conditions, it is of central importance to know the predictability horizons of climate variability on different time scales. We address this issue in this final chapter, again from a stochastic dynamical

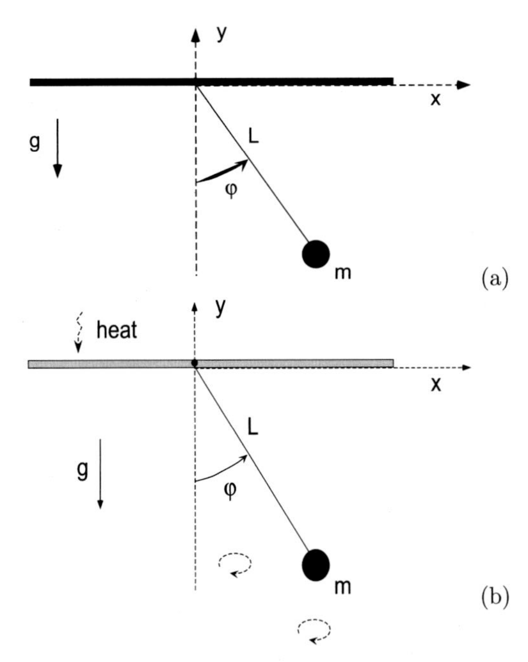

Figure 12.1 (a) Sketch of a single pendulum in vacuum, consisting of a mass m at the end of a rope with length L which is connected to a fixed surface. (b) The same pendulum except that it now moves in air that is heated at the upper surface.

systems point of view. In Section 12.1, the essence of the prediction problem is illustrated using a simple example. Then in Section 12.2, measures of predictability, such as predictive power and relative entropy, are presented. This is followed by the methodology (Section 12.3) to study the behaviour of nearby trajectories needed to assess error propagation. In the next section (12.4), methods to incorporate observations into models to 'steer' trajectories are discussed. As we see, many dynamical systems concepts and methods are used in the study of the predictability of phenomena in the climate system.

12.1 The prediction problem

It is often questioned in the popular media: how can we ever aim to predict the climate state for 2050 with some confidence if we cannot predict the weather more than a few days ahead with a reasonable skill? This question is at the heart of the prediction problem, and we address it in the next section.

12.1.1 A simple example

Consider, first, a single pendulum (Fig. 12.1) where a mass m is connected to the end of a wire of length L, which is connected to a fixed surface. The position of the mass can

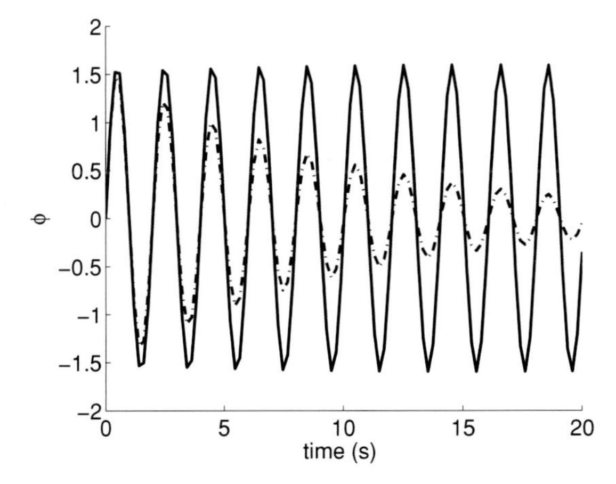

Figure 12.2 Plot of the amplitude of the angle $\phi(t)$ of the mass of a simple pendulum for $L = 1$ m, $g = 9.8$ ms^{-2} versus time (in s) without (drawn) and with (dash-dotted) the presence of air friction ($\gamma = 0.1$ s^{-1}). The initial conditions are $\phi(0) = 0$ and $d\phi/dt = 5$.

be completely specified by the angle ϕ. The prediction problem can be loosely formulated as follows: if at a certain time we release the mass of the pendulum at a specific angle (and with a specific speed), what is the position of the mass at a later time?

As a first approach, we idealise the situation as being in a vacuum (Fig. 12.1a). We can use Newton's second law with gravity as the only force to give the equation for the angle position as

$$\frac{d^2\phi}{dt^2} + \frac{g}{L}\sin\phi = 0, \tag{12.1}$$

where g is the gravitational acceleration. If we know both ϕ and $d\phi/dt$ at the initial time t_0, then (12.1) provides a model to calculate $\phi(t_1)$ from $\phi(t_0)$. For small angles ϕ, we can approximate $\sin\phi \approx \phi$ and (12.1) can be solved as

$$\phi(t) = A\sin\omega_0 t + B\cos\omega_0 t;\ \omega_0 = \sqrt{\frac{g}{L}},$$

where A and B are determined from the initial conditions. A typical solution of $\phi(t)$ is shown as the drawn curve in Fig. 12.2; once the initial conditions are known accurately, we can make an accurate prediction of the position and velocity of the mass at a later time.

If, instead, the pendulum motion is considered to occur in air, the problem is suddenly much more complicated (Fig. 12.1b). The air undergoes a complex motion, interacts with the mass of the pendulum and introduces rapid fluctuations in the position of the mass of the pendulum. We can call these fast motions of the pendulum the 'weather' of the pendulum and the slow motion of the mass of the pendulum its 'climate'.

For an exact prediction of the pendulum's weather, one would need to represent the interactions of the air with the pendulum mass. We can divide, for example, the air volume into grid boxes and compute the motion of the air parcels using the Navier-Stokes equations. As the motion of the air is chaotic on the small time scale (much smaller than the time scale of the period $2\pi/\omega_0$ of the pendulum), we are certainly not able to predict the detailed position of the pendulum mass on this time scale.

However, we may predict the effect of the air on the pendulum's 'climate' relatively easily. For example, the effect of the collisions of the air molecules with the pendulum mass can be represented as a simple net frictional force, giving a modified equation (12.1), again for small ϕ as

$$\frac{d^2\phi}{dt^2} + \gamma\frac{d\phi}{dt} + \frac{g}{L}\phi = 0. \tag{12.2}$$

In this case, we use a so-called parameterisation and do not need to compute the positions of the air parcels. However, the use of this parameterisation has its toll. There is now a friction coefficient γ in the model that is unknown, and it needs to be determined empirically.

The behaviour of the position of the mass of the pendulum (for a specific value of γ) is shown as the dash-dotted curve in Fig. 12.2, and it is clear that it approaches zero for $t \to \infty$ due to the presence of the friction. When the air around the pendulum is slowly heated at the boundary of the container (Fig. 12.1b), the motion of the air parcels will be affected (the viscosity of the air is temperature dependent) and hence also the slow variations in the position of the pendulum. This effect could be taken into account by a temperature-dependent friction coefficient γ.

In summary, because the 'climate' of the pendulum is determined by very different processes than its 'weather', detailed predictions of the position of the pendulum on the long time scale (the 'climate') still may be made despite the fact that the air motion (and the pendulum's 'weather') is unpredictable on the short time scale.

However, the processes responsible for the 'climate' of the pendulum also may lead to unpredictable behaviour. An example is the slightly more complicated system of the double pendulum (Fig. 12.3a). In vacuum, Newton's law now leads to a coupled set of nonlinear equations given by

$$\ddot{\phi}_1 + \frac{g}{L_1}\sin\phi_1 + \frac{m_1}{m_1+m_2}\frac{L_2}{L_1}\left[\cos(\phi_2-\phi_1)\ddot{\phi}_2 - \sin(\phi_2-\phi_1)\dot{\phi}_2^2\right] = 0,$$

$$\ddot{\phi}_2 + \frac{g}{L_2}\sin\phi_2 + \frac{L_1}{L_2}\left[\cos(\phi_2-\phi_1)\ddot{\phi}_1 + \sin(\phi_2-\phi_1)\dot{\phi}_1^2\right] = 0,$$

where a dot indicates differentiation to t. A plot of the variable $\cos\phi_1(t)$ is shown in Fig. 12.3b, displaying irregular oscillations. It is well known that this double pendulum system displays chaotic behaviour, with the largest Lyapunov exponent being positive

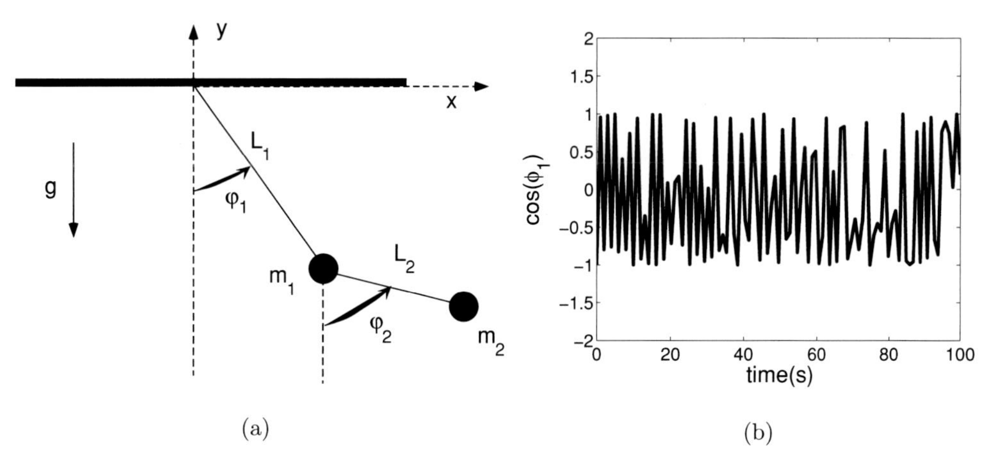

Figure 12.3 (a) The double pendulum consisting of two masses m_1 and m_2 connected by strings of length L_1 and L_2. (b) Solution $\cos\phi_1$ for the case $m_1 = m_2 = 1$ kg, $L_2 = 2$ m, $L_1 = 1$ m and $g = 9.8$ ms^{-2}. The initial conditions are $\phi_1(0) = \phi_2(0) = \pi$ and $d\phi_1/dt = 0$, $d\phi_2/dt = 5$.

(Stachowiak and Okada, 2006). The new element here is that this pendulum system already displays chaotic motion on the 'climate' time scale, limiting predictability of the system, even if no air molecules are considered.

A pendulum system having even more complexity is the double pendulum placed in air that is heated from the boundaries. Here both the 'weather' and the 'climate' of the pendulum have very different limits of predictability and interact with each other, whereas there are also slowly varying boundary influences affecting the motion of the air and the pendulum.

12.1.2 Prediction using climate models

As seen in Chapter 6, a climate model is the recipe that relates future states to past states, similarly to the equation of the pendulum in the previous section. Such models necessarily have many approximations to the real world as a spatially and temporally discrete representation of the state vector is used and parameterisations are incorporated (with uncertain parameters). With a given initial condition, the model can be used to compute future states; this is usually referred to as a simulation.

There are several reasons why it is not so useful for predictions to perform a single simulation with a particular climate model over a certain time period: (i) the initial conditions are poorly known, (ii) there are many uncertainties in the representation of physical and chemical processes (Section 6.4) and (iii) the presence of internal variability may introduce a random element in the model solutions, as in the double pendulum motion (Section 12.1.1).

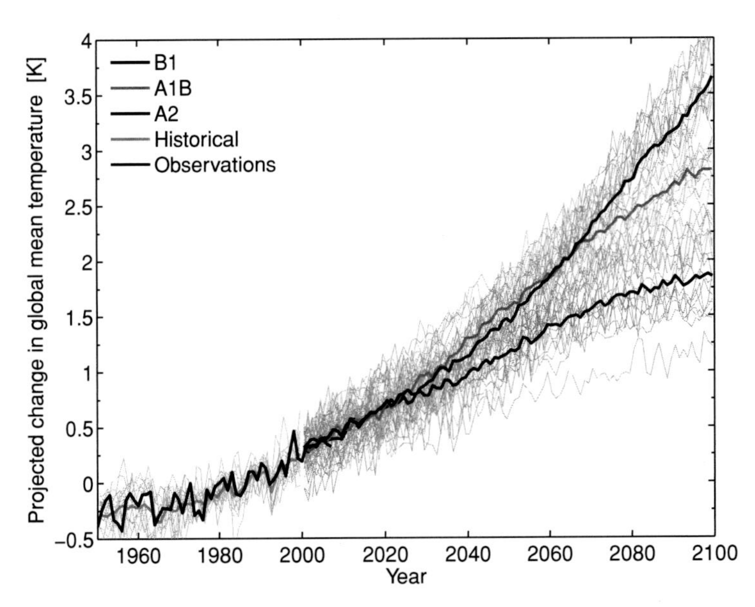

Figure 12.4 Global mean, annual mean and surface air temperature projections from fifteen different global climate models under three different greenhouse emission scenarios from 2000 to 2100 (thin lines): SRES A2 (red), A1B (green), and B1 (blue), designated as high-, medium-, and low-emission paths, respectively. The same models forced with historical forcings are shown as the thin grey lines, and the observed global mean temperatures from 1950 to 2007 are shown as the thick black line. The multimodel mean for each emission scenario is shown with thick coloured lines (figure from Hawkins and Sutton, 2009). (See Colour Plate.)

This has motivated the use of the following types of multiple simulations, or ensembles:

- The standard ensemble: in these simulations, one considers the sensitivity of the climate model solutions to the initial conditions and provides an ensemble of 'equally likely climate development'.
- The perturbed-parameter ensemble: here one investigates the uncertainties due to the representation of physical and chemical processes by varying parameters in the climate model; prominent examples can be found on http://www.climateprediction.net.
- The multimodel ensemble: here one investigates the uncertainties due to a different representation of physical processes in different models. The different models serve as a perturbed parameter ensemble and the different simulations with a single model as a standard ensemble.

The final aim with these ensemble methods is to estimate a range of possible trajectories from an uncertain initial condition with an imperfect model.

An example of the use of ensemble simulations is given in Fig. 12.4 from simulation efforts carried out for the IPCC-AR4 assessment (Hawkins and Sutton, 2009), where one is interested in, for example, the global mean surface temperature up to 2100.

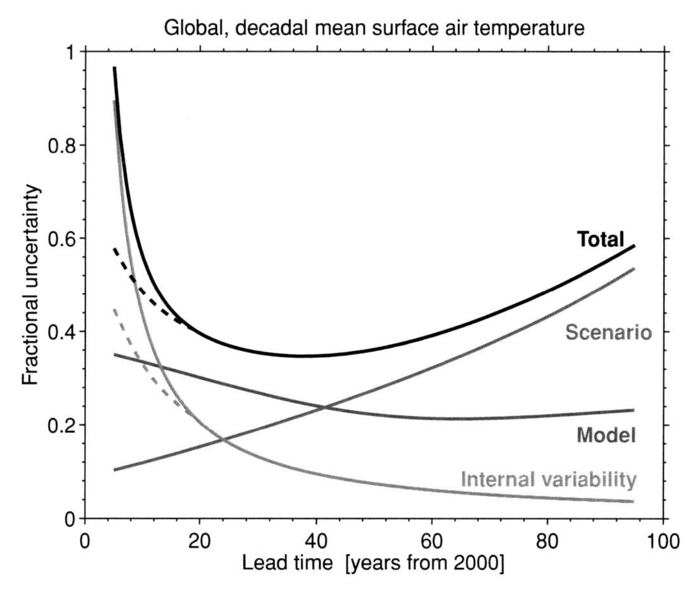

Figure 12.5 The relative importance of each source of uncertainty in decadal mean surface air temperature preductions is shown by the fractional uncertainty (the 90% confidence level divided by the mean prediction), for the global mean, relative to the warming since the year 2000 (i.e., a lead of zero years) (figure from Hawkins and Sutton, 2009).

Here, an additional element of uncertainty comes from the future forcing conditions, for example, future emissions, and hence simulations for different so-called emission scenarios are performed. Hence, predictions are usually referred to as projections. In Fig. 12.4, the annual mean global mean surface temperature from fifteen different Global Climate Models is plotted for three different emission scenarios (low: B1; medium: A1b; and high: A2). The spread around the multimodel mean provides here an estimate of effect of model uncertainty on future climate under different emission scenarios.

The different contributions to the uncertainty (due to internal variability, model uncertainty and forcing scenario uncertainty) for decadal scale global mean temperature predictions is shown in Fig. 12.5; the initial condition uncertainty is not important on the longer time scales. On the shorter lead times, the internal variability is the most important component, whereas the contribution of the forcing scenario uncertainty is relatively small. Indeed, in Fig. 12.4, the model results for the different scenarios are very similar up to 2040. On the longer time scale, however, scenario uncertainty becomes dominant.

12.2 Concepts of predictability

Objective measures of predictability have been developed since the work of Lorenz (1969). Predictability studies of the first kind address how uncertainties in the initial state affect the prediction at a later stage. Indeed, initial uncertainties can amplify as the prediction lead time increases, thus limiting the predictability of the first kind. Examples are weather prediction and El Niño prediction (which is a climate prediction of the first kind). For these studies, a standard ensemble is suited from which it can be seen how nearby trajectories behave.

Predictability studies of the second kind address the predictability of the response of the system to changes in boundary conditions. Examples are the response of atmospheric flows due to changes in sea-surface temperature and the response of climate to changes in orbital insolation variations. For these kinds of studies, a parameter ensemble is suited as it shows how trajectories for nearby parameter values behave.

In this section, after a more abstract discussion on the density of trajectories, a measure of predictability is presented for predictability of the first kind, that is, the concept of predictive power.

12.2.1 The Liouville equation

For a one-dimensional stochastic dynamical system, as discussed in Section 3.5, the probability density function $p(x,t)$ of the random variable X_t, evolving according to the Itô SDE,

$$X_t = X_0 + \int_0^t a(X_s, s)ds + \int_0^t b(X_s, s)\, dW_s, \tag{12.4}$$

is determined by the Fokker-Planck equation (3.73), that is,

$$\frac{\partial p}{\partial t} + \frac{\partial (ap)}{\partial x} - \frac{1}{2}\frac{\partial^2 (pb^2)}{\partial x^2} = 0. \tag{12.5}$$

For a deterministic system (with $b = 0$), this reduces to the so-called Liouville equation,

$$\frac{\partial p}{\partial t} = -\frac{\partial (ap)}{\partial x}. \tag{12.6}$$

Given an uncertainty in the initial conditions, represented by an initial probability density function $p(x, 0) = f(x)$, the Liouville equation provides the development of the probability density function in time. For an N-dimensional deterministic system

$$\frac{d\mathbf{x}}{dt} = \mathbf{F}(\mathbf{x}), \tag{12.7}$$

with initial condition $\mathbf{x}(0) = \mathbf{x}_0$, the probability density function is usually indicated by a density $\rho(\mathbf{x}, t)$, and the Liouville equation (12.6) generalises to

$$\frac{\partial \rho}{\partial t} + \sum_{k=1}^{N} \frac{\partial (\rho \mathbf{F}_k)}{\partial x_k} = 0. \tag{12.8}$$

Equation (12.8) can be formally solved as (Ehrendorfer, 1994)

$$\rho(\mathbf{x}, t) = f(\mathbf{x}_0)\, e^{-\int_0^t \psi(\mathbf{x}(\mathbf{x}_0,s))ds}, \tag{12.9}$$

where $f(\mathbf{x})$ is again the initial density and ψ is given by

$$\psi(\mathbf{x}) = \sum_{k=1}^{N} \frac{\partial \mathbf{F}_k}{\partial x_k}, \tag{12.10}$$

which is the trace of the Jacobian matrix.

Example 12.1 Solution of the 1D Liouville equation In Ehrendorfer (1994), the one-dimensional example

$$\frac{dx}{dt} = ax^2 + bx + c\ ,\ \Delta = \frac{b^2}{4} - ac > 0, \tag{12.11}$$

is considered. With $x(0) = \eta$, the solution of (12.11) can be determined as

$$x(\eta, t) = \frac{r_1(a\eta + r_2)e^{\gamma t} - r_2(a\eta + r_1)}{-a(a\eta + r_2)e^{\gamma t} + a(a\eta + r_1)}, \tag{12.12}$$

with $r_1 = b/2 + \sqrt{\Delta}$, $r_2 = b/2 - \sqrt{\Delta}$ and $\gamma = r_1 - r_2$, as can be verified by direct substitution. To obtain the density $\rho(x, t)$, we determine $\psi = 2ax + b$ and use it to obtain

$$\rho(x, t) = f(\eta)e^{-\int_0^t (2ax(\eta,s)+b)ds}, \tag{12.13}$$

which eventually provides an expression of ρ in terms of η. The latter is implicitly a function of x and t through inversion of (12.12), giving

$$\eta = \eta(x, t) = \frac{1}{a}\left[\frac{ax(r_2 e^{\gamma t} - r_1) + r_1 r_2(e^{\gamma t} - 1)}{ax(1 - e^{\gamma t}) - r_1 e^{\gamma t} + r_2}\right] \tag{12.14}$$

to finally give

$$\begin{aligned}\rho(x, t) = \frac{f(\eta)}{\gamma^2} \exp(bt &+ \frac{2r_1}{\gamma} \ln[a\eta(e^{-\gamma t} - 1) - r_2 + r_1 e^{-\gamma t}] \\ &- \frac{2r_2}{\gamma} \ln[a\eta(1 - e^{\gamma t}) + r_1 - r_2 e^{\gamma t}]).\end{aligned} \tag{12.15}$$

In Fig. 12.6, the development of the density ρ over time (six values in the interval $t \in [0, 0.3]$) is shown for $a = -1$, $b = 1$ and $c = 2$. The initial density is Gaussian with a mean $\bar{x} = -2$ and variance 0.1. In this case, the fixed points of the dynamical system (12.11) are determined by $x^2 - x - 2 = 0$, of which $\bar{x} = -1$ is unstable and $\bar{x} = 2$ is stable. In time, the density moves away from the unstable fixed point to give a substantial spread. ■

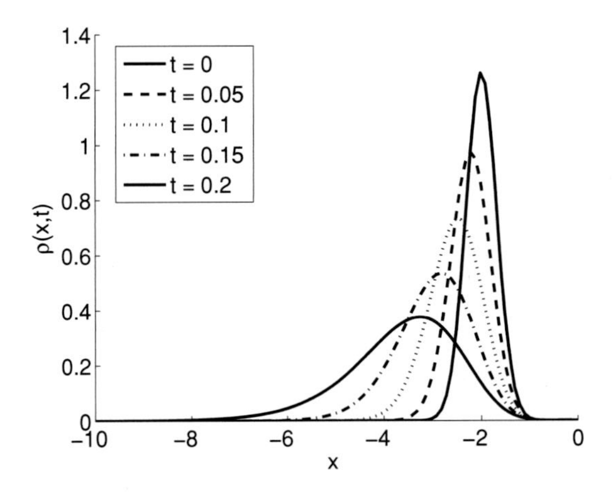

Figure 12.6 Development of the probability density function $\rho(x, t)$ in time for the problem (12.11) with $a = -1, b = 1$ and $c = 2$ (redrawn from Ehrendorfer, 1994).

12.2.2 Predictive power

Several measures have been developed of the predictability of nonlinear systems, such as predictive power (Schneider and Griffies, 1999) and prediction utility (Kleeman, 2002). Of those, we discuss only the predictive power next.

Consider the random n-dimensional state vector of a system $\mathbf{X}_\nu$ where ν indicates a time index. Indicate a particular realisation by $\mathbf{x}_\nu$ and denote its prediction by $\hat{\mathbf{x}}_\nu$. Because of the stochastic nature of the system, there is a prediction error

$$\mathbf{e}_\nu = \mathbf{x}_\nu - \hat{\mathbf{x}}_\nu \rightarrow \mathbf{x}_\nu = \hat{\mathbf{x}}_\nu + \mathbf{e}_\nu, \tag{12.16}$$

and hence in terms of random variables, this can be written as

$$\mathbf{X}_\nu = \hat{\mathbf{X}}_\nu + \mathbf{E}_\nu, \tag{12.17}$$

where the predictor $\hat{\mathbf{X}}_\nu$ is the random vector of which $\hat{\mathbf{x}}_\nu$ is a realisation.

The probability distribution of the state $\mathbf{X}_\nu$ is the climatological distribution, which reflects the prior (before any predictive information is available besides the climatological mean) uncertainty. The statistical properties of $\mathbf{X}_\nu$ can, for example, be obtained by analysing a long data set of observations or a long control simulation of a particular model. The probability distribution of the prediction error $\mathbf{E}_\nu$ reflects the a posteriori uncertainty that remains after the prediction has become available. Intuitively, in particular for univariate processes, one would expect that the predictability of the system is low if the variance of the prediction error is as large (or larger) than the climatological variance.

For multivariate processes, measures based on information theory can be used to quantify predictability. The degree of uncertainty associated with a probability density function $p_{\mathbf{x}}(\mathbf{x})$ of a random variable $\mathbf{x}$ is given by the entropy (Cover and Thomas, 2006)

$$S_{\mathbf{x}} = -k \int p_{\mathbf{x}}(\mathbf{x}) \ln p_{\mathbf{x}}(\mathbf{x}) d\mathbf{x}, \tag{12.18}$$

where k is a constant. The quantity $S_{\mathbf{x}}$ can be seen as the additional information, say, the average number of bits, which is necessary to completely specify $\mathbf{x}$.

The prior entropy $S_{\mathbf{X}_\nu}$ is the average missing information when only the climatological distribution is known. The posterior entropy $S_{\mathbf{E}_\nu}$ is the average missing information after the prediction has become available. The predictive information R_ν, given by

$$R_\nu = S_{\mathbf{X}_\nu} - S_{\mathbf{E}_\nu}, \tag{12.19}$$

is the average information about the state contained in the prediction; it should always be positive. The predictive power $\alpha_\nu \in [0, 1]$ is then defined as (Schneider and Griffies, 1999)

$$\alpha_\nu = 1 - e^{-R_\nu}. \tag{12.20}$$

When the probability distribution of an n-dimensional state vector $\mathbf{X}$ is Gaussian (see Section 3.2), with covariance matrix Σ, then it can be shown (Schneider and Griffies, 1999) that

$$S_{\mathbf{x}} = \frac{k}{2}(n + n \log 2\pi + \ln \det(\Sigma)), \tag{12.21}$$

which can be easily verified for $n = 1$ for which $\Sigma = \sigma^2$ and p_X given by (3.8). When the climatological covariance matrix is indicated by Σ_ν and the covariance matrix of the prediction error by C_ν, the predictive information is given by

$$R_\nu = -\frac{k}{2} \ln \frac{\det C_\nu}{\det \Sigma_\nu}, \tag{12.22}$$

and with the choice $k = 1/n$, the predictive power α_ν is finally determined as (using the product rule of determinants)

$$\alpha_\nu = 1 - e^{-R_\nu} = 1 - (\det C_\nu \Sigma_\nu{}^{-1})^{\frac{1}{2n}}. \tag{12.23}$$

For univariate processes with climatological variance σ_c^2 and prediction error variance σ^2, α_ν reduces to

$$\alpha_\nu = 1 - \frac{\sigma^2}{\sigma_c^2}. \tag{12.24}$$

When the prediction error variance is equal to the climatological variance, the predictive power is indeed zero. In this case, the prediction does not add any additional information than already available through the climatology.

In practice, the matrices C_ν and Σ_ν can be determined by performing a long control and a standard ensemble simulation. From the long control integration, we obtain a trajectory $\mathbf{x}^c_\nu$, $\nu = 1, \ldots N$, where the superscript c refers to the control simulation. From this time series, the mean and (time-independent) covariance matrix of the climatological distribution can be estimated as

$$\bar{\mathbf{x}} = \frac{1}{N}\sum_{\nu=1}^{N} \mathbf{x}^c_\nu, \tag{12.25a}$$

$$\hat{\Sigma}_\nu = \frac{1}{N-1}\sum_{\nu=1}^{N} (\mathbf{x}^c_\nu - \bar{\mathbf{x}})(\mathbf{x}^c_\nu - \bar{\mathbf{x}})^T. \tag{12.25b}$$

Next, consider a standard ensemble of the same model in which M $(N \gg M)$ initial conditions $\mathbf{x}^1_0, \ldots, \mathbf{x}^M_0$ are integrated forward in time. At time t_ν, these initial conditions have evolved to states $\mathbf{x}^1_\nu, \ldots, \mathbf{x}^M_\nu$. The mean of the M-member ensemble

$$\hat{\mathbf{x}}_\nu = \frac{1}{M}\sum_{i=1}^{M} \mathbf{x}^i_\nu \tag{12.26}$$

is a prediction of the model state $\mathbf{x}_\nu$ at lead time t_ν.

The residuals

$$\mathbf{e}^i_\nu = \mathbf{x}^i_\nu - \hat{\mathbf{x}}_\nu \tag{12.27}$$

form a sample of the prediction error distribution and are unbiased as they have zero (ensemble) mean. The sample covariance matrix of the residuals, given by

$$\begin{aligned}\hat{C}_\nu &= \frac{1}{M-1}\sum_{i=1}^{M} \mathbf{e}^i_\nu (\mathbf{e}^i_\nu)^T, \\ &= \frac{1}{M-1}\sum_{i=1}^{M} (\mathbf{x}^i_\nu - \hat{\mathbf{x}}_\nu)(\mathbf{x}^i_\nu - \hat{\mathbf{x}}_\nu)^T,\end{aligned} \tag{12.28}$$

is then an estimate of the prediction error covariance matrix.

Example 12.2 North Atlantic multidecadal predictability An example of the use of the predictive power was given in Schneider and Griffies (1999) using the ensemble simulation results of Griffies and Bryan (1997) regarding North Atlantic multidecadal variability. The data consist of twelve ensemble simulations over a period of 30 years in addition to the 200-year control simulation with the GFDL-R15 model. A principal component analysis of the dynamic topography of the control simulation provides the first two EOF patterns as shown in Fig. 12.7; the EOFs together explain about 35% of the variance of the dynamic topography field. EOF1represents variations in the

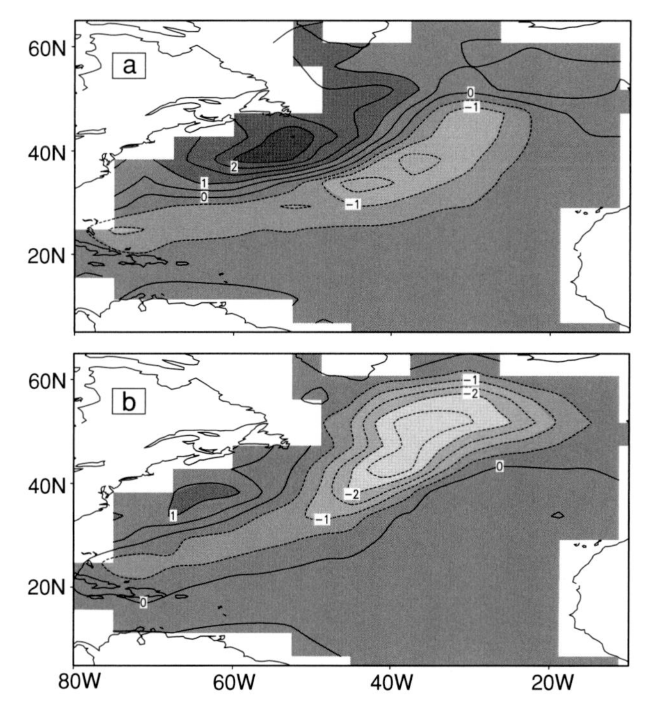

Figure 12.7 First (a) and second (b) EOF of North Atlantic dynamic topography (cm). The amplitude is scaled by the standard deviations of the associated principal components (figure from Schneider and Griffies, 1999).

strength of the North Atlantic Current's northeastward drift, whereas EOF2 displays gyre-shaped variations in the North Atlantic circulation, with the strongest current variations located in the central portion of the basin.

The data from the ensemble integrations are projected onto the space formed by the two EOFs, and estimates of covariance matrix $\hat{C}_\nu$ are obtained for each lead time $\nu = 1, \ldots, 30$ years. Fig. 12.8 shows the predictive power (here indicated as PP) as a function of lead time ν with 95% confidence intervals obtained by Monte Carlo simulation of 1,000 samples (cf. Section 5.2). The bias of the PP estimate is small enough that the PP can be considered significantly greater than zero when the 95% confidence interval does not include zero.

The null hypothesis is that the climatological covariance matrix Σ_ν and the prediction error covariance matrix $\hat{C}_\nu$ are equal, in which case there are no predictable components in the state space of the two EOFs. This null hypothesis is rejected at the 95% significance level for PPs greater than 0.28 (which is the dash-dotted line in Fig. 12.8). The overall PP decays rapidly over the first ten years of the forecasting lead time, remains marginally significant up to about year 17 and becomes insignificant beyond year 17. ■

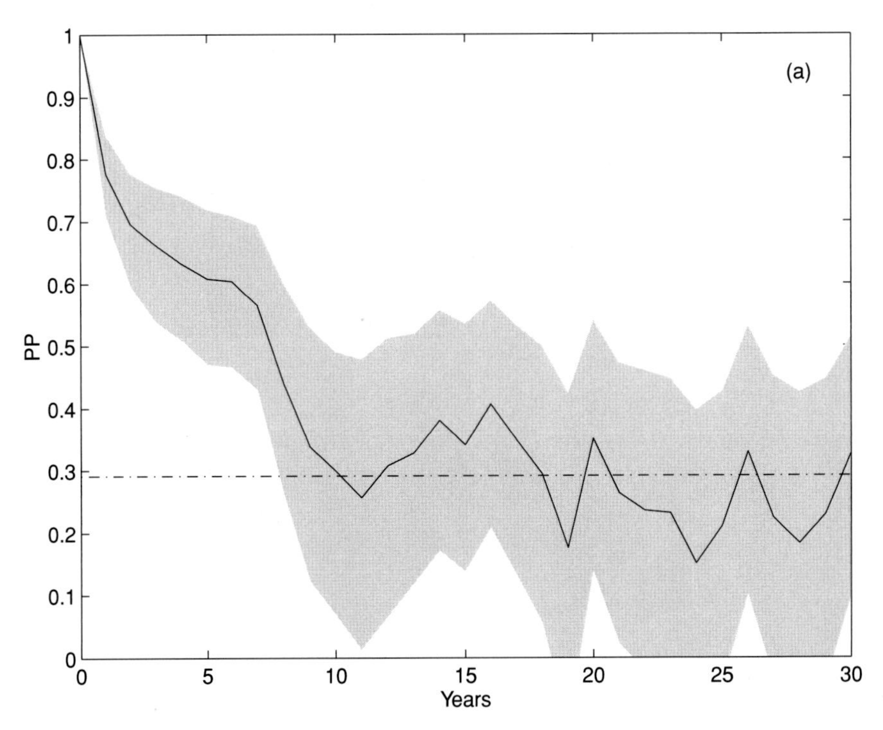

Figure 12.8 Overall predictive power (PP) for the first two EOFs for North Atlantic dynamic topography as a function of forecast lead time (figure from Schneider and Griffies, 1999).

12.3 Behaviour of nearby trajectories

From the previous sections, it is clear that the spread of trajectories for different initial conditions is crucial for predictability of the first kind. The predictive power measure is based on this spread. One of the central measures of the spread of trajectories is the Lyapunov exponent of the model (cf. Section 2.5). However, these exponents are not easy to determine for complicated climate models. What is usually only available is a climate model with specific choices for parameters for which we have to specify initial conditions and then integrate these in time. In this section, we discuss methods addressing the spread of trajectories in such climate models.

12.3.1 Linear view: singular vectors

We write the model equations of an autonomous dynamical system as

$$\frac{d\mathbf{x}}{dt} = \mathbf{F}(\mathbf{x}), \tag{12.29}$$

and indicate the solution $\mathbf{x}$ at time t of a trajectory starting at $\mathbf{x}(t_0)$ for time t_0 as (Kalnay, 2003)

$$\mathbf{x}(t) = \mathcal{M}(\mathbf{x}(t_0)), \tag{12.30}$$

where $\mathcal{M}$ is often referred to as the propagator.

A nearby trajectory $\mathbf{y}$ can be determined by looking at the evolution of the initial conditions $\mathbf{x}(t_0) + \mathbf{y}(t_0)$, where, for example, $\mathbf{y}(t_0) = \epsilon\mathbf{v}$ with $||\mathbf{v}|| = 1$ and $\epsilon \ll 1$. By Taylor expansion, we find

$$\mathcal{M}(\mathbf{x}(t_0) + \mathbf{y}(t_0)) = \mathcal{M}(\mathbf{x}(t_0)) + \frac{\partial \mathcal{M}}{\partial \mathbf{x}}\mathbf{y}(t_0) + \mathcal{O}(\epsilon^2) \approx \mathbf{x}(t) + \mathbf{y}(t), \tag{12.31}$$

where $\mathbf{y}$ satisfies

$$\frac{d\mathbf{y}}{dt} = \mathbf{J}\,\mathbf{y}, \tag{12.32}$$

with $\mathbf{J}$ is the Jacobian matrix of $\mathbf{F}$. The linear system of equations (12.32) is called the tangent linear model, and its solutions are indicated by (using the notation $\mathcal{L}(t_0, t) = \partial\mathcal{M}/\partial\mathbf{x}$)

$$\mathbf{y}(t) = \mathcal{L}(t_0, t)\mathbf{y}(t_0), \tag{12.33}$$

indicating the propagation of the initial perturbation $\mathbf{y}(t_0)$ to the perturbation $\mathbf{y}(t)$ at time t.

When the interval (t_0, t) is split into two intervals (t_0, t_1) and (t_1, t), the evolution of the perturbation is given by (integrating first up to t_1 and then to t)

$$\mathbf{y}(t) = \mathcal{L}(t_1, t)\mathcal{L}(t_0, t_1)\mathbf{y}(t_0) \rightarrow \mathcal{L}(t_0, t) = \mathcal{L}(t_1, t)\mathcal{L}(t_0, t_1). \tag{12.34}$$

With the standard inner product $\langle , \rangle$ on $\mathbb{R}^n$, the adjoint of the tangent linear model $\mathcal{L}^T$ is defined as $\langle \mathcal{L}\mathbf{v}, \mathbf{w}\rangle = \langle \mathbf{v}, \mathcal{L}^T\mathbf{w}\rangle$, and it follows that

$$\mathcal{L}^T(t_0, t) = (\mathcal{L}(t_1, t)\mathcal{L}(t_0, t_1))^T = \mathcal{L}^T(t_0, t_1)\mathcal{L}^T(t_1, t). \tag{12.35}$$

This shows that the adjoint tangent linear model can be viewed as starting at time t and following the trajectory backwards in time.

To determine the action of the (adjoint) tangent linear model on vectors, we use the notation $\mathbf{L} \equiv \mathcal{L}(t_0, t_1)$. For any matrix $n \times n$ matrix $\mathbf{L}$, there exist unitary matrices $\mathbf{U}$ and $\mathbf{V}$ (with $\mathbf{U}^T\mathbf{U} = \mathbf{V}^T\mathbf{V} = \mathbf{I}$, the identity matrix) such that

$$\mathbf{U}^T\mathbf{L}\mathbf{V} = \mathbf{S}, \tag{12.36}$$

where $\mathbf{S}$ is a diagonal matrix containing the singular values σ_i, $i = 1, \ldots n$ of $\mathbf{L}$.

Multiplying (12.36) from the left with $\mathbf{U}$ gives

$$\mathbf{L}\mathbf{V} = \mathbf{U}\mathbf{S} \rightarrow \mathbf{L}\mathbf{v}_i = \sigma_i\mathbf{u}_i \tag{12.37}$$

and multiplying (12.36) from the right with $\mathbf{V}^T$ gives, similarly,

$$\mathbf{U}^T\mathbf{L} = \mathbf{S}\mathbf{V}^T \rightarrow \mathbf{L}^T\mathbf{U} = \mathbf{V}\mathbf{S}^T = \mathbf{V}\mathbf{S} \rightarrow \mathbf{L}^T\mathbf{u}_i = \sigma_i\mathbf{v}_i. \tag{12.38}$$

The $\mathbf{v}_i$ are called the right (or initial) singular vectors and the $\mathbf{u}_i$ the left (or final) singular vectors. Combining both (12.38) and (12.37) gives

$$\mathbf{L}^T\mathbf{L}\mathbf{v}_i = \sigma_i\mathbf{L}^T\mathbf{u}_i = \sigma_i^2\mathbf{v}_i, \tag{12.39}$$

and hence the initial singular vectors can be determined as eigenvectors of $\mathbf{L}^T\mathbf{L}$.

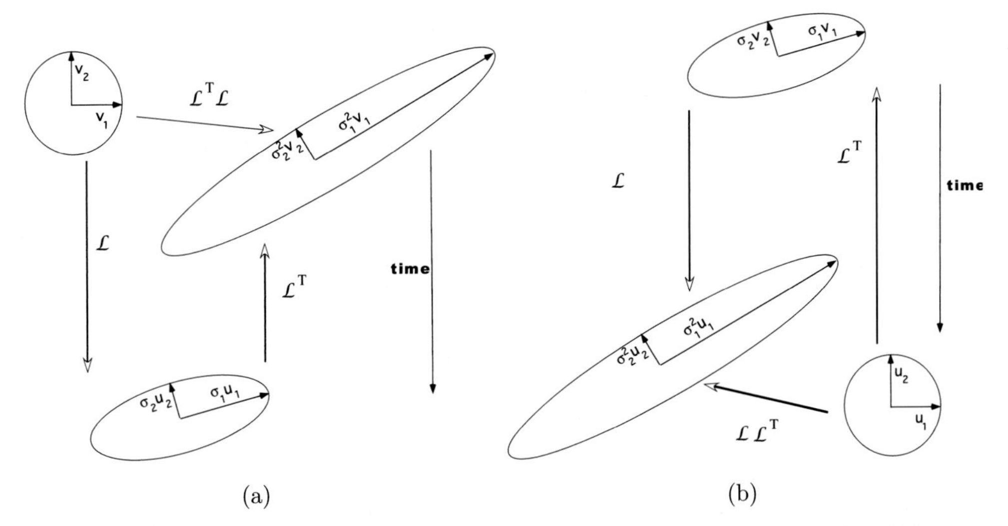

Figure 12.9 (a) Application of the tangent linear model forward in time and the adjoint tangent linear model backward in time to a two-dimensional sphere at initial time. (b) Application of the adjoint tangent linear model backward in time and the tangent linear model forward in time to a two-dimensional sphere at final time (figure based on Kalnay, 2003).

By applying $\mathbf{L}$ to each initial singular vector $\mathbf{v}_i$, the norm of this vector is changed by a factor σ_i, and its direction will be rotated to that of the vector $\mathbf{u}_i$. Similarly, by applying $\mathbf{L}^T$ to each $\mathbf{u}_i$, the norm of this vector is changed by a factor σ_i, and its direction will be rotated to that of the vector $\mathbf{v}_i$ (Fig. 12.9). If $\mathbf{L}$ is applied first to $\mathbf{v}_i$ and then $\mathbf{L}^T$ to $\mathbf{u}_i$, the norm of the vector $\mathbf{v}_i$ is changed by a factor σ_i^2. As the σ_i are real and ordered as $\sigma_1 > \sigma_2 > \ldots > \sigma_n$, the norm of the vector $\mathbf{v}_1$ is affected most.

Having understood the action of $\mathbf{L}$ and $\mathbf{L}^T$, we can now consider the behaviour of nearby trajectories associated with the initial perturbation vector $\mathbf{y}(t_0)$. These can behave very differently, and for the prediction problem, we are in particular interested in which directions; the distance between $\mathbf{x}(t)$ and the trajectory starting at $\mathbf{x}(t_0) + \mathbf{y}(t_0)$ is maximal, say in the norm $\|, \|$ associated with standard inner product, at time $t = t_1$. Using the tangent linear model, with $\mathbf{y}(t_1) = \mathbf{L}\mathbf{y}(t_0)$, we find

$$\|\mathbf{y}(t_1)\|^2 = \langle \mathbf{L}\mathbf{y}(t_0), \mathbf{L}\mathbf{y}(t_0)\rangle = \langle \mathbf{L}^T\mathbf{L}\mathbf{y}(t_0), \mathbf{y}(t_0)\rangle, \tag{12.40}$$

which suggests that this optimisation problem is solved by the singular vectors, and hence the first singular vector is also called the first optimal vector.

In Xue et al. (1997), the singular values and vectors are determined from the tangent linear model of the Zebiak-Cane ENSO model (Section 8.2). The method of constructing the tangent linear model makes use of principle component reduction and a large ensemble of two-year simulations. In Fig. 12.10, the first singular vector optimised over a time interval of six months from January of model year 26 is shown. The thermocline is deeper over much of the western part of the basin, and the westerly

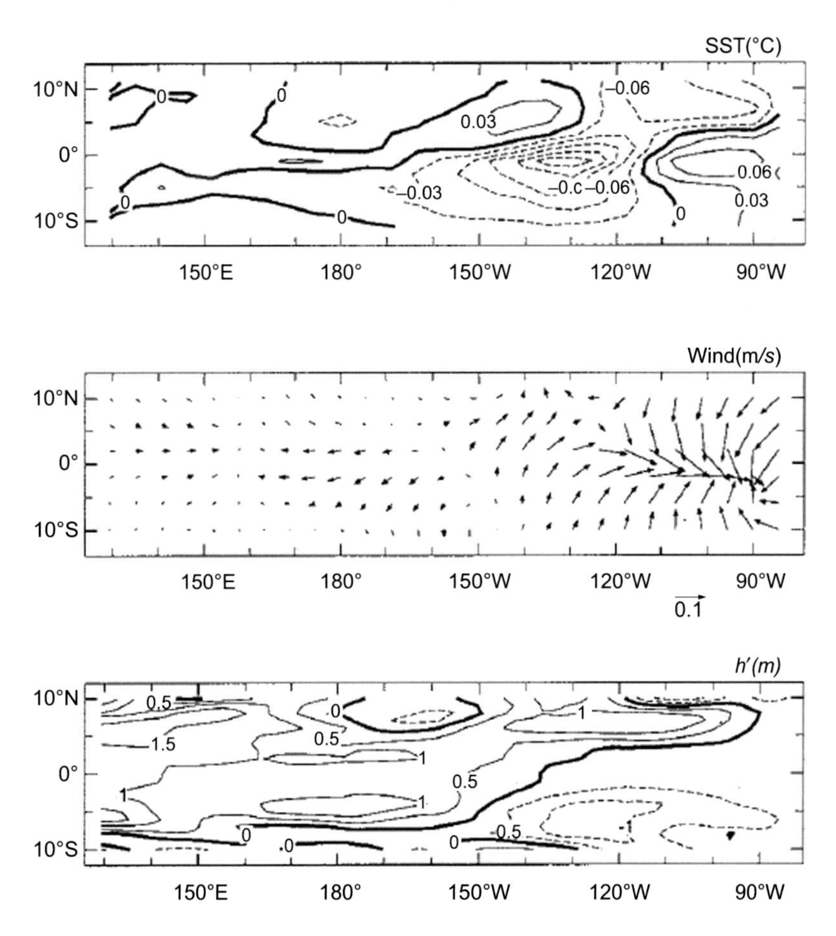

Figure 12.10 The first singular vector (SST, wind and thermocline depth) of the local TLM of the Zebiak-Cane model of ENSO optimised at 6 months from January of year 26 (figure from Xue et al., 1997).

wind anomalies of the singular vector are dominant in the eastern Pacific. With this perturbation on the model state, it develops into a substantial El Niño–like anomaly pattern after six months, with SST anomalies reaching up to 2°C (Xue et al., 1997).

12.3.2 Nonlinear view: CNOPs

When it is aimed to take nonlinear aspects of the spread of trajectories into account, one has to resort to nonlinear optimisation methods, maximising optimal growth of some norm of the solution. To study nonlinear mechanisms of trajectory divergence, Mu (2000) proposed the concept of nonlinear singular vectors and nonlinear singular values, and it was applied in Mu and Wang (2001) to shallow-water flows. In Mu and

Duan (2003), the concept of the conditional nonlinear optimal perturbation (CNOP) was introduced, which we discuss next.

In general, assume that the equations governing the evolution of perturbations can be written as

$$\begin{cases} \dfrac{\partial \boldsymbol{x}}{\partial t} + F(\boldsymbol{x};\bar{\boldsymbol{x}}) = 0, \\ \boldsymbol{x}|_{t=0} = \boldsymbol{x}_0, \end{cases} \quad \text{in} \quad \Omega \times [0, t_e], \tag{12.41}$$

where t is time, $\boldsymbol{x}(t) = (x_1(t), x_2(t), \ldots, x_n(t))$ is the perturbation state vector on a basic state $\bar{\mathbf{x}}$ and F is a nonlinear differentiable operator. Furthermore, $\boldsymbol{x}_0$ is the initial perturbation, $(\boldsymbol{x}, t) \in \Omega \times [0, t_e]$ with Ω a domain in R^n, and $t_e < +\infty$.

Suppose the initial value problem (12.41) is well posed, and the nonlinear propagator $\mathcal{M}$ is defined as the evolution operator of (12.41), which determines a trajectory from the initial time $t = 0$ to time t_e. Hence, for fixed $t_e > 0$, the solution $\boldsymbol{x}(t_e) = \mathcal{M}(\boldsymbol{x}_0;\bar{\boldsymbol{x}})(t_e)$ is well defined, that is,

$$\boldsymbol{x}(t_e) = \mathcal{M}(\boldsymbol{x}_0;\bar{\boldsymbol{x}})(t_e). \tag{12.42}$$

For a chosen norm $\|\cdot\|$, the perturbation $\boldsymbol{x}_{0\delta}$ is called the conditional nonlinear optimal perturbation (CNOP) with constraint condition $\|\boldsymbol{x}_0\| \leq \delta$ if and only if

$$J(\boldsymbol{x}_{0\delta}) = \max_{\|\boldsymbol{x}_0\| \leq \delta} J(\boldsymbol{x}_0), \tag{12.43}$$

where the 'cost function' J is given by

$$J(\boldsymbol{x}_0) = \|\mathcal{M}(\boldsymbol{x}_0;\bar{\boldsymbol{x}})(t_e)\|. \tag{12.44}$$

The CNOP is the initial perturbation whose nonlinear evolution attains the maximal value of the functional J at time t_e with the constraint conditions. The CNOP can be regarded as the most (nonlinearly) unstable initial perturbation superposed on the basic state.

In Mu et al. (2004), the CNOP approach was applied to the two-box model (Stommel, 1961) of the thermohaline circulation described by the dimensionless equations (cf. Example 6.1)

$$\frac{dT}{dt} = \eta_1 - T(1+ \mid T - S \mid), \tag{12.45a}$$

$$\frac{dS}{dt} = \eta_2 - S(\eta_3+ \mid T - S \mid), \tag{12.45b}$$

where $T = T_e - T_p, \;\; S = S_e - S_p$ are the dimensionless temperature and salinity difference between the equatorial and polar box and $\Psi = T - S$ is the dimensionless meridional overturning streamfunction. The initial perturbation is written as $\boldsymbol{x}_0 = (T_0', S_0') = (\delta\cos\theta, \delta\sin\theta)$, where δ indicates the magnitude and θ the direction of the vector.

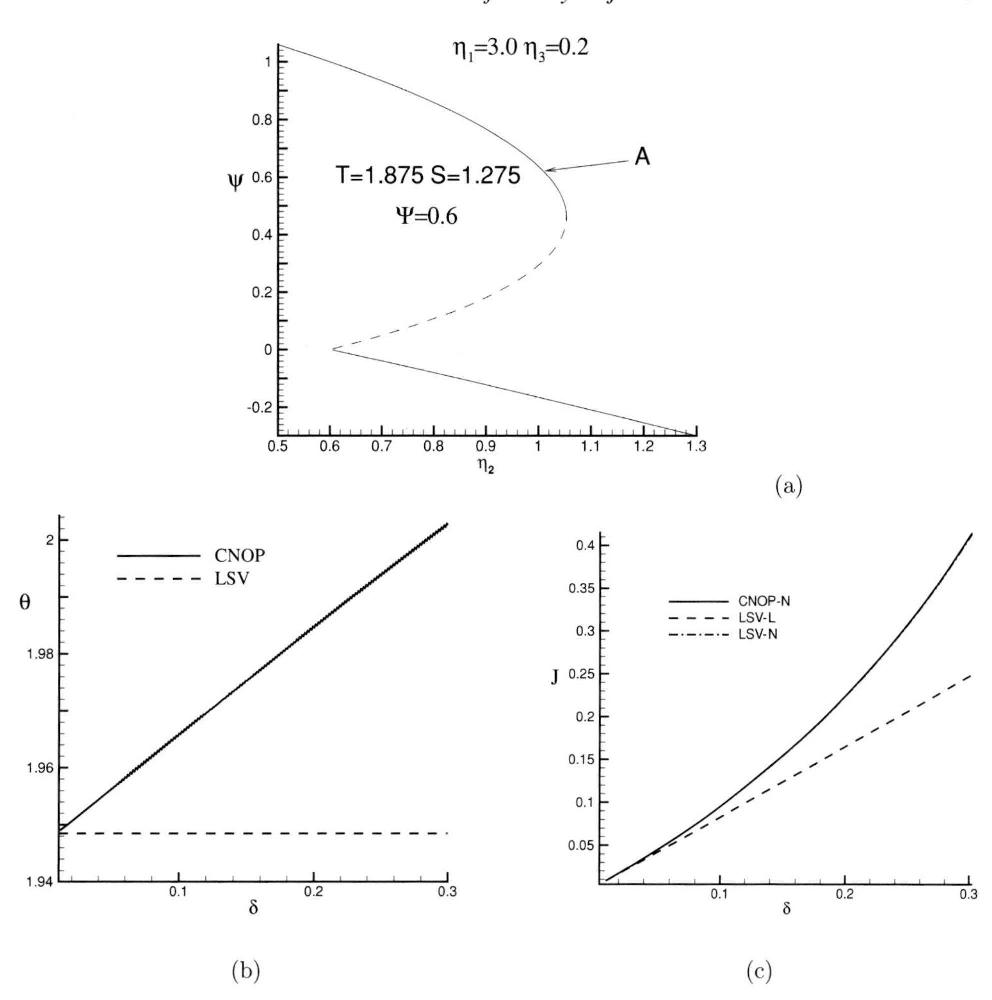

Figure 12.11 (a) Bifurcation diagram for the two-box model (12.45) with η_2 as control parameter and $\eta_1 = 3.0$, $\eta_3 = 0.2$. (b) Values of θ the CNOP and the first linear singular vector (LSV) for the two-box model. (c) Evolution of the cost function J of the time-dependent solution (figure from Mu et al., 2004).

For a thermally dominant (stable) steady state, the state $\bar{T} = 1.875$, $\bar{S} = 1.275$, $\bar{\Psi} = 0.6$ for $\eta_2 = 1.02$ as indicated as point A in Fig. 12.11a is chosen. Furthermore, $\delta = 0.3$ is used as a maximum amplitude of the perturbations. The time $t_e = 2.5$ is about half the time the solution takes to equilibrate to steady state from a particular initial perturbation. The amplitude $\delta = 0.3$ is about 10% of the typical amplitude of the steady state of temperature and salinity $(\bar{T}, \bar{S})$. For θ in the range $\pi/4 < \theta < 5\pi/4$, the initial perturbation flow has $\Psi'(0) < 0$ and hence weakens the thermally dominated flow. For other values of θ, the initial perturbation flow has $\Psi'(0) > 0$, which strengthens the thermally dominated flow.

The results for the CNOP and the first linear singular vectors (LSV) are shown in Fig. 12.11b. The directions of the LSVs, being independent of δ, have constant values of $\theta_1 = 1.948$ (dashed line) and $\theta_2 = 5.089$ (not shown). The directions of the CNOPs (solid curve) increase monotonously, with δ varying from 0.01 to 0.3. The difference between the CNOP and (first) LSV is relatively small when δ is small. Integrating the model with CNOPs and LSVs as initial conditions, respectively, we obtain their value at time t_e, which are denoted as CNOP-N and LSV-N in Fig. 12.11c. To make a comparison, the linear evolution (by the linearised model) of LSVs is also shown (LSV-L) in Fig. 12.11c. It is clear that CNOPs increase nonlinearly as the initial perturbation constraint increases, whereas LSV-Ls only increases linearly. The line of LSV-N is between CNOP-N and LSV-L, but the difference between LSV-N and CNOP-N is hardly distinguishable in Fig. 12.11c.

12.3.3 Nonlinear view: Lyapunov techniques

In the previous sections, all the measures of growth or decay of vectors along a trajectory were dependent on the norm chosen, and hence the results can only be coupled to the physics of the problem when combined with other information of the model system or observations. As we saw in Section 2.4.3, the Lyapunov exponents of an attractor indicate whether exponential divergence of trajectories occurs. When there is a positive Lyapunov exponent, there is sensitivity to initial conditions. These Lyapunov exponents describe the long-term exponential rate of stretching or contraction in the attractor and are norm independent. They reduce to the Floquet multipliers when the trajectory is periodic and to the growth factors of the normal modes when the trajectory is steady. Lyapunov vectors $\boldsymbol{\phi}_i$ are the generalisation of normal mode vectors of a stable steady state and Floquet vectors for a stable periodic orbit to a general time-dependent trajectory. In this section, we discuss methods to obtain information on the Lyapunov vectors directly from the model simulations.

Direct computation

From the Oseledec theorem (see Section 5.5.2), it follows that the Lyapunov exponents $\lambda^{\pm}$, which characterise the asymptotic evolution of linear disturbances to bounded trajectories of arbitrary time dependence, can be calculated as $\lambda_i^{\pm} = \ln s_{\pm}$, where $s_{\pm}$ are the eigenvalues of the matrices

$$\mathcal{S}_{\pm} = \lim_{t \to \pm\infty} (\mathcal{L}^T(t_0, t)\mathcal{L}(t_0, t))^{\frac{1}{2(t-t_0)}}, \qquad (12.46)$$

where $\mathcal{L}(t_0, t)$ is again the tangent linear model. The eigenvalues are norm independent, independent of the initial time t_0 and $\lambda^- = -\lambda^+$. A norm-independent set of Lyapunov vectors $\boldsymbol{\phi}_i$, such that $\boldsymbol{\phi}_i$ grows at a rate $\lambda_i^{\pm}$ as $t \to \pm\infty$, can also be defined using the Oseledec theorem (Eckmann and Ruelle, 1985) using nested subspaces and can be shown to reduce to Floquet vectors and normal modes when the trajectory is periodic and steady, respectively.

As can be anticipated from the discussion on singular vectors in the Section 12.3.1, the set of singular vectors with large optimisation intervals $t - t_0$ are orthogonalisations of the Lyapunov vectors. Let the evolution of the initial singular vectors $\mathbf{v}_j(t_1, t_2)$ using an optimisation interval $t_2 - t_1$ be indicated by

$$\boldsymbol{\xi}_j(t; t_1, t_2) = \mathcal{L}(t_1, t)\mathbf{v}_j(t_1, t_2), \tag{12.47}$$

then the backward singular vectors $\hat{\boldsymbol{\eta}}_j(t)$ and forward singular vectors $\hat{\boldsymbol{\xi}}_j(t)$ are defined as

$$\hat{\boldsymbol{\eta}}_j(t) = \lim_{t_1 \to -\infty} \boldsymbol{\xi}_j(t; t_1, t), \tag{12.48a}$$

$$\hat{\boldsymbol{\xi}}_j(t) = \lim_{t_2 \to \infty} \boldsymbol{\xi}_j(t; t, t_2). \tag{12.48b}$$

To determine a backwards singular vector, the evolved singular vector that was initialised in the distant past is determined at time t. Similarly, the forward singular vector is determined by the initial singular vector with an optimisation time far into the future. The relation between these singular vectors and the Lyapunov vectors is given by

$$\hat{\boldsymbol{\eta}}_j(t) = \sum_{i=1}^{j} \hat{p}_{ji}\boldsymbol{\phi}_i(t), \tag{12.49a}$$

$$\hat{\boldsymbol{\xi}}_j(t) = \sum_{i=j}^{N} \hat{q}_{ji}\boldsymbol{\phi}_i(t), \tag{12.49b}$$

with coefficients $\hat{p}_{ji}$ and $\hat{q}_{ji}$ and where the Lyapunov vectors are ordered with respect to the magnitude of the Lyapunov exponents.

The relations (12.49) provide a method to compute the Lyapunov vectors (Wolfe and Samelson, 2007). It follows that $\langle \boldsymbol{\phi}_i, \hat{\boldsymbol{\xi}}_j \rangle = 0$ for $i < j$ and $\langle \boldsymbol{\phi}_i, \hat{\boldsymbol{\eta}}_j \rangle = 0$ for $i > j$, and hence the $\boldsymbol{\phi}_n$ can be expressed into the vectors $\hat{\boldsymbol{\xi}}_j$ and $\hat{\boldsymbol{\eta}}_j$ through

$$\boldsymbol{\phi}_n = \sum_{i=n}^{N} \langle \boldsymbol{\phi}_n, \hat{\boldsymbol{\xi}}_i \rangle \hat{\boldsymbol{\xi}}_i, \tag{12.50a}$$

$$\boldsymbol{\phi}_n = \sum_{j=1}^{n} \langle \boldsymbol{\phi}_n, \hat{\boldsymbol{\eta}}_j \rangle \hat{\boldsymbol{\eta}}_j, \tag{12.50b}$$

where the dependence on time has been suppressed. Taking the inner product of (12.50a) with $\hat{\boldsymbol{\eta}}_k$ and of (12.50b) with $\hat{\boldsymbol{\xi}}_k$ and eliminating the $\langle \hat{\boldsymbol{\xi}}_k, \boldsymbol{\phi}_n \rangle$ term leads to the linear system of equations

$$\langle \hat{\boldsymbol{\eta}}_k, \boldsymbol{\phi}_n \rangle = \sum_{j=1}^{n} \left[\sum_{i=n}^{N} \langle \hat{\boldsymbol{\eta}}_k, \hat{\boldsymbol{\xi}}_i \rangle \langle \hat{\boldsymbol{\xi}}_i, \hat{\boldsymbol{\eta}}_j \rangle \right] \langle \hat{\boldsymbol{\eta}}_j, \boldsymbol{\phi}_n \rangle \ , \ k \leq n. \tag{12.51}$$

The solution of this system of equations provides the expansion coefficients of the Lyapunov vectors in terms of the backwards singular vectors.

In Wolfe and Samelson (2007), it was realised that because the $\hat{\boldsymbol{\eta}}_i$ and $\hat{\boldsymbol{\xi}}_j$ are orthonormal sets of vectors, it holds that

$$\sum_{i=1}^{N} \langle \hat{\boldsymbol{\eta}}_k, \hat{\boldsymbol{\xi}}_i \rangle \langle \hat{\boldsymbol{\xi}}_i, \hat{\boldsymbol{\eta}}_j \rangle = \delta_{kj}, \tag{12.52}$$

and hence using this in (12.51), it follows that

$$\sum_{j=1}^{n} \left[\sum_{i=1}^{n-1} \langle \hat{\boldsymbol{\eta}}_k, \hat{\boldsymbol{\xi}}_i \rangle \langle \hat{\boldsymbol{\xi}}_i, \hat{\boldsymbol{\eta}}_j \rangle \right] \langle \hat{\boldsymbol{\eta}}_j, \boldsymbol{\phi}_n \rangle = 0 \ , \ k \leq n. \tag{12.53}$$

With the notation $y_k^n = \langle \hat{\boldsymbol{\eta}}_k, \boldsymbol{\phi}_n \rangle$, $k = 1, \ldots, n$ and

$$D_{kj}^n = \sum_{i=1}^{n-1} \langle \hat{\boldsymbol{\eta}}_k, \hat{\boldsymbol{\xi}}_i \rangle \langle \hat{\boldsymbol{\xi}}_i, \hat{\boldsymbol{\eta}}_j \rangle, \ k, j \leq n \tag{12.54}$$

(12.53) can be written as (with D indicating the matrix with elements D_{kj})

$$\mathrm{D}^n \mathbf{y}^n = 0. \tag{12.55}$$

The Lyapunov vectors can hence be determined from only the first $n-1$ forward singular vectors and n backward singular vectors. As we saw in Section 12.2.3, these singular vectors can be obtained from a singular value decomposition for large optimisation times.

Breeding techniques

A method to determine the leading Lyapunov vector that is often applied in operational practice is the calculation of so-called Bred vectors. The original idea (Toth and Kalnay, 1997) was to construct an ensemble of optimal perturbations to carry out an ensemble forecast by selecting the most important growing error. This essentially nonlinear method consists of the following steps (Fig. 12.12a):

(i) Add to the initial state $\mathbf{x}(t_0)$ a small random perturbation $\delta\mathbf{x}(t_0)$ to obtain a perturbed initial state $\tilde{\mathbf{x}}(t_0)$.
(ii) Integrate the model from both the unperturbed and the perturbed initial state for a given time $T = t_1 - t_0$.
(iii) Measure the distance $||\delta\mathbf{x}(t_1)||$ between the two trajectories at time t_1 and rescale this distance to have the same size as the initial one $||\delta\mathbf{x}(t_0)||$.

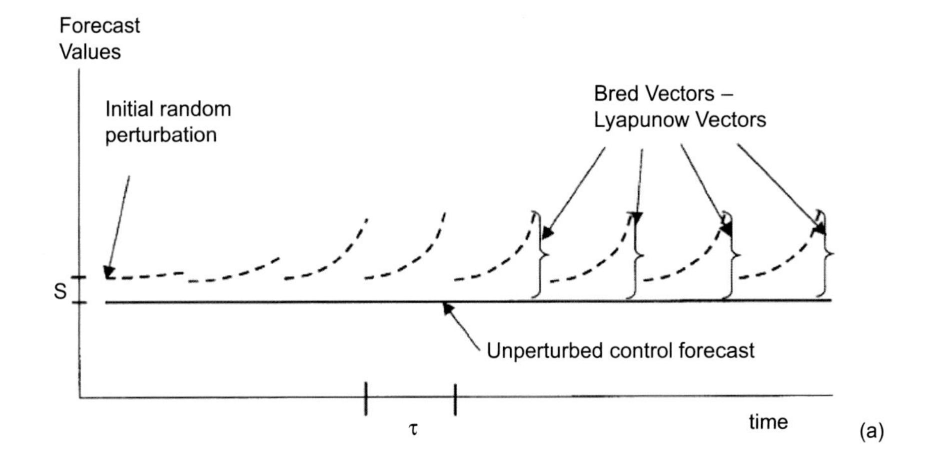

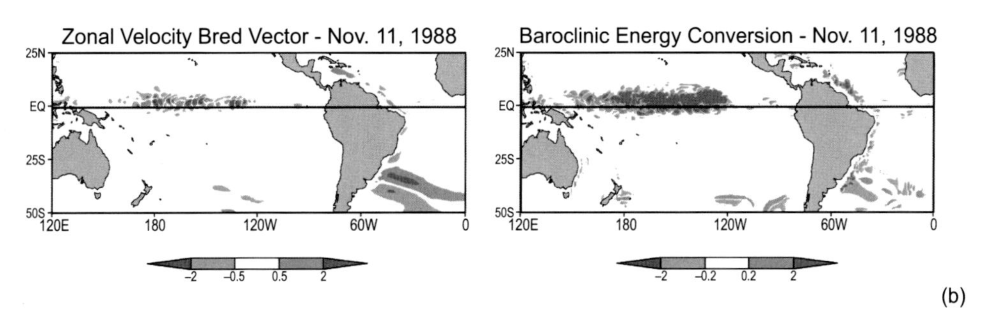

Figure 12.12 (a) Sketch of the procedure to determine Bred vectors. (b) Bred vector (*left panel*) of the zonal velocity starting at 11 November 1988 from the GFDL MOM2 model using a ten-day breeding interval and corresponding baroclinic energy conversion (figure from Hoffman et al., 2009). (See Colour Plate.)

(iv) Add this rescaled perturbation to the 'control simulation' $\mathbf{x}(t_1)$ and repeat from step (ii) until convergence.

After several rescaling steps, the perturbation evolves towards the leading Lyapunov vector (Toth and Kalnay, 1997), and hence the method 'breeds' the nonlinear perturbation that grows fastest. An example of a Bred vector determined from simulations with a global version of the GFDL MOM2 model ($1^\circ \times 1^\circ$ horizontal resolution in midlatitudes reducing to $1^\circ \times 1/2^\circ$ near the equator) with twenty vertical levels and forced by the NCEP reanalysis data was presented in Hoffman et al. (2009). The zonal velocity of the Bred vector for a ten-day breeding interval (Fig. 12.12b) starting on 11 November 1988 shows a dipole pattern near South America and a wave pattern in the Tropical Pacific; the latter could be identified with tropical instability waves (Hoffman et al., 2009).

12.4 Data assimilation

The path of the trajectories of climate models can be 'improved' by combining the model outcomes with available observations in so-called data-assimilation methods. Central to this improvement is an adequate state estimation. Here a distinction is made among a so-called filter, a smoother and a predictor. Suppose that discrete measurements $x_1, \ldots, x_n$ are available, and we are interested in a state estimation $\hat{x}_k$. For a predictor we use only the measurements up to x_{k-1}, for a filter we use measurements up to x_k and for a smoother measurements in the future $(x_l, l > k)$ are also used.

There are basically two classes of data-assimilation methods: (i) those based on optimisation methods using variational techniques and (ii) those based on ensemble approaches. The central ideas of all data assimilation methods, however, go back to those behind the Kalman filter, which we therefore discuss in detail in the next subsection.

12.4.1 The Kalman Filter

To present the Kalman filter in its simplest form, a one-dimensional linear stochastic discrete dynamical system of the form, with $x_j = x(t_j)$,

$$x_j = cx_{j-1} + w_j, \tag{12.56}$$

is considered, with $c \in \mathbb{R}$ and w_j representing Gaussian noise having a $N(0, \sigma_w)$ distribution. Suppose that we measure the state of the system through the output (Fig. 12.13a)

$$y_j = hx_j + v_j, \tag{12.57}$$

where $h \in \mathbb{R}$ and v_j again represent Gaussian noise with a $N(0, \sigma_v)$ distribution; v_j and w_j are independent.

The question that is addressed by the Kalman filter is the following: can we use the y_j to optimally (to minimise the effect of the noise) determine an estimate $\hat{x}_j$ of the state x_j?

Assuming that an estimate $\hat{x}_{j-1}$ is available (Fig. 12.13b), the Kalman filter at time t_j proceeds as follows. First, an a priori estimate $\hat{x}_j^-$ is taken as

$$\hat{x}_j^- = c\hat{x}_{j-1} \tag{12.58}$$

and used to estimate the output $\hat{y}_j = h\hat{x}_j^-$. The difference between this estimate and the measured y_j is, in a second step, used to correct the a priori estimate $\hat{x}_{j-1}^-$ according to

$$\hat{x}_j = \hat{x}_j^- + k_j(y_j - \hat{y}_j) = \hat{x}_j^- + k_j(y_j - h\hat{x}_j^-), \tag{12.59}$$

where k_j is the Kalman gain at time t_j.

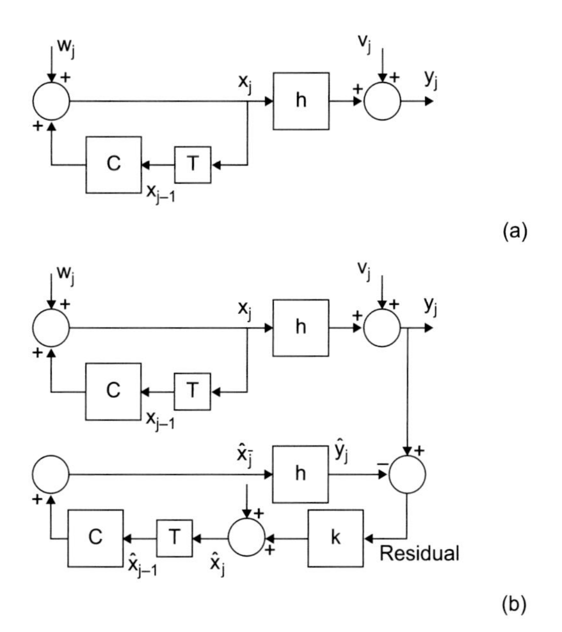

Figure 12.13 (a) Block diagram of the original system where the block 'T' indicates the time delay (slightly modified from http://www.swarthmore.edu/NatSci/echeeve1/Ref/Kalman/ScalarKalman.html). (b) Block diagram of the Kalman filter construction.

To determine k_j, a priori e_j^- and a posteriori e_j errors and variances p_j^- and p_j are defined as

$$e_j^- = x_j - \hat{x}_j^- \ ; \ p_j^- = E[(e_j^-)^2], \tag{12.60a}$$

$$e_j = x_j - \hat{x}_j \ ; \ p_j = E[(e_j)^2], \tag{12.60b}$$

where, as in Chapter 3, $E[x]$ denotes the expectation value of x. The a priori error variance can be written as (using 12.56)

$$p_j^- = E[(x_j - \hat{x}_j^-)^2] = E[(cx_{j-1} + w_j - \hat{x}_j^-)^2] =$$
$$E[(cx_{j-1} + w_j - c\hat{x}_{j-1})^2] = c^2 p_{j-1} + \sigma_w^2, \tag{12.61}$$

where the last equality arises because w_j is uncorrelated to both output and the a priori estimate. The variance p_{j-1} (computed at t_{j-1}) is known at time t_j.

In the Kalman filter, the gain k_j is determined such that p_j is minimised, hence, using (12.59),

$$\frac{\partial p_j}{\partial k_j} = \frac{\partial E[(x_j - \hat{x}_j)^2]}{\partial k_j} = \frac{\partial E[(x_j - \hat{x}_j^- + k_j(y_j - h\hat{x}_j^-))^2]}{\partial k_j} = 0, \tag{12.62}$$

from which it follows that

$$k_j = \frac{E[(y_j - h\hat{x}_j^-)(x_j - \hat{x}_j^-)]}{E[(y_j - h\hat{x}_j^-)^2]}. \tag{12.63}$$

This expression can be simplified using (12.57) according to

$$E[(y_j - h\hat{x}_j^-)(x_j - \hat{x}_j^-)] = E[h(x_j - \hat{x}_j^-)^2] + E[v_j(x_j - \hat{x}_j^-)] = hE[(e_j^-)^2] = hp_j^-,$$
$$E[(y_j - h\hat{x}_j^-)^2] = E[(h(x_j - \hat{x}_j^-) + v_j)^2] = h^2 p_j^- + \sigma_v^2,$$

because v_j is uncorrelated to both output and the a priori estimate. As a consequence, (12.63) reduces to

$$k_j = \frac{hp_j^-}{h^2 p_j^- + \sigma_v^2} = \frac{h(c^2 p_{j-1} + \sigma_w^2)}{h^2(c^2 p_{j-1} + \sigma_w^2) + \sigma_v^2}, \tag{12.64}$$

where in the last equation the expression (12.61) for p_j^- was used.

Finally, the a posteriori error variance p_j can be determined from

$$\begin{aligned} p_j &= E[(x_j - \hat{x}_j)^2] = E[(x_j - (\hat{x}_j^- + k_j(hx_j + v_j - h\hat{x}_j^-)))^2] \\ &= E[((x_j - \hat{x}_j^-)(1 - hk_j) - k_j v_j)^2] \\ &= (1 - hk_j)^2 p_j^- + k_j^2 \sigma_v^2. \end{aligned}$$

Expressing σ_v^2 into p_j^- using (12.64), that is,

$$\sigma_v^2 = p_j^- \frac{h(1 - hk_j)}{k_j}, \tag{12.65}$$

then eventually gives

$$p_j = (1 - hk_j)p_j^- = (1 - hk_j)(c^2 p_{j-1} + \sigma_w^2). \tag{12.66}$$

We now summarise the two-step process of the Kalman filter and simultaneously provide its multidimensional extension. In this case, the system and measurement is described by

$$\mathbf{x}_j = \mathbf{C}\mathbf{x}_{j-1} + \mathbf{W}_j, \tag{12.67a}$$

$$\mathbf{y}_j = \mathbf{H}\mathbf{x}_j + \mathbf{V}_j, \tag{12.67b}$$

with obvious extensions from the preceding one-dimensional case, and let Σ_W and Σ_V indicate the covariance matrices of the noise $\mathbf{W}$ and $\mathbf{V}$, respectively.

The predictor and corrector steps in the Kalman filter are then given by

$$\hat{\mathbf{x}}_j^- = \mathbf{C}\hat{\mathbf{x}}_{j-1}, \tag{12.68a}$$

$$\hat{\mathbf{x}}_j = \hat{\mathbf{x}}_j^- + \mathbf{K}_j(\mathbf{y}_j - \mathbf{H}\hat{\mathbf{x}}_j^-). \tag{12.68b}$$

The a priori and a posteriori error covariances are given by (compare with [12.60])

$$\mathbf{P}_j^- = E[(\mathbf{x} - \hat{\mathbf{x}}_j^-)(\mathbf{x} - \hat{\mathbf{x}}_j^-)^T], \tag{12.69a}$$

$$\mathbf{P}_j = E[(\mathbf{x} - \hat{\mathbf{x}}_j)(\mathbf{x} - \hat{\mathbf{x}}_j)^T], \tag{12.69b}$$

and $\mathbf{K}_j$ is determined by minimising $\mathbf{P}_j$ which results in (compare with [12.64])

$$\mathbf{K}_j = \frac{\mathbf{P}_j^- \mathbf{H}^T}{\mathbf{H}\mathbf{P}_j^- \mathbf{H}^T + \Sigma_V}, \tag{12.70}$$

where the a priori covariance matrix is found from (compare with [12.61])

$$\mathbf{P}_j^- = \mathbf{C}\mathbf{P}_{j-1}\mathbf{C}^T + \Sigma_W, \tag{12.71}$$

and finally the a posteriori covariance matrix is given by (compare with [12.66])

$$\mathbf{P}_j = (\mathbf{I} - \mathbf{K}_j \mathbf{H})\mathbf{P}_j^-. \tag{12.72}$$

In the climate data assimilation literature, one often uses a slightly different terminology. Here a background model state $\mathbf{x}^b$ (instead of $\hat{\mathbf{x}}^-$) serves as the prior estimate with covariance error matrix $\mathbf{B}$ (instead of $\mathbf{P}^-$). The estimated state is called the analysis $\mathbf{x}^a$ (instead of $\hat{\mathbf{x}}$), and the a posteriori covariance matrix is usually indicated by $\mathbf{A}$, the analysis errors (instead of $\mathbf{P}$). The covariance matrix of the observational errors is usually indicated by $\mathbf{R}$ (instead of Σ_V), and often the model error covariance Σ_W is neglected (perfect model). From the Kalman filter, the analysis $\mathbf{x}^a$ then follows from

$$\mathbf{x}^a = \mathbf{x}^b + \mathbf{K}(\mathbf{y} - \mathbf{H}\mathbf{x}^b), \tag{12.73}$$

and the Kalman-gain matrix $\mathbf{K}$ and the analysis errors follow from

$$\mathbf{K} = \frac{\mathbf{B}\mathbf{H}^T}{\mathbf{H}\mathbf{B}\mathbf{H}^T + \mathbf{R}}, \tag{12.74a}$$

$$\mathbf{A} = (\mathbf{I} - \mathbf{K}\mathbf{H})\mathbf{B}. \tag{12.74b}$$

Although the Kalman filter is a basis for many of the data-assimilation methods in oceanography and meteorology (and hence its basic technique was presented in this section), the multidimensional version becomes prohibitively computationally expensive for large-dimensional dynamical systems, and hence many approximative Kalman filter methods are in use (Evensen, 2009).

12.4.2 Ensemble-based methods

As mentioned in Section 12.1, ensemble methods are often used in climate modelling. Here, many simulations are performed with different initial conditions, and using data assimilation techniques, observations can be used to limit the spread of the ensemble.

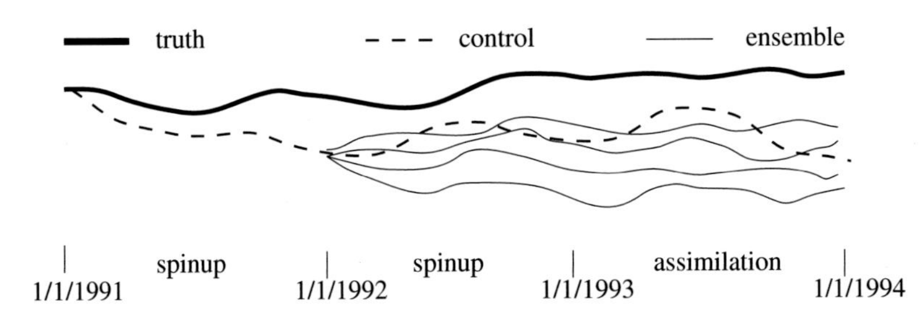

Figure 12.14 Schematic of the experiment set-up in Leeuwenburgh (2005).

Next, the state vector of ensemble member i is indicated by $\mathbf{x}_i$, and we indicate the ensemble mean operator with an overbar.

Ensemble Kalman Filter

In the Ensemble Kalman filter (EnKF) method with given observations $\mathbf{y}$ at a time t, these observations first are perturbed (Burgers et al., 1998) to obtain

$$\mathbf{y}_i = \mathbf{y} + \mathbf{e}_i \rightarrow \mathbf{R}_e = \overline{\mathbf{e}\mathbf{e}^T}, \tag{12.75}$$

and the analysis for each ensemble member follows from (12.73) as

$$\mathbf{x}_i^a = \mathbf{x}_i^b + \mathbf{K}_e\left(\mathbf{y}_i - \mathbf{H}\mathbf{x}_i^b\right). \tag{12.76}$$

The Kalman gain matrix $\mathbf{K}_e$ is now computed using ensemble mean quantities, similar to (12.74),

$$\mathbf{K}_e = \frac{\mathbf{B}_e\mathbf{H}^T}{\mathbf{H}\mathbf{B}_e\mathbf{H}^T + \mathbf{R}_e}, \tag{12.77}$$

where

$$\mathbf{B}_e = \overline{(\mathbf{x}^b - \overline{\mathbf{x}^b})(\mathbf{x}^b - \overline{\mathbf{x}^b})^T}, \tag{12.78}$$

and the analysis error is computed from

$$\mathbf{A} = \overline{(\mathbf{x}^a - \overline{\mathbf{x}^a})(\mathbf{x}^a - \overline{\mathbf{x}^a})^T} = (\mathbf{I} - \mathbf{K}_e\mathbf{H})\mathbf{B}_e + \mathcal{O}(N^{-1/2}), \tag{12.79}$$

if N is the number of ensemble members. The EnKF is used in many prediction and data-assimilation studies in climate dynamics (Evensen, 2009), and a website with up-to-date information and codes can be found at http://enkf.nersc.no/.

An example of the use of the EnKF is provided in the identical twin study of Leeuwenburgh (2005), where SSH data are assimilated into a global ocean model. The 'truth' was created by forcing the ocean model with NCEP reanalysis fields, whereas during the control and assimilation runs, the model was forced by ERA40 fields (Fig. 12.14). SSH 'observations' from the truth were assimilated into along

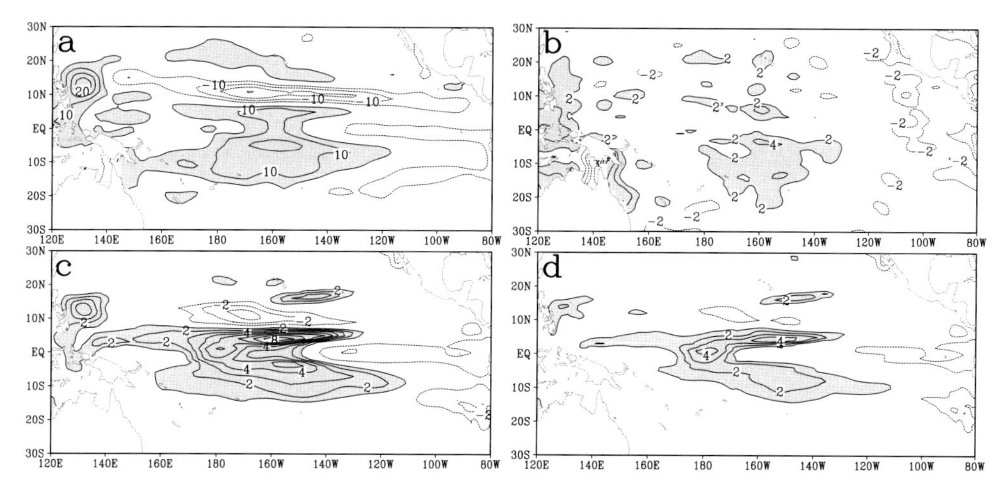

Figure 12.15 Errors in the (a–c) forecast (without data assimilation) and (b–d) analysis at 8 January 1993, for sea-level height (cm) (a–b) and temperature (°C) (c–d) at 50 depth (figure from Leeuwenburgh, 2005).

TOPEX/POSEIDON tracks over the equatorial Pacific every ten days. The results indicate that the assimilation of SSH leads to a significant improvement along the equator in all subsurface fields relative to the unconstrained control simulation (Fig. 12.15).

Particle filtering

With an ensemble of trajectories (or particles), we obtain at a certain time a sampling of the probability density function of the background (or prior estimate), and through the Kalman filter, we obtain an estimate of the probability density function of the analysis (or posterior estimate). When the model is nonlinear and/or the prior probability density function is non-Gaussian, the EnKF methodology is no longer appropriate. One of the possibilities is to use Bayes's theory to determine the posterior probability density function from the prior probability density function using the extra information provided by additional observations. This requires that the probability density functions are properly sampled and ensemble methods are ideally suited to do so; it leads to so-called particle filter methods (Van Leeuwen, 2009).

To illustrate the basics of a particle filter method, assume that the model is represented by the equations

$$\mathbf{x}_{k+1} = \mathcal{M}_k(\mathbf{x}_k, \mathbf{W}_k), \tag{12.80}$$

where $\mathbf{x}_k$ is the state vector at time t_k, $\mathcal{M}$ is the propagator and $\mathbf{W}_k$ is the noise in the model. At time t_k, the observation vector is indicated by $\mathbf{y}_k$. To determine the probability density function of the state vector $\mathbf{x}_k$, N independent particles or trajectories are available (from a model simulation over the interval $[t_{k-1}, t_k]$), which provide states $\mathbf{x}_k^i$, $i = 1, \ldots, N$. The probability density function of the model, $p_N(\mathbf{x}_k)$, can

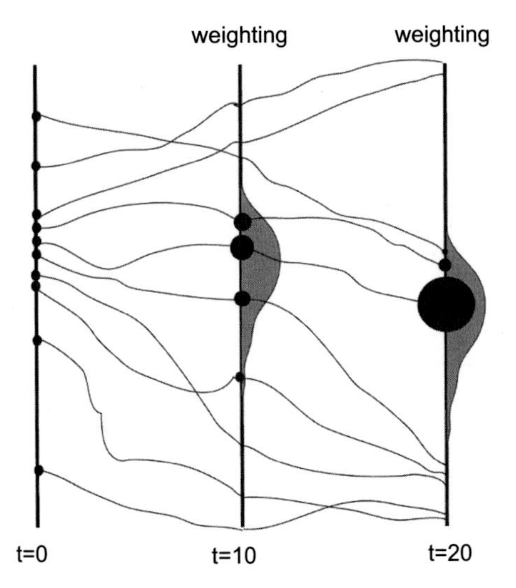

Figure 12.16 Schematic of the standard particle filter method with importance sampling (figure from Van Leeuwen, 2009).

then be represented by

$$p_N(\mathbf{x}_k) = \frac{1}{N}\sum_{i=1}^{N}\delta(\mathbf{x}_k - \mathbf{x}_k^i), \tag{12.81}$$

where δ is the Dirac function.

Suppose now that at an initial time $t = t_0$, N trajectories are computed. As observations are not yet taken into account, the particles are given an equal weight $w_0^i = 1/N$, as any of these trajectories could be close to later observations and $p_N(\mathbf{x}_0)$ is chosen (Fig. 12.16). After time $\Delta t = t_1 - t_0$, the prior probability density function of the model $p_N(\mathbf{x}_1)$ can be computed through (12.81), which can be written as

$$p_N(\mathbf{x}_1) = \sum_{i=1}^{N} w_0^i \delta\big(\mathbf{x}_1 - \mathbf{x}_1^i\big). \tag{12.82}$$

Next, the observations $\mathbf{y}_1$ are taken into account to compute a posterior probability density function. The joint probability of two events a and b can be written in two ways using conditional probabilities

$$P(a \in A, b \in B) = P(a \in A|b \in B)P(b \in B), \tag{12.83a}$$

$$P(a \in A, b \in B) = P(b \in B|a \in A)P(a \in A), \tag{12.83b}$$

and, combining these two relations, we find Bayes's theorem

$$P(a \in A|b \in B) = \frac{P(b \in B|a \in A)P(a \in A)}{P(b \in B)}. \tag{12.84}$$

In the particle filter methodology, Bayes's theorem is used in the form

$$p_N(\mathbf{x}_1|\mathbf{y}_1) = \frac{p_N(\mathbf{y}_1|\mathbf{x}_1)p_N(\mathbf{x}_1)}{p_N(\mathbf{y}_1)} = \sum_{i=1}^{N} w_1^i \delta(\mathbf{x}_1 - \mathbf{x}_1^i), \tag{12.85}$$

where the new weights are found from (using [12.82])

$$w_1^i = \frac{p_N(\mathbf{y}_1|\mathbf{x}_1)}{p_N(\mathbf{y}_1)} w_0^i. \tag{12.86}$$

The probability $p_N(\mathbf{y}_1|\mathbf{x}_1)$ of the observation $\mathbf{y}_1$ given the model state $\mathbf{x}_1$, is directly linked to the observational error. For example, with a univariate measurement y_1 having a Gaussian distribution with a variance σ_{obs}^2, the probability follows from

$$p_N(y_1|\mathbf{x}_1) \sim e^{-\frac{(H(\mathbf{x}_1)-y_1)^2}{2\sigma_{obs}^2}}, \tag{12.87}$$

where $H(\mathbf{x}_1)$ is the model equivalent of the observation calculated using the observational operator H. The probability $p_N(\mathbf{y}_1)$ serves as a normalisation that can be calculated by the constraint

$$\sum_{i=1}^{N} w_1^i = \sum_{i=1}^{N} \frac{p_N(\mathbf{y}_1|\mathbf{x}_1)}{p_N(\mathbf{y}_1)} w_0^i = 1. \tag{12.88}$$

Once the new weights have been determined, the posterior probability density function is known. In the particle filter procedure, the weights of the different particles increase when the trajectories are closer to observations (see Fig. 12.16).

A typical example, where horizontal mixing coefficients in a global ocean model are estimated from sea-level observations using a particle filter method, can be found in Vossepoel and Van Leeuwen (2007). Let the horizontal diffusion coefficient for heat and salt be indicated by K_H and the horizontal viscosity by A_H. In a 128-member ensemble starting from the same initial conditions, the ocean model is integrated forward in time. For each ensemble member, the diffusivity and viscosity are taken as $A_H = cA_H^0$ and $K_H = cK_H^0$, where c is chosen from a uniform distribution on the interval (0, 1) and A_H^0 and K_H^0 are standard values.

From the simulations, the global sea-surface height field is determined, and the spread of this quantity in the ensemble is shown in Fig. 12.17a. In the regions where the spread of the ensemble is largest, the sea level is most sensitive to differences in lateral mixing parameterization. Synthetic sea-level observations are next taken from a simulation of the model with a value $c = 0.2$ (and also from $c = 0.8$) and random uncorrelated noise has been added with a standard deviation of 5 cm. These 'observations' are assimilated using the particle filter, and the weights w_i from (12.86) are shown for both cases ($c = 0.2$ and $c = 0.8$) in Fig. 12.17c–d. The narrow weights for the case $c = 0.2$, compared with those for $c = 0.8$, indicate a high sensitivity of sea level to horizontal mixing when the global horizontal mixing (the 'truth' from a low mixing model case) is low. The shape of the weights for $c = 0.8$ demonstrates the strong nonlinear dependence of sea level on the horizontal mixing coefficients. The resulting sea level in the inverse estimation for the case $c = 0.8$ is close to the synthetic truth (Fig. 12.17b).

12.4.3 Variational data assimilation

The Kalman filter result (12.74) is optimal, and it can be shown that the analysis $\mathbf{x}^a$ in (12.73) is a solution of the optimisation problem (Lorenc, 1988)

$$\mathbf{x}^a = \min_{\mathbf{x}} J(\mathbf{x}), \tag{12.89}$$

where the cost function J is given by

$$J(\mathbf{x}) = (\mathbf{x} - \mathbf{x}^b)^T \mathbf{B}^{-1} (\mathbf{x} - \mathbf{x}^b) + (\mathbf{y} - \mathbf{H}\mathbf{x})^T \mathbf{R}^{-1} (\mathbf{y} - \mathbf{H}\mathbf{x}). \tag{12.90}$$

The variational data assimilation methods used differ in (i) the cost function J that is minimised, (ii) which control variable is used in the minimisation and (iii) how errors are taken into account. As an example, we provide a brief description of 4D-Var. Here the control variable is the initial state $\mathbf{x}(t_0)$ at time $t = t_0$. Suppose the initial condition $\mathbf{w}^b(t_0)$ of the background model is given. The analysis is the model trajectory that simultaneously minimises the distance to the initial background $\mathbf{w}^b(t_0)$ and the observations $\{\mathbf{y}_i : i = 0, \cdots, N-1\}$. This is an optimisation problem, which in the incremental 4D-Var formulation (Courtier et al., 1994) is stated as

$$\delta\mathbf{w}^a = \min_{\delta\mathbf{w}} J(\delta\mathbf{w}), \tag{12.91a}$$

$$J(\delta\mathbf{w}) = \delta\mathbf{w}^T \mathbf{B}^{-1} \delta\mathbf{w} + \sum_{i=0}^{N-1} \mathbf{d}_i^T \mathbf{R}_i^{-1} \mathbf{d}_i, \tag{12.91b}$$

$$\mathbf{d}_i = \mathbf{y}_i - H_i \mathcal{M}(t_0, t_i)(\mathbf{w}^b(t_0)) - \mathbf{H}_i \mathbf{L}(t_0, t_i)\delta\mathbf{w}. \tag{12.91c}$$

In the preceding equations, J is the cost function that measures the distance to the observations and the initial conditions, $\delta\mathbf{w}^a$ is the optimal increment on the initial background $\mathbf{w}^b(t_0)$ state and $\mathbf{d}_i$ is the departure of the model trajectory from observation $\mathbf{y}_i$ (Fig. 12.18). The operators $\mathcal{M}$ and H_i are the evolution operator and the observation operator with $\mathbf{L}$ and $\mathbf{H}_i$ their linearisations around the background trajectory $\mathbf{w}^b(t_i)$. The matrices $\mathbf{B}$ and $\mathbf{R}_i$ are the covariance matrices for the background errors and observational errors. Given an optimum $\delta\mathbf{w}^a$ of (12.91), the analysis $\mathbf{w}^a(t_i)$ is given by

$$\mathbf{w}^a(t_i) = \mathcal{M}(t_0, t_i)(\mathbf{w}^b(t_0) + \delta\mathbf{w}^a). \tag{12.92}$$

A minimum of the cost function J is computed, for example, with a quasi-Newton conjugate gradient method in which the gradient ∇J is needed. For (12.91), the

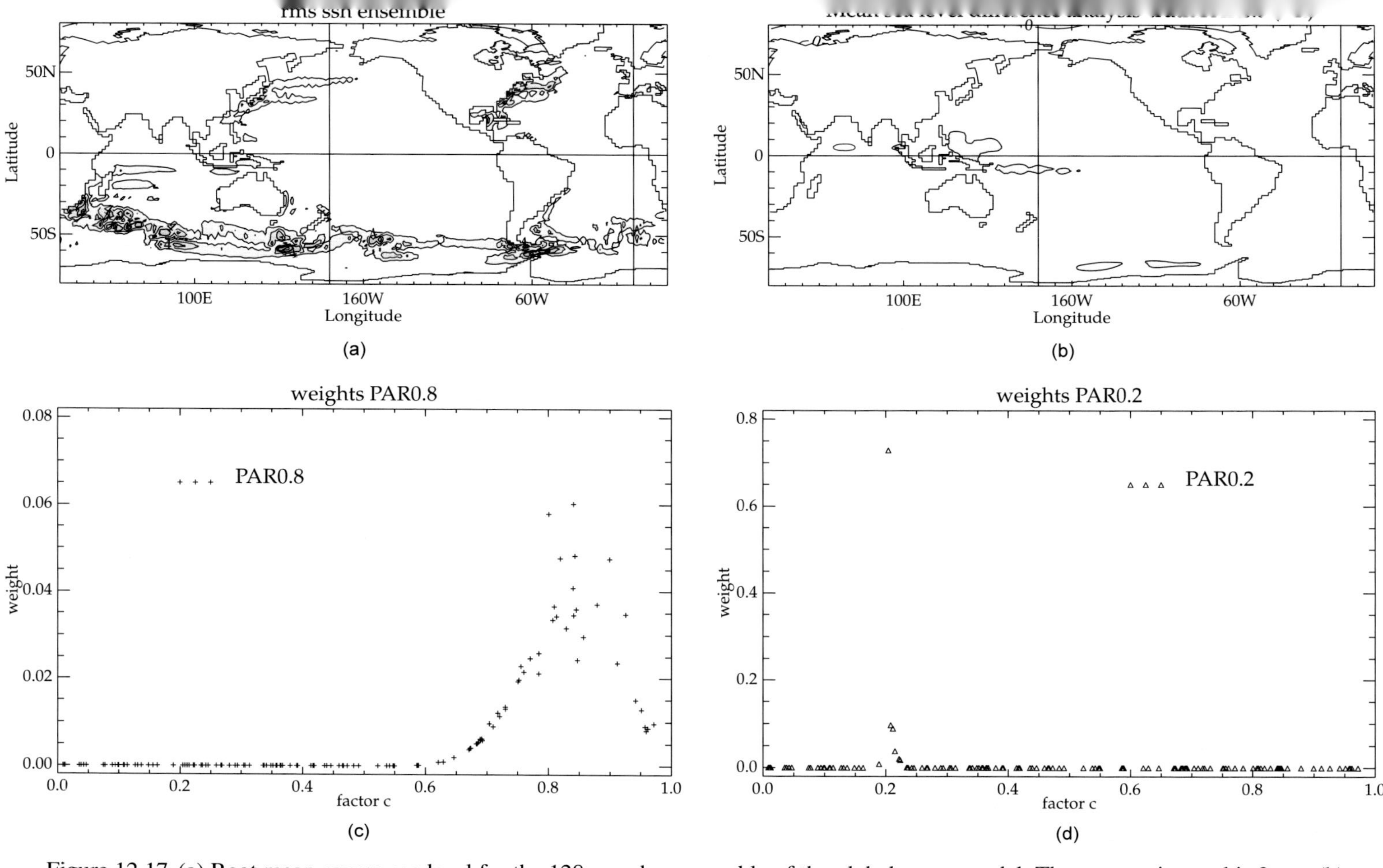

Figure 12.17 (a) Root mean square sea level for the 128-member ensemble of the global ocean model. The contour interval is 2 cm. (b) Difference between synthetic observations and inverse estimate for $c = 0.8$. (c) Values of w_i for each of the ensemble members as a function of c, with 'observations' for the case $c = 0.2$. (d) Same as (c) with 'observations' for the case $c = 0.8$ (figure from Vossepoel and Van Leeuwen, 2007).

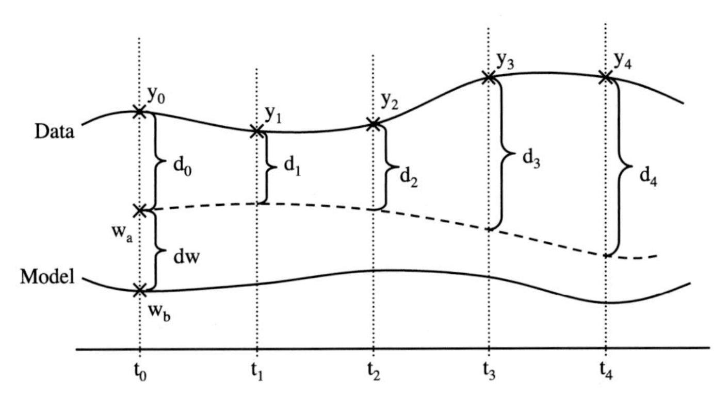

Figure 12.18 Sketch of the 4D-Var method, showing an assimilation interval with 4 points and indicating the model trajectory, the observations and the analysis.

gradient is given by

$$\nabla J(\delta \mathbf{w}) = 2\mathbf{B}^{-1}\delta \mathbf{w} - 2\sum_{i=0}^{N-1} \mathbf{L}(t_0, t_i)^T \mathbf{H}_i^T \mathbf{R}_i^{-1} \mathbf{d}_i. \tag{12.93}$$

For explicit time integration models, the procedure is to evaluate the cost function by forward time stepping, whereas the gradient is evaluated by integrating backward in time using the adjoint model $\mathbf{L}(t_0, t_i)^T$. For 4D-Var with implicit time stepping, only forward time stepping is used to evaluate the gradient (Terwisscha van Scheltinga and Dijkstra, 2005).

The 4D-Var method is used in many applications in the geosciences, such as numerical weather prediction (Lorenc, 2007), atmospheric reanalysis (Kalnay et al., 1996; Compo et al., 2011) and ocean state estimation (Wunsch and Heimbach, 2007; Moore et al., 2011). Comparisons between the use of 4D-Var and EnKF methods on several atmospheric models are provided in Kalnay et al. (2007) and Buehner et al. (2010).

12.5 Outlook

The aim of this chapter was to show the role of concepts of (stochastic) dynamical systems theory in climate prediction. As we have seen, one of the central elements in prediction is the amplification of errors (perturbations) along trajectories. Many systems, even on climate time scales, are nonlinear and display non-Gaussian statistics, and concepts of CNOPs, Lyapunov vectors and Bred vectors are better suited in real applications than those (e.g., singular vectors) based on linear systems and Gaussian statistics. The concepts of dynamical systems are now also widely and successfully applied in prediction systems where, by incorporating observations through data assimilation, adequate state estimations and useful predictions can be obtained for highly nonlinear systems.

The skill of numerical weather prediction has improved over the years, and forecasts a few days ahead over many regions over the globe have reasonable skill. The growth of perturbations associated with instabilities in the atmospheric flow field limits the predictability horizon to about ten days. On the seasonal time scale, skill is improving, and regions of high-potential predictability (Rowell, 2010) are being identified. El Niño prediction is hampered by the existence of the spring predictability barrier, with a large growth of errors during the northern hemispheric spring due to a particular sensitivity of the equatorial ocean-atmosphere system (Duan et al., 2009). There is much activity in the area of decadal predictability (Keenlyside et al., 2008), but the processes controlling the predictability horizons are not very clear yet. In the IPCC-AR4 and also for the next IPCC assessment, attempts are made to determine the range of possible future climate states in 2100 from ensembles of simulations with a suite of global climate models (Meehl et al., 2007). As mentioned in Section 12.1, the additional uncertainty here is the forcing of the system, in particular the concentration of atmospheric greenhouse gas concentrations. Hence, in the IPCC-AR4 assessment, one only talks about projections instead of predictions.

Important issues in climate research where the potential of dynamical systems has not been fully explored are the detection of critical behaviour and the a priori determination of extremes. How do we sense that the climate state vector is close to a transition point such as a Hopf bifurcation or a saddle-node bifurcation? Can we predict the approximation of these points from the methods used in the previous sections with a useful skill? In recent years, there has been substantial effort to detect transition behaviour from time series (Held and Kleinen, 2004; Livina and Lenton, 2007; Scheffer et al., 2009; Lenton, 2011; Thompson and Sieber, 2011), although in many cases, the available data series are too short to provide convincing answers. Extreme behaviour is centrally important in climate change issues: eventually, it is the extremes in temperature (heat waves) and precipitation (flooding) that provide most impact and damage to nature and society. The analysis of extreme behaviour in dynamical systems is currently an active research area (Sterl et al., 2008; Holland et al., 2012).

The main challenge of the application of all the techniques in this chapter to problems in climate predictability is related to the 'curse of dimensionality'. The spatial resolution and the number of processes represented in climate models will increase with time (cf. Chapter 6), and a model in which all dynamically relevant scales of motion are represented will have an estimated 10^{12} degrees of freedom. With the speed of processors being stagnant and the computer hardware switching to graphical processors units (GPUs) and multicore platforms, enormous challenges lie ahead to be able to apply data-assimilation techniques to future climate models on the algorithm aspects, the data-handling aspects and the high-performance computing aspects. Definitely, major technical hurdles must be cleared before reliable forecasts of future climate conditions can be produced using the best climate models available.

Finally, although this may not be realised by the general public and policy makers, climate prediction provides a historic opportunity for humankind to determine at least one aspect of life relatively far into the future with a specified quantitative uncertainty. The impact of such predictions cannot be underestimated, as is now already the case for numerical weather predictions a few days ahead.

References

Abarbanel, H. (1996). *Analysis of Observed Chaotic Data*. Springer, New York, U.S.A.

Allen, M. R. and Robertson, A. W. (1996). Distinguishing modulated oscillations from coloured noise in multivariate datasets. *Clim. Dyn.*, **12**, 775–784.

Alley, R. B., Marotzke, J., Nordhaus, W. D., Overpeck, J. T., Peteet, D. M., Pielke Jr, R. A., Pierrehumbert, R. T., Rhines, P. B., Stocker, T. F., Talley, L. D., and Wallace, J. M. (2003). Abrupt climate change. *Nature*, **299**, 2005–2010.

Alvarez-Garcia, F., Latif, M., and Biastoch, A. (2008). On multidecadal and quasidecadal North Atlantic variability. *J. Clim.*, **21**(14), 3433–3452.

Ambaum, M., Hoskins, B. J., and Stephenson, D. B. (2001). Arctic oscillation or North Atlantic oscillation? *J. Clim.*, **14**(16), 3495–3507.

Ambaum, M. H. P. (2008). Unimodality of wave amplitude in the northern hemisphere. *J. Atmos. Sci.*, **65**(3), 1077–1086.

Andersen, K. and coauthors (2004). High-resolution record of Northern Hemisphere climate extending into the last interglacial period. *Nature*, **431**, 147–151.

Archer, D., Winguth, A., and Lea, D. (2000). What caused the glacial/interglacial atmospheric pCO2 cycles. *Rev. Geophys.*, **38**, 159–189.

Arnold, L. (1998). *Random Dynamical Systems*. Springer Verlag, New York, U.S.A.

Athanasiadis, P. J. and Ambaum, M. H. P. (2010). Do high-frequency eddies contribute to low-frequency teleconnection tendencies?. *J. Atmos. Sci.*, **67**(2), 419–433.

Barsugli, J. J. and Battisti, D. S. (1998). The basic effects of atmosphere-ocean thermal coupling on midlatitude variability. *J. Atmos. Sci.*, **55**, 477–493.

Battisti, D. and Hirst, A. (1989). Interannual variability in a tropical atmosphere-ocean model: influence of the basic state, ocean geometry and nonlinearity. *J. Atmos. Sci.*, **46**, 1687–1712.

Belkin, I. M., Levitus, S., Antonov, J. I., and Malmberg, S.-A. (1998). "Great Salinity Anomalies" in the North Atlantic. *Prog. Oceanogr.*, **41**, 1–68.

Benedict, J. J., Lee, S., and Feldstein, S. B. (2004). Synoptic view of the North Atlantic Oscillation. *J. Atmos. Sci.*, **61**, 121–144.

Benzi, R., Parisi, G., Sutera, A., and Vulpiani, A. (1982). Stochastic resonance in climatic change. *Tellus*, **34**, 10–16.

Benzi, R., Parisi, G., Sutera, A., and Vulpiani, A. (1983). A theory of stochastic resonance in climatic change. *SIAM J. Appl. Math.*, **43**, 565–578.

Benzi, R., Speranza, A., and Sutera, A. (1986). A minimal baroclinic model for the statistical properties low-frequency variability. *J. Atmos. Sci.*, **43**(23), 2962–2967.

Berner, J. and Branstator, G. (2007). Linear and nonlinear signatures in the planetary wave dynamics of an AGCM: Probability density functions. *J. Atmos. Sci.*, **64**(1), 117–136.

Blanke, B., Neelin, J., and Gutzler, D. (1997). Estimating the effect of stochastic wind stress forcing on ENSO irregularity. *J. Clim.*, **10**, 1473–1486.

Bond, G., Showers, W., Cheseby, M., Lotti, R., Almasi, P., deMenocal, P., Priori, P., Cullen, H., Hajdas, I., and Bonani, G. (1997). A pervasive millennial-scale cycle in the North Atlantic Holocene and glacial climates. *Science*, **278**, 1257–1265.

Braun, H., Christl, M., Rahmstorf, S., Ganopolski, A., Mangini, A., Kubatzki, C., Roth, K., and Kromer, B. (2005). Possible solar origin of the 1,470-year glacial climate cycle demonstrated in a coupled model. *Nature*, **438**, 208–211.

Braun, H., Ganopolski, A., Christl, M., and Chialvo, D. R. (2007). A simple conceptual model of abrupt glacial climate events. *Nonlin. Process. Geophys.*, **14**(6), 709–721.

Bretherton, F. (1988). *Earth System Science: A Closer View*. NASA, Washington, DC, U.S.A.

Budyko, M. I. (1969). The effect of solar radiations on the climate on the Earth. *Tellus*, **21**, 611–619.

Buehner, M., Houtekamer, P. L., Charette, C., Mitchell, H. L., and He, B. (2010). Intercomparison of variational data assimilation and the ensemble Kalman filter for global deterministic NWP. Part I: description and single-observation experiments. *Month. Weather Rev.*, **138**(5), 1550–1566.

Burgers, G. (1999). The El Niño stochastic oscillator. *Clim. Dyn.*, **15**(7), 521–531.

Burgers, G., van Leeuwen, P., and Evensen, G. (1998). Analysis scheme in the ensemble Kalman filter. *Month. Weather Rev.*, **126**, 1719–1724.

Cane, M. A. and Sarachik, E. S. (1977). Forced baroclinic ocean motions. II: the linear equatorial bounded case. *J. Marine Res.*, **35**, 395–432.

Cane, M. A., Munnich, M. M., and Zebiak, S. E. (1990). Study of self-excited oscillations of the tropical ocean-atmosphere system. Part I: linear analysis. *J. Atmos. Sci.*, **47**, 1562–1577.

Cayan, D. R. (1992). Latent and sensible heat flux anomalies over the northern oceans: the connection to monthly atmospheric circulation. *J. Clim.*, **5**, 354–369.

Cessi, P. (1994). A simple box model of stochastically forced thermohaline flow. *J. Phys. Oceanogr.*, **24**, 1911–1920.

Cessi, P. and Speranza, A. (1985). Orographic instability of nonsymmetric baroclinic flows and nonpropagating planetary waves. *J. Atmos. Sci.*, **42**, 2585–2596.

Chang, P., Wang, B., Li, T., and Ji, L. (1994). Interactions between the seasonal cycle and the Southern Oscillation – frequency entrainment and chaos in a coupled ocean-atmosphere model. *Geophys. Res. Lett.*, **21**, 2817–2820.

Charney, J. and DeVore, J. (1979). Multiple flow equilibria in the atmosphere and blocking. *J. Atmos. Sci.*, **36**, 1205–1216.

Chatfield, C. (2004). *The Analysis of Time Series: An Introduction, 6th edition*. CRC Press, Boca Raton, U.S.A.

Chekroun, M. D., Simonnet, E., and Ghil, M. (2011). Stochastic climate dynamics: random attractors and time-dependent invariant measures. *Phys. D: Nonlin. Phenom.*, **240**(21), 1685–1700.

Chen, F. and Ghil, M. (1996). Interdecadal variability in a hybrid coupled ocean-atmosphere model. *J. Phys. Oceanogr.*, **26**, 1561–1578.

Cheng, W., Bleck, R., and Rooth, C. (2004). Multi-decadal thermohaline variability in an ocean-atmosphere general circulation model. *Clim. Dyn.*, **22**, 573–590.

Cheng, X. and Wallace, J. M. (1993). Cluster analysis of the Northern Hemisphere wintertime 500 pHa heigh field: spatial patterns. *J. Atmos. Sci.*, **50**, 2674–2696.

Chylek, P., Folland, C. K., Dijkstra, H. A., Lesins, G., and Dubey, M. K. (2011). Ice-core data evidence for a prominent near 20 year time-scale of the Atlantic Multidecadal Oscillation. *Geophys. Res. Lett.*, **38**, L13704.

Claussen, M. and coauthors (2002). Earth system models of intermediate complexity: closing the gap in the spectrum of climate system models. *Clim. Dyn.*, **18**, 579–586.
Clement, A. C. and Peterson, L. C. (2008). Mechanisms of abrupt climate change of the last glacial period. *Rev. Geophys.*, **46**, RG4002.
Colin de Verdière, A. (2007). A simple model of millennial oscillations of the thermohaline circulation. *J. Phys. Oceanogr.*, **37**(5), 1142–1155.
Colin de Verdière, A., Ben Jelloul, M. and Sévellec, F. (2006). Bifurcation structure of thermohaline millennial oscillations. *J. Clim.*, **19**, 5777–5795.
Colin de Verdière, A. and Huck, T. (1999). Baroclinic instability: an oceanic wavemaker for interdecadal variability. *J. Phys. Oceanogr.*, **29**, 893–910.
Compo, G. and coauthors (2011). The Twentieth Century Reanalysis Project. *Quart. J. R. Meteor. Soc.*, **137**, 1–28.
Corti, S., Molteni, F., and Palmer, T. N. (1999). Signature of recent climate change in frequencies of natural atmospheric circulation regimes. *Nature*, **398**(6730), 799–802.
Courtier, P., Thépaut, J., and Hollingsworth, A. (1994). A strategy for operational implementation of 4D Var, using an incremental approach. *Quart. J. R. Meteor. Soc.*, **120**, 1367–1387.
Cover, T. M. and Thomas, J. A. (2006). *Elements of Information Theory*, John Wiley, New York, U.S.A.
Crommelin, D. (2003). Regime transitions and heteroclinic connections in a barotropic atmosphere. *J. Atmos. Sci.*, **60**(2), 229–246.
Crommelin, D., Opsteegh, J., and Verhulst, F. (2004). A mechanism for atmospheric regime behavior. *J. Atmos. Sci.*, **61**(12), 1406–1419.
Crucifix, M. (2012). Oscillators and relaxation phenomena in Pleistocene climate theory. *Phil. Trans. R. Soc. A.* **370**, 1140–1165.
Cunningham, S. A., Kanzow, T., Rayner, D., Baringer, M. O., Johns, W. E., Marotzke, J., Longworth, H. R., Grant, E. M., Hirschi, J. J.-M., Beal, L. M., Meinen, C. S., and Bryden, H. L. (2007). Temporal variability of the Atlantic meridional overturning circulation at 26.5 N. *Science*, **317**(5840), 935–938.
Da Costa, E. D. and Colin de Verdière, A. C. (2004). The 7.7 year North Atlantic oscillation. *Quart. J. R. Meteor. Soc.*, **128**, 797–817.
De Swart, H. (1989). Analysis of a six-component atmospheric spectral model: chaos, predictability and vacillation. *Phys. D: Nonlin. Phenom.*, **36**(3), 222–234.
De Vries, P. and Weber, S. L. (2005). The Atlantic freshwater budget as a diagnostic for the existence of a stable shut down of the meridional overturning circulation. *Geophys. Res. Lett.*, **32**, No. 9, L09606.
Delworth, T. L. and Greatbatch, R. G. (2000). Multidecadal thermohaline circulation variability driven by atmospheric surface flux forcing. *J. Clim.*, **13**, 1481–1495.
Delworth, T. L. and Mann, M. E. (2000). Observed and simulated multidecadal variability in the Northern Hemisphere. *Clim. Dyn.*, **16**, 661–676.
Delworth, T. L., Manabe, S., and Stouffer, R. J. (1993). Interdecadal variations of the thermohaline circulation in a coupled ocean-atmosphere model. *J. Clim.*, **6**, 1993–2011.
Dickey, J. O., Ghil, M., and Marcus, S. L. (1991). Extratropical Aspects of the 40–50 day oscillation in length-of-day and atmospheric angular momentum. *J. Geophys. Res.*, **96**(D12), 22643–22658.
Dijkstra, H. A. (2005). *Nonlinear Physical Oceanography: A Dynamical Systems Approach to the Large Scale Ocean Circulation and El Niño*, 2nd revised and enlarged edition. Springer, New York, U.S.A.
Dijkstra, H. A. (2006). Interaction of SST Modes in the North Atlantic Ocean. *J. Phys. Oceanogr.*, **36**, 286–299.
Dijkstra, H. A. and Burgers, G. (2002). Fluid dynamics of El Niño variability. *Ann. Rev. Fluid Mech.*, **34**, 531–558.

Dijkstra, H. A. and Ghil, M. (2005). Low-frequency variability of the large-scale ocean circulation: a dynamical systems approach. *Rev. Geophys*, **43**(3).
Dijkstra, H. A. and Molemaker, M. J. (1997). Symmetry breaking and overturning oscillations in thermohaline-driven flows. *J. Fluid Mech.*, **331**, 195–232.
Dijkstra, H. A. and Neelin, J. (1995). Coupled ocean-atmosphere models and the tropical climatology. II: why the cold tongue is in the east. *J. Clim.*, **8**, 1343–1359.
Dijkstra, H. A. and Weijer, W. (2005). Stability of the global ocean circulation: basic bifurcation diagrams. *J. Phys. Oceanogr.*, **35**, 933–948.
Dima, M. and Lohmann, G. (2007). A hemispheric mechanism for the Atlantic multidecadal oscillation. *J. Clim.*, **20**, 2706–2719.
Doedel, E. J. (1980). AUTO: A program for the automatic bifurcation analysis of autonomous systems. In *Proc. 10th Manitoba Conf. on Numerical Math. and Comp.*, pages 265–274.
Dommenget, D. and Latif, M. (2002). Analysis of observed and simulated SST spectra in the midlatitude. *Clim. Dyn.*, **19**, 277–288.
Dong, B. and Sutton, R. (2005). Mechanism of interdecadal thermohaline circulation variability in a coupled ocean-atmosphere GCM. *J. Clim.*, **18**(8), 1117–1135.
Dorfle, M. and Graham, R. (1983). Probablity density of the Lorenz model. *Phys. Rev. A*, **27**(2), 1096–1105.
Duan, W., Liu, X., Zhu, K., and Mu, M. (2009). Exploring the initial errors that cause a significant "spring predictability barrier" for El Niño events. *J. Geophys. Res.*, **114**(C4).
Eckert, C. and Latif, M. (1997). Predictability of a stochastically forced hybrid coupled model of the Tropical Pacific ocean-atmosphere system. *J. Clim.*, **10**, 1488–1504.
Eckmann, J. and Ruelle, D. (1985). Ergodic theory of chaos and strange attractors. *Rev. Mod. Phy.*, **57**, 617–658.
Eden, C. and Jung, T. (2001). North Atlantic interdecadal variability: oceanic response to the North Atlantic oscillation (1865–1997). *J. Clim.*, **14**, 676–691.
Ehrendorfer, M. (1994). The Liouville equation and its potential usefulness for the prediction of forecast skill. Part I: theory. *Monthly Weather Rev.*, **122**(4), 703–713.
Enfield, D. B., Mestas-Nunes, A. M., and Trimble, P. (2001). The Atlantic multidecadal oscillation and its relation to rainfall and river flows in the continental US. *Geophys. Res. Lett.*, **28**, 2077–2080.
EPICA (2006). One-to-one coupling of glacial climate variability in Greenland and Antarctica. *Nature*, **444**, 195–198.
Evensen, G. (2009). *Data Assimilation: The Ensemble Kalman Filter*. Springer, New York, U.S.A.
Farrell, B. F. and Ioannou, P. J. (1996). Generalized stability theory. Part I: autnomous operators. *J. Atmos. Sci.*, **53**, 2025–2040.
Fedorov, A. V. and Philander, S. G. (2000). Is El Niño changing? *Science*, **288**, 1997–2002.
Feng, S. and Hu, Q. (2008). How the North Atlantic multidecadal oscillation may have influenced the Indian summer monsoon during the past two millennia. *Geophys. Res. Lett.*, **35**, L01707.
Frankcombe, L. M. and Dijkstra, H. A. (2009). Coherent multidecadal variability in North Atlantic sea level. *Geophys. Res. Lett.*, **36**, L15604.
Frankcombe, L. M. and Dijkstra, H. A. (2011). The role of Atlantic-Arctic exchange in North Atlantic multidecadal climate variability. *Geophys. Res. Lett.*, **38**(16).
Frankcombe, L. M., Dijkstra, H. A., and von der Heydt, A. (2008). Sub-surface signatures of the Atlantic Multidecadal Oscillation. *Geophys. Res. Lett.*, **35**, L19602.
Frankcombe, L. M., Dijkstra, H. A., and von der Heydt, A. (2009). Noise-induced multidecadal variability in the North Atlantic: excitation of normal modes. *J. Phys. Oceanogr.*, **39**(1), 220–233.

Frankcombe, L. M., von der Heydt, A., and Dijkstra, H. A. (2010a). North Atlantic multidecadal climate variability: an investigation of dominant time scales and processes. *J. Clim.*, **23**(13), 3626–3638.

Frankcombe, L. M., Dijkstra, H. A., and von der Heydt, A. S. (2010b). The Atlantic Multidecadal Oscillation: a stochastic dynamical systems view. In T. Palmer and P. Williams, editors, *Stochastic Physics and Climate Modelling*. Cambridge University Press, Cambridge, U.K.

Franzke, C., Crommelin, D., Fischer, A., and Majda, A. J. (2008). A hidden Markov model perspective on regimes and metastability in atmospheric flows. *J. Clim.*, **21**(8), 1740–1757.

Galanti, E. and Tziperman, E. (2000). ENSO's phase locking to the seasonal cycle in the fast-SST, fast-wave and mixed-mode regimes. *J. Atmos. Sci.*, **57**, 2936–2950.

Ganopolski, A. and Rahmstorf, S. (2002). Abrupt glacial climate change due to stochastic resonance. *Phys. Rev. Lett.*, **88**, 038501–1–4.

Ganopolski, A., Calov, R., and Claussen, M. (2010). Simulation of the last glacial cycle with a coupled climate ice-sheet model of intermediate complexity. *Clim. Past*, **6**(2), 229–244.

Ganopolsky, A. and Rahmstorf, S. (2001). Rapid changes of glacial climate simulated in a coupled climate model. *Nature*, **409**, 153–158.

Gardiner, C. W. (2002). *Handbook of Stochastic Methods*, 2nd edition. Springer, New York, U.S.A.

Gargett, A. (1984). Vertical eddy diffusivity in the ocean interior. *J. Marine Res.*, **42**, 359–393.

Gent, P. and McWilliams, J. C. (1990). Isopycnal mixing in ocean circulation models. *J. Phys. Oceanogr.*, **20**, 150–155.

Ghil, M. (1994). Cryothermodynamics: the chaotic dynamics of paleoclimate. *Phys. D: Nonlin. Phenom.*, **77**, 130–159.

Ghil, M. and Childress, S. (1987). *Topics in Geophysical Fluid Dynamics: Atmospheric Dynamics, Dynamo Theory, and Climate Dynamics*. Springer Verlag, Berlin, Germany.

Ghil, M. and Le Treut, H. (1981). A climate model with cryodynamics and geodynamics. *J. Geophys. Res.*, **86**, 5262–5270.

Ghil, M. and Robertson, A. W. (2002). "Waves" vs. "particles" in the atmosphere's phase space: a pathway to long-range forecasting? *Proc. Natl. Acad. Sci.* U.S.A., **99** (Suppl. 1), 2493–2500.

Ghil, M. and Vautard, R. (1991). Interdecadal oscillations and the warming trend in global temperature time series. *Nature*, **350**, 324–327.

Ghil, M., Allen, M. R., Dettinger, M. D., Ide, K., Kondrashov, D., Mann, M. E., Robertson, A. W., Saunders, A., Tian, Y., Varadi, F., and Yiou, P. (2002). Advanced spectral methods for climatic time series. *Rev. Geophys.*, **40**, 3.1–3.41.

Ghil, M., Chekroun, M., and Simonnet, E. (2008). Climate dynamics and fluid mechanics: natural variability and related uncertainties. *Phys. D: Nonlin. Phenom.*, **237**(14-17), 2111–2126.

Gildor, H. and Tziperman, E. (2000). Sea ice as the glacial cycles climate switch: role of seasonal and orbital forcing. *Paleoceanography*, **15**, 605–615.

Gildor, H. and Tziperman, E. (2001). A sea ice climate switch mechanism for the 100-kyr glacial cycles. *J. Geophys. Res.*, **106**, 9117–9133.

Gildor, H., Tziperman, E., and Toggweiler, J. (2002). Sea ice switch mechanism and glacial-interglacial CO_2 variations. *Global Biogeochem. Cycles*, **16**(3), 1032.

Gill, A. E. (1980). Some simple solutions for heat induced tropical circulation. *Quart. J. R. Meteor. Soc.*, **106**, 447–462.

Greatbatch, R. and Zhang, S. (1995). An interdecadal oscillation in an idealized ocean basin forced by constant heat flux. *J. Clim.*, **8**, 82–91.

Gregory, J. M. (2005). A model intercomparison of changes in the Atlantic thermohaline circulation in response to increasing atmospheric CO_2 concentration. *Geophys. Res. Lett.*, **32**(12), L12703.

Griffies, S. and Bryan, K. (1997). Predictability of North Atlantic multidecadal climate variability. *Science*, **275**, 181–184.

Guckenheimer, J. and Holmes, P. (1990). *Nonlinear Oscillations, Dynamical Systems and Bifurcations of Vector Fields*, 2nd edition. Springer Verlag, Berlin, Germany.

Guilyardi, E. (2006). El Niño–mean state–seasonal cycle interactions in a multi-model ensemble. *Clim. Dyn.*, **26**, 329–348.

Guilyardi, E., Wittenberg, A., Fedorov, A., Collins, M., Wang, C., Capotondi, A., van Oldenborgh, G. J., and Stockdale, T. (2009). Understanding El Niño in ocean-atmosphere general circulation models: progress and challenges. *Bull. Am. Meteor. Soc.*, **90**(3), 325–340.

Haak, H., Jungclaus, J., Mikolajewicz, U., and Latif, M. (2003). Formation and propagation of great salinity anomalies. *Geophys. Res. Lett.*, **30**(9), 26.1–26.4.

Hansen, A. and Sutera, A. (1995). The probability density distribution of the planetary-scale atmospheric wave amplitude revisited. *J. Atmos. Sci.*, **52**(13), 2463–2472.

Hartmann, D. L. (1994). *Global Physical Climatology*. Academic Press, San Diego, U.S.A.

Hasselmann, K. (1976). Stochastic climate models. I: theory. *Tellus*, **28**, 473–485.

Hawkins, E. and Sutton, R. (2009). The potential to narrow uncertainty in regional climate predictions. *Bull. Am. Meteor. Soc.*, **90**(8), 1095–1107.

Hawkins, E. and Sutton, R. T. (2007). Variability of the Atlantic thermohaline circulation described by three-dimensional empirical orthogonal functions. *Clim. Dyn.*, **29**, 745–762.

Hawkins, E., Smith, R. S., Allison, L. C., Gregory, J. M., Woollings, T. J., Pohlmann, H., and De Cuevas, B. (2011). Bistability of the Atlantic overturning circulation in a global climate model and links to ocean freshwater transport. *Geophys. Res. Lett.*, **38**(10), L10605.

Held, H. and Kleinen, T. (2004). Detection of climate system bifurcations by degenerate fingerprinting. *Geophys. Res. Lett.*, **31**, L23207.

Hendon, H. H., Wheeler, M. C., and Zhang, C. (2007). Seasonal dependence of the MJO-ENSO relationship. *J. Clim.*, **20**(3), 531–543.

Hendon, H. H., Lim, E., Wang, G., Alves, O., and Hudson, D. (2009). Prospects for predicting two flavors of El Niño. *Geophys. Res. Lett.*, **36**, L19713.

Higham, D. J. (2001). An algorithmic introduction to numerical solution of stochastic differential equations. *SIAM Rev.*, **43**, 525–546.

Hirst, A. C. (1986). Unstable and damped equatorial modes in simple coupled ocean-atmosphere models. *J. Atmos. Sci.*, **43**, 606–630.

Hoffman, M. J., Kalnay, E., Carton, J. A., and Yang, S.-C. (2009). Use of breeding to detect and explain instabilities in the global ocean. *Geophys. Res. Lett.*, **36**(12), L12608.

Holland, M. P., Vitolo, R., Rabassa, P., Sterk, A. E., and Broer, H. W. (2012). Extreme value laws in dynamical systems under physical observables. *Phys. D: Nonlin. Phenom.*, **241**(5), 497–513.

Holton, J. R. (1992). *An Introduction to Dynamic Meteorology*. Academic Press, New York, U.S.A.

Huang, R., Luyten, J. R., and Stommel, H. M. (1992). Multiple equilibrium states in combined thermal and saline circulation. *J. Phys. Oceanogr.*, **22**, 231–246.

Huck, T., Colin de Verdière, A., and Weaver, A. J. (1999). Interdecadal variability of the thermohaline circulation in box-ocean models forced by fixed surface fluxes. *J. Phys. Oceanogr.*, **29**, 865–892.

Huck, T. and Vallis, G. (2001). Linear stability analysis of the three-dimensional thermally-driven ocean circulation: application to interdecadal oscillations. *Tellus*, **53A**, 526–545.

Huck, T., Vallis, G., and de Verdière, C. (2001). On the robustness of interdecadal oscillations of the thermohaline circulation. *J. Clim.*, **14**, 940–963.
Hurrell, J. W. (1995). Decadal trends in the North Atlantic oscillation: regional temperatures and precipitation. *Science*, **269**, 676–680.
Hurrell, J. W. and Deser, C. (2010). North Atlantic climate variability: the role of the North Atlantic Oscillation. *J. Marine Syst.*, 231–244.
Hurrell, J. W., Kushnir, Y., Ottersen, G., and Visbeck, M. (2003). An overview of the North Atlantic Oscillation. In J. W. Hurrell, Y. Kushnir, G. Ottersen, and M. Visbeck, editors, *The North Atlantic Oscillation: Climatic Significance and Environmental Impact*. American Geophysical Union, Washington, DC.
Huybers, P. (2006). Early Pleistocene glacial cycles and the integrated summer insolation forcing. *Science*, **313**(5786), 508–511.
Imbrie, J. and Imbrie, K. P. (1986). *Ice Ages: Solving the Mystery*, 2nd edition. Harvard Univ. Press, Cambridge, Mass., U.S.A.
Itoh, H. and Kimoto, M. (1996). Multiple attractors and chaotic itinerancy in a quasigeostrophic model with realistic topography: implications for weather regimes and low-frequency variability. *J. Atmos. Sci.*, **53**(15), 2217–2231.
Jansen, M. F., Dommenget, D., and Keenlyside, N. (2009). Tropical atmosphere-ocean interactions in a conceptual framework. *J. Clim.*, **22**(3), 550–567.
Jiang, N., Neelin, J., and Ghil, M. (1995). Quasi-quadrennial and quasi-biennial varibility in the equatorial Pacific. *Clim. Dyn.*, **12**, 101–112.
Jin, F. and Ghil, M. (1990). Intraseasonal oscillations in the extratropics – Hopf-bifurcation and topographic instabilities. *J. Atmos. Sci.*, **47**(24), 3007–3022.
Jin, F.-F. (1997a). An equatorial recharge paradigm for ENSO. I: Conceptual model. *J. Atmos. Sci.*, **54**, 811–829.
Jin, F.-F. (1997b). An equatorial recharge paradigm for ENSO. II: A stripped-down coupled model. *J. Atmos. Sci.*, **54**, 830–8847.
Jin, F.-F. (2010). Eddy-induced instability for low-frequency variability. *J. Atmos. Sci.*, **67**(6), 1947–1964.
Jin, F.-F. and An, S. I. (1999). Thermocline and zonal advective feedbacks within the equatorial ocean recharge oscillator model for ENSO. *Geophys. Res. Lett.*, **26**, 2989–2992.
Jin, F.-F. and Neelin, J. (1993). Modes of interannual tropical ocean-atmosphere interaction – a unified view. I: numerical results. *J. Atmos. Sci.*, **50**, 3477–3503.
Jin, F.-F., Neelin, J., and Ghil, M. (1994). El Niño on the devil's staircase: annual subharmonic steps to chaos. *Science*, **264**, 70–72.
Jin, F.-F., Neelin, J., and Ghil, M. (1996). El Niño/Southern Oscillation and the annual cycle: subharmonic frequency-locking and aperiodicity. *Phys. D: Nonlin. Phenom.*, **98**, 442–465.
Johns, W., Baringer, M., and Beal, L. (2011). Continuous, array-based estimates of Atlantic Ocean heat transport at 26.5 N. *J. Clim.*, **24**, 2429–2449.
Jouzel, J. and coauthors. (2007). Orbital and millennial Antarctic climate variability over the past 800,000 years. *Science*, **317**(5839), 793–796.
Jungclaus, J. H., Haak, H., Latif, M., and Mikolajewicz, U. (2005). Arctic-North Atlantic interactions and multidecadal variability of the meridional overturning circulation. *J. Clim.*, **18**, 4013–4031.
Jungclaus, J. H., Keenlyside, N., Botzet, M., Haak, H., Luo, J.-J., Marotzke, J., Mikolajewicz, U., and Roeckner, E. (2006). Ocean circulation and tropical variability in the coupled model ECHAM5/MPI-OM. *J. Clim.*, **19**, 3952–3972.
Källén, E., Crafoord, C., and Ghil, M. (1979). Free oscillations in a climate model with ice-sheet dynamics. *J. Atmos. Sci.*, **36**(12), 2292–2303.
Kalnay, E. (2003). *Atmospheric Modeling, Data Assimilation, and Predictability*. Cambridge University Press, Cambridge, U.K.

Kalnay, E. and coauthors (1996). The NCEP/NCAR reanalysis 40-year project. *Bull. Am. Meteor. Soc.*, **77**, 437–471.

Kalnay, E., Li, H., Miyoshi, T., Yang, S.-C., and Ballabrera-Poy, J. (2007). 4-D-Var or ensemble Kalman filter? *Tellus A Dyn. Meteor. Oceanogr.*, **59**(5), 758–773.

Kantelhardt, J., Koscielny-Bunde, E., Rego, H., Havlin, S., and Bunde, A. (2001). Detecting long-range correlations with detrended fluctuation analysis. *Phys. A*, **295**, 441–454.

Kaplan, A., Kushnir, Y., Cane, M. A., and Blumenthal, M. B. (1997). Reduced space optimal analysis for historical data sets: 136 years of Atlantic sea surface temperatures. *J. Geophys. Res.*, **102**(C13), 27835–27860.

Keenlyside, N. S., Latif, M., Jungclaus, J., Kornblueh, L., and Roeckner, E. (2008). Advancing decadal-scale climate prediction in the North Atlantic sector. *Nature*, **453**(7191), 84–88.

Keller, H. B. (1977). Numerical solution of bifurcation and nonlinear eigenvalue problems. In P. H. Rabinowitz, editor, *Applications of Bifurcation Theory*. Academic Press, New York, U.S.A.

Kemfert, C. (2005). An integrated assessment model of economy-energy-climate – The model Wiagem. *Integr. Assess.*, **3**(4), 281–298.

Keppenne, C. L., Marcus, S. L., Kimoto, M., and Ghil, M. (2000). Intraseasonal variability in a two-layer model and observations. *J. Atmos. Sci.*, **57**, 1010–1028.

Kerr, R. A. (2000). A North Atlantic climate pacemaker for the centuries. *Science*, **288**, 1984–1986.

Kessler, W. S., Spillane, M. C., McPhaden, M. J., and Harrison, D. E. (1996). Scales of variability in the Equatorial Pacific inferred from the tropical atmosphere-ocean array. *J. Clim.*, **9**, 2999–3024.

Kirtman, B. P. and Schopf, P. S. (1998). Decadal variability in ENSO predictability and prediction. *J. Clim.*, **11**, 2804–2822.

Kleeman, R. (1993). On the dependence of hindcast skill on ocean thermodynamics in a coupled ocean-atmosphere model. *J. Clim.*, **6**, 2012–2033.

Kleeman, R. (2002). Measuring dynamical prediction utility using relative entropy. *J. Atmos. Sci.*, **59**, 2057–2072.

Kleeman, R. (2008). Stochastic theories for the irregularity of ENSO. *Phil. Transact. R. Soc. A Mathe. Phys. Engineer. Sci.*, **366**(1875), 2509–2524.

Kloeden, P., Platen, E., and Schurz, H. (1994). *Numerical Solution of Stochastic Differential Equations through Computer Experiments*. Springer Verlag, New York, U.S.A.

Kloeden, P. E. and Platen, E. (1999). *Numerical Solution of Stochastic Differential Equations*. Springer, New York, U.S.A.

Knight, J. R., Folland, C. K., and Scaife, A. A. (2006). Climate impacts of the Atlantic Multidecadal Oscillation. *Geophys. Res. Lett.*, **33**, L17706.

Kondrashov, D., Ide, K., and Ghil, M. (2004). Weather regimes and preferred transition paths in a three-level quasigeostrophic model. *J. Atmos. Sci.*, **61**(5), 568–587.

Kravtsov, S. and Ghil, M. (2004). Interdecadal variability in a hybrid coupled ocean-atmosphere-sea ice model. *J. Phys. Oceanogr.*, **34**, 1756–1775.

Kuehn, C. (2011). A mathematical framework for critical transitions: Bifurcations, fast–slow systems and stochastic dynamics. *Phys. D: Nonlin. Phenom.*, **240**(12), 1020–1035.

Kuhlbrodt, T., Griesel, A., Montoya, M., Levermann, A., Hofmann, M., and Rahmstorf, S. (2007). On the driving processes of the Atlantic meridional overturning circulation. *Rev. Geophys.*, **45**, RG2001, doi:10.1029/2004RG000166.

Kushnir, Y. (1994). Interdecadal variations in North Atlantic sea surface temperature and associated atmospheric conditions. *J. Phys. Oceanogr.*, **7**, 141–157.

Kuznetsov, Y. A. (1995). *Elements of Applied Bifurcation Theory*. Springer Verlag, New York, U.S.A.

Latif, M. (1998). Dynamics of interdecadal variability in coupled ocean-atmosphere models. *J. Clim.*, **11**, 602–624.

Le Treut, H., Portes, J., Jouzel, J., and Ghil, M. (1988). Isotopic modeling of climatic oscillations: implications for a comparative study of marine and ice-core records. *J. Geophys. Res.*, **93**, 9365–9383.

Leeuwenburgh, O. (2005). Assimilation of along-track altimeter data in the tropical Pacific region of a global OGCM ensemble. *Quart. J. R. Meteor. Soc.*, **131**(610), 2455–2472.

Legras, B. and Ghil, M. (1983). Ecoulements atmosphériques stationnaires, périodiques et apériodiques. *J. Méc. Théor. Appl.*, 45–82. Special Issue (Two-Dimensional Turbulence, R. Moreau (Ed.), Gauthier-Villars, Paris).

Legras, B. and Ghil, M. (1985). Persistent anomalies, blocking and variations in atmospheric predictability. *J. Atmos. Sci.*, **42**, 433–471.

Lenton, T. M. (2011). Early warning of climate tipping points. *Nat. Pub. Group*, **1**(4), 201–209.

Lisiecki, L. E. and Raymo, M. (2005). A Pliocene-Pleistocene stack of 57 globally distributed benthic 18O records. *Paleoceanography*, **20**(1), PA1003.

Livina, V. N. and Lenton, T. M. (2007). A modified method for detecting incipient bifurcations in a dynamical system. *Geophys. Res. Lett.*, **34**(3).

Lorenc, A. (2007). 4D-Var and the butterfly effect: Statistical four-dimensional data assimilation for a wide range of scales. *Quart. J. R. Soc.*, **133**, 607–614.

Lorenc, A. C. (1988). Optimal nonlinear objective analysis. *Quart. J. R. Meteor. Soc.*, **114**(479), 205–240.

Lorenz, E. N. (1963). Deterministic nonperiodic flow. *J. Atmos. Sci.*, **20**, 130–141.

Lorenz, E. N. (1969). The predictability of a flow which possesses many scales of motion. *Tellus*, **21**(3), 289–307.

Lucarini, V., Calmanti, S., Dell'Aquila, A., Ruti, P. M., and Speranza, A. (2007). Intercomparison of the northern hemisphere winter mid-latitude atmospheric variability of the IPCC models. *Clim. Dyn.*, **28**, 829–848.

Maas, L. R. M. (1994). A simple model for the three-dimensional, thermally and wind-driven ocean circulation. *Tellus Ser. A Dyn. Meteor. Oceanogr.*, **46**(5), 671–680.

Madden, R. A. and Julian, P. R. (1994). Observations of the 40-50-day tropical oscillation – a review. *Month. Weather Rev.*, **122**, 814–835.

Majda, A. J. and Wang, X. (2006). *Non-Linear Dynamics and Statistical Theories for Basic Geophysical Flows*. Cambridge University Press, Cambridge.

Majda, A. J., Franzke, C. L., Fischer, A., and Crommelin, D. (2006). Distinct metastable atmospheric regimes despite nearly Gaussian statistics: a paradigm model. *Proc. Nat. Acad. Sci.* U.S.A., **103**(22), 8309–8314.

Malkus, W. V. R. (1972). Non-periodic convection at high and low Prandtl number. *Mem. Soc. R. Sci. Liege*, **6e Serie, Tome IV**, 125–128.

Mantua, N. J. and Battisti, D. S. (1995). Aperiodic variability in the Zebiak-Cane coupled ocean-atmosphere model: air-sea interaction in the western equatorial Pacific. *J. Clim.*, **8**, 2897–2927.

McGuffie, K. and Henderson-Sellers, A. (2006). *A Climate Modeling Primer*, 3rd edition. Wiley, New York, U.S.A.

McPhaden, M. J. and coauthors (1998). The Tropical Ocean-Global Atmosphere observing system: a decade of progress. *J. Geophys. Res.*, **103**, 14,169–14,240.

Mechoso, C., Neelin, J., and Yu, J. (2003). Testing simple models of ENSO. *J. Atmos. Sci.*, **60**(2), 305–318.

Meehl, G. A., and coauthors (2007). Global climate projections. In *Climate Change 2007: The Physical Science Basis. Contribution of Working Group I to the Fourth Assessment Report of the Intergovernmental Panel on Climate Change* [Solomon, S., D. Qin, M. Manning, Z. Chen, M. Marquis, K.B. Averyt, M. Tignor and H.L. Miller (cds.)]. Cambridge University Press, Cambridge, United Kingdom and New York, NY, U.S.A.

Mikosch, T. (2000). *Elementary Stochastic Calculus*. World Scientific, New York, U.S.A.

Milankovitch, M. (1930). Mathematische Klimalehre und astronomische Theorie der Klimaschwankungen. In W. Koppen and R. Geiger, editors, *Handbuch der Klimatologie*. Borntraeger, Berlin, Germany.

Miller, L. and Douglas, B. C. (2007). Gyre-scale atmospheric pressure variations and their relation to 19th and 20th century sea level rise. *Geophys. Res. Lett.*, **34**, L16602.

Mitchell, J. M. (1976). An overview of climate variability and its causal mechanisms. *Quatern. Res.*, **6**, 481–493.

Mo, K. and Ghil, M. (1988). Cluster-analysis of multiple planetary flow regimes. *J. Geophys. Res. Atmos.*, **93**, 10927–10952.

Moore, A. and Kleeman, R. (2001). The differences between the optimal perturbations of coupled models of ENSO. *J. Clim.*, **14**(2), 138–163.

Moore, A., Arango, H., and Broquet, G. (2011). The Regional Ocean Modeling System (ROMS) 4-dimensional variational data assimilation systems: Part I – system overview and formulation. *Prog. Oceanogr.*, **91**, 34–49.

Moore, A. M. and Kleeman, R. (1999). Stochastic forcing of ENSO by the Intraseasonal Oscillation. *J. Clim.*, **12**, 1199–1220.

Moore, D. W. (1968). *Planetary-Gravity Waves in an Equatorial Ocean.* Harvard University, Cambridge, Mass., U.S.A.

Mu, M. (2000). Nonlinear singular vectors and nonlinear singular values. *Sci. China Ser. D Earth Sci.*, **43**, 375–385.

Mu, M. and Duan, W. (2003). Conditional nonlinear optimal perturbation and its applications. *Nonlin. Process. Geophys.*, **2003**, 493–501.

Mu, M. and Wang, J. (2001). Nonlinear fastest growing perturbation and the first kind of predictability. *Sci. China*, **44D**, 1128–1139.

Mu, M., Liang, S., and Dijkstra, H. A. (2004). The sensitivity and stability of the ocean's thermohaline circulation to finite amplitude perturbations. *J. Phys. Oceanogr.*, **34**, 2305–2315.

Mukougawa, H. (1988). A dynamical model of 'quasi-stationary' states in large-scale atmospheric motions. *J. Atmos. Sci.*, **45**, 2868–2888.

Munnich, M. M., Cane, M., and Zebiak, S. E. (1991). A study of self-excited oscillations of the tropical ocean-atmosphere system. II: nonlinear cases. *J. Atmos. Sci.*, **48**, 1238–1248.

Mysak, L. A., Stocker, T. F., and Huang, F. (1993). Century-scale variability in a randomly forced, two-dimensional thermohaline ocean circulation model. *Clim. Dyn.*, **8**, 103–116.

Neelin, J. (1991). The slow sea surface temperature mode and the fast-wave limit: analytic theory for tropical interannual oscillations and experiments in a hybrid coupled model. *J. Atmos. Sci.*, **48**, 584–606.

Neelin, J. (2011). *Climate Change and Climate Modeling*. Cambridge University Press, Cambridge, U.K.

Neelin, J. and Dijkstra, H. A. (1995). Coupled ocean-atmosphere models and the tropical climatology. I: the dangers of flux-correction. *J. Clim.*, **8**, 1325–1342.

Neelin, J., Latif, M., and Jin, F.-F. (1994). Dynamics of coupled ocean-atmosphere models: the tropical problem. *Ann. Rev. Fluid Mech.*, **26**, 617–659.

Neelin, J., Battisti, D. S., Hirst, A. C., Jin, F.-F., Wakata, Y., Yamagata, T., and Zebiak, S. E. (1998). ENSO Theory. *J. Geophys. Res.*, **103**, 14,261–14,290.

Neelin, J., Jin, F.-F., and Syu, H.-H. (2000). Variations of ENSO phase locking. *J. Clim.*, **13**, 2570–2590.

Nicolis, C. (1982). Stochastic aspects of climatic transitions-response to a periodic forcing. *Tellus*, **34**(1), 1–9.

Nitsche, G., Wallace, J., and Kooperberg, C. (1994). Is there evidence of multiple equilibria in planetary wave amplitude statistics. *J. Atmos. Sci.*, **51**(2), 314–322.

North, G., Bell, T., Cahalan, R., and Moeng, F. (1982). Sampling errors in the estimation of empirical orthogonal functions. *Month. Weather Rev.*, **110**, 699–706.

North, G. R., Cahalan, R. F., and Coakley, J. A. (1981). Energy balance climate models. *Rev. Geophys. Space Phys.*, **19**, 19–121.
Oerlemans J. and Van der Veen, C. J. (1984). *Ice Sheets and Climate*, D. Reidel, Dordrecht, Holland.
Oksendal, B. (1995). *Stochastic Differential Equations*. Springer Verlag, Berlin, Germany.
Paillard, D. (1998). The timing of Pleistocene glaciations from a simple multiple-state climate model. *Nature*, **391**(6665), 378–381.
Palmer, T., Doblas-Reyes, F., and Weisheimer, A. (2008). Toward seamless prediction. *Bull. Am. Meteor. Soc.*, **90**, 459–470.
Pedlosky, J. (1987). *Geophysical Fluid Dynamics*, 2nd edition. Springer Verlag, New York, U.S.A.
Peixoto, J. P. and Oort, A. H. (1992). *Physics of Climate*. AIP Press, New York, U.S.A.
Penland, C. and Sardeshmukh, P. D. (1995). The optimal growth of tropical sea surface temperature anomalies. *J. Clim.*, **8**, 1999–2024.
Penland, C., Flugel, M., and Chang, P. (2000). Identification of dynamical regimes in an intermediate coupled ocean-atmosphere model. *J. Clim.*, **13**(12), 2105–2115.
Philander, S. G. H. (1990). *El Niño and the Southern Oscillation*. Academic Press, New York, U.S.A.
Philander, S. G. H., Yamagata, T., and Pacanowski, R. C. (1984). Unstable air-sea interactions in the tropics. *J. Atmos. Sci.*, **41**, 604–613.
Philander, S. G. H., Gu, D., Halpern, D., Lambert, G., Lau, N. C., Li, T., and Pacanowski, R. C. (1996). Why the ITCZ is mostly north of the equator. *J. Clim.*, **9**, 2958–2972.
Pikovsky, A. and Kurths, J. (1997). Coherence resonance in a noise-driven excitable system. *Phys. Rev. Lett.*, **78**(5), 775–778.
Plaut, G. and Vautard, R. (1994). Spells of low-frequency oscillations and weather regimes in the Northern Hemisphere. *J. Atmos. Sci.*, **51**, 210–236.
Plaut, G., Ghil, M., and Vautard, R. (1995). Interannual and interdecadal variability in 335 years of Central England Temperature. *Science*, **268**, 710–713.
Polyakov, I. V. and Johnson, M. A. (2000). Arctic decadal and interdecadal variability. *Geophys. Res. Lett.*, **27**, 4097–4100.
Polyakov, I. V., Walsh, D., Dmitrenko, I., Colony, R. L., and Timokhov, L. A. (2003). Arctic Ocean variability derived from historical observations. *Geophys. Res. Lett.*, **30**, 31.1–31.4.
Preisendorfer, R. W. (1988). *Principal Component Analysis in Meteorology and Oceanography*. Elsevier, Amsterdam, The Netherlands.
Quon, C. and Ghil, M. (1995). Multiple equilibria and stable oscillations in thermosolutal convection at small aspect ratio. *J. Fluid Mech.*, **291**, 33–56.
Rahmstorf, S., Crucifix, M., Ganopolski, A., Goosse, H., Kamenkovich, I., Knutti, R., Lohmann, G., March, R., Mysak, L., Wang, Z., and Weaver, A. J. (2005). Thermohaline circulation hysteresis: A model intercomparison. *Geophys. Res. Lett.*, **L23605**, 1–5.
Raymo, M., Lisiecki, L. and Nisancioglu, K. H. (2006). Plio-Pleistocene ice volume, Antarctic climate, and the global δ^{18}O record. *Science*, **313**, 492–495.
Raymo, M. E. and Huybers, P. (2008). Unlocking the mysteries of the ice ages. *Nature*, **451**(7176), 284–285.
Rayner, N. A., Parker, D. E., Horton, E. B., Folland, C. K., Alexander, L. V., Rowell, D. P., Kent, E. C., and Kaplan, A. (2003). Global analyses of sea surface temperature, sea ice, and night marine air temperature since the late nineteenth century. *J. Geophys. Res.*, **108**(D14).
Rayner, N. A., Brohan, P., Parker, D. E., Folland, C. K., Kenndy, J. J., Vanicek, M., Ansell, T. J., and Tett, S. F. B. (2006). Improved analysis of changes and uncertainties in sea surface temperature measured in situ since the mid-ninteenth century: The HadSST2 dataset. *J. Clim.*, **19**, 446–469.

Risken, H. (1989). The Fokker-Planck Equation. 2nd ed. Springer Verlag, 488 pp.
Roulston, M. and Neelin, J. (2000). The response of an ENSO model to climate noise, weather noise and intraseasonal forcing. *Geophys. Res. Lett.*, **27**, 3723–3726.
Rowell, D. P. (2010). Assessing potential seasonal predictability with an ensemble of multidecadal GCM simulations. *J. Clim.*, **11**, 109–120.
Ruddiman, W. F. (2001). *Earth's Climate: Past and Future*. W.H. Freeman & Company, New York, U.S.A.
Ruelle, D. and Takens, F. (1970). On the nature of turbulence. *Comm. Math. Phys.*, **20**, 167–192.
Ruti, P., Lucarini, V., Dell'Aquila, A., Calmanti, S., and Speranza, A. (2006). Does the subtropical jet catalyze the midlatitude atmospheric regimes? *Geophys. Res. Lett.*, **33**, L06814.
Sakai, K. and Peltier, W. R. (1995). A simple model of the Atlantic thermohaline circulation: Internal and forced variability with paleoclimatological implications. *J. Geophys. Res.*, **100**, 13,455–13,479.
Sakai, K. and Peltier, W. R. (1996). A multi basin reduced model of the global thermohaline circulation: peleoceanographic analyses of the origins of ice-age climate variability. *J. Geophys. Res.*, **101**, 22,535–22,262.
Sakai, K. and Peltier, W. R. (1997). Dansgaard-Oeschger oscillations in a coupled atmosphere-ocean climate model. *J. Clim.*, **10**, 949–970.
Sakai, K. and Peltier, W. R. (1999). A dynamical systems model of the Dansgaard-Oeschger oscillation and the origin of the Bond cycle. *J. Clim.*, **12**, 2238–2255.
Saltzman, B. (1984). On the role of equilibrium atmospheric climate models in the theory of long-period glacial variations. *J. Atmos. Sci.*, **41**, 2263–2266.
Saltzman, B. and Maasch, K. (1988). Carbon cycle instability as a cause of the late Pleistocene ice age oscillations: modeling the asymmetric response. *Global Biogeochem. Cycles*, **2**, 177–184.
Saltzmann, B. (2001). *Dynamical Paleoclimatology*. Academic Press, New York, U.S.A.
Saravanan, R. and McWilliams, J. C. (1998). Advective ocean-atmosphere interaction: an analytical stochastic model with implications for decadal variability. *J. Clim.*, **11**, 165–188.
Sarmiento, J. L. and Gruber, N. (2006). *Ocean Biogeochemical Dynamics*. Princeton Univ. Press, Princeton, N.J., U.S.A.
Scheffer, M. and coauthors (2009). Early warning signals for critical transitions, *Nature*, **461**, 53–59.
Schlesinger, M. E. and Ramankutty, N. (1994). An oscillation in the global climate system of period 65-70 years. *Nature*, **367**, 723–726.
Schneider, T. and Griffies, S. (1999). A conceptual framework for predictability studies. *J. Clim.*, **12**(10), 3133–3155.
Schultz, M. (2002). On the 1470-year pacing of Dansgaard-Oeschger warm events. *Paleoceanography*, **17**(2), 4.1–4.9.
Sellers, W. D. (1969). A global climate model based on the energy balance of the earth-atmosphere system. *J. Appl. Meteor.*, **8**, 392–400.
Selten, F. (1995). An efficient description of the dynamics of barotropic flow. *J. Atmos. Sci.*, **52**(7), 915–936.
Selten, F. and Branstator, G. (2004). Preferred regime transition routes and evidence for an unstable periodic orbit in a baroclinic model. *J. Atmos. Sci.*, **61**, 2267–2282.
Sevellec, F., Huck, T., and Ben Jelloul, M. (2006). On the mechanism of centennial thermohaline oscillations. *J. Marine Res.*, **64**(3), 355–392.
Sevellec, F., Huck, T., Ben Jelloul, M., and Vialard, J. (2009). Nonnormal multidecadal response of the thermohaline circulation induced by optimal surface salinity perturbations. *J. Phys. Oceanogr.*, **39**(4), 852–872.

Shilnikov, L. P. (1965). A case of the existence of a denumerable set of periodic motions. *Sov. Math. Dokl.*, **6**, 163–166.

Siegenthaler, U., Stocker, T., Monnin, E., and Lüthi, D. (2005). Stable carbon cycle–climate relationship during the late Pleistocene. *Science*, **310**, 1313–1317.

Simonnet, E., Dijkstra, H. A., and Ghil, M. (2009). Bifurcation analysis of ocean, atmosphere and climate models. In R. Temam and J. Tribbia, editors, *Handbook of Numerical Analysis*, pages 187–230. North-Holland, Amsterdam, The Netherlands.

Smyth, P., Ide, K., and Ghil, M. (1999). Multiple regimes in Northern Hemisphere height fields via mixture model clustering. *J. Atmos. Sci.*, **56**, 3704–3723.

Stachowiak, T. and Okada, T. (2006). A numerical analysis of chaos in the double pendulum. *Chaos*, **29**, 417–422.

Stephenson, D. B., Hannachi, A., and O'Neill, A. (2004). On the existence of multiple climate regimes. *Quart. J. R. Meteor. Soc.*, **130**(597), 583–605.

Sterk, A. E., Vitolo, R., Broer, H. W., Simo, C., and Dijkstra, H. A. (2010). New nonlinear mechanisms of midlatitude atmospheric low-frequency variability. *Phys. D: Nonlin. Phenom.*, **239**(10), 702–718.

Sterl, A., Severijns, C., Dijkstra, H. A., Hazeleger, W., van Oldenborgh, G. J., van den Broeke, M., Burgers, G., van den Hurk, B., van Leeuwen, P. J., and van Velthoven, P. (2008). When can we expect extremely high surface temperatures? *Geophys. Res. Lett.*, **35**, L14703.

Stommel, H. (1961). Thermohaline convection with two stable regimes of flow. *Tellus*, **2**, 244–230.

Stouffer, R. J., Yin, J., Gregory, J. M., Dixon, K. W., Spelman, M. J., Hurlin, W., Weaver, A. J., et al. (2006). Investigating the causes of the response of the thermohaline circulation to past and future climate changes. *J. Clim.*, **19**, 1365–1387.

Strong, C., Jin, F.-F., and Ghil, M. (1995). Intraseasonal oscillations in a barotropic model with annual cycle, and their predictability. *J. Atmos. Sci.*, **52**, 2627–2642.

Stuiver, M. and Grootes, P. M. (2000). GISP2 Oxygen Isotope Ratios. *Quatern. Res.*, **53**(3), 277–284.

Suarez, M. and Schopf, P. S. (1988). A delayed action oscillator for ENSO. *J. Atmos. Sci.*, **45**, 3283–3287.

Sutton, R. T. and Hodson, D. L. (2005). Atlantic ocean forcing of North American and European summer climate. *Science*, **309**, 115–118.

Te Raa, L. A. and Dijkstra, H. A. (2002). Instability of the thermohaline ocean circulation on interdecadal time scales. *J. Phys. Oceanogr.*, **32**, 138–160.

Te Raa, L. A. and Dijkstra, H. A. (2003a). Modes of internal thermohaline variability in a single-hemispheric ocean basin. *J. Marine Res.*, **61**, 491–516.

Te Raa, L. A. and Dijkstra, H. A. (2003b). Sensitivity of North Atlantic multidecadal variability to freshwater flux forcing. *J. Clim.*, **32**, 138–160.

Te Raa, L. A., Gerrits, J., and Dijkstra, H. A. (2004). Identification of the mechanism of interdecadal variability in the North Atlantic Ocean. *J. Phys. Oceanogr.*, **34**, 2792–2807.

Terwisscha van Scheltinga, A. and Dijkstra, H. A. (2005). Nonlinear data-assimilation using implicit models. *Nonlin. Process.*, **12**, 515–525.

Thompson, D. W. J. and Wallace, J. M. (2000). Annular modes in the extratropical circulation. Part I: month-to-month variability. *J. Clim.*, **13**, 1000–1016.

Thompson, D. W. J. and Wallace, J. M. (2001). Regional climate impacts of the Northern Hemisphere annular mode. *Science*, **293**, 85–89.

Thompson, D. W. J., Wallace, J. M., and Hegerl, G. C. (2000). Annular modes in the extratropical circulation. Part II: trends. *J. Clim.*, **13**, 1018–1036.

Thompson, J. M. T. and Sieber, J. (2011). Predicting climate tipping as a noisy bifurcation: a review. *Int. J. Bifurcation Chaos*, **21**(2), 399–423.

Tian, Y., Weeks, E., Ide, K., Urbach, J., Baroud, C., Ghil, M., and Swinney, H. (2001). Experimental and numerical studies of an eastward jet over topography. *J. Fluid Mech.*, **438**, 129–157.

Timmermann, A., Jin, F.-F., and Abshagen, J. (2003a). A nonlinear theory of El Niño bursting. *J. Atmos. Sci.*, **60**, 165–176.

Timmermann, A., Gildor, H., Schulz, M., and Tziperman, E. (2003b). Coherent resonant millennial-scale climate oscillations triggered by massive meltwater pulses. *J. Clim.*, **16**, 2569–2585.

Toggweiler, J. (1999). Variation of atmospheric CO_2 by ventilation. *Paleoceanography*, **14**, 571–588.

Toth, Z. and Kalnay, E. (1997). Ensemble forecasting at NCEP and the breeding method. *Month. Weather Rev.*, **125**, 3297–3319.

Trauth, M. H. (2007). *MATLAB Recipes for Earth Sciences*. Springer, Berlin, Germany.

Trenberth, K. (1997). El Niño and climate change. *Geophys. Res. Lett.*, **24**, 3057–3060.

Tziperman, E., Stone, L., Cane, M. A., and Jarosh, H. (1994a). El Niño chaos: overlapping of resonances between the seasonal cycle and the Pacific ocean-atmosphere oscillator. *Science*, **264**, 72–74.

Tziperman, E., Toggweiler, J. R., Feliks, Y., and Bryan, K. (1994b). Instability of the thermohaline circulation with respect to mixed boundary conditions: is it really a problem for realistic models? *J. Phys. Oceanogr.*, **24**, 217–232.

Tziperman, E., Cane, M. A., and Zebiak, S. E. (1995). Irregularity and locking to the seasonal cycle in an ENSO prediction model as explained by the quasi-periodicity route to chaos. *J. Atmos. Sci.*, **52**, 293–306.

Tziperman, E., Zebiak, S. E., and Cane, M. A. (1997). Mechanisms of seasonal-ENSO interaction. *J. Atmos. Sci.*, **54**, 61–71.

Tziperman, E., Cane, M. A., Zebiak, S. E., Xue, Y., and Blumenthal, B. (1998). Locking of El Niño's peak time to the end of the calendar year in the delayed oscillator picture of ENSO. *J. Clim.*, **11**, 2191–2199.

Tziperman, E., Raymo, M. E., Huybers, P., and Wunsch, C. (2006). Consequences of pacing the Pleistocene 100 kyr ice ages by nonlinear phase locking to Milankovitch forcing. *Paleoceanography*, **21**(4), PA4206.

Van der Vaart, P. C. F., Dijkstra, H. A., and Jin, F.-F. (2000). The Pacific Cold Tongue and the ENSO mode: unified theory within the Zebiak-Cane model. *J. Atmos. Sci.*, **57**, 967–988.

Van Kampen, N. (1981). *Stochastic Processes in Physics and Chemistry*. North-Holland, New York, U.S.A.

Van Leeuwen, P. J. (2009). Particle filtering in geophysical systems. *Month. Weather Rev.*, **137**(12), 4089–4114.

Van Oldenborgh, G. J. (2000). What caused the onset of the 1997/98 El Niño? *Month. Weather Rev.*, **128**, 2601–2607.

Van Oldenborgh, G. J., te Raa, L. A., Dijkstra, H. A., and Philip, S. Y. (2009). Frequency- or amplitude dependent effects of the Atlantic meridional overturning on the tropical Pacific Ocean. *Ocean Sci.*, **5**, 293–301.

Vautard, R. and Ghil, M. (1989). Singular spectrum analysis in nonlinear dynamics with applications to paleoclimatic time series. *Phys. D: Nonlin. Phenom.*, **35**, 395–424.

Vautard, R., Yiou, P., and Ghil, M. (1992). Singular spectrum analysis: a toolkit for short, noisy chaotic signals. *Phys. D: Nonlin. Phenom.*, **58**, 95–126.

Vecchi, G. A. and Harrison, D. E. (2000). Tropical Pacific sea surface temperature anomalies, El Niño, and equatorial westerly wind events. *J. Clim.*, **13**, 1814–1830.

Velez-Belchi, P., Alvarez, A., Colet, P., Tintore, J., and Haney, R. L. (2001). Stochastic resonance in the thermohaline circulation. *Geophys. Res. Lett.*, **28**, 2053–2056.

Vellinga, M. and Wu, P. (2004). Low-latitude freshwater influence on centennial variability of the Atlantic thermohaline circulation. *J. Clim.*, **17**, 4498–4511.

Venegas, S. and Mysak, L. (2000). Is there a dominant timescale of natural climate variability in the Arctic? *J. Clim.*, **13**, 3412–3434.

Vinje, T., Loyning, T. B., and Polyakov, I. (2002). Effects of melting and freezing in the Greenland sea. *Geophys. Res. Lett.*, **29**, 44.1–44.4.

Visbeck, M., Chassignet, E. P., Curry, R. G., Delworth, T. L., Dickson, R. R., and Krahmann, G. (2003). The ocean's response to North Atlantic oscillation variability. In J. W. Hurrell, Y. Kushnir, G. Ottersen, and M. Visbeck, editors, *The North Atlantic Oscillation: Climatic Significance and Environmental Impact*. American Geophysical Union, Washington DC, U.S.A.

Von Storch, H. and Zwiers, F. W. (1999). *Statistical Analysis in Climate Research.* Cambridge University Press, Cambridge, U.K.

Vossepoel, F. C. and Van Leeuwen, P. J. (2007). Parameter estimation using a particle method: inferring mixing coefficients from sea level observations. *Month. Weather Rev.*, **135**(3), 1006–1020.

Walin, G. (1985). The thermohaline circulation and the control of ice ages. *Paleogeogr. Paleoclim. Paleoecol.*, **50**, 323–332.

Wallace, J. and Gutzler, D. S. (1981). Teleconnections in the geopotential height field during the Northern Hemisphere winter. *Month. Weather Rev.*, **109**, 784–812.

Wang, C. (2001). A unified oscillator model for the El Niño-Southern Oscillation. *J. Clim.*, **14**(1), 98–115.

Wang, C. and Picaut, J. (2004). Understanding ENSO physics – a review. *Earth's Climate: The Ocean-Atmosphere Interaction. AGU Geophysical Monograph Series*, **147**, 21–48.

Weare, B. C. and Nasstrom, J. N. (1982). Examples of extended empirical orthogonal function analyses. *Month. Weather Rev.*, **110**, 481–485.

Weber, S. L., Drijfhout, S. S., Abe-Ouchi, A., Crucifix, M., Eby, M., Ganopolski, A., Murakami, S., Otto-Bliesner, B., and Peltier, W. R. (2007). The modern and glacial overturning circulation in the Atlantic ocean in PMIP coupled model simulations. *Clim. Past Discuss.*, **3**, 51–64.

Weeks, E., Tian, Y., Urbach, J., Ide, K., Swinney, H., and Ghil, M. (1997). Transitions between blocked and zonal flows in a rotating annulus with topography. *Science*, **278**(5343), 1598–1601.

Weijer, W. and Dijkstra, H. A. (2003). Multiple oscillatory modes of the global ocean circulation. *J. Phys. Oceanogr.*, **33**, 2197–2213.

Weijer, W., Maltrud, M. E., Hecht, M. W., Dijkstra, H. A., and Kliphuis, M. A. (2012). Response of the Atlantic Ocean circulation to Greenland Ice Sheet melting in a strongly-eddying ocean model. *Geophys. Res. Lett.*, **39**(9).

Welander, P. (1982). A simple heat-salt oscillator. *Dyn. Atmos. Oceans*, **6**, 233–242.

Welander, P. (1986). Thermohaline effects in the ocean circulation and related simple models. In J. Willebrand and D. L. T. Anderson, editors, *Large-Scale Transport Processes in Oceans and Atmosphere*, pages 163–200. D. Reidel, Dordrecht, Holland.

Wiggins, S. (1994). *Global Bifurcations and Chaos: Analytical Methods*. Springer Verlag (Applied Mathematical Sciences), New York, U.S.A.

Winton, M. and Sarachik, E. S. (1993). Thermohaline oscillations induced by strong steady salinity forcing of ocean general circulation models. *J. Phys. Oceanogr.*, **23**, 1389–1410.

Wolfe, C. L. and Samelson, R. M. (2007). An efficient method for recovering Lyapunov vectors from singular vectors. *Tellus Ser. A Dyn. Meteor. Oceanogr.*, **59**(3), 355–366.

Wunsch, C. (2000). On sharp spectral lines in the climate record and the millenial peak. *Paleoceanography*, **15**, 417–424.

Wunsch, C. (2010). Towards understanding the Paleocean. *Quatern. Sci. Rev.*, **29**(17–18), 1960–1967.
Wunsch, C. and Heimbach, P. (2007). Practical global oceanic state estimation. *Phys. D: Nonlin. Phenom.*, **230**, 197–208.
Xie, S. P. and Philander, S. G. H. (1994). A coupled ocean-atmosphere model of relevance to the ITCZ in the eastern Pacific. *Tellus*, **46A**, 340–350.
Xue, Y., Cane, M. and Zebiak, S. E. (1997). Predictability of a coupled model of ENSO using singular vector analysis. Part I: optimal growth in seasonal background and ENSO cycles. *Month. Weather Rev.*, **125**(9), 2043–2056.
Zebiak, S. E. and Cane, M. A. (1987). A model El Niño-Southern Oscillation. *Month. Weather Rev.*, **115**, 2262–2278.
Zhang, R. (2008). Coherent surface-subsurface fingerprint of the Atlantic meridional overturning circulation. *Geophys. Res. Lett.*, **35**, L20705.
Zhang, R. and Delworth, T. L. (2006). Impact of Atlantic multidecadal oscillations on India/Sahel rainfall and Atlantic hurricanes. *Geophys. Res. Lett.*, **33**, L17712.
Zhang, R. and Vallis, G. K. (2006). Impact of great salinity anomalies on the low-frequency variability of the North Atlantic climate. *J. Clim.*, **19**(3), 470–482.
Zhang, R., Delworth, T. L., and Held, I. M. (2007). Can the Atlantic Ocean drive the observed multidecadal variability in Northern Hemisphere mean temperature? *Geophys. Res. Lett.*, **34**, L02709.
Zhang, Y., Wallace, J. M., and Battisti, D. S. (1997). ENSO-like interdecadal variability: 1900-1993. *J. Clim.*, **10**, 1004–1020.
Zhu, X. and Jungclaus, J. H. (2008). Interdecadal variability of the meridional overturning circulation as an ocean internal mode. *Clim. Dyn.*, **31**(6), 731–741.

Copyright Acknowledgements

I gratefully acknowledge the following copyright holders who have kindly provided permission to reproduce the figures indicated. Sources of all figures are also referenced in each figure caption. I thank my colleagues Ed Hawkins, Christian Franzke, Florian Sévellec, Adam Phillips, Jim Hurrell, Daan Crommelin, Andrey Ganopolski, Didier Paillard, Alain Colin de Verdiére, Alef Sterk, Hugues Goosse, Mike Roulston, David Neelin, Richard Kleeman, Eli Tziperman, Femke Vossepoel, Olwijn Leeuwenburgh, Stephen Griffies, Michel Crucifix, Eugenia Kalnay, Matthew Hoffman, Pedro Velez-Belchi and Tapio Schneider for sending me .eps files of the figures so promptly (and I also thank those who tried but could not find these files anymore).

Chapter 1: Fig. 1.2 by AGU (Dijkstra and Ghil, 2005). ■ Chapter 2: Fig. 2.7 and Fig. 2.8 by Elsevier (Kuehn, 2011). ■ Chapter 3: Fig. 3.6 by Elsevier (Kuehn, 2011). ■ Chapter 4: Fig. 4.4 by Elsevier (Chekroun et al., 2011). ■ Chapter 5: Fig. 5.14 by Elsevier (Kantelhardt et al., 2001). ■ Chapter 6: Fig. 6.4, courtesy by Hugues Goosse. ■ Chapter 7: Fig. 7.1 by U.S. National Academy of Sciences (Ghil and Robertson, 2002). ♯ Fig. 7.2 by AMS (Plaut and Vautard, 1994). ♭ Fig. 7.3 and Fig. 7.4 by AGU (Hurrell et al., 2003). ♯ Fig. 7.5, Fig. 7.6 and Fig. 7.14 by AMS (Crommelin, 2003). ♭ Fig. 7.7. Fig. 7.8, Fig. 7.11 and Fig. 7.13 by AMS (Crommelin et al., 2004). ♯ Fig. 7.15, Fig. 7.16 and Fig. 7.17 and AMS (Itoh and Kimoto, 1996). ♭ Fig. 7.18 by Elsevier (Sterk et al., 2010). ♯ Fig. 7.19, Fig. 7.20 and Fig. 7.21 by AMS (Franzke et al., 2008). ■ Chapter 8: Fig. 8.3 by AMS (Neelin et al., 2000). ♭ Fig. 8.6 and Fig. 8.7 by AMS (Van der Vaart et al., 2000). ♯ Fig. 8.8 by Science Publishers (Fedorov and Philander, 2000). ♭ Fig. 8.9 and Fig. 8.10 by AMS (Jin, 1997a). ♯ Fig. 8.11, Fig. 8.12 and Fig. 8.13 by AGU (Roulston and Neelin, 2000). ♭ Fig. 8.14 by AMS (Moore and Kleeman, 1999). ♯ Fig. 8.16 by AMS (Neelin et al., 2000). ♭ Fig. 8.17 by AMS (Moore and Kleeman, 1999). ♯ Fig. 8.18 by AMS (Moore and Kleeman, 1999). ♭ Fig. 8.19 by Science Publisher (Tziperman et al., 1994b). ♯ Fig. 8.20 and Fig. 8.21 by AMS

(Timmermann et al., 2003a). ■ Fig. 9.15, Fig. 9.16 and Fig. 9.17 by Journal of Marine Research (Sevellec et al., 2009). ♯ Fig. 9.19 by AMS (Saravanan and McWilliams, 1998). ♭ Fig. 9.20 by AGU (Frankcombe and Dijkstra, 2011). ■ Chapter 10: Fig. 10.1 by AGU (Clement and Peterson, 2008). ♯ Fig. 10.2 by AGU (Kuhlbrodt et al., 2007). ♭ Fig. 10.3 by AGU (Weijer et al., 2012). ♯ Fig. 10.5c by AGU (Rahmstorf et al., 2005). Fig. 10.6 by AGU (Held and Kleinen, 2004). ♯ Fig. 10.7 by Nature Publishing (Lenton, 2011). ♭ Fig. 10.9 by Journal of Marine Research (Sevellec et al., 2009). ♯ Fig. 10.11 by AGU (Weijer and Dijkstra, 2003). ♭ Fig. 10.13 by AGU (Colin de Verdière, 2007). ♯ Fig. 10.17 and Fig. 10.18 by AGU (Velez-Belchi et al., 2001). ♭ Fig. 10.20, Fig. 10.21 and Fig. 10.22 by AMS (Timmermann et al., 2003a). ♯ Fig. 10.24, Fig. 10.25 and Fig. 10.26 by Nature Publishing (Ganopolsky and Rahmstorf, 2001). ♭ Fig. 10.27 and Fig. 10.28 by APS (Ganopolski and Rahmstorf, 2002). ♯ Fig. 10.29 by AMS (Timmermann et al., 2003b). ■ Chapter 11: Fig. 11.10 and Fig. 11.11 by Nature Publishing (Paillard, 1998). ♭ Fig. 11.12 and Fig. 11.13 by AGU (Le Treut et al., 1988). ♯ Fig. 11.14 by AGU (Gildor and Tziperman, 2000). ♭ Fig. 11.16 by AGU (Tziperman et al., 2006). ■ Chapter 12: Fig. 12.4 and Fig. 12.5 by AMS (Hawkins and Sutton, 2009). ♭ Fig. 12.6 by AMS (Ehrendorfer, 1994). ♯ Fig. 12.7 and Fig. 12.8 by AMS (Schneider and Griffies, 1999). ♯ Fig. 12.10 by AMS (Xue and Cane, 1997). ♭ Fig. 12.11 by AMS (Mu et al., 2004). ♯ Fig. 12.12 by AGU (Hoffman et al., 2009). ♭ Fig. 12.14 and Fig. 12.15 by the Royal Meteorological Society (Leeuwenburgh, 2005). ♯ Fig. 12.16 by AMS (Van Leeuwen, 2009). ♭ Fig. 12.17 by AMS (Vossepoel and Van Leeuwen, 2007). ■

Index